AF344503

Genetics and Improvement of Forest Trees

Genetics and Improvement of Forest Trees

Editor

Yuji Ide

MDPI • Basel • Beijing • Wuhan • Barcelona • Belgrade • Manchester • Tokyo • Cluj • Tianjin

Editor
Yuji Ide
The University of Tokyo
Japan

Editorial Office
MDPI
St. Alban-Anlage 66
4052 Basel, Switzerland

This is a reprint of articles from the Special Issue published online in the open access journal *Forests* (ISSN 1999-4907) (available at: https://www.mdpi.com/journal/forests/special_issues/tree_genetics_improvement).

For citation purposes, cite each article independently as indicated on the article page online and as indicated below:

LastName, A.A.; LastName, B.B.; LastName, C.C. Article Title. *Journal Name* **Year**, *Volume Number*, Page Range.

ISBN 978-3-0365-1242-6 (Hbk)
ISBN 978-3-0365-1243-3 (PDF)

Cover image courtesy of Yuji Ide.

Contents

About the Editor

Yuji Ide, PhD., Professor Emeritus, The University of Tokyo, Japan, retired in 2018. He has been involved in forest tree improvement research for more than forty years, and his achievements include the evaluation of selected individuals by progeny tests, tissue culture, and population genetics. He has served as president of the Japanese Forest Society and the Japanese Society of Forest Genetics and Tree Breeding.

Editorial

Genetics and Improvement of Forest Trees

Yuji Ide

Graduate School of Agricultural and Life Sciences, The University of Tokyo, 1-1-1 Bunkyo ku, Tokyo 113-8657, Japan; ide@es.a.u-tokyo.ac.jp

Citation: Ide, Y. Genetics and Improvement of Forest Trees. *Forests* **2021**, *12*, 182. https://doi.org/10.3390/f12020182

Academic Editor: Carol A. Loopstra
Received: 1 February 2021
Accepted: 3 February 2021
Published: 5 February 2021

Publisher's Note: MDPI stays neutral with regard to jurisdictional claims in published maps and institutional affiliations.

Forest tree improvement has mainly been implemented to enhance the productivity of artificial forests. However, given the drastically changing global environment, improvement of various traits related to environmental adaptability is more essential than ever.

Plant genome research has revealed the genetic background of various useful traits. Abundant genomic and genetic information has accumulated for several important forest tree species. Such information not only enables rapid improvement of new traits, but also greatly contributes to forest tree breeding programs, including interpretation of the results thereof. However, there are few cases in which genetic information has been used effectively for forest tree improvement.

Tree improvement involves many processes, from determining the target tree species to setting breeding goals, establishing breeding strategies, and producing seeds. Each of these processes requires specific genetic information. Research on the phylogeny and genetic diversity of target species can provide information of fundamental importance at the beginning of a tree improvement program and for evaluating its results. Knowledge of the genes controlling the heritability and physiology of traits is essential. Accumulated genetic information can be used directly for marker-assisted selection (MAS) or genomic selection (GS) of target traits, facilitating more rapid tree improvement.

This special issue focuses on genetic information, including trait heritability and the physiological mechanisms thereof, which facilitates tree improvement. Nineteen papers are included reporting genetic approaches to improving various species, including conifers, broad-leaf trees, and bamboo.

Of the 19 papers, 5 deal with species phylogeny and the phylogeography of woody species. Li et al. [1] developed retrotransposon-based markers for Bamboo (*Phyllostachys* spp.), which are important natural resources, especially in Asia. The taxonomy and genetics of this group are complex. Therefore, this research should aid future studies. Research on within-species genetic diversity focuses mainly on conservation issues. However, this is also an essential issue for forest tree improvement, especially the derivation of efficacious improvement strategies. Kitamura et al. [2] (*Abies sachalinensis* F.Schmidt) and Chen et al. [3] (*Larix kaempferi* (Lamb.) Carr.) both discuss genetic variation in marginal populations. Understanding the genetic diversity of such populations will be useful for estimating their adaptability under changing environmental conditions. Inanaga et al. [4] report on the genetic diversity of *Thujopsis dolabrata* (Thunb. ex L. f.) Siebold & Zucc., which is an important traditional species in Japanese forestry. Cai et al. [5] report on the genetic diversity and structure of *Cryptomeria japonica* var. *sinensis* (*Cryptomeria fortune* (Hooibrenk)), which is endemic to China. These studies not only reveal the phylogeny of these species, but are also useful for evaluating the genetic resources available for further improvement thereof. They also provide useful information for determining the genetic consequences of tree improvement. Mukasyaf et al. [6] revealed the population genetic effects of adding *Pinus thunbergii* Parl. seedlings from pine wood nematode-resistant stock to existing forests.

Several papers discuss the inheritance of trees and relationship between genotype and environment, which is important to understand for forest tree improvement. Carignato et al. [7] report genetic variability in drought resistance among *Eucalyptus* clones and indicate the possibility of early selection in the nursery. Two papers evaluate selected *P. thunbergii* clones in terms of pine wood nematode resistance [8,9]. Iki et al. [8] report that the cumulative temperature after nematode infection does not affect the resistance ranking of progeny seedlings of nematode-resistant pine wood clones. Matsunaga et al. [9] examine the stability of traits in progeny seedlings moved to a different area from that in which they were selected. Tsuyama et al. [10] report the results of a 10-year provenance trial of *A. sachalinensis* using multivariate random forests, a machine learning method. Matsushita et al. [11] examine the growth-promoting effects of the light environment and girdling treatment on female cones in larch seed orchards. Takashima et al. [12] develop an infrared thermography/chlorophyll fluorescence technology for rapid evaluation of the response to drought stress of *C. japonica*. All of these papers not only provide specific selection criteria, but may also stimulate further research on the genes controlling these traits.

Four papers comprehensively analyze the physiological pathway of trait expression; this was made possible by recent advances in metabolomics and transcriptome analysis. Kurita et al. [13] examine genes promoting male strobilus formation after gibberellin treatment in *C. japonica*. Yang et al. [14] reveal candidate genes related to wood formation in *C. fortunei*. Cao et al. [15] study genes controlling wood properties in *Cunninghamia lanceolate* (Lamb.) Hook., especially heartwood color. Yang et al. [16] assess genes associated with the accumulation of tea oil in the leaves of Camellia oleifera. These four papers show the usefulness of genomic research for expediting genetic improvement of tree species with long lifespans.

Among the major targets of contemporary genetics research are MAS for specific traits and GS for quantitative trait loci. Lebedev et al. [17] provide a detailed review of GS and emphasize the need for advanced research in this field. Finally, two papers report on practical research. Moriguchi et al. [18] present the successful results of MAS of male sterility in *C. japonica*. Nagano et al. [19] report advanced results of GS in the same species. Both studies were successful because they used vast amounts of accumulated genetic information. Molecular breeding requires intensive research on the target species.

All of the papers published in this special issue provide cutting-edge genetic information on tree genetics and suggest research directions for future tree improvement. Tree improvement efforts are expected to be facilitated by research progress in various research fields.

Funding: This research received no external funding.

Institutional Review Board Statement: Not applicable.

Informed Consent Statement: Not applicable.

Data Availability Statement: Not applicable.

Acknowledgments: I thank all of the authors who submitted papers to this special issue, and the reviewers thereof. I also thank the editorial board and staff for their valuable support of this special issue.

Conflicts of Interest: The author declares no conflict of interest.

References

1. Li, S.; Ramakrishnan, M.; Vinod, K.; Kalendar, R.; Yrjälä, K.; Zhou, M. Development and Deployment of High-Throughput Retrotransposon-Based Markers Reveal Genetic Diversity and Population Structure of Asian Bamboo. *Forests* **2020**, *11*, 31. [CrossRef]
2. Kitamura, K.; Uchiyama, K.; Ueno, S.; Ishizuka, W.; Tsuyama, I.; Goto, S. Geographical Gradients of Genetic Diversity and Differentiation among the Southernmost Marginal Populations of *Abies sachalinensis* Revealed by EST-SSR Polymorphism. *Forests* **2020**, *11*, 233. [CrossRef]
3. Chen, S.; Ishizuka, W.; Hara, T.; Goto, S. Complete Chloroplast Genome of Japanese Larch (*Larix kaempferi*): Insights into Intraspecific Variation with an Isolated Northern Limit Population. *Forests* **2020**, *11*, 884. [CrossRef]
4. Inanaga, M.; Hasegawa, Y.; Mishima, K.; Takata, K. Genetic Diversity and Structure of Japanese Endemic Genus *Thujopsis* (Cupressaceae) Using EST-SSR Markers. *Forests* **2020**, *11*, 935. [CrossRef]
5. Cai, M.; Wen, Y.; Uchiyama, K.; Onuma, Y.; Tsumura, Y. Population Genetic Diversity and Structure of Ancient Tree Populations of *Cryptomeria japonica* var. sinensis Based on RAD-seq Data. *Forests* **2020**, *11*, 1192. [CrossRef]

6. Mukasyaf, A.; Matsunaga, K.; Tamura, M.; Iki, T.; Watanabe, A.; Iwaizumi, M. Reforestation or Genetic Disturbance: A Case Study of *Pinus thunbergii* in the Iki-no-Mastubara Coastal Forest (Japan). *Forests* **2021**, *12*, 72. [CrossRef]
7. Carignato, A.; Vázquez-Piqué, J.; Tapias, R.; Ruiz, F.; Fernández, M. Variability and Plasticity in Cuticular Transpiration and Leaf Permeability Allow Differentiation of *Eucalyptus* Clones at an Early Age. *Forests* **2020**, *11*, 9. [CrossRef]
8. Iki, T.; Matsunaga, K.; Hirao, T.; Ohira, M.; Yamanobe, T.; Iwaizumi, M.; Miura, M.; Isoda, K.; Kurita, M.; Takahashi, M.; et al. Effects of Temperature Factors on Resistance against Pine Wood Nematodes in *Pinus thunbergii*, based on Multiple Location Sites Nematode Inoculation Tests. *Forests* **2020**, *11*, 922. [CrossRef]
9. Matsunaga, K.; Iki, T.; Hirao, T.; Ohira, M.; Yamanobe, T.; Iwaizumi, M.; Miura, M.; Isoda, K.; Kurita, M.; Takahashi, M.; et al. Do Seedlings Derived from Pinewood Nematode-Resistant *Pinus thunbergii* Parl. Clones Selected in Southwestern Region Perform Well in Northern Regions in Japan? Inferences from Nursery Inoculation Tests. *Forests* **2020**, *11*, 955. [CrossRef]
10. Tsuyama, I.; Ishizuka, W.; Kitamura, K.; Taneda, H.; Goto, S. Ten Years of Provenance Trials and Application of Multivariate Random Forests predicted the Most Preferable Seed Source for Silviculture of *Abies sachalinensis* in Hokkaido, Japan. *Forests* **2020**, *11*, 1058. [CrossRef]
11. Matsushita, M.; Nishikawa, H.; Tamura, A.; Takahashi, M. Effects of Light Intensity and Girdling Treatments on the Production of Female Cones in Japanese Larch (*Larix kaempferi* (Lamb.) Carr.): Implications for the Management of Seed Orchards. *Forests* **2020**, *11*, 1110. [CrossRef]
12. Takashima, Y.; Hiraoka, Y.; Matsushita, M.; Takahashi, M. Evaluation of Responsivity to Drought Stress using Infrared Thermography and Chlorophyll Fluorescence in Potted Clones of *Cryptomeria japonica*. *Forests* **2021**, *12*, 55. [CrossRef]
13. Kurita, M.; Mishima, K.; Tsubomura, M.; Takashima, Y.; Nose, M.; Hirao, T.; Takahashi, M. Transcriptome Analysis in Male Strobilus induction by Gibberellin Treatment in *Cryptomeria japonica* D. Don. *Forests* **2020**, *11*, 633. [CrossRef]
14. Yang, J.; Guo, Z.; Zhang, Y.; Mo, J.; Cui, J.; Hu, H.; He, Y.; Xu, J. Transcriptomic Profiling of *Cryptomeria fortunei* Hooibrenk Vascular Cambium Identifies Candidate Genes involved in Phenylpropanoid Metabolism. *Forests* **2020**, *11*, 766. [CrossRef]
15. Cao, S.; Zhang, Z.; Sun, Y.; Li, Y.; Zheng, H. Profiling of Widely Targeted Metabolomics for the Identification of Secondary Metabolites in Heartwood and Sapwood of the Red-Heart Chinese Fir (*Cunninghamia lanceolata*). *Forests* **2020**, *11*, 897. [CrossRef]
16. Yang, S.; Liang, K.; Wang, A.; Zhang, M.; Qiu, J.; Zhang, L. Physiological Characterization and Transcriptome Analysis of *Camellia oleifera* Abel. during Leaf Senescence. *Forests* **2020**, *11*, 812. [CrossRef]
17. Lebedev, V.; Lebedeva, T.; Chernodubov, A.; Shestibratov, K. Genomic Selection for Forest Tree Improvement: Methods, Achievements and Perspectives. *Forests* **2020**, *11*, 1190. [CrossRef]
18. Moriguchi, Y.; Ueno, S.; Hasegawa, Y.; Tadama, T.; Watanabe, M.; Saito, R.; Hirayama, S.; Iwai, J.; Konno, Y. Marker-Assisted Selection of Trees with MALE STERILITY 1 in *Cryptomeria japonica* D. Don. *Forests* **2020**, *11*, 734. [CrossRef]
19. Nagano, S.; Hirao, T.; Takashima, Y.; Matsushita, M.; Mishima, K.; Takahashi, M.; Iki, T.; Ishiguri, F.; Hiraoka, Y. SNP Genotyping with Target Amplicon Sequencing Using a Multiplexed Primer Panel and Its Application to Genomic Prediction in Japanese Cedar, *Cryptomeria japonica* (L.f.) D.Don. *Forests* **2020**, *11*, 898. [CrossRef]

 *forests*

Review

Genomic Selection for Forest Tree Improvement: Methods, Achievements and Perspectives

Vadim G. Lebedev [1,*], Tatyana N. Lebedeva [2], Aleksey I. Chernodubov [3] and Konstantin A. Shestibratov [1,3]

[1] Branch of the Shemyakin-Ovchinnikov Institute of Bioorganic Chemistry of the Russian Academy of Sciences, Prospekt Nauki 6, 142290 Pushchino, Moscow Region, Russia; schestibratov.k@yandex.ru

[2] Institute of Physicochemical and Biological Problems in Soil Science of the Russian Academy of Sciences, Instituskaya Str. 2, 142290 Pushchino, Moscow Region, Russia; tanyaniko@mail.ru

[3] Department of Forest Crops, Selection and Forest Reclamation, Voronezh State University of Forestry and Technologies Named after G.F. Morozov, 8 Timiryazeva Str. 394087 Voronezh, Russia; chernodubov2010@yandex.ru

* Correspondence: vglebedev@mail.ru

Received: 25 October 2020; Accepted: 8 November 2020; Published: 11 November 2020

Abstract: The breeding of forest trees is only a few decades old, and is a much more complicated, longer, and expensive endeavor than the breeding of agricultural crops. One breeding cycle for forest trees can take 20–30 years. Recent advances in genomics and molecular biology have revolutionized traditional plant breeding based on visual phenotype assessment: the development of different types of molecular markers has made genotype selection possible. Marker-assisted breeding can significantly accelerate the breeding process, but this method has not been shown to be effective for selection of complex traits on forest trees. This new method of genomic selection is based on the analysis of all effects of quantitative trait loci (QTLs) using a large number of molecular markers distributed throughout the genome, which makes it possible to assess the genomic estimated breeding value (GEBV) of an individual. This approach is expected to be much more efficient for forest tree improvement than traditional breeding. Here, we review the current state of the art in the application of genomic selection in forest tree breeding and discuss different methods of genotyping and phenotyping. We also compare the accuracies of genomic prediction models and highlight the importance of a prior cost-benefit analysis before implementing genomic selection. Perspectives for the further development of this approach in forest breeding are also discussed: expanding the range of species and the list of valuable traits, the application of high-throughput phenotyping methods, and the possibility of using epigenetic variance to improve of forest trees.

Keywords: forest tree breeding; genomic selection; molecular markers; high-throughput phenotyping; epigenetics; genotyping; genomic prediction models; quantitative trait locus; breeding cycle

---

1. Forest Tree Breeding

1.1. Traditional Breeding

In contrast to agricultural crops with their breeding history of centuries or even millennia, the history of forest tree breeding is relatively recent. It was only in the late 1950s in the USA and Europe that the increasing demand for wood for construction, fuel, and paper production, and the reduction in natural forests generated the need for forest tree domestication using modern breeding methods [1]. The breeding of woody plants is much more difficult, longer, and expensive than that of agricultural crops. Trees have a long juvenile period (late age flowering and seed production) and a large physical size; their progeny testing requires large areas and long-term observations.

Some important forest tree traits, e.g., tolerance to biotic and abiotic stresses, are difficult to assess in the field. For forest trees in general and coniferous trees in particular, the stages of reproduction and progeny testing may take up 15 years each [2]. In the USA, one cycle of pine species improvement with classical methods can take about 30 years, including breeding (at least 10 years), field testing (at least 8 years), and propagation (at least 8 years) [3]. Even for the fast-growing *Eucalyptus globulus* Labill., improvement programs take 12 years and are quite expensive [4]. Furthermore, it is nearly impossible to assess the most economically important traits (wood quality, growth rate) in adult trees (aged 30–50 years), and therefore, selection has to be based on measurements made in young trees (aged 5–15 years). Normally, selection is made at about one-third of the rotation age, e.g., 2–3 years for *Eucalyptus* and 6–12 years for *Pinus* spp. [5]. However, it is known that tree traits at juvenile and mature ages poorly correlate with each other [6]. Moreover, most of the complex tree traits (growth, stem form, and branching) have low or moderate heritability, which limits the selection response and thereby hampers the genetic gain [7]. Unlike annual crops and fruit trees, forest trees were practically not subject to domestication, which would have reduced their genetic diversity, and they actually represent wild populations. Their nearly absent pedigree makes it difficult to estimate their breeding values. Moreover, possible errors in the determined progeny relationships among open-pollinated species (pines, spruces) may lead to lower accuracy of estimated breeding values (EBVs) [8]. EBV is the genetic merit of individual plants that refers to its value in a breeding program for a particular trait. Thus, forest trees have a slow accumulation of genetic gain per unit time.

Due to the above features, many forest tree breeding programs are now only in the first or second cycle of breeding, and only the most advanced programs are in their third or fourth cycle [2]. For example, it took 55 years to complete three breeding cycles in a loblolly pine (*Pinus taeda* L.) breeding program in the USA [9]. The classical approach based on phenotypic assessment helped improve the tree genotypes and reduce the rotation time, yet the speed of breeding was absolutely unsatisfactory. Advanced methods of breeding (e.g., top-grafting) and propagation (e.g., somatic embryogenesis) can shorten the stages in coniferous trees from 10 to 4 years, and from 8 to 1 year, respectively [3], but the duration of field testing still remains unchanged. Long breeding cycles of forest trees hinder the growing demand for new valuable genotypes for forest plantations that are increasingly expanding around the world. Ongoing global climate change also requires the accelerated development of new genotypes of forest trees, especially stress-resistant ones. Therefore, breeders are looking for new ways to improve the efficiency of forest breeding, i.e., to accelerate the selection of valuable genotypes.

1.2. Marker-Assisted Selection

There were attempts to speed up the selection process by using genotypic rather than phenotypic evaluation of genotypes, i.e., using molecular markers. The first use of molecular markers in traditional phenotypic selection led to the development of marker-assisted selection (MAS). This method has been used in plant breeding since the 1990s [10]. The principle of MAS consists of exploiting linkage disequilibrium (LD) between markers and quantitative trait loci (QTLs), i.e., non-random association between marker and QTL alleles [5]. MAS is most effective for simple traits controlled by a few QTLs, each responsible for a relatively large part of the overall phenotypic variability. It was for simply inherited traits, e.g., disease resistance, that some success was achieved with MAS in forest trees [11]. However, the most economically important tree traits (wood properties, trunk straightness, growth rate) have quantitative, complex inheritance patterns, and the use MAS was much less effective for them. Such traits are linked with multiple QTLs, each having a small effect, and MAS is not designed to track a large number of loci. For example, as shown on two populations of *Pinus taeda*, QTLs explained as little as 5.3% to 15.7% of the phenotypic variance in several wood traits, and these populations, in turn, contained only part of the genetic variability present in the entire population of loblolly pine [12]. Errors in defining a stringent statistical threshold for declaring a significant effect of a particular QTL can lead to a high percentage of false-positive or false-negative results. Being undomesticated, forest trees are also characterized by low levels of LD [13]. Finally, successful use of MAS in forest trees was hindered by strong QTL-environment and QTL-genetic background interactions, and the fluctuation of alleles frequency over generations [14,15]. On the whole, it was shown that QTLs do not explain enough genetic variations to enable any effective implementation of MAS for complex traits in forest trees [11].

1.3. Genomic Selection

About 20 years ago, Meuwissen et al. [16] proposed an alternative approach based on analysis of all QTL effects, regardless of their significance. This technology began developing after the costs of genotyping had dropped due to advances in next generation sequencing (NGS) and the use of large amounts of SNP markers, as well as the development of statistical methods for large dataset analysis. Genomic selection (GS) first became widely used in animal breeding, and then in crop breeding. GS proved to be thrice as efficient as MAS in maize and twice as efficient as MAS in wheat [17]. GS is also based on molecular markers, but, unlike MAS, preliminary information about phenotype-marker linkage, localizations of QTLs in the genome and their relative effect on the phenotype are not prerequisite for GS [18]. In simple words, the key difference is as follows: MAS uses few markers linked with large-effect QTLs, whereas GS uses lots of markers linked with minor-effect QTLs. The use of high-density markers is one of the fundamental features of GS: each locus of a trait has likelihood to be in LD with at least one marker locus in the entire target population [10]. To date, GS has demonstrated promising results in plants, with prediction accuracy higher than that of MAS [19].

Compared to phenotypic selection-based traditional breeding, GS can significantly shorten the breeding cycle because the marker-based assessment of genotypes can be done at a very early stage, when DNAs can be isolated without harming the plants. This eliminates the need for lengthy and expensive field testing of progeny, otherwise necessary for phenotypic evaluation. The possibility of early assessment is especially important for traits that manifest at late stages of development (particularly in long-living species) or are difficult to assess with an adequate accuracy (e.g., pest- and disease-resistance). Ultimately, this will significantly increase genetic gain per unit time [20,21].

GS is relatively seldom used in forest tree breeding, in contrast to crop breeding. The large and poorly studied genomes of trees, conifers in particular, make their genotyping difficult. However, GS can significantly reduce the breeding cycle, which is particularly important for slow-growing boreal coniferous species with a breeding cycle of 20 to 30 years. For example, the use of GS reduced the field testing stage to several months [3]. As a result, GS nearly halved the breeding cycle of loblolly pine: it made phenotype-based assessment unnecessary and allowed marker-based genotyping as soon as seeds were available. A similar effect was obtained in deciduous trees: GS reduced the breeding cycle of eucalyptus from 10 to 5 years [11]. One of the first GS studies already showed that, for eucalyptus species, a 50% shortening of the breeding cycle would increase efficiency gains by 50–100%, and a 75% shortening would increase the efficiency by 200–300% [22]. Stimulation of early flowering in individual plants can shorten the breeding cycle even more: eucalyptus breeding programs achieve flowering at 1–4 years compared to 4–8 years in natural conditions, and in *Pinus taeda*, by grafting, flowering can be achieved at 3 years [23]. In addition, GS allows simultaneous and early selection for several traits among a large number of individuals—an impossible task for conventional tree breeding now mostly based on tandem selection [24]. Thus, GS can provide a higher genetic gain per unit time, a faster succession of generations in breeding programs, and a faster creation of genotypes better adapted to environmental changes (the global climate change, the spread of diseases), and can better meet the industrial demand for wood. Thus, although GS can potentially improve the efficiency of tree breeding programs, the outcomes would strongly depend on species and traits of interest. Therefore, cost-benefit analysis is essential.

Due to the low degree of forest tree domestication, long traditional breeding cycles, and a large genetic variability of almost any trait, GS will probably be much more efficient in forest breeding than the traditional approach and, perhaps, even more efficient than in crops and animals [2]. Since the first simulation [23,25] and empirical studies [22,26] in the early 2010s, GS has attracted considerable interest among forest breeders around the world: about 60 studies on this topic have been conducted to date (Tables 1 and 2). As shown in Table 2, more than 80% of them were performed in trees from the genera *Eucalyptus*, *Picea*, and *Pinus*, the main tree plantation species in the world. In the last 3–4 years alone, the list was expanded by adding less common forest species, such as Douglas-fir, Japanese cedar, rubber tree, etc. As in traditional breeding, phenotype was mainly assessed based on growth and wood traits (mainly the physical ones), less often, based on tree architecture and pulp yield, and very rarely, based on species-specific traits, e.g., the content of essential oils and their components in *Eucalyptus polybractea* R.T.Baker [8] or rubber production in *Hevea brasiliensis* (Willd. ex A. Juss.) Müll. Arg. [27]. More recently, there appeared studies on GS for resistance to stresses, such as diseases [28,29], pests [30,31], and drought [32].

Table 1. Summary of the published simulation studies on GS in forest trees.

Species	Population Size	Effective Population Size	Heritability of Trait	Number of QTL	Marker Density, Marker/cM	Prediction Model [1]	Scenario [2]	Reference
forest trees	200–8000	10–1000	0.2–0.6	1–200	1–20	BLUP	PS GS	[23]
Cryptomeria japonica (Thunb. ex L.f.) D.Don	2000	-	0.1–0.5	50 or 150	0.5–5	linear	PS GS PS + GS	[25]
Eucalyptus	-	100	0.1 or 0.6	44	0.1	BLASSO	PS GS GS + dominance	[33]
Pinus taeda	1000	-	0.25	30 or 1000	8.33	BRR, Bayes A, Bayes B, BLASSO	GS GS + dominance	[34]
conifers	800–3000	-	0.2 or 0.5	350	4–53.3	Bayes C	PS PS + clonal archive GS GS + top grafting	[35]
forest trees	2050	5–25	0.2 or 0.6	100–200	1–10	ABLUP, GBLUP	GS	[36]
Hevea brasiliensis (Willd. ex A. Juss.) Müll. Arg.	100–5000	-	-	-	-	linear	PS PS + GS	[27]

[1] BLUP—best linear unbiased prediction, BLASSO—Bayesian least absolute shrinkage and selection operator, BRR—Bayesian ridge regression, ABLUP—additive BLUP, GBLUP—genomic BLUP. [2] PS—phenotypic selection, GS—genomic selection.

Table 2. Summary of the published experimental studies on GS in forest trees.

Species	Population Structure [1]	Genotyping Technology	Marker Type	Number of Markers	Traits [2]	Phenotyping Age, Years [3]	Prediction Model [4]	Breeding Cycle in PS/GS, Years [5]	Reference
Castanea dentata (Marshall) Borkh.	1230 (HS)	GBS	SNP	71,507	disease resistance	5–16	ABLUP, HBLUP, Bayes C		[28]
Cryptomeria japonica	476 (WP)	Axiom array	SNP	32,036	growth, physWP, male fecundity	10	GBLUP, Bayes B, RF		[37]
C. japonica	547 (FS), 29 (HS)	Axiom array	SNP	3034	growth, physWP	18	GBLUP, RF		[13]

Table 2. *Cont.*

Species	Population Structure [1]	Genotyping Technology	Marker Type	Number of Markers	Traits [2]	Phenotyping Age, Years [3]	Prediction Model [4]	Breeding Cycle in PS/GS, Years [5]	Reference
Eucalyptus benthamii Maiden and Cambage *E. pellita* F. Muell.	505 (HS) 732 (HS)	EUChip60K	SNP	13,787 19,506	growth	56 months 42 months	ABLUP, GBLUP, Bayes A, Bayes B, Bayes Cπ, BLASSO, BRR		[38]
E. cladocalyx F. Muell.	480 (49 HS)	EUChip60K	SNP	3879	growth, physWP, treeArc, flowering intensity	17 or 18	Bayes A, Bayes B, Bayes C, BRR		[39]
E. globulus	310 (40 FS, 13 HS)	EUChip60K	SNP	~12,000	growth	5–11	GBLUP, Bayes B, Bayes C, BLASSO		[4]
E. globulus	647 (62 FS, 3 HS)	EUChip60K	SNP	14,442	growth, treeArc	4	RR-BLUP, RR-BLUP-B, Bayes A, Bayes B, BLASSO, PCR, S-PCR		[6]
E. globulus	646 (FS, HS)	EUChip60K	SNP	14,422	growth, physWP, treeArc	6	Bayes A, Bayes B, Bayes C, BLASSO, BRR		[40]
E. grandis W. Hill × *E. urophylla* S.T. Blake *E. grandis, E. urophylla, E. globulus* and hybrids	738 (43 FS) 920 (75 FS)	DArT Pty array	DArT	3129 3564	growth, physWP, pulp yield	3 3 or 3.7	RR-BLUP	18/9	[22]
E. grandis	187 (131 HS)	DArT Pty array	DArT	7680	growth	5	ABLUP, HBLUP		[41]
E. grandis × *E. urophylla*	768 (37 FS)	EUChip60K	SNP	24,806	growth, physWP, pulp yield	3	GBLUP (+ D)	18/14 or 9	[42]
E. grandis	187 (131 HS)	DArT Pty array	DArT	2816	growth, treeArc	5	ABLUP, HBLUP (+ MT)		[43]
E. grandis, E. grandis × *E. camaldulensis* Dehnh.	995 (69 FS)	sequence capture, EUChip60K	SNP	55,722, 40,932	growth, physWP, chemWP	24 and 36 months	RR-BLUP, BayesB		[44]
E. grandis × *E. urophylla*	999 (45 FS)	EUChip60K	SNP	33,398	growth, physWP, chemWP	5	ABLUP, GBLUP, HBLUP		[45]

Table 2. *Cont.*

Species	Population Structure [1]	Genotyping Technology	Marker Type	Number of Markers	Traits [2]	Phenotyping Age, Years [3]	Prediction Model [4]	Breeding Cycle in PS/GS, Years [5]	Reference
E. grandis	1548 (FS, HS)	EUChip60K	SNP	15,040	growth, physWP, chemWP	7	ABLUP, GBLUP	8/4	[46]
E. nitens (H. Deane and Maiden) Maiden	691 (72 HS)	EUChip60K	SNP	12,236	growth, physWP	6 or 7	GBLUP, BLUP		[47]
E. pellita	468 (28 HS)	marker panel	SNP	2023	growth, pulp yield	54 months	ABLUP, GBLUP, HBLUP		[48]
E. polybractea R.T. Baker	480 (40 HS)	whole genome sequencing	SNP	10,000, 97,000 and 502,000	growth, terpene, and essential oil content	1 and 2	ABLUP, GBLUP, Bayes B		[8]
E. robusta Sm.	415 (WP)	DArTseq	SNP	2919	growth, chemWP	49 months or 20 years	GBLUP (+ MT)		[49]
E. urophylla × *E. grandis*	1130 (69 FS)	DArTseq	SNP	3303	growth	32 months	PBLUP (+ D, E)		[50]
E. urophylla × *E. grandis*	958 (338 FS)	EUChip60K	SNP	41,304	growth, physWP, pulp yield	3 and 6	ABLUP, GBLUP, RR-BLUP, BLASSO, RKHS		[51]
E. urophylla × *E. grandis*	958 (338 FS)	EUChip60K	SNP	41,304	growth, physWP, pulp yield	3 and 6	GBLUP (+ D, E)		[52]
E. urophylla × *E. grandis*	1130 (69 FS)	DArTseq	SNP	3303	growth, chemWP, water use efficiency	55 months	GBLUP (+ MT)		[32]
Fraxinus excelsior L.	1250 (WP, HS)	whole genome sequencing	SNP	100–50,000	disease resistance	6	RR-BLUP		[29]
Hevea brasiliensis	435 (FS)	GBS	SNP	107,294	growth	1, 2, 3, and 4	GBLUP	10/3	[53]
H. brasiliensis	330 (FS)		SSR	332	rubber production	4	RR-BLUP (+ D), BLASSO, RKHS		[27]
Picea abies (L.) H. Karst.	1370 (128 FS)	sequence capture	SNP	116,765	growth, physWP	17 or 30	ABLUP, GBLUP, BLASSO, BRR, RKHS	25/15	[54]
P. abies	1370 (128 FS)	sequence capture	SNP	116,765	growth, physWP	17 or 30	ABLUP, GBLUP (+ D, E)		[55]

Table 2. *Cont.*

Species	Population Structure [1]	Genotyping Technology	Marker Type	Number of Markers	Traits [2]	Phenotyping Age, Years [3]	Prediction Model [4]	Breeding Cycle in PS/GS, Years [5]	Reference
P. abies	726 (40 FS)	Infinium iSelect array	SNP	5660	growth, physWP pest resistance	15 or 16 10 and 15	ABLUP, GBLUP, TGBLUP, Bayes Cπ, BRR (+ MT)		[30]
P. abies	484 (62 HS)	exome capture	SNP	130,269	physWP	19 or 22	ABLUP, GBLUP, RR-BLUP, Bayes B, RKHS		[56]
P. glauca (Moench) Voss,	1748 (FS, HS)	Infinium array PgLM3	SNP	6932	growth, physWP	17	ABLUP, RR-BLUP, BLASSO		[57]
P. glauca	1694 (214 HS)	Infinium array PgAS1	SNP	7338	growth, physWP	27	BLUP	30/10	[58]
P. glauca × *P. engelmannii* Parry ex Engelm.	769 (25 HS)	GBS	SNP	34,570–50,803	growth	3, 6, 10, 15, 30, and 40	RR-BLUP, Bayes Cπ, GRR		[14]
P. glauca × *P. engelmannii*	1126 (25 HS)	GBS	SNP	8868–62,618	growth, physWP	38	ABLUP, GBLUP, RR-BLUP, GRR (+ MT)		[15]
P. glauca × *P. engelmannii*	1694 (214 HS)	Infinium array PgAS1	SNP	7338	growth, physWP	38	ABLUP, GBLUP (+ D, E)		[59]
P. glauca	1694 (214 HS)	Infinium array PgAS1	SNP	6716	growth, physWP	22	HBLUP		[60]
P. glauca × *P. engelmannii*	1126 (25 HS)	GBS	SNP	~30,000	growth, physWP	38	ABLUP, GBLUP (+ D, E)		[61]
P. glauca	1516 (136 FS)	Infinium iSelect array	SNP	4148	growth, physWP, pest resistance	16–28	ABLUP, GBLUP, Bayes Cπ (+ D)		[31]
P. glauca	1513 (54 FS), 892 (HS)	Infinium array PgLM3	SNP	4092	growth, physWP,	19 or 20	ABLUP, GBLUP		[62]
P. glauca × *P. engelmannii*	1126 (25 HS)	GBS	SNP	62,190	growth, physWP	38	ABLUP, RR-BLUP		[63]
P. mariana (Mill.) Britton, Sterns, and Poggenb.	734 (34 FS)	Infinium iSelect array	SNP	4993	growth, physWP	25	BLUP	28/9	[64]

Table 2. *Cont.*

Species	Population Structure [1]	Genotyping Technology	Marker Type	Number of Markers	Traits [2]	Phenotyping Age, Years [3]	Prediction Model [4]	Breeding Cycle in PS/GS, Years [5]	Reference
P. sitchensis (Bong.) Carr.	498 (3 FS)	RADseq	SNP	from ~2000–3000 to ~56,000	growth, timing of budburst	5 or 6	GBLUP		[65]
Pinus contorta Douglas ex Loudon	732 (FS), 697 (HS)	Axiom array	SNP	19,584	growth, physWP	10	ABLUP, GBLUP, Bayes B, Bayes C		[66]
P. pinaster Aiton	818 (35 HS)	Infinium array	SNP	4436	growth, treeArc	8 or 12	ABLUP, BLUP, BLASSO		[7]
P. pinaster	661 (FS, HS)	Infinium array	SNP	2500	growth, treeArc	6–15	GBLUP, BLASSO, BRR		[18]
P. radiata D. Don	1103	exome capture	SNP	67,168	treeArc, internal checking, external resin bleeding	7, 8, or 9	ABLUP, GBLUP	17/9	[67]
P. sylvestris L.	694 (183 FS)	GBS	SNP	8719	growth, physWP	10 and 30/30 and 36	ABLUP, GBLUP, BLASSO, BRR	36/18 and 21/11	[68]
P. taeda	951 (61 FS)	Infinium array	SNP	4853	growth, physWP, chemWP, treeArc, disease resistance (greenhouse), rooting ability (greenhouse)	4 or 6	RR-BLUP, RR-BLUP B, Bayes A, Bayes Cπ, BLASSO		[69]
P. taeda	926 (61 FS)	Infinium array	SNP	4825	growth	1, 2, 3, 4, and 6/3, 4, and 6	BLUP		[26]
P. taeda	149 (13 FS)	Infinium array	SNP	3406	growth, chemWP, treeArc	5	BLUP	15/7.5	[12]
P. taeda	165 (9 FS)	Infinium array	SNP	3461	growth	5	ABLUP, GBLUP		[9]
P. taeda	951 (61 FS)	Infinium array	SNP	4853	growth		PBLUP, GBLUP (+ D, E)		[70]
P. taeda	923 (71 FS)	Infinium array	SNP	7216	growth, disease resistance (greenhouse)	6	Bayes A, Bayes B, BLASSO, BRR		[34]

Table 2. *Cont.*

Species	Population Structure [1]	Genotyping Technology	Marker Type	Number of Markers	Traits [2]	Phenotyping Age, Years [3]	Prediction Model [4]	Breeding Cycle in PS/GS, Years [5]	Reference
Populus deltoides W. Bartram ex Marshall	473 (WP)	sequence capture	SNP	92,000	growth	1, 2, 3, 4 and 5 (field)1–15 weeks (greenhouse)	GBLUP		[71]
Pseudotsuga menziesii (Mirb.) Franco	1372 (37 FS)	exome capture	SNP	69,551	growth physWP	12 and 35 38	ABLUP, RR-BLUP, GRR		[72]
P. menziesii	1321 (37 FS)	exome capture	SNP	69,551	growth	12	ABLUP, RR-BLUP, Bayes B, GRR		[19]
P. menziesii	1321 (37 FS)	exome capture	SNP	56,454	growth, physWP	35 or 38	ABLUP, RR-BLUP		[63]
Shorea platyclados Sloot. ex Foxw.	356 (HS)	whole genome sequencing	SNP	5900	growth, physWP, treeArc	11	Bayes A, Bayes B, Bayes C, BL, BRR		[73]

[1] HS—half-sibs; FS—full-sibs; WP—wild population. [2] physWP—physical wood properties; chemWP—chemical wood properties; treeArc—tree architecture. [3] x or y—different traits at different age; x and y—all traits at different age. [4] BLUP—best linear unbiased prediction; ABLUP—additive BLUP; GBLUP—genomic BLUP; TGBLUP—threshold genomic BLUP; HBLUP—single-step genomic BLUP; PBLUP—parent BLUP; RR-BLUP—ridge regression BLUP; BLASSO—Bayesian least absolute shrinkage and selection operator; BRR—Bayesian ridge regression; RKHS—reproducing kernel Hilbert space; GRR—generalized ridge regression; PCR—principal component regression; S-PCR—supervised PCR; D—dominance effect; E—epistatic effects; MT—multi-trait models. [5] PS—phenotypic selection; GS—genomic selection.

14

2. Methodology of Genomic Selection

GS is usually implemented in three or four steps: (1) genotyping and phenotyping of a "training population" from the breeding population of trees; (2) development of genomic prediction models where marker effects are simultaneously estimated for the alleles of all marker loci and which can predict phenotype from genotype; (3) validation of predictive models (on a "test population", i.e., a group of individuals not included in model training; (4) application of the models to predict GEBV of non-phenotyped individuals and selection for further purposes based on these values [19]. Some authors, e.g., Tan [51], unite the first and second steps. Thus, a GS program uses two populations: a training population, for which genotypic and phenotypic data are determined, and a test population, where the breeding value of each individual is estimated based on the genotypic data.

Establishment of a training population is a critical step in GS [74]. A successful approach to establishing a training population in forest genomic breeding is to use available progeny tests from classical breeding programs. Such tests usually involve thousands of trees from several dozen (rarely hundreds) half-sib or full-sib families. Full-sibs produced from controlled pollination are preferable to open-pollinated half-sibs with only one known parent. However, half-sibs are widely used because they can offer some advantages: (1) fast and inexpensive production due to the lack of cross-breeding stage; (2) screening of a large number of parents with minimal efforts; (3) better genetic sampling due to a large number of pollen donors [61]. For minor tree species, genotypes from wild populations are also used. Training populations of forest trees usually include over a thousand individuals, i.e., much more than in GS of agricultural crops [11].

The plants of a training population are used for whole-genome genotyping through a large number (thousands to tens of thousands) of SNP markers and phenotyping for as many economically valuable traits as possible. The use of genome-wide markers is one of the fundamental features of GS. For each trait locus, there is a probability of being in LD with the marker locus, hence there is no need to look for specific significant QTL-marker associations, as in MAS [10]. In GS, phenotyping is only needed for building and validating genomic prediction models. Furthermore, with the same genomic profiles, the models can be easily recalibrated for other traits of interest in accordance with new breeding objectives [30].

Genomic prediction models are developed based on genotyping and phenotyping data. Such models are validated on a test population, which is only genotyped. The test population is usually related to the training population. Models with higher accuracy, i.e., correlation between observed and predicted breeding values, can be used to predict BVs in non-phenotyped individuals based on their genotypic data and ranking by this parameter. Plants with a high GEBV can be used for further breeding, and those with a high genomic estimated genotypic value (GEGV; where the prediction took account of non-additive genetic effects)—for clonal propagation and subsequent cultivation [24]. GS is focused on predicting general breeding value rather than identifying specific genes associated with a particular trait, and is therefore less limited by polygenic heritability of many traits in agricultural and forest species [12].

3. Genotyping Forest Trees

The development of high-throughput genotyping technologies has paved the way for a wide use of GS. A successful GS requires a dense and genome-wide distribution of genetic markers because in this case, each locus of a trait of interest is likely to be in LD with at least one marker locus in the entire target population [10]. In addition to the reduction in genotyping costs, the GS breakthrough was also facilitated by the development of statistical methods for accurate prediction of marker effects [75]. Efficient GS requires a low-cost, flexible, and accurate high-density genotyping platform [44]. GS of forest trees applies a number of genotyping technologies: SNP chip/array, diversity array technology (DArT and DArTseq), genotyping by sequencing (GBS), restriction site-associated DNA sequencing (RAD-seq), sequence capture, and genome-wide sequencing.

3.1. Single Nucleotide Polymorphic Marker Arrays

A single-nucleotide polymorphic marker (SNP) is a sequence with a single nucleotide replacement that does not change the overall length of the DNA sequence. SNPs have many applications in plants, including positional cloning, whole-genome association studies, mapping of QTLs, and determination of genetic relationships between individuals [40]. SNP array-based genotyping, with detection of the incorporated nucleotide on a two-dimensional surface (the chip), has become the most commonly used method in tree GS. Fixed SNP arrays provide the current gold standard for data reproducibility in forest tree GS; they are also breeder friendly, available from multiple service providers, and easily managed and stored [24].

Although two microarray-based genotyping platforms—Infinium (Illumina) and Axiom (Affymetrix)—were used for tree populations, most parts of GS research in trees were done using the Illumina Infinium platform. This genotyping platform for a non-model species (*Pinus taeda*) was first applied by Eckert et al. [76]. It contains 7216 SNPs, each representing a unique pine expressed sequence tag (EST) contig, and was used for GS of *Pinus taeda* populations [26,70]. For maritime pine, the Illumina Infinium SNP array with a large number of markers (9K and 12K) was used [7,18]. The genotyping of *Picea glauca* was done using two iSelect Infinium (Illumina) SNP arrays: PgAS1 [59] and PgLM3 [57]. Both had a similar number of assayed SNPs (13,162 and 14,139, respectively), but the first array was mainly designed for population genetics and genetic association studies, whereas the second one was constructed for population genetics, genomic prediction, and linkage mapping purposes [77]. SNP arrays were also developed for genotyping of minor species, e.g., the Infinium iSelect SNP genotyping array containing 5300 SNPs representing as many distinct black spruce gene contigs [64].

Based on the sequencing of 12 eucalyptus species, the high-density Illumina Infinium EuCHIP60K was developed, which contains probes for 60,904 SNPs [78]. According to the system's developers, EUChip60K represents an outstanding tool for GS, genome-wide association study (GWAS), and the broader study of complex trait variation in eucalypts. The chip was used in GS of various eucalyptus species and hybrids: *Eucalyptus globulus* [4], *E. grandis* × *E. urophylla* [42], *E. nitens* [47]. However, population screening for traits with oligogenic effects, such as disease resistance genes, can supposedly be done with a low-density SNP chip [1].

The Axiom system was used later and much less often: e.g., for genotyping *Cryptomeria japonica* [37] (more than 73,000 SNPs) and *Pinus contorta* [66] (more than 51,000 SNPs).

3.2. Diversity Array Technology

Diversity Arrays Technology (DArT) is yet another microchip-based marker system. This cost-effective sequence-independent ultra-high-throughput marker system was developed in 2001 [79]. The DArT technology is based on detection of polymorphic DNA fragments and, unlike SNPs, its chip development does not need genome sequence data. Hence, GS can also be used for species with an unsequenced or totally unstudied genome. This technology is also used for large-genome species such as grain crops [80]. Comparative evaluation of SNPs versus DArT markers in GS of wheat (*Triticum durum* Desf.) showed similar prediction accuracies of the two marker systems, although the number of DArTs was 3 times that of the SNPs (16,383 vs. 5649, respectively) [81]. Later on, there appeared an improved technology, DArTseq, which has largely replaced the original DArT. DArTseq is a DArT marker platform combined with next generation sequencing (NGS) platforms, which allows reduction in genome complexity mediated by restriction enzymes and sequencing of restriction fragments [82]. Comparative evaluation of various array-based platforms—DArT, Illumina Infinium BeadChip wheat iSelect, and DArTseq—in GS of spring wheat showed that DArTseq—given the low cost per SNP—was the best platform for genomic prediction [83].

SNPs and DArT markers were not compared on forest trees, but DArT was used on *Eucalyptus grandis* [41,43] and various eucalyptus species and hybrids [22], whereas DArTSeq was used on *E. robusta* [49] and *E. urophylla* × *E. grandis* [50].

3.3. Complex Genome Genotyping

Coniferous species are characterized by large genomes of about 20 Gb or more (e.g., the *Pinus taeda* genome is 22 Gb [84]), which many times exceed the genomes of deciduous trees (e.g., 485 Mb for poplar [85]). The huge genomes of coniferous species impede their GS. Although the number of SNP markers in tree GS widely varies (from 2–3 to several dozens of thousands), most studies use no more than 10 thousand. This is quite enough for deciduous tree genomes, but not for conifers, where such a small number of markers may not be able to capture most of the QTL effects in large breeding populations [54]. To achieve the same coverage density as in deciduous species, one would need at least 100,000 to 200,000 markers. To save costs and time on conifer genotyping, Neves et al. [86] proposed focusing on the coding region. This method—exome capture—is a target enrichment method for sequencing the protein coding regions in a genome, which greatly improves the analysis efficiency. Genetic variation is not limited to exomes, but sequence capture is a cost-effective alternative to whole-genome sequencing. Sequence capture consists of targeted hybridization and allows genotyping almost any number and density of genetic markers. The method has become popular quite recently, after Suren et al. [87] showed the utility of sequence capture for re-sequencing in complex genomes of interior spruce (*Picea glauca* × *P. engelmanii*) and lodgepole pine (*Pinus contorta*). Sequence capture genotyping may be a useful alternative to DNA chips in GS of forest trees. It was used for genotyping *Picea abies* [54–56] and *Pinus radiata* [67], and was the only method applied in all GS studies on Douglas-fir [19,63,72]. A comparison of two genotyping methods—sequence capture followed by next generation sequencing (NGS) versus EucHIP60K.br—showed their equivalence in terms of genomic prediction of the traits of interest in eucalyptus [44]. The authors concluded that sequence capture could be a good alternative for species where SNP arrays are not available or too expensive to develop. This was later confirmed by Ballesta et al. [39], where the efficiency of EUChip60K on *Eucalyptus cladocalyx* was as low as 6% (about 3900 SNPs), probably due to a distant relatedness with the species on which the chip was developed.

An alternative to genotyping complex genomes may be provided by DNA sequencing associated with the restriction site (RAD-seq), which genotypes regions in proximity to endonuclease restriction sites [44]. This method allows a relatively cheap identification of a large number of SNPs required for GS in species with an unsequenced genome, and the number of loci selected depends on the number of these restriction sites. Fuentes-Utrillo [65] applied this method for GS of *Picea sitchensis*, with four restriction enzymes and the number of markers for mapping ranging from ~2000 to ~56,000, depending on the enzyme.

3.4. Next Generation Sequencing Technologies

Yet another method—genotyping-by-sequencing (GBS)—is based on advances in next generation sequencing technologies for obtaining SNP data and has very low per sample costs [80]. The GBS employs restriction enzymes to reduce genome complexity, and a barcoding system for multiplex sequencing. GBS does not require a decoded genome and is suitable for non-model species, such as forest trees [14]. Furthermore, GBS uses the genotyped population to detect markers, thus minimizing the ascertainment bias. A comparison of DArT versus GBS marker platforms in winter wheat showed that 38,412 GBS SNPs provided a higher GS accuracy than 1544 DArT markers. Noteworthy, the per sample cost of DArT genotyping was 2.5 times as high as that of GBS. The authors suggested that the higher accuracy of GBS was mainly due to the large increase in the number of markers. GBS was used in GS of both coniferous—*Picea engelmannii* × *P. glauca* [14,15], *Pinus sylvestris* [68]—and deciduous species—rubber tree [53], *Castanea dentata* [28].

Finally, the use of whole-genome sequencing (WGS) in GS is quite ambiguous: its high cost is not always compensated by a higher accuracy of GEBV due to increased marker density. This method was used for genotyping *Eucalyptus polybractea* containing a high concentration of desired oil composition in leaves [8]. Increased SNP density from 10K to 500K generally results in increased predictive ability for most traits tested, although for height, chip-based genotyping may be more cost-effective than

WGS. Apart from its use on eucalyptus, the same method has been recently employed to genotype *Shorea platyclados*, a tropical tree from the Dipterocarpaceae family with an excellent timber quality. This was the first use of GS for wood species from the Southeast Asian rainforest [73].

The use of high-density molecular markers can uncover hidden relatedness in open-pollinated populations and correct potential pedigree errors. For example, the average pairwise estimates of genetic relationship among individuals were substantially lower using SNP data than expectations based on pedigree information of eucalyptus hybrids [51]. The authors suggest that these inconsistencies likely derived from pollen contamination and/or mislabeling in the process of generating the full- and half-sib families. Pedigree errors are common in breeding programs and result in incorrect estimates of heritabilities and decreased breeding value accuracies. The accuracy of GEBVs was 0.55–0.75 when using the documented pedigree of the radiata pine training population and 0.61–0.80 when using the SNP-corrected pedigree [67]. The documented pedigree was corrected using a subset of 704 SNPs and about 50% of parents were reassigned. This pedigree error was significantly higher than that reported in livestock and plant breeding programs (about 10%) [67]. Errors were also found during the verification of the lodgepole pine's population structure [66]. Correct pedigree information is essential for accurate selection of suitable individuals as parents of the next generation.

The advent of NGS technologies has sharply reduced the cost of sequencing and allowed decoding the genomes of not only model species and important agricultural crops, but of some forest trees as well. With the exception of the genome of the model species *Populus trichocarpa* Torr. and A. Gray ex. Hook [85], all other sequenced tree genomes were published in the last decade. In total, the genomes of less than 10 coniferous and less than 20 deciduous forest tree species have been sequenced so far. They do not include Scots pine (*Pinus sylvestris*), an economically important species dominant in forested areas [88]. The availability of a sequenced genome is not a prerequisite for GS, although it would simplify the genotyping process. On the other hand, there are several important tree species with known genomes that have not yet been subject to GS (e.g., birch, oak), and they may probably be used to develop genotyping systems for forest breeding.

4. Phenotyping Forest Trees

4.1. Problems of Classical Phenotyping

The mechanisms of tree phenotype formation have their own specific features: during their long lives, trees are exposed to alternating dormancy/growth periods and lots of stresses including those never experienced by annual plants, e.g., autumn and late spring frosts. Poor understanding of phenotype formation mechanisms in forestry can lead to unpredictable behavior of some genotypes under certain environmental conditions [89]. Tree phenotyping should be done in different environments because the phenotype is influenced by both environmental and genetic factors.

Phenotyping has always been a bottleneck in classical breeding programs, since the size of breeding population is limited by the evaluable number of plants for which the phenotype can be determined. Genomic selection is no exception, and the phenotyping of the training population is as important as its genotyping. Precise phenotypic data are key to accurate GEBV prediction for the training population [90]. While plant genotyping once used to be the bottleneck, the development of NGS has reduced the costs of genotyping and now, the limiting factor is phenotyping. The current technical challenge for implementing GS in crop plants is the reliability of phenotypic data [20]. Traditionally, plant phenotypic traits were determined by time-consuming, labor-intensive, and often destructive manual measurements, which were also prone to researcher bias. The situation has changed with the recent advent of high-throughput phenotyping (HTP) methods that are an integral part of plant phenomics and are based on obtaining images in various spectral regions and their subsequent computer processing [91]. It is the phenomic approach that, for the first time, has provided an opportunity to deduce the patterns of changes in plant phenotypes in response to changing environmental factors. These methods allow non-invasive precise phenotyping of large number of plants over long periods of time. Plants can be studied at a whole plant level (holistic phenotypes) or at an organ level, e.g., leaves and stems (component phenotypes) [92]. These methods assess various parameters automatically, thus eliminating the issues of researcher bias and inadequate statistical processing.

4.2. High-Throughput Phenotyping

A variety of equipment has been developed for HTP: individual devices, including robotic ones, as well as automated greenhouse systems that can quickly scan and record accurate data for thousands of plants by means of non-invasive image capture techniques in various spectral regions. Special equipment has been also developed for plant phenotyping in the field: moving platforms ("phenomobiles") and aerial systems (drones, unmanned aerial vehicles, balloons). Thus, the traditional manual techniques of plant phenotyping are now giving way to high-precision non-destructive imaging techniques. Furthermore, specialized image processing software has been developed to assess plant growth and development both at specific time points and throughout their life. Computer image analysis has significantly increased the phenotyping throughput that used to be the limiting factor in the analysis of genotypes or their behavior [93]. For instance, the Quantitative Plant database [94], formerly called Plant Image Analysis, contains nearly 200 plant image analysis tools.

HTP is most often based on visible-range images that provide much information about the plant's height, the structure of its parts, and the size, color, and spatial orientation of its leaves. The use of infrared (IR) cameras to capture night images allows round-the-clock measurements. According to some studies, the spectral reflection of leaves closely correlates with their nitrogen and chlorophyll content, and it can be remotely measured with hyper- and multispectral devices, as well as a chlorophyll meter (SPAD meter) [95]. Therefore, it is possible to determine not only the rate of plant growth but also their physiological state, resistance to diseases, etc. Studies of the relationship between drought severity and associated physiological changes in plants showed the particular importance of long-term experiments [96]. Obtaining a true picture would require collecting phenotypic data at regular intervals throughout the plant's life cycle.

For several years, HTP methods have been used in the GS of annual crops, grains in particular. Rutkoski [97] observed that aerial measurement of canopy temperature and vegetation index, as secondary traits, with a thermal and hyperspectral camera, increased the prediction accuracies for wheat grain yield by 56% in pedigree, and by 70% in genomic prediction models. Juliana et al. [98] used aerial HTP platforms in the wheat genomic selection for grain yield under stress conditions (drought, heat) and found that it increased selection intensity due to the large populations used. An unmanned aerial vehicle (UAV) was first used to assess the heritability of vegetation index traits measured on small unreplicated plots and to estimate the extent to which they are predictive of wheat grain yield in replicated yield trials [99]. In that study, aerial HTP provided a substantially better response to selection for grain yield than conventional visual selection. Not only remote phenotyping but also other methods were used in GS of wheat. For example, a robotic field phenotyping platform measured plant height [100]. In another study, grain yield was assessed from spectral reflectance data collected with a handheld multiple spectral radiometer [101]. HTP was also used in GS of other annual crops. Greenhouse-based HTP platforms were applied to measure shoot biomass and water use in GS of rice [102]. Processed images of cassava roots were used as a source of phenotypic data in genomic selection [103].

The developing technologies of precision phenotyping, remote sensing, robotics, and artificial intelligence enable breeders to perform high-throughput, low-cost, and labor-saving precision phenotyping. This can help scale up the experiments, reduce labor costs, and eliminate human errors in manual measurements [104]. However, HTP methods are still rarely used in GS of tree species.

4.3. High-Throughput Phenotyping in GS of Forest Trees

Large training populations (about 1000 trees) of woody plants make their phenotyping a long, complex, and expensive process. Apple breeders note that the development of gene technologies has brought about the big problem of HTP of large populations [1]. Thus, the main challenge in GS of trees is field phenotyping of large numbers of plants under various environmental conditions. The challenge cannot be met without HTP. Furthermore, tree breeding populations do not differ much from the wild ones, i.e., forest species are in early breeding stages. Meanwhile, selection is most effective at early breeding cycles, when the frequencies of favorable alleles in the target population are low, and here, the precision of phenotyping is critical. Otherwise, it will lower genetic gains because of the very low frequency of favorable alleles [105].

To the best our knowledge, GS of forest tree species, in contrast to annual crops, hardly ever uses direct HTP techniques, nor are they often applied in forestry. UAV phenotyping was first used on a black poplar (*Populus nigra* L.) population of 503 genotypes on an area of 1.67 hectares, where two water treatments (well-watered and moderate drought) were compared [106]. The study showed that such a phenotyping technique can be an important tool for improving the efficiency of forest tree selection for climate change tolerance. Later on, UAV was used to estimate the total height, intra-annual height growth, and phenology of individual 15-year-old Norway spruce trees (*Picea abies*) in a dense field trial [107]. The method proved to be a cost- and time-effective alternative to manual height measurements, and its precision was high enough for selection with total tree height as the target trait. An automated phenotyping platform was successfully used to measure a variety of growth parameters and response to dry-down period in seedlings of two oak species (*Quercus bicolor* Willd. and *Q. prinoides* Willd.) under greenhouse conditions [108]. The use of image processing demonstrated changes in the leaf shape and size in some transgenic lines of aspen (*Populus tremula* L.) carrying the recombinant xyloglucanase gene in the second year of vegetation under semi-natural conditions [109]. Of particular interest are image-based methods that allow reconstructing of the 3D structures of the trunk and branches. Tree architecture is one of the main traits in GS of forest trees (Table 2), but its traditional assessment is very labor-intensive and imprecise.

There are hardly any known studies on direct HTP of forest species, whereas indirect phenotyping methods, such as near-infrared reflection spectroscopy (NIRS) and X-ray diffraction, are used in GS for high-throughput measurements of chemical and physical properties of wood. Although data obtained with such methods may not reflect the state of the whole tree, they are usually quite sufficient for GS [11]. For example, NIRS measurements of chest-level samples of *Acacia crassicarpa* A. Cunn. ex Benth. wood correlated well enough with those of samples taken along the entire tree height, so that tree breeders would still effectively select the best-ranked genotypes [110].

NIRS combines spectroscopic techniques and mathematical algorithms for indirect measurement of concentrations of OH-, NH-, CH-, or SH-containing compounds and is widely used to determine the content of nitrogen, moisture, carbohydrates, amino acids, and some other plant compounds in several crop species [90]. This method allows fast and precise assessment of several phenotypes at a time, thus saving wet chemistry costs [111]. In one of the first GS studies on trees, NIRS was used for indirect measurement of eucalyptus pulp yield [22], and later, it was also used to measure physical properties of wood, such as fiber length, coarseness, and number of fibers per gram [44], basic density [51], as well as chemical properties of wood, such as α-cellulose content, syringyl to guaiacyl lignin monomer ratio [46].

Density is one of the main characteristics of wood and is traditionally assessed by extracting samples of increment cores from trees and then, measuring their volume and weight. The method is relatively simple and accurate, but also time-consuming and labor-intensive [112]. An alternative method for measuring wood density is X-ray densitometry, which was often used to phenotype various spruce species in GS [58,59,64]. Another method of X-ray structure analysis, X-ray diffractometry, is used for microfibril angle (MFA) measurement [113]. Wood modulus of elasticity (MOE) can be calculated from wood density and MFA [58].

4.4. Importance of Age-Related Phenotyping

The age of phenotyping is very important for trees, which is not so for agricultural crops. According to published data, phenotyping age can vary in a wide range (Table 2): from 1–2 [8] to 40 years [14]. It is known that correlation between the same traits in young and mature trees can differ significantly depending on trait, species, environment, and age [114]. This was also observed in GS studies. Studies on *Pinus taeda* showed that models developed for growth traits based on data from 1–2-year-old plants had a limited accuracy in predicting phenotypes at 6 years [26]. Ratcliffe et al. [14] compared the heights of interior spruce trees at the ages of 3, 6, 10, 15, 30, and 40 years. The prediction accuracy of GS models based on 30-year height was nearly equivalent to those based on 40-year, but it significantly decreased with an increasing age difference between the training and test populations. On the other hand, quite opposite results have been obtained lately. In order to reduce the breeding cycle, Alves et al. [71] integrated indirect phenotypic selection based on greenhouse phenotypes with traditional GS. Plants of the same genotypes of *Populus deltoides* were grown in the field and in a greenhouse, with plant height measurements collected from week 1 to 15 in the greenhouse and at years 1 to 5 in the field. The study showed a moderate correlation (0.39–0.42) between greenhouse height measurements at weeks 13 and 15 and field measurements at years 3–5. By combining multiple greenhouse phenotypes into a selection index, a relative efficiency of ~0.48 was achieved, which is comparable with the results of GS models based on field phenotypes only [71].

The results of the study offer promising prospects. In GS of trees, phenotyping is almost always performed in the field. Exceptions are very rare and relate to specific traits, e.g., the rooting ability of *Pinus taeda* cuttings [69]. The phenotype is influenced by both the genotype and the environment and therefore, data obtained for naturally perennial plants grown under controlled conditions cannot be extrapolated to their behavior in the nature. However, the manifestation of some traits, such as tolerance to biotic and abiotic factors, are difficult to assess in the field, because it is difficult to provide a large variability of stress factors in order to evaluate the resistance of genotypes. Although pest and disease resistance can still be evaluated using artificial infection, practical, GS uses indirect methods. In particular, the levels of acetophenone aglycones (piceol and pungenol) that are known to be active ingredients against spruce budworm were analyzed in white spruce needles [31], and the growth traits in Norway spruce plants were assessed in areas with different pest abundance [30]. A trait such as disease resistance was evaluated under greenhouse conditions [34,69]. It is problematic to assess drought tolerance in plants, which, unlike some agricultural crops, are normally not artificially irrigated. Drought in field conditions is hardly predictable, and even more so is its duration and intensity. In a study on rubber tree [53], plants were cultivated on sites with different water availability conditions, but such sites would differ not only in the amount of precipitation, but also in temperature, soil composition, and many other factors. Meanwhile, drought tolerance is becoming an increasingly valuable trait for forest trees due to ongoing global climate change. The first study on GS for drought tolerance in trees was conducted not long ago on a eucalyptus hybrid population using water use efficiency (WUE) as a selective trait [32] and there is no doubt that research in this area will expand.

We think that GS for tolerance to abiotic stresses can use clonally propagated planting material from training populations to assess the effects of external factors on plant growth and development (in a greenhouse or, even better, outdoors). The level of correlation between the manifestations of such traits in juvenile and mature plants is unknown but can be roughly estimated from the growth history of the adult population and climate data. Traditional assessment of the physiological state of plants is long and labor-intensive. Bouvet et al. [32] assessed WUE in eucalyptus by measuring stable carbon isotope ratio (δ^{13}C) in plant tissues by isotopic mass ratio spectrometry, which allows large-scale screening of plants. An alternative method is HTP using various spectral cameras. These techniques can effectively evaluate the growth, biophysical, and biochemical performance of plants [115] and will help reduce the cost of phenotyping and improve breeding value prediction accuracy for forest trees.

5. Genomic Prediction Models

5.1. Parametric and Nonparametric Models

GEBV estimation is no less important in GS than genotyping and phenotyping of individuals in a population. A genomic prediction method should provide a high accuracy and preferably capture LD between markers and QTLs rather than relatedness for higher long-term stability [11]. In addition, it must be a simple, reliable, and efficient tool for estimation of various traits. There are a number of statistical methods developed for use in GS. They mainly differ in the assumptions of the distribution and variances of marker effects. These models usually contain a huge amount of genotypic data (markers) and a limited amount of phenotypic data. All these methods can be divided into two groups: parametric and nonparametric models [10,111].

Parametric models were the first to be applied in GS. In the first GS study, Meuwissen et al. [16] used three such models: Ridge Regression Best Linear Unbiased Predictor (RR-BLUP), Bayes A, and Bayes B. RR-BLUP assumes that all marker effects are normally distributed and that all markers have equal variations with small but non-zero effect. The other two models are based on Bayesian estimation. In contrast to normal distribution in RR-BLUP, marker effects in the Bayes models are a priori assumed to follow Student's t-distribution [111]. Later on, modifications of RR-BLUP were developed (GBLUP (genomic BLUP), HBLUP (single-step genomic BLUP), as were a number of Bayesian models (Bayes Cπ, Bayes LASSO (least absolute shrinkage and selection operator), etc.).

GEBV prediction accuracy depends, among other things, on the statistical methods used. They differ in their assumptions and algorithms regarding the variance of complex traits with different genetic architectures. There are two types of genetic architecture: (1) genetic effects follow a mixed inheritance process where there are few genetic variants of large effects and many variants of very small effects, or (2) each genetic effect contributes only a very small fraction of the total genetic variance [67]. Bayesian models are better suited for traits, with the first type of genetic architecture, where marker effects are modeled to follow a prior distribution. The basic difference between various Bayesian methods is that they have different prior distributions and produce different degrees of shrinkage [116]. In particular, Bayes A assumes that genetic variance follows an inverted chi-square distribution and therefore, this model is suitable for traits controlled by a moderate number of genes; Bayes B assumes the variance of markers is equal to zero with probability π and is better suited when the trait is strongly influenced by certain loci; Bayes Cπ assumes that probability π has a prior uniform distribution and therefore, it is better suited for analyzing real data [69,116]. The drawback of the Bayesian methods is the need to set prior values, but this requirement is circumvented in the Bayesian LASSO that requires less data [117].

For traits with the second type of genetic architecture, i.e., influenced by a large number of minor genes, better prediction accuracies are achieved with models like GBLUP and RR-BLUP, which assume that the effects of all loci have a common variance. GBLUP is equivalent to RR-BLUP and uses a genomic relationship matrix (GRM) generated by evaluating marker covariance across all individuals, providing greater resolution in genetic relationships among individuals [111]. This model can be applied even with simple or absent pedigrees and is therefore preferred for breeding programs of forest trees without a long record of commercial cultivation [67].

In one of the first GS studies, Bayesian approaches outperformed RR-BLUP for disease resistance traits controlled by a limited number of loci [69]. However, GS for disease resistance is very rare, and most studies of complex quantitative traits failed to find any advantages of certain models. For example, both RR-BLUP and Bayes Cπ performed consistently well in a study on interior spruce [14]. A study on eucalyptus [4] showed comparable prediction accuracies of GBLUP and Bayesian models (Bayes B, Bayes C, BLASSO). The absence of marked differences between GBLUP and Bayesian methods in the evaluation of growth and wood quality traits was also observed in *Pinus pinaster* [18] and *Pinus contorta* [66]. In *Cryptomeria japonica*, however, GBLUP provided higher prediction accuracies than Bayes B for almost all assessed traits, which suggests their control by many QTLs [37].

Unlike phenotypic mass selection based on an ancestral relationship matrix (matrix A), genomic prediction relies on a marker-based relationship matrix (matrix G), which provides a more accurate assessment of genetic similarity [56]. Generally, the studies showed the superiority of marker-based models over pedigree-based models. In a eucalyptus hybrid population [50], prediction accuracy and stability were improved by using marker-based instead of pedigree-based relationship matrices. In *Picea glauca*, marker-based models (GBLUP) showed better prediction accuracies compared to pedigree-based models (ABLUP) [59]. However, the advantage of such models was not always observed. In Norway spruce, ABLUP had higher accuracy for all four traits than four genomic selection methods (GBLUP, BRR, BLASSO, and RKHS (reproducing kernel Hilbert space) [54]. In a recent study on radiata pine [67], the accuracy of GEBVs (from a GBLUP model) was higher than that of EBVs (from an ABLUP model) for branch cluster frequency, but lower for stem straightness, internal checking, and external resin bleeding.

In addition to parametric models, GS also uses a group of nonparametric and semi-parametric models. Examples of such models are random forests (RF), support vector regression (SVR), and neural networks (NN), nonparametric machine learning methods, and RKHS, a semi-parametric method where the genomic relationship matrix used in GBLUP is replaced by a kernel matrix, which enables nonlinear regression in a higher-dimensional feature space [118]. Unlike parametric models that evaluate additive genetic effects, these models can also capture non-additive ones (e.g., dominance, epistasis) [111]. Thus, they can predict phenotypes better than the parametric models, especially where non-additive effects are important.

Nonparametric models are rarely used in forest tree GS, and only in recent years. In eucalyptus, RKHS demonstrated slightly better predictive abilities than four other models for traits with lower heritabilities (such as trunk CBH, height, and volume), but it was the worst for pulp yield [51]. In other studies, RKHS did not differ in accuracy from parametric statistical methods (GBLUP and BLASSO on Norway spruce [54] and BLASSO and RR-BLUP on rubber tree [27]). A far as we know, RF has only been used in studies on *Cryptomeria japonica* [13,37]. The prediction accuracy depended on where the plants were grown, but on the whole, GBLUP and RF were better models than Bayes B [37].

In each GS model, prediction is based on different analytical assumptions, hence there is no universally applicable statistical method for any traits in any population. Heslot et al. [119] compared 11 different parametric and nonparametric models for predicting various quantitative traits on wheat, barley, maize, and *Arabidopsis* and no model was better than the others for all traits. Normally, one should start with the use of RR-BLUP or GBLUP, include Bayesian models for large-effect loci, and machine learning methods for important non-additive effects [11].

5.2. Non-Additive Genetic Effects

Additive effects are considered the most important in breeding, since only they can be inherited, as opposed to non-additive effects associated with specific genotypes. However, dominance and epistasis can be confused with additive or random environmental effects, and if they are significant, ignoring them will lead to bias in genetic estimation. On the other hand, their inclusion in models can improve the accuracy of prediction [34]. Dominance assessment is important for crossed populations commonly used in the breeding of perennial species. Parametric models can also be used to estimate non-additive effects by replacing pedigree-based relationship matrices due to non-additive effects with their marker-based counterpart [70].

Early studies in forest tree GS evaluated only additive effects, but later simulation studies on eucalyptus [33] and loblolly pine [34] showed that the inclusion of dominance in the GS prediction model improved its accuracy. This was verified in practice by some studies. Bouvet et al. [50] showed that non-additive variance explained a significant part of the total genetic variance, although epistatic variance could not be clearly estimated. The authors attributed this to their use of

Eucalyptus grandis × *E. urophylla* hybrids that could have enhanced heterosis. In another study on eucalyptus, inclusion of dominance effects in the model increased the accuracy of GS for traits with a large dominance variance (height and CBH) [52]. The role of non-additive effects was also confirmed on coniferous trees: El-Dien et al. [59] reported a high proportion of epistatic variance in wood density of white spruce. Comparing GBLUP-A with GBLUP-AD on interior spruce showed the superiority of the latter model, which was more pronounced for height than for wood density due to the observed dominance variance [61]. In contrast to wood density, height assessments showed no differences between GBLUP-AD and GBLUP-ADE due to the lack of epistatic genetic variances. Thus, in interior spruce, tree height was best assessed with GBLUP-AD, whereas wood density with GBLUP-ADE, reflecting the presence of significant additive × additive genetic variances.

The role of non-additive effects was not always noted and often depended on trait. For instance, inclusion of dominance effects in BLR and RR-BLUP had no effect on GS accuracy in rubber tree [27]. Compared to an additive-only model, the use of an additive + dominance model in eucalyptus improved the predictive abilities for growth (mean annual increment) but not for wood quality traits [42]. Inclusion of dominance effects increased R2 for tree height and acoustic velocity in Norway spruce both in the pedigree-based (ABLUP) and the genomic-based (GBLUP) models [55]. Yet, R2 for Pilodyn penetration (a surrogate for the trait of wood density) and MOE did not change in either model, which was consistent with the zero estimates of dominance variations for both traits. Furthermore, the full genomic-based model with additive, dominance, and epistatic effects (GBLUP-ADE) almost did not differ from the GBLUP-AD model for all four traits (height, acoustic velocity, Pilodyn penetration, and MOE), which indicates the absence of three kinds of epistatic interactions in Norway spruce. Finally, as shown on eucalyptus hybrids, 0% to 30% of the phenotypic variance for growth traits could be attributed to epistatic variation depending on the plant age, and these findings are consistent with classical breeding studies in *Eucalyptus* [52]. However, epistasis showed no effect on basic density and pulp yield in eucalyptus, in contrast to El-Dien et al. [59], where the epistatic effect on wood density was noted. Thus, the contribution of non-additive effects may depend on plant species, trait, and age.

5.3. Multi-Trait and Multi-Environment GS

In classical breeding, tests under various environmental conditions are essential for assessing the influence of environmental factors on genotype behavior and for studying genotype by environment interactions (G × E). G × E interactions are quite common in forest trees and are used to estimate the effectiveness of the same lines under different environmental conditions in genotype stability studies [11]. However, most GS studies use estimates obtained in only one environment, which prevents the use of information about environmental factors and thus, limits the predictive abilities of the GS model. Meanwhile, inclusion of genotype-by-environment interactions in GS models can improve their prediction accuracy. The effect of a genomic marker can be considered as a function of environmental covariates that can be evaluated by GS methods [120].

Genotype-by-environment interactions in trees are usually studied on contrasting sites [114]. Resende et al. [26] planted loblolly pine on four sites and found that the equation for any single site provided good prediction accuracy (0.64–0.74) within that site and a lower accuracy (0.18–0.66) for other sites. Site location was also important: where cross-validation was performed between sites within the same state (Florida or Georgia), the accuracy was higher compared with sites located in different states (0.54 and 0.37, on average). This finding was later confirmed on interior spruce planted on three sites: any of the two GS models, RR-BLUP and Generalized Ridge Regression (GRR), designed for a specific site predicted GEBV for other sites with a worse accuracy (0.41–0.59 vs. 0.00–0.39) [15]. Thus, G × E interaction was evident even though all three sites were located within one breeding zone. The researchers also used the multi-site population for prediction of the GEBV for each individual site. Although the accuracies varied by site, they were higher (0.42–0.49) than in the cross-site validation study [15]. A GS study on Norway spruce planted on two sites in northern Sweden showed that for all four traits, the accuracies of within-site training and selection were always higher than those of cross-site training and selection [54]. A recent study on rubber tree demonstrated the superiority of multi-environment GS models over single-environment ones [53].

The effect of G×E may also depend on the trait. As known from classical breeding, G×E usually has a significant effect on growth traits in coniferous species, while having no influence on their wood quality [54]. The GS studies on white spruce planted on two sites in different ecoclimatic regions of Canada confirmed that growth traits were more sensitive to G×E than wood traits [57]. Similar findings were reported for Norway spruce [54]. Thus, a genomic model developed for one site can be used to predict GEBV for wood traits in another site, but one should bear in mind that G×E can have a stronger impact in a more heterogeneous environment. In general, the accuracy of a predictive model decreases when applied to a different environment, but the degree of the decrease depends on the trait. GS should also take G×E into account in order to select genotypes adaptable to changes in environmental conditions.

In addition to multi-environmental trials, multi-trait selection is also of considerable interest in GS. GS models are usually designed to predict only one trait. However, multi-trait models were also developed and helped improve the accuracy of BV predictions in animals [43], crops [116], and forest trees. Studies on *Eucalyptus grandis* demonstrated the superiority of the multiple-trait combined approach in predicting BVs over the single-trait combined approach, in particular for a low-heritability trait (height) [43]. As shown on a eucalyptus hybrid, positive genetic gains can be achieved by associating biomass, a proxy of WUE, and wood chemical traits (cellulose and lignin) [32]. Interesting results were obtained on interior spruce with multi-trait GS prediction models based on Principal Component Analysis (PCA) [15]. There is a known negative genetic correlation between wood yield and wood quality, which was confirmed by the results from PC1. However, it turned out that the use of PC2 and PC3 allowed a concurrent selection of traits with different phenotypic optima, i.e., PC2 and PC3 accessed different combinations of SNPs (i.e., causal genes) that work in the same direction [15]. All these studies show the feasibility of GS for several traits at a time, even with a negative correlation between them.

5.4. Epigenetic Effects

Along with genetic effects, there are also epigenetic effects, and there is growing evidence that epigenetics has the potential to contribute to important traits in many plant species. Epigenetic trait is a heritable change in gene expression without changes in DNA sequence. There are several epigenetic mechanisms, and the best studied one is the methylation of cytosines within CG dinucleotides [121]. These changes can occur much faster than the genetic ones and they respond to external stresses; therefore, they may be particularly important in the context of a rapid climate change [122]. The role of epigenetic changes in phenotypic plasticity has been increasingly studied in recent years. Studies of DNA methylation in response to climate change are of special importance for long-living trees because their long generation period limits their ability to respond to rapid environmental changes through genetic mechanisms [123]. For example, studies revealed an association between DNA methylation and climatic conditions in natural populations of valley oak (*Quercus lobata* Nee) [123] and a correlation of DNA methylation levels and biomass productivity of poplar plants grown under different water availability conditions [124]. In traditional plant breeding, there is such a phenomenon as transgressive segregation: the hybrid progeny has a wider phenotypic variation than its parents. Unlike heterotic phenotypes, extreme phenotypes caused by transgressive segregation are heritably stable, but the precise molecular mechanisms of this phenomenon remain unclear [125]. It is assumed that transgressive segregants in plant breeding that are likely due to genetic and/or epigenetic effects are more common for adaptive traits such as tolerance to abiotic stresses, and water and nutrient use efficiency that strengthen the plant viability [126].

Although epigenetic effects are considered to be inheritable (additive), they can also be non-additive. Such a phenomenon as heterosis, where the progeny (F1 hybrids) surpasses its parents in a number of traits, is often used in plant breeding, including forest breeding. However, despite the wide practical use of heterosis, its underlying biological mechanisms are not well understood [127]. Epigenetic changes were shown to be involved in the generation of heterotic phenotype in annual plants [128]. Later, Gao et al. [129] investigated the role of DNA methylation changes in F1 hybrids of *Populus deltoides* and found non-additive levels of methylation, suggesting a role of DNA methylation in heterosis of forest trees. Thus, the stability of epigenetic changes (their heritability or non-heritability) may be of great importance in GS, as it will affect the degree of LD of epialleles with nearby SNPs [130].

Statistical discrimination between genetic and epigenetic changes underlying phenotype changes has not been studied well enough yet. The existence of epigenetic inheritance means that matrix A, which is used to describe inheritance in the calculation of estimated breeding values (EBVs), does not exactly describe the genetic similarity between relatives [131]. A recent application of Bayesian statistical models to assess the epigenetic architecture of complex traits in clinical practice showed that BayesRR distinguish between the variance explained by genetic markers and methylation probes better than LASSO or ridge regression [132].

It is already known that epigenetics can influence all aspects of the phenotypic variance in plants, including genetic variance, environmental variance, G×E, etc. [121]. Clinical studies showed that DNA methylation profiles have the potential to significantly improve complex trait prediction over and above that of SNP markers [133]. Recent advances in genomic technologies and bioinformatics have made genome-wide high-resolution methylome analysis feasible even for very large genomes [134]. Thus, the use of these technologies in the GS of trees to detect epigenetic polymorphisms (markers) helps understand the role of epigenetics in phenotypic variation and improve predictive abilities.

6. Accuracy Drivers in Genomic Predictions

The efficiency of GS in plants was initially evaluated using computer simulations. The first study of this kind was performed by Bernardo et al. [135] in maize, and was soon followed by a simulation study on a woody plant, oil palm [136]. According to the latter, response to GS was higher than to MAS for all population sizes, and higher than to phenotypic selection for a population of 50 or 70 individuals. Simulation studies on forest trees [23,25] showed that GS was more efficient than traditional breeding. According to these simulations and the subsequent empirical studies, the accuracy of GS in forest trees depended on several key factors [19,51]: (1) the extent of LD between markers and QTLs, (2) effective population size (Ne), (3) marker density, (4) training population size, (5) relatedness between training and test populations, (6) genetic architecture of a trait (number of loci and effect size), (7) trait heritability. Of these, only the inheritance and the genetic architecture of traits cannot be controlled by the breeder, while the others can be, more or less.

6.1. Linkage Disequilibrium, Effective Population Size, and Marker Density

The first three of the above-mentioned factors are interrelated. Linkage disequilibrium (LD) is particularly important in the context of GS in forest trees [23]. GS uses the LD between markers and QTLs by employing a large number of DNA markers: with dense coverage, any QTL associated with a trait would be in LD with at least one marker [18,137]. Thus, GS prediction accuracy depends on the degree of LD, which, in turn, depends on the effective population size (Ne) of training population and the marker density.

Most tree genomes have a low level of LD, which is due to high outcrossing rates, long-distance propagule dispersal, and large effective population sizes [138]. Grattapaglia et al. [23] suggested that the low LD levels in most forest tree species could be increased by reducing the effective population size and increasing the marker density. Ne is one of the most important parameters in population genetics because it determines the effectiveness of natural selection [139]. Ne is also important in GS: the smaller Ne, the higher the predictive ability, since a lower Ne is associated with a higher LD, and hence, a stronger association between markers and QTLs over long distances throughout the genome [66]. It is believed that the success of GS in trees mostly depends on the interplay of two factors: (1) Ne that determines the marker density and (2) LD that can be controlled by selecting Ne (it can be increased by reducing Ne) [140].

A simulation study on forest trees revealed a non-proportional inverse relationship between marker density and Ne: an adequate GS accuracy (≥ 0.68) was achieved with a marker density of about 2 markers per cM for Ne = 30, about 10 markers per cM with Ne = 60, and up to 20 markers per cM with Ne = 100 [23]. Similar results were obtained in a simulation in conifers (*Cryptomeria japonica*): efficient GS was achieved with one marker per cM [25]. Forest tree genomes are about 1000–2000 cM long: from 919 to 1814 cM for *Eucalyptus* species [141], 1637 cM for *Pinus taeda* [142], 1859 cM for Douglas-fir [143]. Thus, with a small Ne, the GS of an average genome would require as little as 3000–5000 markers, but even with a 2–3-fold larger Ne it would require several dozens or hundreds of times more markers, depending on the genome size.

Practically, non-domesticated forest trees are characterized by a large genetic diversity and a large Ne. Ne can be reduced by using populations of related individuals (half-sibs or full-sibs), as done in some tree breeding programs. Meuwissen et al. [144] found that adequate accuracy of predicted breeding values in populations of unrelated individuals can be achieved with a minimum of 10×Ne×L, where L is the total genome length in Morgans. Within-family GS requires much less markers, as they would track only the large chromosome segments shared by family members. For example, GS in a biparental population of apple trees showed a good accuracy (>0.7) with as little as 2500 markers [145], whereas theoretically it would have required 130,000 (10 × 1000 × 13) markers in an outbreeding population of apples [111]. A study on eucalyptus showed that a reduction in Ne from 51 to 11 improved the accuracy from 0.65 to 0.80 [22]. On the other hand, a large effective population size may lead to greater recombination and genetic diversity within the population due to high outcrossing rates in wind-pollinated species, which is favorable for a long-term genetic gain [35].

Experimental studies in forest tree species demonstrate quite fair predictive abilities at relatively moderate genotyping densities (2500–10,000 SNPs), probably due to the impact of relatedness as a driver of accuracy [11]. For example, in a study on *Picea abies*, the accuracy reached a plateau at 4000–8000 SNPs [55]. A subset of 3000–4000 markers was sufficient to reach the same predictive abilities and accuracies as the full set of 8719 markers in Scots pine [68]. If possible, however, higher marker densities should be used to increase the prediction accuracy. Furthermore, higher marker densities may be necessary where the training and test populations do not originate genetically from the same primary population [144].

6.2. Size and Structure of Tree Populations in GS

A larger training population allows more accurate assessment of marker effects and, hence, a higher accuracy of GEBV for candidate selection. Unlike some agricultural crops and even more so animals, for most forest trees, the size of training population is usually not a limiting factor. A simulation study showed that with a population size of more than 2000 individuals, the prediction accuracy leveled off to a plateau regardless of Ne and marker density [23]. However, with a high marker density or Ne < 30 individuals, 1000 individuals were enough to achieve accuracies of at least 0.7. Therefore, the training populations included 800–1200 trees in most studies (Table 2). A training population is usually sampled from an existing progeny trial derived from interbreeding (open or controlled pollinated) a few dozen elite parents. The costs of genotyping have dropped in recent years and it is no longer the limiting factor in the analysis of large populations. Moreover, genotyping of a specific individual has to be done only once. On the other hand, the costs of phenotyping remain high because it requires ample manual labor and should preferably be repeated in different years, environments, and in several replications. Therefore, it is phenotyping that is becoming the limiting factor for increasing the size of training population. The use of HTP methods will help remove this limitation.

In theory, a small Ne (related individuals) should provide higher prediction accuracy than unrelated genotypes. Calculations show that the use of 1000 eucalyptus half-sibs potentially has a very high predicted GEBV accuracy (>0.9) owing to the relatively small effective number of independent chromosome segments within the half-sib families [111]. To test this statement, the effectiveness of GS was evaluated in a white spruce (*Picea glauca*) population of a large effective size [58]. The study showed that the accuracy of GEBVs obtained with individuals of unknown relatedness was lower, with about half of the accuracy achieved with half-sibs.

GS studies on forest trees generally reported moderate to high accuracies of selection models with correlations from 0.6 to 0.8 for full-sib-families, from 0.3 to 0.5 in half-sibs, and a very low predictive ability for unrelated individuals [64]. Furthermore, the full-sib family structure required significantly fewer markers than did the half-sib family structure to achieve the same effect. For example, in the full-sib family of Norway spruce, 250 markers provided the same accuracy, and 750 markers provided the same predictive ability for all four traits, as 100,000 markers in the half-sib family structure [54].

The degree of relatedness in a training population affects the prediction accuracy and is an important factor in the relationships between training and test populations, as was demonstrated in a number of tree studies. In these studies, the progeny of full- and/or half-sibs within one generation was usually divided into training and test populations [56]. This improved the accuracy of GEBV prediction by increasing the likelihood of the same chromosome segments being present in both populations [111]. On the contrary, the use of unrelated populations reduced the accuracy of predictive models. For example, models built with a half-sib structure of *Picea mariana* (Mill.) Britton, Sterns, and Poggenb. led to a large decrease in accuracy, predictability, and genetic gain compared with a full-sib design [64]. In a study on eucalyptus, the average values of the realized genomic relationships among full-sibs, half-sibs, and unrelated individuals consistently decreased (0.309, 0.131, and 0.0056, respectively).

In addition to the degree of relatedness, the ratio of training to test population was also important. In a study on Norway spruce, Chen et al. [54] compared the effects of five different training to test ratios (1:1, 3:1, 5:1, 7:1, and 9:1) on the accuracy of statistical models. The GS accuracy increased with the increasing ratio, although it also depended on the evaluated trait: for tree height, the maximum accuracy was achieved at the ratio of 5:1, whereas for wood quality traits, only minor improvements were observed after the ratio had been increased to 3:1. In a similar study on a deciduous species, eucalyptus genotypes were divided into five different size groups with the training to test ratios of 1:1, 2:1, 3:1, 4:1, or 9:1 [51]. In contrast to the findings of Chen et al. [54], here, the predictive ability significantly improved after the ratio had been increased from 1:1 to 9:1 (which supports the importance of an adequate size of the training set in GS), but it did not depend on trait. However, the ratio increase from 1:1 to 2:1 caused a greater increase in predictive ability than the ratio increase from 2:1 to 9:1. Thus, it may be more appropriate to improve the predictive accuracy by increasing the size of training population rather than the marker density.

6.3. Heritability and Genetic Architecture of Traits

Trait heritability and genetic architecture also affect model accuracy, but, unlike the above key factors, they are natural features and cannot be controlled by the breeder. To some extent, they can be influenced by choosing an adequate statistical model. Traits with a simpler architecture (e.g., disease resistance) have fewer loci that control large proportions of phenotypic variance. These are the features that are best suited for MAS applications and are better predictable in GS. More complex quantitative traits (growth, wood properties) may be controlled by dozens or hundreds of QTLs with weaker effects [11]. Given these differences in marker effects, calculations can be done using models that provide for different or equal contributions of all markers to the observed variability. For instance, the growth and wood attributes of interior spruce were more accurately predicted by RR-BLUP than by GRR, thus suggesting the complex genetic architecture of the traits [15]. De Almeida Filho et al. [34] compared predictions of polygenic (height, 1000 QTLs) and oligogenic (disease resistance, 30 QTLs) traits in a simulated population of loblolly pine (*Pinus taeda*). Computations showed that the models Bayes A and Bayes B were more accurate than BL and BRR for oligogenic traits in all scenarios. Thus, RR-BLUP (frequentist version of BRR) is better suited for predicting complex traits than simple ones.

The heritability of a trait can be defined as the fraction of phenotype variability which is due to genetic variation. Narrow-sense heritability considers only additive genetic effects, while disregarding the non-additive effects (dominance, epistasis) and genotype-environment interaction [111]. Traits with a higher heritability are more accurately described by marker effects and are less dependent on other factors. Heritability was shown to have a relatively small impact on the accuracy where the training population was large enough for adequate assessment of marker effects [11]. However, in the same way as for genetic architecture, modeling of dominance effects can improve the GS models for some traits [66]. The heritability of economically valuable traits is usually high enough to assure accurate GS of plants, provided there are sufficient markers and large training populations [111].

7. Economic Efficiency of GS in Tree Breeding

The current common breeding approaches—traditional, MAS, and GS—differ in cost and efficiency. Breeding of new varieties should consider the economic aspect in the process of cost–benefit analysis. Traditional breeding is based on phenotypic selection and has a long cycle and a low efficiency when selecting for complex traits [17]. Owing to early evaluation of the genotype, MAS can shorten the breeding cycle, which could otherwise reach 25 or more years in boreal conifers [146], but it is still ineffective for selection of polygenic traits. GS can shorten the breeding process even further, primarily by reducing long and expensive field trials, but this technology is less than 10 years old and its use is currently limited due to poorly studied genomes of many tree species. Nearly all studies that compared GS and traditional breeding of forest trees showed the superiority of GS, which significantly reduced the breeding cycle (Table 2). This was mainly achieved due to the possibility to evaluate plants at a very young age (down to seedlings). In loblolly pine, the efficiency of GS was 53–112% higher than with traditional breeding, and the breeding cycle shortened by 50% [26]. In radiata pine, the efficiency of GS was 37–115% higher, and the breeding cycle shortened from 17 to 9 years [67] compared with traditional breeding. The results in interior spruce were more modest than in pine species: efficiency increased by 6–33% and the breeding cycle reduced by 25% [14]. A twofold reduction in the breeding cycle was achieved in eucalyptus [22,46] and even threefold reduction in some Picea species [58,64] and rubber tree [53]. At the same time, GS relies on expensive genotypic analysis, and this must be taken into account when comparing breeding methods in terms of efficiency.

The economic efficiency of a breeding method may depend on such factors as population size, trait heritability, costs of plant phenotyping and genotyping, etc. For example, a simulation study on herbaceous plants showed that GS was more cost-effective than phenotypic selection if the following conditions were met: the heritability of traits of interest was below 0.25, and the cost of one plant phenotyping did not exceed that of genotyping [147]. Forest trees species differ in genome organization, duration of juvenile period, vegetative propagation abilities, and different rotation periods and requirements for growing conditions. The traits of interest for selection are also very diverse: quantitative (growth rate), qualitative (wood properties), and those responsible for tolerance to biotic and abiotic stresses. All the above can influence genotyping and phenotyping, propagation and cultivation during the breeding process, and makes each GS program unique.

The cost-effectiveness of GS in forestry depends on many factors and can vary greatly in each particular case. This technology is, perhaps, hardly suitable for small breeding programs with minor tree species and a poorly studied genome, but when used on major industrial forest species, its considerable time gain may be very important from a commercial perspective [11]. Resende et al. [22] calculated that the costs of GS of 20,000 eucalyptus seedlings would be paid back at least 20 times due to a 1% increase in pulp yield, and 9 years earlier than in the case of conventional breeding. On the other hand, genetic technologies are rapidly developing, and in the coming years, genome exploration may stop being a limiting factor.

The breeding of trees, even using GS methods, is a long and expensive process, and therefore, a detailed cost-benefit analysis is absolutely necessary before its practical implementation [11]. The very first GS study on woody plants (oil palm) [136] estimated that the cost per unit gain with GS was

26–57% lower than with phenotypic selection when markers cost USD 1.50 per data point, and 35–65% lower when markers cost USD 0.15 per data point. Only recently, however, there appeared full-scale studies on forest trees, with the assessment of various scenarios of breeding, propagation, plantation establishment, forest management, and harvesting. Chang et al. [148] conducted a stand-level financial benefit-cost analysis to compare GS and traditional breeding of two major commercial tree species in Western Canada, white spruce and lodgepole pine, if grown for up to 250 years. According to the results, GS could shorten the breeding cycle for these species in Canada from 33 to 18 years, but under current market conditions, traditional breeding should still remain the main tree breeding strategy for producing improved seedlings of white spruce and lodgepole pine for reforestation. Yet, GS would become a promising approach in the following scenarios: (1) an increased log price premium at harvest; (2) reduced seedling costs; (3) achieving higher genetic gain; and (4) planting on high-productivity sites. This study also shows the importance of choosing an appropriate reference year for comparing different breeding strategies, as this can significantly influence the conclusions made [148]. The authors emphasize that their calculations were made for the current market conditions and breeding strategies for white spruce and lodgepole pine in the province of Alberta, Canada, and cannot be extrapolated to other species and regions of the world. For example, a shorter rotation period and a lower cost of pine seedlings in the southern USA, New Zealand, or Chile may make GS financially more attractive.

Another study compared the financial performance of various breeding and deployment scenarios, with or without GS, in the context of intensively managed plantations of white spruce in Quebec (Canada) [149]. The duration of a classical breeding cycle, 34 years, was similar to the previous study, but the rotation period lasted for up to 60 years, and weeds were controlled with herbicides. According to the results of the study, the best scenario used GS with somatic embryogenesis (SE), followed by a scenario with GS in combination with top-grafting for the production of improved seedlings. As was already reported earlier, the most significant breeding cycle reduction with maximum increase in genetic gain could be achieved by combining GS and SE [150]. Li et al. [35] reported that the use of SE for propagation, even after traditional breeding, led to an additional 8.11% genetic gain compared with propagation by seeds of the selected individuals. On the other hand, combining GS with top-grafting resulted in a significant increase in additional genetic gain per year in conifers due to the reduced age of coning from 5 to 3 years [35]. Thus, top-grafting could be an interesting alternative to increase genetic gain and shorten breeding cycles for tree breeding programs where SE at an operational scale is yet not available [149]. Finally, the continuing reduction in genotyping costs is making GS increasingly financially attractive for use on forest trees.

8. Perspectives of GS in Forestry

Studies of GS in forest trees have been conducted for less than 10 years, but they have already demonstrated the promising prospects of this approach in forest tree improvement. A few years ago, Isik et al. [1] named three challenges for GS in trees: (1) lack of reliable, repeatable, and cost-efficient genotyping platforms; (2) lack of infrastructure to store, retrieve, and analyze large numbers of markers; (3) lack of well-designed and well-tested breeding populations for many species. The latest advances in genomics and bioinformatics can help solve the first two problems rather quickly, but the establishment of breeding population takes much longer. Nevertheless, the list of tree genera used in GS studies has significantly expanded over the last 2–3 years and is likely to continue expanding in the future. Among boreal species, good candidates for inclusion in the list are representatives of the genera *Betula* and *Quercus*. Russian breeders have been working with such species as *B. pendula* Roth, *B. pubescens* Ehrh., and *Q. robur* L. for several decades, and the breeding populations, including full-sib families, can be used for GS. Genotyping of these species is facilitated by the availability of the recently sequenced reference genomes of birch (*B. pendula*) [151] and oak (*Q. robur*) [152]. Their small genomes, 440 and 740 Mb, respectively, are ten times smaller than those of pine and spruce (~20 Gb), which is also an advantage. Birch and oak have valuable wood and they are very promising for plantation forestry in southern regions of European Russia. Of tropical species, *Tectona grandis* L.f., for which a draft genome was recently published [153], can be considered as the best candidates for GS. It was already shown that the breeding cycle in coniferous trees can be further shortened by using grafting and somatic embryogenesis. In deciduous species, flowering can be accelerated with the fast-track breeding system using genetic transformation by flowering inducing genes. The system was successfully applied on fruit trees—apple [154], plum [155], and trifoliate orange [156]—and made it possible to reduce the juvenile period several times.

The list of traits of interest for GS will also expand. Global climate change is drastically changing the cultivation conditions of forest trees, especially in the boreal regions. On the one hand, increasing temperature can improve forest productivity. On the other hand, the duration and intensity of various stresses also increase, and traditional breeding fails to respond timely to these challenges. GS programs would probably also include such traits as tolerance to abiotic stresses, primarily drought. It is difficult and time-consuming to assess the state of plants exposed to stress factors using traditional approaches, hence the need for wider adoption of HTP techniques in GS, because they allow such assessment to be made remotely by capturing images in various spectral regions. Stresses are environmental factors that occur unpredictably and usually continue for a limited period of time, and can affect economically important traits of trees. Considering the above-said factors, GS programs should make a wider use of multi-environment, multi-age, and multi-trait statistical models.

In addition, the proper use of epigenetic variance can open up new opportunities for improvement of forest trees. This would require a detailed understanding of how to predict the stability of epigenetic variants (their additive or non-additive nature) so that epigenetics could be used to improve valuable heritably stable traits, primarily stress tolerance. In particular, changes in methylation of transposable elements can be responsible for the variability of traits [157]. This is especially important for conifers, whose large genomes contain a high number of transposable elements, which provide vast potential genetic resources [158].

Finally, we can expect more extensive use of GS simulations. The GS of forest trees began with simulation studies and such studies are still occasionally conducted. As shown by the recent cost–benefit studies, the economic efficiency of GS implementation depends on a large number of factors, including those difficult to predict (e.g., the market value of wood). GS programs are expensive and lengthy, although much shorter than traditional breeding, and their implementation should be preceded by simulation studies to select optimal strategies for specific species, traits, and growing conditions.

9. Conclusions

Despite its short history, GS has already proved to be a powerful tool in plant breeding. It is especially useful for the prediction of complex quantitative traits, including productivity and wood quality, the main characteristics of woody plants for cultivation on forest plantations. In each specific case, it is essential to consider a number of factors and their combinations with the biggest impact on the selection efficacy. Clarification of the pedigrees of individuals from breeding populations using molecular markers will improve the accuracy of predicting their breeding values. In general, studies on GS of forest trees confirm that it can significantly reduce the duration of the breeding cycle and increase the genetic gain. The reduction in genotyping and phenotyping costs is likely to contribute to a wider use of this method in forest breeding.

Author Contributions: Conceptualization, V.G.L.; writing—original draft preparation, V.G.L. and T.N.L.; writing—review and editing, V.G.L. and K.A.S.; funding acquisition, A.I.C. and K.A.S. All authors have read and agreed to the published version of the manuscript.

Funding: This research was carried out within the state program of Ministry of Science and High Education of the Russian Federation (theme "Plant molecular biology and biotechnology: their cultivation, pathogen and stress protection (BIBCH)" (No 0101-2019-0037).

Conflicts of Interest: The authors declare no conflict of interest.

References

1. Isik, F.; Kumarx, S.; Martínez-García, P.J.; Iwatajj, H.; Yamamoto, T. Acceleration of forest and fruit tree domestication by genomic selection. *Adv. Bot. Res.* **2015**, *74*, 93–124. [CrossRef]
2. Isik, F. Genomic selection in forest tree breeding: The concept and an outlook to the future. *New For.* **2014**, *45*, 379–401. [CrossRef]
3. Harfouche, A.; Meilan, R.; Kirst, M.; Morgante, M.; Boerjan, W.; Sabatti, M.; Mugnozza, G.S. Accelerating the domestication of forest trees in a changing world. *Trends Plant Sci.* **2012**, *17*, 64–72. [CrossRef] [PubMed]
4. Durán, R.; Isik, F.; Zapata-Valenzuela, J.; Balocchi, C.; Sofía Valenzuela, S. Genomic predictions of breeding values in a cloned Eucalyptus globulus population in Chile. *Tree Genet. Genomes* **2017**, *13*, 74. [CrossRef]
5. Muranty, H.; Jorge, V.; Bastien, C.; Lepoittevin, C.; Bouffier, L.; Sanchez, L. Potential for marker-assisted selection for forest tree breeding: Lessons from 20 years of MAS in crops. *Tree Genet. Genomes* **2014**, *10*, 1491–1510. [CrossRef]
6. Ballesta, P.; Serra, N.; Guerra, F.P.; Hasbún, R.; Mora, F. Genomic prediction of growth and stem quality traits in *Eucalyptus globulus* Labill. at its southernmost distribution limit in Chile. *Forests* **2018**, *9*, 779. [CrossRef]
7. Bartholomé, J.; van Heerwaarden, J.; Isik, F.; Boury, C.; Vidal, M.; Plomion, C.; Bouffier, L. Performance of genomic prediction within and across generations in maritime pine. *BMC Genom.* **2016**, *17*, 604. [CrossRef]
8. Kainer, D.; Stone, E.A.; Padovan, A.; Foley, W.J.; Külheim, C. Accuracy of genomic prediction for foliar terpene traits in *Eucalyptus polybractea*. *G3 Genes Genomes Genet.* **2018**, *8*, 2573. [CrossRef]
9. Zapata-Valenzuela, J.; Whetten, R.W.; Neale, D.; McKeand, S.; Isik, F. Genomic estimated breeding values using genomic relationship matrices in a cloned population of loblolly pine. *G3 Genes Genomes Genet.* **2013**, *3*, 909–916. [CrossRef]
10. Desta, Z.A.; Ortiz, R. Genomic selection: Genome-wide prediction in plant improvement. *Trends Plant Sci.* **2014**, *19*, 592–601. [CrossRef]
11. Grattapaglia, D. Status and perspectives of genomic selection in forest tree breeding. In *Genomic Selection for Crop Improvement*; Varshney, R.K., Roorkiwal, L., Sorrells, M.E., Eds.; Springer International Publishing AG: Cham, Switzerland, 2017; pp. 199–249. [CrossRef]
12. Zapata-Valenzuela, J.; Isik, F.; Maltecca, C.; Wegrzyn, J.; Neale, D.; McKeand, S.; Whetten, R. SNP markers trace familial linkages in a cloned population of *Pinus taeda*—Prospects for genomic selection. *Tree Genet. Genomes* **2012**, *8*, 1307–1318. [CrossRef]
13. Nagano, S.; Hirao, T.; Takashima, Y.; Matsushita, M.; Mishima, K.; Takahashi, M.; Iki, T.; Ishiguri, F.; Hiraoka, Y. SNP genotyping with target amplicon sequencing using a multiplexed primer panel and its application to genomic prediction in Japanese cedar, *Cryptomeria japonica* (L.f.) D.Don. *Forests* **2020**, *11*, 898. [CrossRef]
14. Ratcliffe, B.; El-Dien, O.G.; Klápště, J.; Porth, I.; Chen, C.; Jaquish, B.; El-Kassaby, Y.A. A comparison of genomic selection models across time in interior spruce (*Picea engelmannii* × *glauca*) using unordered SNP imputation methods. *Heredity* **2015**, *115*, 547–555. [CrossRef] [PubMed]
15. El-Dien, O.G.; Ratcliffe, B.; Klápště, J.; Chen, C.; Porth, I.; El-Kassaby, Y.A. Prediction accuracies for growth and wood attributes of interior spruce in space using genotyping-by-sequencing. *BMC Genom.* **2015**, *16*, 370. [CrossRef]
16. Meuwissen, T.H.E.; Hayes, B.J.; Goddard, M.E. Prediction of total genetic value using genome-wide dense marker maps. *Genetics* **2001**, *157*, 1819–1829.
17. Heffner, E.L.; Lorenz, A.J.; Jannink, J.-L.; Sorrells, M.E. Plant breeding with genomic selection: Gain per unit time and cost. *Crop Sci.* **2010**, *50*, 1681–1690. [CrossRef]
18. Isik, F.; Bartholomé, J.; Farjat, A.; Chancerel, E.; Raffina, A.; Sanchez, L.; Plomion, C.; Bouffiera, L. Genomic selection in maritime pine. *Plant Sci.* **2016**, *242*, 108–119. [CrossRef]
19. Thistlethwaite, F.R.; Ratcliffe, B.; Klápště, J.; Porth, I.; Chen, C.; Stoehr, M.U.; El-Kassaby, Y.A. Genomic selection of juvenile height across a single-generational gap in Douglas-fir. *Heredity* **2019**, *122*, 848–863. [CrossRef]
20. Bhat, J.A.; Ali, S.; Salgotra, R.K.; Mir, Z.A.; Dutta, S.; Jadon, V.; Tyagi, A.; Mushtaq, M.; Jain, N.; Singh, P.K.; et al. Genomic selection in the era of next generation sequencing for complex traits in plant breeding. *Front. Genet.* **2016**, *7*, 221. [CrossRef]

21. Crossa, J.; Pérez-Rodríguez, P.; Cuevas, J.; Montesinos-López, O.; Jarquín, D.; de los Campos, G.; Burgueño, J.; Camacho-González, J.M.; Pérez-Elizalde, S.; Beyene, Y.; et al. Genomic selection in plant breeding: Methods, models, and perspectives. *Trends Plant Sci.* **2017**, *22*, 961–975. [CrossRef]

22. Resende, M.D.V.; Resende, M.F.R., Jr.; Sansaloni, C.P.; Petroli, C.D.; Missiaggia, A.A.; Aguiar, A.M.; Abad, J.M.; Takahashi, E.K.; Rosado, A.M.; Faria, D.A.; et al. Genomic selection for growth and wood quality in *Eucalyptus*: Capturing the missing heritability and accelerating breeding for complex traits in forest trees. *New Phytol.* **2012**, *194*, 116–128. [CrossRef] [PubMed]

23. Grattapaglia, D.; Resende, M.D.V.; Resende, M.R.; Sansaloni, C.P.; Petroli, C.D.; Missiaggia, A.A.; Takahashi, E.K.; Zamprogno, K.C.; Kilian, A. Genomic Selection for growth traits in *Eucalyptus*: Accuracy within and across breeding populations. *BMC Proc.* **2011**, *5*, O16. [CrossRef]

24. Grattapaglia, D.; Silva-Junior, O.B.; Resende, R.T.; Cappa, E.P.; Müller, B.S.F.; Tan, B.; Isik, F.; Blaise Ratcliffe, B.; Yousry, A. El-Kassaby, Y.A Quantitative genetics and genomics converge to accelerate forest tree breeding. *Front. Plant Sci.* **2018**, *9*, 1693. [CrossRef] [PubMed]

25. Iwata, H.; Hayashi, T.; Tsumura, Y. Prospects for genomic selection in conifer breeding: A simulation study of *Cryptomeria japonica*. *Tree Genet. Genomes* **2011**, *7*, 747–758. [CrossRef]

26. Resende, M.F.R., Jr.; Munoz, P.; Acosta, J.J.; Peter, G.F.; Davis, J.M.; Grattapaglia, D.; Resende, M.D.V.; Kirst, M. Accelerating the domestication of trees using genomic selection: Accuracy of prediction models across ages and environments. *New Phytol.* **2012**, *193*, 617–624. [CrossRef]

27. Cros, D.; Mbo-Nkoulou, L.; Bell, J.M.; Oum, J.; Masson, A.; Soumahoro, M.; Tran, D.M.; Achour, Z.; Le Guen, V.; Clement-Demange, A. Within-family genomic selection in rubber tree (*Hevea brasiliensis*) increases genetic gain for rubber production. *Ind. Crops Prod.* **2019**, *138*, 111464. [CrossRef]

28. Westbrook, J.W.; Zhang, Q.; Mandal, M.K.; Jenkins, E.V.; Barth, L.E.; Jenkins, J.W.; Grimwood, J.; Schmutz, J.; Holliday, J.A. Optimizing genomic selection for blight resistance in American chestnut backcross populations: A trade-off with American chestnut ancestry implies resistance is polygenic. *Evol. Appl.* **2020**, *13*, 31–47. [CrossRef]

29. Stocks, J.J.; Metheringham, C.L.; Plumb, W.J.; Lee, S.J.; Kelly, L.J.; Nichols, R.A.; Buggs, R.J.A. Genomic basis of European ash tree resistance to ash dieback fungus. *Nat. Ecol. Evol.* **2019**, *3*, 1686–1696. [CrossRef]

30. Lenz, P.R.N.; Nadeau, S.; Mottet, M.-J.; Perron, M.; Isabel, N.; Beaulieu, J.; Bousquet, J. Multi-trait genomic selection for weevil resistance, growth, and wood quality in Norway spruce. *Evol. Appl.* **2020**, *13*, 76–94. [CrossRef]

31. Beaulieu, J.; Nadeau, S.; Ding, C.; Celedon, J.M.; Azaiez, A.; Carol Ritland, C.; Laverdière, J.-P.; Marie Deslauriers, M.; Adams, G.; Fullarton, M.; et al. Genomic selection for resistance to spruce budworm in white spruce and relationships with growth and wood quality traits. *Evol. Appl.* **2020**, 1–19. [CrossRef]

32. Bouvet, J.-M.; Ekomonod, C.G.M.; Brendelf, O.; Laclaug, J.-P.; Bouillet, J.-P.; Epron, D. Selecting for water use efficiency, wood chemical traits and biomass with genomic selection in a *Eucalyptus* breeding program. *For. Ecol. Manag.* **2020**, *465*, 118092. [CrossRef]

33. Denis, M.; Bouvet, J.-M. Efficiency of genomic selection with models including dominance effect in the context of *Eucalyptus* breeding. *Tree Genet. Genomes* **2013**, *9*, 37–51. [CrossRef]

34. de Almeida Filho, J.E.; Guimarães, J.F.R.; de Silva, F.F.; de Resende, M.D.V.; Muñoz, P.; Kirst, M.; Resende, M.F.R., Jr. The contribution of dominance to phenotype prediction in a pine breeding and simulated population. *Heredity* **2016**, *117*, 33–41. [CrossRef] [PubMed]

35. Li, Y.; Dungey, H.S. Expected benefit of genomic selection over forward selection in conifer breeding and deployment. *PLoS ONE* **2018**, *13*, e0208232. [CrossRef] [PubMed]

36. Stejskal, J.; Lstibůrek, M.; Klápště, J.; Čepl, J.; El-Kassaby, Y.A. Effect of genomic prediction on response to selection in forest tree breeding. *Tree Genet. Genomes* **2018**, *14*, 74. [CrossRef]

37. Hiraoka, Y.; Fukatsu, E.; Mishima, K.; Hirao, T.; Teshima, K.M.; Tamura, M.; Tsubomura, M.; Iki, T.; Kurita, M.; Takahashi, M.; et al. Potential of genome-wide studies in unrelated plus trees of a coniferous species, *Cryptomeria japonica* (Japanese cedar). *Front. Plant Sci.* **2018**, *9*, 1322. [CrossRef]

38. Müller, B.S.F.; Neves, L.G.; de Almeida Filho, J.E.; Resende, M.F., Jr.; Muñoz, P.R.; dos Santos, P.E.T.; Filho, E.P.; Kirst, M.; Dario Grattapaglia, D. Genomic prediction in contrast to a genome-wide association study in explaining heritable variation of complex growth traits in breeding populations of *Eucalyptus*. *BMC Genom.* **2017**, *18*, 524. [CrossRef]

39. Ballesta, P.; Bush, D.; Silva, F.F.; Freddy Mora, F. Genomic predictions using low-density SNP markers, pedigree and GWAS information: A case study with the non-model species *Eucalyptus cladocalyx*. *Plants* **2020**, *9*, 99. [CrossRef]

40. Ballesta, P.; Maldonado, C.; Pérez-Rodríguez, P.; Mora, F. SNP and haplotype-based genomic selection of quantitative traits in *Eucalyptus globulus*. *Plants* **2019**, *8*, 331. [CrossRef]

41. Cappa, E.P.; El-Kassaby, Y.A.; Muñoz, F.; Garcia, M.N.; Villalba, P.V.; Klápště, J.; Poltri, S.N.M. Improving accuracy of breeding values by incorporating genomic information in spatial-competition mixed models. *Mol. Breed.* **2017**, *37*, 125. [CrossRef]

42. Resende, R.T.; Resende, M.D.V.; Silva, F.F.; Azevedo, C.F.; Takahashi, E.K.; Silva-Junior, O.B.; Grattapaglia, D. Assessing the expected response to genomic selection of individuals and families in *Eucalyptus* breeding with an additive-dominant model. *Heredity* **2017**, *119*, 245–255. [CrossRef] [PubMed]

43. Cappa, E.P.; El-Kassaby, Y.A.; Muñoz, F.; Garcia, M.N.; Villalba, P.V.; Klápště, J.; Poltri, S.N.M. Genomic-based multiple-trait evaluation in *Eucalyptus grandis* using dominant DArT markers. *Plant Sci.* **2018**, *271*, 27–33. [CrossRef] [PubMed]

44. de Moraes, B.F.X.; dos Santos, R.F.; de Lima, B.M.; Aguiar, A.M.; Missiaggia, A.A.; da Costa Dias, D.; Rezende, G.D.P.S.; Gonçalves, F.M.A.; Acosta, J.J.; Kirst, M.; et al. Genomic selection prediction models comparing sequence capture and SNP array genotyping methods. *Mol. Breed.* **2018**, *38*, 115. [CrossRef]

45. Cappa, E.P.; de Lima, B.M.; da Silva-Junior, O.B.; Garcia, C.C.; Shawn, D.; Mansfield, S.D.; Grattapaglia, D. Improving genomic prediction of growth and wood traits in *Eucalyptus* using phenotypes from non-genotyped trees by single-step GBLUP. *Plant Sci.* **2019**, *284*, 9–15. [CrossRef] [PubMed]

46. Mphahlele, M.M.; Isik, F.; Mostert-O'Neill, M.M.; Reynolds, S.M.; Hodge, G.R.; Myburg, A.A. Expected benefits of genomic selection for growth and wood quality traits in *Eucalyptus grandis*. *Tree Genet. Genomes* **2020**, *16*, 49. [CrossRef]

47. Suontama, M.; Klápště, J.; Telfer, E.; Graham, N.; Stovold, T.; Low, C.; McKinley, R.; Dungey, H. Efficiency of genomic prediction across two *Eucalyptus nitens* seed orchards with different selection histories. *Heredity* **2019**, *122*, 370–379. [CrossRef] [PubMed]

48. Thavamanikumar, S.; Arnold, R.J.; Luo, J.; Thumma, B.R. Genomic studies reveal substantial dominant effects and improved genomic predictions in an open-pollinated breeding population of *Eucalyptus pellita*. *G3 Genes Genomes Genet.* **2020**, *10*, 3751–3763. [CrossRef]

49. Rambolarimanana, T.; Ramamonjisoa, L.; Verhaegen, D.; Tsy, J.-M.L.P.; Jacquin, L.; Cao-Hamadou, T.-V.; Makouanzi, G.; Bouvet, J.-M. Performance of multi-trait genomic selection for *Eucalyptus robusta* breeding program. *Tree Genet. Genomes* **2018**, *14*, 71. [CrossRef]

50. Bouvet, J.-M.; Makouanzi, G.; Cros, D.; Vigneron, P. Modeling additive and non-additive effects in a hybrid population using genome-wide genotyping: Prediction accuracy implications. *Heredity* **2016**, *116*, 146–157. [CrossRef]

51. Tan, B.; Grattapaglia, D.; Martins, G.S.; Ferreira, K.Z.; Sundberg, B.; Ingvarsson, P.K. Evaluating the accuracy of genomic prediction of growth and wood traits in two *Eucalyptus* species and their F1 hybrids. *BMC Plant Biol.* **2017**, *17*, 110. [CrossRef]

52. Tan, B.; Grattapaglia, D.; Wu, H.X.; Ingvarsson, P.K. Genomic relationships reveal significant dominance effects for growth in hybrid *Eucalyptus*. *Plant Sci.* **2018**, *267*, 84–93. [CrossRef] [PubMed]

53. Souza, L.M.; Francisco, F.R.; Gonçalves, P.S.; Junior, E.J.S.; Le Guen, V.; Fritsche-Neto, R.; Souz, A.P. Genomic selection in rubber tree breeding: A comparison of models and methods for managing G×E interactions. *Front Plant Sci.* **2019**, *10*, 1353. [CrossRef] [PubMed]

54. Chen, Z.-Q.; Baison, J.; Pan, J.; Karlsson, B.; Andersson, B.; Westin, J.; García-Gil1, M.R.; Wu, H.X. Accuracy of genomic selection for growth and wood quality traits in two controlpollinated progeny trials using exome capture as the genotyping platform in Norway spruce. *BMC Genom.* **2018**, *19*, 946. [CrossRef]

55. Chen, Z.-Q.; Baison, J.; Pan, J.; Westin, J.; Gil, M.R.G.; Wu, H.X. Increased prediction ability in Norway spruce trials using a marker × environment interaction and non-additive genomic selection model. *J. Hered.* **2019**, *110*, 830–843. [CrossRef]

56. Zhou, L.; Chen, Z.; Olsson, L.; Grahn, T.; Karlsson, B.; Wu, H.W.; Lundqvist, S.-O.; García- Gil, M.R. Effect of number of annual rings and tree ages on genomic predictive ability for solid wood properties of Norway spruce. *BMC Genom.* **2020**, *21*, 323. [CrossRef]

57. Beaulieu, J.; Doerksen, T.K.; MacKay, J.; Rainville, A.; Bousquet, J. Genomic selection accuracies within and between environments and small breeding groups in white spruce. *BMC Genom.* **2014**, *15*, 1048. [CrossRef]

58. Beaulieu, J.; Doerksen, T.; Clement, S.; MacKay, J.; Bousquet, J. Accuracy of genomic selection models in a large population of open-pollinated families in white spruce. *Heredity* **2014**, *113*, 343–352. [CrossRef] [PubMed]

59. El-Dien, O.G.; Ratcliffe, B.; Klápště, J.; Chen, C.; Porth, I.; El-Kassaby, Y.A. Implementation of the realized genomic relationship matrix to open-pollinated white spruce family testing for disentangling additive from nonadditive genetic effects. *G3 Genes Genomes Genet.* **2016**, *6*, 743–753. [CrossRef]

60. Ratcliffe, B.; El-Dien, O.G.; Cappa, E.P.; Porth, I.; Klápste, J.; Chen, C.; El-Kassaby, Y.A. Single-step BLUP with varying genotyping effort in open-pollinated *Picea glauca*. *G3 Genes Genomes Genet.* **2017**, *7*, 935–942. [CrossRef] [PubMed]

61. El-Dien, O.G.; Ratcliffe, B.; Klápště, J.; Chen, C.; Porth, I.; El-Kassaby, Y.A. Multienvironment genomic variance decomposition analysis of open-pollinated Interior spruce (*Picea glauca* x *engelmannii*). *Mol. Breed.* **2018**, *38*, 26. [CrossRef] [PubMed]

62. Lenz, P.R.N.; Nadeau, S.; Azaiez, A.; Gérardi, S.; Deslauriers, M.; Perron, M.; Isabel, N.; Beaulieu, J.; Bousquet, J. Genomic prediction for hastening and improving efficiency of forward selection in conifer polycross mating designs: An example from white spruce. *Heredity* **2020**, *124*, 562–578. [CrossRef] [PubMed]

63. Thistlethwaite, F.R.; El-Dien, O.G.; Ratcliffe, B.; Klarpstě, J.; Porth, I.; Chen, C.; Michael, U.; Stoehr, M.U.; Ingvarsson, P.K.; El-Kassaby, Y.A. Linkage disequilibrium vs. pedigree: Genomic selection prediction accuracy in conifer species. *PLoS ONE* **2020**, *15*, e0232201. [CrossRef] [PubMed]

64. Lenz, P.R.N.; Beaulieu, J.; Mansfield, S.D.; Clément, S.; Desponts, M.; Bousquet, J. Factors affecting the accuracy of genomic selection for growth and wood quality traits in an advanced-breeding population of black spruce (*Picea mariana*). *BMC Genom.* **2017**, *18*, 335. [CrossRef] [PubMed]

65. Fuentes-Utrilla, P.; Goswami, C.; Cottrell, J.E.; Pong-Wong, R.; Law, A.; A'Hara, S.W.; Lee, S.J.; Woolliams, J.A. QTL analysis and genomic selection using RADseq derived markers in Sitka spruce: The potential utility of within family data. *Tree Genet. Genomes* **2017**, *13*, 33. [CrossRef]

66. Ukrainetz, N.K.; Mansfield, S.D. Assessing the sensitivities of genomic selection for growth and wood quality traits in lodgepole pine using Bayesian models. *Tree Genet. Genomes* **2020**, *16*, 14. [CrossRef]

67. Li, Y.; Klápště, J.; Telfer, E.; Wilcox, P.; Graham, N.; Macdonald, L.; Dungey, H.S. Genomic selection for non-key traits in radiata pine when the documented pedigree is corrected using DNA marker information. *BMC Genom.* **2019**, *20*, 1026. [CrossRef]

68. Calleja-Rodriguez, A.; Pan, J.; Funda, T.; Chen, Z.-Q.; Baison, J.; Isik, F.; Abrahamsson, S.; Wu, H.X. Genomic prediction accuracies and abilities for growth and wood quality traits of Scots pine, using genotyping-by-sequencing (GBS) data. *bioRxiv* **2019**. [CrossRef]

69. Resende, M.F.R., Jr.; Muñoz, P.; Resende, M.D.V.; Garrick, D.J.; Fernando, R.L.; Davis, J.M.; Jokela, E.J.; Martin, T.A.; Peter, G.F.; Kirst, M. Accuracy of genomic selection methods in a standard data set of loblolly pine (*Pinus taeda* L.). *Genetics* **2012**, *190*, 1503–1510. [CrossRef]

70. Muñoz, P.R.; Resende, M.F.R., Jr.; Gezan, S.A.; Resende, M.D.V.; de los Campos, G.; Kirst, M.; Huber, D.; Peter, G.F. Unraveling additive from nonadditive effects using genomic relationship matrices. *Genetics* **2014**, *198*, 1759–1768. [CrossRef]

71. Alves, F.C.; Balmant, K.M.; Resende, M.F.R.; Kirst, M.; de los Campos, G. Accelerating forest tree breeding by integrating genomic selection and greenhouse phenotyping. *Plant Genome* **2020**, e20048. [CrossRef]

72. Thistlethwaite, F.R.; Ratcliffe, B.; Klápště, J.; Porth, I.; Chen, C.; Stoehr, M.U.; El-Kassaby, Y.A. Genomic prediction accuracies in space and time for height and wood density of Douglas-fir using exome capture as the genotyping platform. *BMC Genom.* **2017**, *18*, 930. [CrossRef] [PubMed]

73. Sawitri, S.; Tani, N.; Na'iem, M.; Widiyatno, W.; Indrioko, S.; Uchiyama, K.; Suwa, R.; Siong Ng, K.K.; Lee, S.L.; Tsumura, Y. Potential of genome-wide association studies and genomic selection to improve productivity and quality of commercial timber species in tropical rainforest, a case study of *Shorea platyclados*. *Forests* **2020**, *11*, 239. [CrossRef]

74. Akdemir, D.; Isidro-Sánchez, J. Design of training populations for selective phenotyping in genomic prediction. *Sci. Rep.* **2019**, *9*, 1446. [CrossRef] [PubMed]

75. Robertsen, C.D.; Hjortshøj, R.L.; Janss, L.L. Genomic selection in cereal breeding. *Agronomy* **2019**, *9*, 95. [CrossRef]

76. Eckert, A.J.; van Heerwaarden, J.; Wegrzyn, J.L.; Nelson, C.D.; Ross-Ibarra, J.; Gonzalez-Martinez, S.C.; Neale, D.B. Patterns of population structure and environmental associations to aridity across the range of loblolly pine (*Pinus taeda* L., Pinaceae). *Genetics* **2010**, *185*, 969–982. [CrossRef] [PubMed]

77. Pavy, N.; Gagnon, F.; Rigault, P.; Blais, S.; Deschênes, A.; Boyle, B.; Pelgas, B.; Deslauriers, M.; Clément, S.; Lavigne, P.; et al. Development of high-density SNP genotyping arrays for white spruce (*Picea glauca*) and transferability to subtropical and nordic congeners. *Mol. Ecol. Resour.* **2013**, *13*, 324–336. [CrossRef]

78. Silva-Junior, O.B.; Faria, D.A.; Grattapaglia, D. A flexible multi-species genome-wide 60K SNP chip developed from pooled resequencing of 240 *Eucalyptus* tree genomes across 12 species. *New Phytol.* **2015**, *206*, 1527–1540. [CrossRef]

79. Jaccoud, D.; Peng, K.M.; Feinstein, D.; Kilian, A. Diversity Arrays: A solid state technology for sequence information independent genotyping. *Nucleic Acids Res.* **2001**, *29*, e25. [CrossRef]

80. Heslot, N.; Rutkoski, J.; Poland, J.; Jannink, J.-L.; Sorrells, M.E. Impact of marker ascertainment bias on genomic selection accuracy and estimates of genetic diversity. *PLoS ONE* **2013**, *8*, e74612. [CrossRef]

81. Merida-Garcia, R.; Liu, G.; He, S.; Gonzalez-Dugo, V.; Dorado, G.; Garlvez, S.; Solís, I.; Zarco-Tejada, P.J.; Reif, J.C.; Hernandez, P. Genetic dissection of agronomic and quality traits based on association mapping and genomic selection approaches in durum wheat grown in Southern Spain. *PLoS ONE* **2019**, *14*, e0211718. [CrossRef]

82. Sansaloni, C.; Petroli, C.; Jaccoud, D.; Carling, J.; Detering, F.; Grattapaglia, D.; Kilian, A. Diversity Arrays Technology (DArT) and next-generation sequencing combined: Genome-wide, high throughput, highly informative genotyping for molecular breeding of *Eucalyptus*. *BMC Proc.* **2011**, *5*, 54. [CrossRef]

83. Liu, C.; Sikumaran, S.; Jarquin, D.; Crossa, J.; Dreisigacker, S.; Sansaloni, C.; Reynolds, M. Comparison of array-and sequencing-based makers for genome-wide association mapping and genomic prediction in spring wheat. *Crop Sci.* **2020**, *60*, 211–225. [CrossRef]

84. Zimin, A.; Stevens, K.A.; Crepeau, M.W.; Holtz-Morris, A.; Koriabine, M.; Marçais, G.; Puiu, D.; Roberts, M.; Wegrzyn, J.L.; de Jong, P.J.; et al. Sequencing and assembly of the 22-Gb loblolly pine genome. *Genetics* **2014**, *196*, 875–890. [CrossRef] [PubMed]

85. Tuskan, G.A.; Difazio, S.; Jansson, S.; Bohlmann, J.; Grigoriev, I.; Hellsten, U. The genome of black cottonwood, *Populus trichocarpa* (Torr. & Gray). *Science* **2006**, *313*, 1596–1604. [CrossRef] [PubMed]

86. Neves, L.G.; Davis, J.M.; Barbazuk, W.B.; Kirst, M. Whole-exome targeted sequencing of the uncharacterized pine genome. *Plant J.* **2013**, *75*, 146–156. [CrossRef] [PubMed]

87. Suren, H.; Hodgins, K.A.; Yeaman, S.; Nurkowski, K.A.; Smets, P.; Rieseberg, L.H.; Aitken, S.N.; Holliday, J.A. Exome capture from the spruce and pine gigagenomes. *Mol. Ecol. Res.* **2016**, *16*, 1136–1146. [CrossRef]

88. Pyhäjärvi, T.; Kujala, S.T.; Savolainen, O. 275 years of forestry meets genomics in *Pinus sylvestris*. *Evol. Appl.* **2020**, *13*, 11–30. [CrossRef]

89. Li, Y.; Suontama, M.; Burdon, R.D.; Dungey, H.S. Genotype by environment interactions in forest tree breeding: Review of methodology and perspectives on research and application. *Tree Genet. Genom.* **2017**, *13*, 60. [CrossRef]

90. Cabrera-Bosquet, L.; Crossa, J.; von Zitzewitz, J.; Maria Dolors Serret, M.D.; Araus, L.J. High-throughput phenotyping and genomic selection: The frontiers of crop breeding converge. *J. Integr. Plant Biol.* **2012**, *54*, 312–320. [CrossRef]

91. Fiorani, F.; Schurr, U. Future scenarios for plant phenotyping. *Annu. Rev. Plant Biol.* **2013**, *64*, 267–291. [CrossRef]

92. Choudhury, S.D.; Bashyam, S.; Qiu, Y.; Awada, T. Holistic and component plant phenotyping using temporal image sequence. *Plant Methods* **2018**, *14*, 35. [CrossRef] [PubMed]

93. Furbank, R.T.; Tester, M. Phenomics—Technologies to relieve the phenotyping bottleneck. *Trends Plant Sci.* **2011**, *16*, 635–644. [CrossRef] [PubMed]

94. Available online: https://www.quantitative-plant.org/ (accessed on 12 October 2020).

95. Wang, Y.; Wang, D.; Shi, P.; Omasa, K. Estimating rice chlorophyll content and leaf nitrogen concentration with a digital still color camera under natural light. *Plant Methods* **2014**, *10*, 36. [CrossRef] [PubMed]

96. Salekdeh, G.H.; Reynolds, M.; Bennett, J.; Boyer, J. Conceptual framework for drought phenotyping during molecular breeding. *Trends Plant Sci.* **2009**, *14*, 488–496. [CrossRef] [PubMed]

97. Rutkoski, J.; Poland, J.; Mondal, S.; Autrique, E.; González Pérez, L.; Crossa, J.; Reynolds, M.; Ravi, S. Predictor traits from high-throughput phenotyping improve accuracy of pedigree and genomic selection for grain yield in wheat. *G3 Genes Genomes Genet.* **2016**, *6*, 2799–2808. [CrossRef]

98. Juliana, P.; Montesinos-López, O.A.; Crossa, J.; Mondal, S.; González Pérez, L.; Poland, J.; Huerta-Espino, J.; Crespo-Herrera, L.; Govindan, V.; Dreisigacker, S.; et al. Integrating genomic-enabled prediction and high-throughput phenotyping in breeding for climate-resilient bread wheat. *Theor. Appl. Genet.* **2019**, *132*, 177–194. [CrossRef] [PubMed]

99. Krause, M.R.; Mondal, S.; Crossa, J.; Singh, R.P.; Pinto, F.; Haghighattalab, A.; Shrestha, S.; Rutkoski, J.; Gore, M.A.; Sorrells, M.E.; et al. Aerial high-throughput phenotyping enabling indirect selection for grain yield at the early-generation seed-limited stages in breeding programs. *Crop Sci.* **2020**. [CrossRef]

100. Lyra, D.H.; Virlet, N.; Sadeghi-Tehran, P.; Hassall, K.L.; Wingen, L.U.; Orford, S.; Griffiths, S.; Hawkesford, M.J.; Gancho, T.; Slavov, G.T. Functional QTL mapping and genomic prediction of canopy height in wheat measured using a robotic field phenotyping platform. *J. Exp. Bot.* **2020**, *71*, 1885–1898. [CrossRef]

101. Lozada, D.N.; Godoy, J.V.; Ward, B.P.; Carter, A.H. Genomic prediction and indirect selection for grain yield in US Pacific Northwest winter wheat using spectral reflectance indices from high-throughput phenotyping. *Int. J. Mol. Sci.* **2020**, *21*, 165. [CrossRef]

102. Baba, T.; Momen, M.; Campbell, M.T.; Walia, H.; Morota, G. Multi-trait random regression models increase genomic prediction accuracy for a temporal physiological trait derived from high-throughput phenotyping. *PLoS ONE* **2020**, *15*, e0228118. [CrossRef]

103. Yonis, B.O.; del Carpio, D.P.; Wolfe, M.; Jannink, J.L.; Kulakow, P.; Ismail Rabbi, I. Improving root characterisation for genomic prediction in cassava. *Sci. Rep.* **2020**, *10*, 8003. [CrossRef] [PubMed]

104. Xu, Y.; Liu, X.; Fu, J.; Wang, H.; Wang, J.; Huang, C.; Prasanna, B.M.; Olsen, M.S.; Wang, G.; Zhang, A. Enhancing genetic gain through genomic selection: From livestock to plants. *Plant Commun.* **2020**, *1*, 100005. [CrossRef]

105. Chawade, A.; van Ham, J.; Blomquist, H.; Bagge, O.; Alexandersson, E.; Ortiz, R. High-throughput field-phenotyping tools for plant breeding and precision agriculture. *Agronomy* **2019**, *9*, 258. [CrossRef]

106. Ludovisi, R.; Tauro, F.; Salvati, R.; Khoury, S.; Mugnozza Scarascia, G.; Harfouche, A. UAV-based thermal imaging for high-throughput field phenotyping of black poplar response to drought. *Front. Plant Sci.* **2017**, *8*, 1681. [CrossRef]

107. Solvin, T.M.; Puliti, S.; Arne Steffenrem, A. Use of UAV photogrammetric data in forest genetic trials: Measuring tree height, growth, and phenology in Norway spruce (*Picea abies* L. Karst.). *Scand. J. For. Res.* **2020**, *35*, 322–333. [CrossRef]

108. Mazis, A.; Choudhury, S.D.; Morgan, P.B.; Stoerger, V.; Jeremy Hiller, J.; Ge, Y.; Awada, T. Application of high-throughput plant phenotyping for assessing biophysical traits and drought response in two oak species under controlled environment. *For. Ecol. Manag.* **2020**, *465*, 118101. [CrossRef]

109. Vidyagina, E.O.; Subbotina, N.M.; Belyi, V.A.; Lebedev, V.G.; Krutovsky, K.V.; Shestibratov, K.A. Various effects of the expression of the xyloglucanase gene from *Penicillium canescens* in transgenic aspen under semi-natural conditions. *BMC Plant Biol.* **2020**, *20*, 251. [CrossRef]

110. Martins, G.S.; Yuliarto, M.; Antes, R.; Sabki; Prasetyo, A.; Unda, F.; Mansfield, S.D.; Hodge, G.R.; Acosta, J.J. Wood and pulping properties variation of *Acacia crassicarpa* A. Cunn. ex Benth. and sampling strategies for accurate phenotyping. *Forests* **2020**, *11*, 1043. [CrossRef]

111. Lin, Z.; Hayes, B.J.; Daetwyler, H.D. Genomic selection in crops, trees and forages: A review. *Crop Pasture Sci.* **2014**, *65*, 1177–1191. [CrossRef]

112. Gao, S.; Wang, X.; Wiemann, M.C. A critical analysis of methods for rapid and nondestructive determination of wood density in standing trees. *Ann. For. Sci.* **2017**, *74*, 27. [CrossRef]

113. Schimleck, L.; Dahlen, J.; Apiolaza, L.A.; Downes, G.; Emms, G.; Evans, R.; Moore, J.; Pâques, L.; Van den Bulcke, J.; Wang, X. Non-destructive evaluation techniques and what they tell us about wood property variation. *Forests* **2019**, *10*, 728. [CrossRef]

114. White, T.L.; Adams, W.T.; Neale, D.B. *Forest Genetics*; CABI Publishing CAB International: Cambridge, UK, 2007.

115. Banerjee, B.P.; Joshi, S.; Thoday-Kennedy, E.; Pasam, R.K.; Tibbits, J.; Hayden, M.; Spangenberg, G.; Kant, S. High-throughput phenotyping using digital and hyperspectral imaging-derived biomarkers for genotypic nitrogen response. *J. Exp. Bot.* **2020**, *71*, 4604–4615. [CrossRef] [PubMed]

116. Wang, X.; Xu, Y.; Hu, Z.; Xu, C. Genomic selection methods for crop improvement: Current status and prospects. *Crop J.* **2018**, *6*, 330–340. [CrossRef]

117. Legarra, A.; Robert-Granie, C.; Croiseau, P.; Guillaume, F.; Fritz, S. Improved LASSO for genomic selection. *Genet. Res.* **2011**, *93*, 77–87. [CrossRef] [PubMed]

118. Gianola, D.; Fernandoand, R.L.; Stella, A. Genomic assisted prediction of genetic value with semi-parametric procedures. *Genetics* **2006**, *173*, 1761–1776. [CrossRef]

119. Heslot, N.; Yang, H.-P.; Sorrells, M.E.; Jannink, J.-L. Genomic selection in plant breeding: A comparison of models. *Crop Sci.* **2012**, *56*, 146–160. [CrossRef]

120. Heslot, N.; Jannink, J.-L.; Sorrells, M.E. Perspectives for genomic selection applications and research in plants. *Crop Sci.* **2015**, *55*, 1–12. [CrossRef]

121. Banta, J.A.; Richards, C.L. Quantitative epigenetics and evolution. *Heredity* **2018**, *121*, 210–224. [CrossRef]

122. Nicotra, A.B.; Atkin, O.K.; Bonser, S.P.; Davidson, A.M.; Finnegan, E.J.; Mathesius, U.; Poot, P.; Purugganan, M.D.; Richards, C.L.; Valladares, F.; et al. Plant phenotypic plasticity in a changing climate. *Trends Plant Sci.* **2010**, *15*, 684–692. [CrossRef]

123. Gugger, P.F.; Fitz-Gibbon, S.T.; Pellegrini, M.; Sork, V.L. Species-wide patterns of DNA methylation variation in *Quercus lobata* and their association with climate gradients. *Mol. Ecol.* **2016**, *25*, 1665–1680. [CrossRef]

124. Le Gac, A.-L.; Lafon-Placette, C.; Chauveau, D. Winter-dormant shoot apical meristem in poplar trees shows environmental epigenetic memory. *J. Exp. Bot.* **2018**, *69*, 4821–4837. [CrossRef]

125. Koide, Y.; Sakaguchi, S.; Uchiyama, T.; Ota, Y.; Tezuka, A.; Nagano, A.J.; Ishiguro, S.; Takamure, I.; Kishima, Y. Genetic properties responsible for the transgressive segregation of days to heading in rice. *G3 Genes Genomes Genet.* **2019**, *9*, 1655–1662. [CrossRef]

126. de Los Reyes, B.G. Genomic and epigenomic bases of transgressive segregation—New breeding paradigm for novel plant phenotypes. *Plant Sci.* **2019**, *288*, 110213. [CrossRef] [PubMed]

127. Fujimoto, R.; Uezono, K.; Ishikura, S.; Osabe, K.; Peacock, W.J.; Elizabeth, S.; Dennis, E.S. Recent research on the mechanism of heterosis is important for crop and vegetable breeding systems. *Breed. Sci.* **2018**, *68*, 145–158. [CrossRef] [PubMed]

128. Groszmann, M.; Greaves, I.K.; Fujimoto, R.; Peacock, W.J.; Dennis, E.S. The role of epigenetics in hybrid vigour. *Trends Genet.* **2013**, *29*, 684–690. [CrossRef] [PubMed]

129. Gao, M.; Huang, Q.; Chu, Y.; Ding, C.; Zhang, B.; Su, X. Analysis of the leaf methylomes of parents and their hybrids provides new insight into hybrid vigor in *Populus deltoides*. *BMC Genet.* **2014**, *15*, S8. [CrossRef]

130. Springer, N.M.; Schmitz, R.J. Exploiting induced and natural epigenetic variation for crop improvement. *Nat. Rev. Genet.* **2017**, *18*, 563–575. [CrossRef]

131. Goddard, M.E.; Emma Whitelaw, E. The use of epigenetic phenomena for the improvement of sheep and cattle. *Front. Genet.* **2014**, *5*, 247. [CrossRef]

132. Banos, D.T.; McCartney, D.L.; Patxot, M.; Anchieri, L.; Battram, T.; Christiansen, C.; Costeira, R.; Walker, R.M.; Morris, S.W.; Archie Campbell, A. Bayesian reassessment of the epigenetic architecture of complex traits. *Nat. Commun.* **2020**, *11*, 2865. [CrossRef]

133. Shah, S.; Bonder, M.J.; Marioni, R.E.; Zhu, Z.; McRae, A.F.; Zhernakova, A.; Harris, S.E.; Liewald, D.; Henders, A.K.; Mendelson, M.M.; et al. Improving phenotypic prediction by combining genetic and epigenetic associations. *Am. J. Hum. Genet.* **2015**, *97*, 75–85. [CrossRef]

134. Yang, X.; Mackenzie, S.A. Approaches to whole-genome methylome analysis in plants. In *Plant Epigenetics and Epigenomics. Methods in Molecular Biology*; Spillane, C., McKeown, P., Eds.; Humana: New York, NY, USA, 2020; Volume 2093. [CrossRef]

135. Bernardo, R.; Yu, J.M. Prospects for genomewide selection for quantitative traits in maize. *Crop Sci.* **2007**, *47*, 1082–1090. [CrossRef]

136. Wong, C.K.; Bernardo, R. Genomewide selection in oil palm: Increasing selection gain per unit time and cost with small populations. *Theor. Appl. Genet.* **2008**, *116*, 815–824. [CrossRef] [PubMed]

137. Hayes, B.; Goddard, M. Genome-wide association and genomic selection in animal breeding. *Genome* **2010**, *53*, 876–883. [CrossRef] [PubMed]

138. Sork, V.L.; Aitken, S.N.; Dyer, R.J.; Eckert, A.J.; Legendre, P.; Neale, D.B. Putting the landscape into the genomics of trees: Approaches for understanding local adaptation and population responses to changing climate. *Tree Genet. Genomes* **2013**, *9*, 901–911. [CrossRef]

139. Charlesworth, B. Fundamental concepts in genetics: Effective population size and patterns of molecular evolution and variation. *Nat. Rev. Genet.* **2009**, *10*, 195–205. [CrossRef]

140. Grattapaglia, D. Breeding forest trees by genomic selection: Current progress and the way forward. In *Genomics of Plant Genetic Resources*; Tuberosa, R., Graner, A., Frison, E., Eds.; Springer: Dordrecht, The Netherlands, 2014; pp. 651–682.

141. Brondani, R.P.V.; Williams, E.R.; Brondani, C.; Grattapaglia, D. A microsatellite-based consensus linkage map for species of Eucalyptus and a novel set of 230 microsatellite markers for the genus. *BMC Plant Biol.* **2006**, *6*, 20. [CrossRef]

142. Neves, L.G.; Davis, J.M.; Barbazuk, W.B.; Kirst, M. A high-density gene map of loblolly pine (*Pinus taeda* L.) based on exome sequence capture genotyping. *G3 Genes Genomes Genet.* **2014**, *4*, 29–37. [CrossRef]

143. Krutovsky, K.V.; Troggio, M.; Brown, G.R.; Jermstad, K.D.; Neale, D.B. Comparative mapping in the *Pinaceae*. *Genetics* **2004**, *168*, 447–461. [CrossRef]

144. Meuwissen, T.H.E. Accuracy of breeding values of 'unrelated' individuals predicted by dense SNP genotyping. *Genet. Sel. Evol.* **2009**, *41*, 35–44. [CrossRef]

145. Kumar, S.; Bink, M.C.A.M.; Volz, R.K.; Bus, V.G.M.; Chagné, D. Towards genomic selection in apple (*Malus·domestica* Borkh.) breeding programmes: Prospects, challenges and strategies. *Tree Genet. Genomes* **2012**, *8*, 1–14. [CrossRef]

146. Mullin, T.J.; Andersson, B.; Bastien, J.C.; Beaulieu, J.; Burdon, R.D.; Dvorak, W.S.; King, J.N.; Kondo, T.; Krakowski, J.; Lee, S.J.; et al. Economic importance, breeding objectives and achievements. In *Genetics, Genomics and Breeding of Conifers*; Plomion, C., Bousquet, J., Kole, C., Eds.; Science Publishers and CRC Press: New York, NY, USA, 2011; pp. 40–127.

147. Rajsic, P.; Weersink, A.; Navabi, A.; Peter Pauls, K. Economics of genomic selection: The role of prediction accuracy and relative genotyping costs. *Euphytica* **2016**, *210*, 259–276. [CrossRef]

148. Chang, W.-Y.; Gaston, C.; Cool, J.; Barb, R.; Thomas, B.R. A financial analysis of using improved planting stock of white spruce and lodgepole pine in Alberta, Canada: Genomic selection versus traditional breeding. *Int. J. For. Res.* **2019**, *92*, 297–310. [CrossRef]

149. Chamberland, V.; Robichaud, F.; Perron, M.; Gélinas, N.; Bousquet, J.; Beaulieu, J. Conventional versus genomic selection for white spruce improvement: A comparison of costs and benefits of plantations on Quebec public lands. *Tree Genet. Genomes* **2020**, *16*, 17. [CrossRef]

150. Park, Y.-S.; Beaulieu, J.; Bousquet, J. Multi-varietal forestry integrating genomic selection and somatic embryogenesis. In *Vegetative Propagation of Forest Trees*; Park, Y.-S., Bonga, J.M., Moon, H.-K., Eds.; National Institute of Forest Science: Seoul, Korea, 2016; pp. 302–322.

151. Salojärvi, J.; Smolander, O.-P.; Nieminen, K.; Rajaraman, S. Genome sequencing and population genomic analyses provide insights into the adaptive landscape of silver birch. *Nat. Genet.* **2017**, *49*, 904–912. [CrossRef]

152. Plomion, C.; Aury, J.; Amselem, J.; Aury, J.-M.; Leroy, T. Oak genome reveals facets of long lifespan. *Nat. Plants* **2018**, *4*, 440–452. [CrossRef] [PubMed]

153. Yasodha, R.; Vasudeva, R.; Balakrishnan, S.; Sakthi, A.R.; Abel, N.; Binai, N.; Rajashekar, B.; Bachpai, V.K.W.; Pillai, C.; Dev, S.A. Draft genome of a high value tropical timber tree, Teak (*Tectona grandis* L. f): Insights into SSR diversity, phylogeny and conservation. *DNA Res.* **2018**, *25*, 409–419. [CrossRef]

154. Wenzel, S.; Flachowsky, H.; Magda-Viola Hanke, M.-V. The Fast-track breeding approach can be improved by heat-induced expression of the FLOWERING LOCUS T genes from poplar (*Populus trichocarpa*) in apple (*Malus domestica* Borkh.). *Plant Cell Tissue Organ Cult.* **2013**, *115*, 127–137. [CrossRef]

155. Graham, T.; Scorza, R.; Wheeler, R.; Smith, B.; Dardick, C.; Dixit, A.; Raines, D.; Ann Callahan, A.; Srinivasan, C.; Spencer, L.; et al. Over-expression of *FT1* in plum (*Prunus domestica*) results in phenotypes compatible with spaceflight: A potential new candidate crop for bioregenerative life support systems. *Gravit. Space Res.* **2015**, *3*, 39–50.

156. Endo, T.; Fujii, H.; Omura, M.; Takehiko Shimada, T. Fast-track breeding system to introduce CTV resistance of trifoliate orange into citrus germplasm, by integrating early flowering transgenic plants with marker-assisted selection. *BMC Plant Biol.* **2020**, *20*, 224. [CrossRef]

157. Vicient, C.M.; Casacuberta, J.M. Impact of transposable elements on polyploid plant genomes. *Ann. Bot.* **2017**, *120*, 195–207. [CrossRef]
158. Wang, J.; Lu, N.; Yi, F.; Xiao, Y. Identification of transposable elements in conifer and their potential application in breeding. *Evol. Bioinform. Online* **2020**, *16*. [CrossRef] [PubMed]

Publisher's Note: MDPI stays neutral with regard to jurisdictional claims in published maps and institutional affiliations.

Article

Development and Deployment of High-Throughput Retrotransposon-Based Markers Reveal Genetic Diversity and Population Structure of Asian Bamboo

Shitian Li [1,†], Muthusamy Ramakrishnan [1,†], Kunnummal Kurungara Vinod [2], Ruslan Kalendar [3,4], Kim Yrjälä [1,5] and Mingbing Zhou [1,6,*]

[1] State Key Laboratory of Subtropical Silviculture, Zhejiang A&F University, Lin'an, Hangzhou 311300, China; lst323517@163.com (S.L.); ramky@zafu.edu.cn (M.R.); kim.yrjala@helsinki.fi (K.Y.)

[2] Division of Genetics, ICAR-Indian Agricultural Research Institute, New Delhi 110012, India; kkvinod@iari.res.in

[3] Department of Agricultural Sciences, Viikki Plant Science Centre and Helsinki Sustainability Centre, University of Helsinki, P.O. Box 27 (Latokartanonkaari 5), FI-00014 Helsinki, Finland; ruslan.kalendar@helsinki.fi

[4] RSE, National Center for Biotechnology, 13/5, Kurgalzhynskoye road, Nur-Sultan 010000, Kazakhstan

[5] Department of Forest Sciences, University of Helsinki, 00014 Helsinki, Finland

[6] Zhejiang Provincial Collaborative Innovation Center for Bamboo Resources and High-Efficiency Utilization, Zhejiang A&F University, Lin'an, Hangzhou 311300, China

* Correspondence: zhoumingbing@zafu.edu.cn; Tel.: +86-571-63743869

† These authors contributed equally to this article.

Received: 31 October 2019; Accepted: 17 December 2019; Published: 24 December 2019

Abstract: Bamboo, a non-timber grass species, known for exceptionally fast growth is a commercially viable crop. Long terminal repeat (LTR) retrotransposons, the main class I mobile genetic elements in plant genomes, are highly abundant (46%) in bamboo, contributing to genome diversity. They play significant roles in the regulation of gene expression, chromosome size and structure as well as in genome integrity. Due to their random insertion behavior, interspaces of retrotransposons can vary significantly among bamboo genotypes. Capitalizing this feature, inter-retrotransposon amplified polymorphism (IRAP) is a high-throughput marker system to study the genetic diversity of plant species. To date, there are no transposon based markers reported from the bamboo genome and particularly using IRAP markers on genetic diversity. *Phyllostachys* genus of Asian bamboo is the largest of the Bambusoideae subfamily, with great economic importance. We report structure-based analysis of bamboo genome for the LTR-retrotransposon superfamilies, *Ty3-gypsy* and *Ty1-copia*, which revealed a total of 98,850 retrotransposons with intact LTR sequences at both the ends. Grouped into 64,281 clusters/scaffold using CD-HIT-EST software, only 13 clusters of retroelements were found with more than 30 LTR sequences and with at least one copy having all intact protein domains such as *gag* and polyprotein. A total of 16 IRAP primers were synthesized, based on the high copy numbers of conserved LTR sequences. A study using these IRAP markers on genetic diversity and population structure of 58 Asian bamboo accessions belonging to the genus *Phyllostachys* revealed 3340 amplicons with an average of 98% polymorphism. The bamboo accessions were collected from nine different provinces of China, as well as from Italy and America. A three phased approach using hierarchical clustering, principal components and a model based population structure divided the bamboo accessions into four sub-populations, PhSP1, PhSP2, PhSP3 and PhSP4. All the three analyses produced significant sub-population wise consensus. Further, all the sub-populations revealed admixture of alleles. The analysis of molecular variance (AMOVA) among the sub-populations revealed high intra-population genetic variation (75%) than inter-population. The results suggest that *Phyllostachys* bamboos are not well evolutionarily diversified, although geographic speciation could have occurred at a limited level. This study highlights the usability of IRAP markers in determining the inter-species variability of Asian bamboos.

Keywords: LTR-retrotransposon; *Ty3-gypsy*; *Ty1-copia*; IRAP; molecular markers; bamboo; *Phyllostachys*; genetic diversity; populations structure; AMOVA

1. Introduction

Bamboo, a monocot and a major grass genera, is a group of evergreen flowering plants belonging to the subfamily Bambusoideae of the family Poaceae [1]. Although the proliferation of bamboo occurs predominantly through rhizomes, most bamboos do reproduce through seeds, flowering at least once in a lifetime. Usually, flowering intervals are long and vary between species, ranging from several to hundreds of years [2–4]. More than 1642 bamboo species from 75 genera are known (htttps://www.inbar.int), among which 100 species are commercially cultivated over 30 million hectares worldwide, particularly in Asia. Several members of the bamboo, including Asian bamboo, are recognised as fast-growing plants, growing up to a height of 35–50 m and up to 30 cm in diameter (https://www.inbar.int). Among cultivated bamboo species, the Asian bamboo can grow at a maximum rate of 100 cm a day and produces huge biomass [5].

Most Asian bamboo species are native to China, although some are known to grow in India, Vietnam and Myanmar. These bamboos account for approximately 0.8% of the forest area worldwide. Some species were introduced to Japan a hundred years ago and became naturalised. More recently, a few naturalized species from Australia, Europe and the Americas have been reported [6]. Most Asian bamboos belong to the genus *Phyllostachys* in the tribe Arundinarieae. They are chiefly temperate woody bamboos and are tetraploids (2n = 4x = 48) with a 2B karyotype pattern.

Bamboo wood is a non-timber natural raw material having notable industrial importance and economic value in South Asia [7]. Asia is the largest producer of bamboo products in the world, with annual international trade amounting to more than 2.5 billion US dollars (https://www.inbar.int). In spite of being an economically important perennial species, commercial bamboo remains mostly confined to natural populations. Further, the genetic diversity of bamboos has not adequately been explored. The major reasons were the difficulty in assessing the phenotypic variability of clones because of their extended growth period, perennial nature, gigantic size, propagation behavior, non-uniformity of age, long flowering cycle and the extensive area of their natural habitat. However, with the advent of molecular marker-based techniques developed in the 1980s, studies on crop genetic diversity have gained momentum. Subsequently, from 1991, a relatively limited number of molecular fingerprinting studies have been carried out to assess the genetic diversity of the Asian bamboo species using restriction fragment length polymorphism (RFLP), [8,9], randomly amplified polymorphic DNA (RAPD), [10,11], amplified fragment length polymorphism (AFLP), [12–16], simple sequence repeats (SSRs), [17,18], expressed sequence tags-SSR (EST-SSR), [19,20], inter-simple sequence repeats (ISSRs), [15,21] and single-nucleotide polymorphisms (SNP) [22].

In this study, we took advantage of the genome wide abundance of transposable elements to assess genetic variability in bamboo. Transposable elements (TEs) are ubiquitous genetic elements in eukaryotic genomes, capable of self-replicative transposition, affecting genome stability [23–28]. Two types of TEs have been identified based on their transposition mechanisms, namely class I retrotransposons and class II DNA transposons [28]. Retrotransposons are RNA-based TEs which duplicate themselves and move within the genome in a semi-conservative manner through a 'copy-and-paste' mechanism of an RNA intermediate [28–30]. DNA transposons, on the other hand, use a conservative style of transposition and move directly by a 'cut-and-paste' mechanism [31–33]. Retrotransposons are found in abundance, particularly in plant genomes, outnumbering DNA transposons, accounting for a significant part of the genome such as 68% in wheat [34] and 49%–78% in maize [35,36]. While exploring TEs from 44 bamboo species belonging to 38 genera, Zhou et al. [37] identified TEs as widespread, abundant and diverse in the bamboo genome. In moso bamboo, retrotransposons are reported to occupy 39%–46% [38–40] of the genome, accounting for about 65% of the total repetitive elements in the genome [38].

Among the two major types of retroelements, long terminal repeat (LTR)-retrotransposons and non-LTR-retrotransposons, LTR-retrotransposons constitute more than 90% of the retrotransposons found in plant genomes [41–43]. They have typical structural features, such as LTR sequences at both ends, transcription and reverse transcription processing signals and target site duplications [24,44]. Additionally, they possess a primer-binding site and a polypurine tract, aiding the synthesis of minus- and plus-strand DNA [45,46]. Based on their characteristics, LTR-retrotransposons are primarily divided into two superfamilies: *Ty1-copia* and *Ty3-gypsy* [42]. Recent estimates of the bamboo genome show that 63.2% of the genome is occupied by TEs [40] and 45.7% of the repeat regions belongs to *Ty3-gypsy* and *Ty1-copia* types, signifying their role in determining genome size [39]. The genome-wide analysis showed that LTR-retrotransposons are transcriptionally active in the bamboo genome and are responsible for generating 30% of small interfering RNAs (siRNAs) [47]. It has been reported that LTR-retroelements get activated by environmental stress [48–50]. Therefore, in the course of genome evolution, LTR-retrotransposon activity could accumulate several variations, making them an ideal source for genome-wide molecular markers [51–54].

The inter-retrotransposon amplified polymorphism (IRAP) technique produces amplified fragments that are characteristic of a dominant marker [55,56]. Amplification of the IRAP fragment between two LTR-retrotransposons is done using outward-facing primers which anneal to LTR sequences. This method needs neither restriction digestion nor ligation enzyme [57]. Due to technical ease, the IRAP method has been utilized in several studies of the genetic diversity of various plant species. Kalendar et al. [51] studied genetic diversity in barley using IRAP markers, proving their usefulness for diversity studies and the method helped to distinguish between Brazilian and Japanese rice genotypes [58]. Furthermore, it has been used in sunflower [59], *Pinus* [60], *Lilium* [61], Persian oak (*Quercus brantii* Lindl.) and in *Bletilla striata* [62], and in wild diploid wheat [63] for diversity studies, as well as for population structure and phylogenetic analyses. Guo et al. [62] reported that the results obtained using IRAP markers were similar to those obtained by start codon-targeted (SCoT) markers.

Although LTR-retrotransposons occupy a significant part of the bamboo genome, the genetic diversity information attributable to them remains mostly unknown. Furthermore, no TE-based bamboo markers have been reported to date. In the current study, we report, for the first time, the development of several IRAP markers based on the moso bamboo genome and the use of these markers to assess the IRAP-based genetic diversity and population structure of *Phyllostachys* bamboo.

2. Materials and Methods

2.1. Plant Material

A total of 58 Asian bamboo accessions were used in the study. The accessions included 47 distinct species belonging to the genus *Phyllostachys*, of which four species had 15 different varieties shared between them. There were nine varieties of *Ph. edulis* and two varieties each from *Ph. nigra*, *Ph. bambusoides* and *Ph. sulphurea*. These materials were collected from the forests of the main Asian bamboo growing regions of China spread over the provinces of Zhejiang, Anhui, Sichuang, Jiangxi, Guangdong, Hunan, Henan, Jiangsu and Taiwan. Three species, one sourced from Italy (*Ph. nidularia*) and two obtained from America (*Ph. elegans* and *Ph. glauca*), were also included. The details are listed in Table 1. The collected plant materials were planted and maintained in red soil of a botanical garden of Fujian province. The conservation site has a subtropical monsoon climate with four distinct seasons in the year, an average rainfall from 1270 and 2030 mm a year, and an annual average temperature of 17.5 °C.

Fresh young leaves of the bamboo clones were randomly collected and surface-cleaned by gently rinsing with 70% ethanol and preserved in a polythene bag containing colour-changing silica gel (Tsingke, China). The leaf bags were stored in a deep freezer at −80 °C for further analysis.

Table 1. List of 58 *Phyllostachys* accessions (Asian bamboo) collected from different geographical regions used for the analysis of genetic diversity using inter-retrotransposon amplified polymorphism (IRAP) markers.

S. No.	Phyllostachys Accessions	Source Location
1.	*Ph. aurea* Carr.	Zhejiang, China
2.	*Ph. acutiligula* G. H. Lai	Anhui, China
3.	*Ph. varioauriculata*	Anhui, China
4.	*Ph. heteroclada*	Sichuang, China
5.	*Ph. hispida*	Zhejiang, China
6.	*Ph. hirtivagina*	Anhui, China
7.	*Ph. edulis* (Carrière) J. Houz	Jiangsu, China
8.	*Ph. edulis* cv. Obliquinoda	Zhejiang, China
9.	*Ph. edulis* cv. Pachyloen	Jiangxi, China
10.	*Ph. edulis* cf. huamaoyhw (Wen.) Wen	Hunan, China
11.	*Ph. edulis* cv. Viridisulcata	Zhejiang, China
12.	*Ph. edulis* (Carr.) Matsumura	Zhejiang, China
13.	*Ph. edulis* Mitford cv.	Zhejiang, China
14.	*Ph. edulis* (Carr.) Mitford cv. Gracilis	Jiangsu, China
15.	*Ph. edulis* cv. Tubaeformis S.Y. Wang	Hunan, China
16.	*Ph. incarnata* T.H. Wen, Bull. Bot. Res	Zhejiang, China
17.	*Ph. nidularia*	Italy
18.	*Ph. sulphurea* (Carr)A. et C. Riv	Anhui, China
19.	*Ph. sulphurea var. viridis*	Henan, China
20.	*Ph. mannii*	Taiwan, China
21.	*Ph. virella*	Zhejiang, China
22.	*Ph. yunhoensxian*	Zhejiang, China
23.	*Ph. rubromarginata*	Guangdong, China
24.	*Ph. elegans* McClure	America
25.	*Ph. fimbriligula*	Zhejiang, China
26.	*Ph. viridiglaucescens*	Zhejiang, China
27.	*Ph. robustiramea*	Zhejiang, China
28.	*Ph. flexuosa*	Anhui, China
29.	*Ph. glauca*	America
30.	*Ph. zhejiangensis*	Zhejiang, China
31.	*Ph. angusta*	Anhui, China
32.	*Ph. meyeri*	Henan, China
33.	*Ph. vivax*	Anhui, China
34.	*Ph. propinqua*	Anhui, China
35.	*Ph. iridescens*	Zhejiang, China
36.	*Ph. purpureociliata*	Anhui, China
37.	*Ph. nuda*	Jiangsu, China
38.	*Ph. primotina*	Zhejiang, China
39.	*Ph. arcana*	Jiangsu, China
40.	*Ph. verrucosa*	Hunan, China
41.	*Ph. bissetii*	Sichuang, China
42.	*Ph. aureosulcata*	Zhejiang, China
43.	*Ph. longiciliata*	Anhui, China
44.	*Ph. rutila*	Zhejiang, China
45.	*Ph. bambusoides*	Henan, China
46.	*Ph. bambusoides* Sieb. et Zucc	Anhui, China
47.	*Ph. nigella*	Zhejiang, China
48.	*Ph. dulcis*	Zhejiang, China
49.	*Ph. aurita*	Henan, China
50.	*Ph. funhuanensis*	Zhejiang, China
51.	*Ph. platyglossa*	Zhejiang, China
52.	*Ph. rubicunda*	Zhejiang, China
53.	*Ph. atrovaginata*	Zhejiang, China
54.	*Ph. parvifolia*	Zhejiang, China
55.	*Ph. corrugata*	Anhui, China
56.	*Ph. nigra*	Hunan, China
57.	*Ph. nigra var.* henonis	Jiangsu, China
58.	*Ph. violascens*	Zhejiang, China

2.2. Isolation of Genomic DNA

A modified cetyltrimethylammonium bromide (CTAB) method [64] was used to extract genomic DNA from the leaf samples. For this, the leaf samples were cut into small pieces of sizes from 3.0 to 5.0 mm. An amount of 200 mg of the cut leaf pieces was ground in a mortar using liquid N, and quickly transferred into a sterile centrifuge tube containing 850 µL preheated (65 °C) 2% CTAB extraction buffer containing 20 mM EDTA, 100 mM Tris of pH 8.0, 1.4 M NaCl, 2% CTAB, 200 mg/mL PVP and 1% β-mercaptoethanol. The tubes' contents were mixed by gentle inversion and incubated at 65 °C for 30 min. The tubes were gently inverted twice every 10 min. After incubation, the tubes were allowed to cool for 15 min at 25 °C. After cooling, an equal volume of ice-cold phenol:chloroform:isoamyl alcohol (25:24:1 *v/v*) mixture was added to the tube; the contents were gently mixed and centrifuged at 12,000 rpm for 10 min. The clear supernatant was collected in another tube and an equal volume of ice-cold chloroform:isoamyl alcohol (24:1 *v/v*) mixture was added. The contents were gently mixed and centrifuged again at 12,000 rpm for 10 min and the supernatant was collected. To the supernatant, an equal volume of ice-cold isopropanol was added, and the tubes were kept at −80 °C for one hour to precipitate the DNA. The mix was centrifuged at 12,000 rpm for 5 min and the supernatant was discarded. To the pellet, 600 µL of 75% ice-cold ethanol was added and the tube was left standing for 10 min. The pellet-ethanol mixture was centrifuged at 12,000 rpm for 5 min. This step was repeated twice and the pellet was air-dried at 40 °C and then dissolved in 40 µL 1× TE (10 mM Tris-Cl pH 8.0 and 1 mM EDTA pH 8.0) buffer. The concentration and purity of DNA were quantified by a Nanodrop-spectrophotometer (ND1000, ThermoScientific, Wilmington, DE, USA).

2.3. Isolation of LTR-Retrotransposons and IRAP-Primer Design

In an earlier study from our lab, a total of 2,004,644 LTR-retrotransposon-related sequences were identified in the moso bamboo genome, accounting for about 40% of the moso bamboo genome [39]. The LTR sequences were identified using LTRharvest and LTR digest software [65], and the terminal repeats were analysed for similarity both at 5′and 3′ LTR regions using CD-HIT software [66]. The LTR sequences were divided into different clusters using an incremental clustering algorithm with 95% similarity criteria. LTR sequence clusters with more than 30 copy numbers were chosen as candidate sequences for IRAP primer designing. The primers were designed using Primer Premier 5.0 software (http://www.premierbiosoft.com/) and synthesized by Bioengineering (Shanghai) Co. Ltd., (Shanghai, China). Following this, IRAP primers were used for IRAP fragment amplification using appropriate polymerase chain reaction (PCR) conditions. Primers that generated a low number of amplicons were subsequently excluded from the analysis. A set of IRAP primers which provided a high proportion of alleles were, thus, finally shortlisted.

2.4. PCR Amplification of IRAP and Electrophoresis

PCR reactions were performed in 20 µL reaction mixture containing 100 ng genomic DNA, 400 nM primer, and 10 µL PCR master mix (Nanjing Nuoweizan Biotechnology Co., Ltd., Nanjing, China) and the final volume was adjusted to 20 µL by adding nuclease-free water. The annealing temperature of each IRAP primer was determined using gradient PCR. The amplification reaction was carried out in a DNA thermal cycler (DNA Engine® Thermal Cycler—Bio-Rad). The PCR reaction was run at an initial denaturation temperature of 94 °C for 5 min, followed by 35 cycles of 30 s denaturation at 94 °C, 30 s annealing and 1 min extension at 72 °C with a final extension at 72 °C for 7 min. The annealing temperature was readjusted for each IRAP primer. The amplified product was electrophoresed in 1.5% (*w/v*) agarose gel at 75–80 V for 2.15 h. The separated alleles were visualised by a gel documentation system (Bio-Rad). The alleles were visually scored as 1 = present; 0 = absent using GelQuest software (https://www.sequentix.de/gelquest/).

2.5. Cloning and Sequencing of IRAP Fragments

Since IRAP markers are dominant, it is important to confirm that it is indeed LTR-retrotransposons that are selectively amplified. Alleles showing clear brightness were randomly excised from the agarose gels and were purified using a DNA gel extraction and purification kit (Simgen, Hangzhou, China). The purified fragment was ligated into the pMD18-T vector between the *Not* I and *EcoR* V restriction sites (TaKaRa, Shiga, Japan). The ligated product was transformed into *Escherichia coli* (*E. coli*) DH5α competent cells. The recombinant *E. coli* clones were obtained on LB agar plates containing ampicillin (100 µg/µL), X-gal (40 mg) and isopropyl β-D-1-thiogalactopyranoside (IPTG) (160 µg) and kept at 37 °C overnight. White *E. coli* colonies were selected by the blue-white screening method, and the insertion was verified by PCR using the corresponding IRAP primer, and the amplified fragment was sequenced (Sino Biological Inc., Beijing, China). The sequences were aligned with the original LTR-retrotransposon sequences using MEGA 7 software [67] to confirm correct amplification of the corresponding LTR sequences. Multiple sequence alignment of the PCR product sequences was also carried out with ClustalW, using the Neighbor-joining method with evolutionary distance for construction of phylogenetic tree created [68] in MEGA software with 500 bootstraps.

2.6. Marker Statistics and Genetic Relations

The total number of alleles, range of allele products, percent of polymorphism and the PIC of each IRAP marker was calculated using the binary data of the corresponding PCR amplicon using the formula PIC = 1 − [f^2 + (1 − f)^2], where 'f' is the frequency of the marker in the data set. PIC for dominant markers is a maximum of 0.5 for 'f' = 0.5 [69]. Marker allele variation and distribution among the bamboo accessions were also worked out.

2.7. Genetic Distance and Diversity of Bamboo Accessions

The distance/similarity matrix of the bamboo accessions was constructed from allele distribution data. The distance was computed based on Jaccard's similarity coefficient [70] and was subjected to hierarchical clustering using the unweighted pair-group method with arithmetic average (UPGMA). The UPGMA dendrogram was produced using Free Tree V. 0. 9.1. 50 software [71]. To access the dendogram reliability, over 10,000 bootstrapping (resampling) values were set. The phylogram was visualized by TreeViewX V.0.5.0 software [72].

2.8. IRAP-Statistical Fitness Analysis

To validate the dendrogram and the genetic diversity a statistical fitness analyses were performed using binary data. The cophenetic correlation coefficient (CCC) was estimated between the dendrogram and the observed dissimilarity matrix [70]. Further, 58 Asian bamboo species were classified into different groups using three-dimensional principal component analysis (PCA) based on PC1, PC2, and PC3. The analyses were performed using PAST v. 3.24 software [73], and the scatter plot of three-dimensional PCA was obtained by three-dimension PCA tool, using the OmicShare online tools (http://www.omicshare.com/tools). Both the Broken Stick model and the Jolliffe cut-off value were used to interpret the number of significant components for the total variation obtained from the PCA analysis [74–76]. Based on the minimum eigenvalue criteria (of more than 1), significant components were used to calculate the accuracy value with respect to the population structure and hierarchical clustering.

2.9. Analysis of Population Structure

An analysis of population structure and gene flow between the 58 Asian bamboo accessions was performed using a model-based clustering approach to divide the species into sub-populations with the help of STRUCTURE v.2.3.4 software [77]. The program uses Bayesian estimates to identify population structure, under assumptions of admixed ancestry and correlated allelic frequencies using unlinked markers [78]. No prior information was ascribed to the IRAP data while estimating sub-populations. The optimal number of sub-populations (K) was determined by running the programme with K values ranging from 1 to 10, with six independent runs for each K value. To determine the most appropriate K value, the length of the burn-in period parameter was configured to 100,000 and the number of Markov Chain Monte Carlo (MCMC) (Bayesian statistics) replications (simulations) after burn-in was set over 500,000 [79]. The optimum K value was found by an ad hoc statistic ΔK based on the percentage of variation in the log probability of the IRAP marker between successive K values using an online tool, Structure Harvester [80].

2.10. Analysis of Molecular Variance

After determining the sub-populations among the accessions tested, the analysis of the molecular variance (AMOVA) between the sub-populations was estimated using GenAlEx 6.5 software [81]. In the parameter set, a nonparametric permutation and standard permute procedure with 999 pairwise-permutations were used. These values were utilised to measure the total molecular variance between and within the populations.

3. Results

3.1. Development of IRAP Makers and Functionality Assay

A total of 98,850 LTR retrotransposons with both ends of intact LTR sequences were identified in the moso bamboo genome in the present study. The incremental clustering divided these LTR sequences into 64,281 clusters with 95% similarity criteria. Only the clusters that contained more than 30 copies of LTR sequences were considered as candidate clusters for IRAP primer development (Supplementary Table S1). Accordingly, 13 clusters with more than 30 LTR copies and at least one copy of all intact protein domains such as *gag* and polyprotein were shortlisted, which accounted only 0.02% of the identified clusters. These 13 clusters had a total of 696 copy numbers of LTR sequences, with an average of 53.5 copies per cluster. The highest copy number of 121 was identified in cluster number 3 followed by cluster 4, cluster 15 and cluster 22. Based on the number of clusters and LTR sequence size, a total of 90 markers were initially designed. Each marker was represented by a primer that acted both forward and reverse primers (Figure 1). Among these, 26 primers (29%) that showed proper amplification with clear allele patterns and high reproducibility were shortlisted. Finally, only 16 primers (18%) those showed clearly distinguishable polymorphism, and were, therefore, chosen for further analysis.

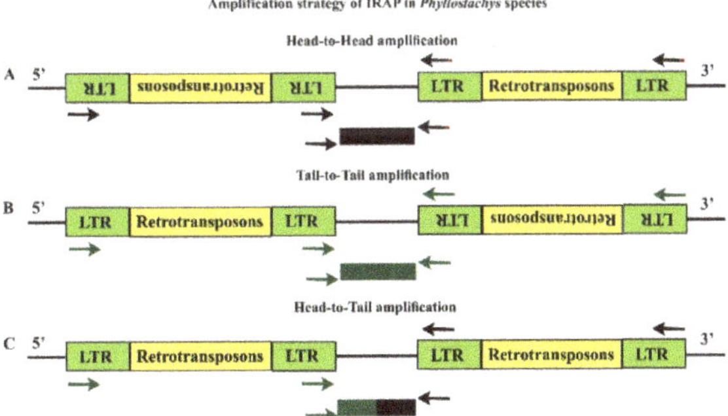

Figure 1. Amplification strategy for inter retrotransposon amplified polymorphism (IRAP) in *Phyllostachys* species (Asian bamboo). LTR stands for long terminal repeat. The single primer acts as both forward and reverse primer in PCR reaction. (**A**), Head-to-Head amplification; (**B**), Tail-to-Tail amplification; (**C**), Head-to-Tail amplification. The arrows and rectangles show the position of the IRAP primers and expected PCR products, respectively.

A functionality check of the amplicons generated by the selected 16 IRAP markers revealed consistently well-resolved and reproducible amplicon patterns among all the 58 Asian bamboo accessions (Figure 2). There was a total of 215 scorable amplicons (alleles) produced of which 214 were polymorphic (99.5%). Polymorphic alleles were produced with an average of 13.3 alleles per marker. Across the test population, a total of 3282 polymorphic amplicons were generated with an average of 56.6 amplicons per accession. The allele numbers produced per marker ranged between 8 (CL54-R) and 16 (CL34-R and CL63-R) with a size variation ranging from 200 bp to 2700 bp (Table 2).

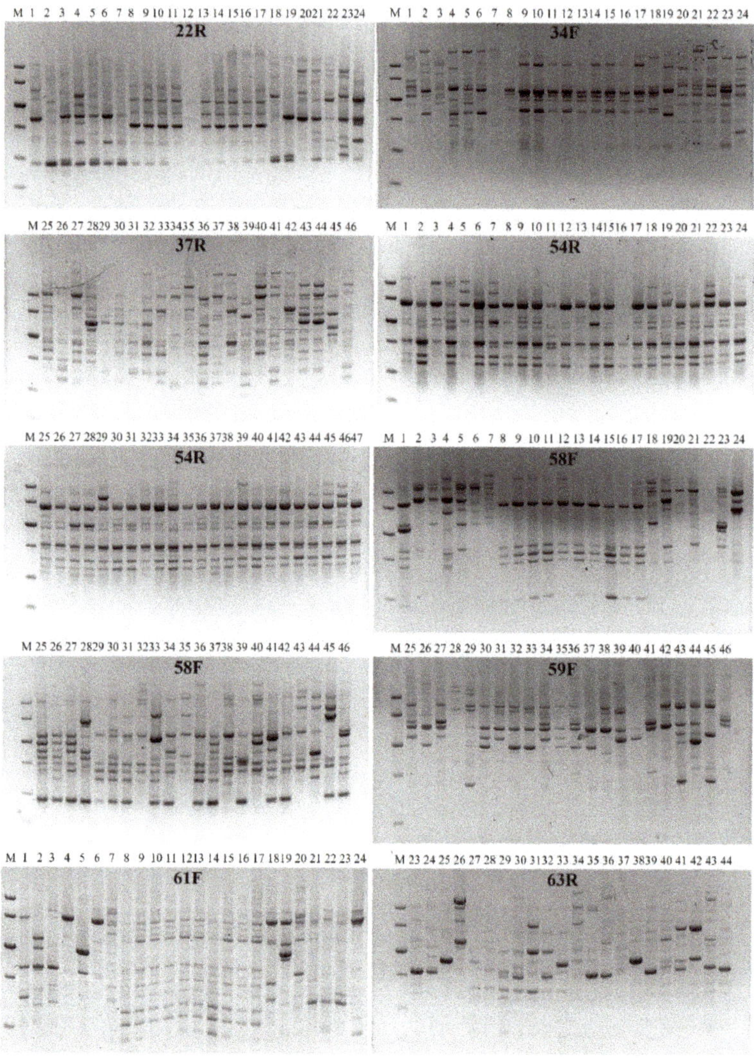

Figure 2. Inter-retrotransposon amplified polymorphism (IRAP) gel fingerprints. Negative agarose gels illustrate the results achieved in different Asian bamboo accessions for different IRAP markers. The bold black letters on top of the gel are the names of the IRAP marker. M above the gel on the left side represents 1 kb DNA Ladder mix (Takara). Numbers 1–47 represent different Asian bamboo species as defined in Table 1.

Table 2. List of inter-retrotransposon amplified polymorphism (IRAP) primers with description of amplicons used for the analysis of the genetic diversity and population structure of different *Phyllostachys* species (Asian bamboo).

IRAP Marker	Primer Sequence (5′-3′)	IRAP Primer Location in LTR Retrotransposons (bp)	Allele Size Range (kb)	No. of Polymorphic Alleles	Total Number of Amplicons in the Population	Average Allele Frequency	PIC Value	Melting Temperature (Tm) °C
CL3-F	TATAAAGGTAGCTTTCGGGTATG	428338_436487	0.65–2.7	13	231	0.306	0.350	52
CL4-R	CTGGTATATAGCTGTTGAGCGACG	82500_84292	0.6–2.4	13	192	0.255	0.351	57
CL15-R	CTCGTGTATTCTCCCTTTGC	221059_224205	0.43–2.5	12	227	0.326	0.400	53
CL22-F	TGATCAGAGAAGAAAGGGGA	49377_57704	0.42–2.7	13	125	0.166	0.256	52
CL22-R	CACGCAGAGAGATTGACACG	49377_57704	0.35–2.7	15	173	0.198	0.319	56
CL34-F	GAACGATTACCTCACAGACA	4137_7249	0.45–2.7	15	183	0.210	0.262	52
CL34-R	GAGCAATAAAGAGAAGCCCG	4137_7249	0.4–2.2	16	273	0.294	0.334	53
CL37-F	AGATTGTTTGATTCGGGGGG	153045_162321	0.3–2.5	12	183	0.263	0.325	54
CL37-R	AGCGGCGTGGAGGAGTTACC	153045_162321	0.5–2.7	13	234	0.310	0.376	61
CL42-R	GCAACAACAAACCCTAAAAA	18144_27070	0.3–2.7	14	230	0.283	0.369	50
CL54-R	GCAAGAACATAAGAACAGAA	56871_64575	0.3–2.0	8	209	0.450	0.336	55
CL58-F	AAGGAATCGTCAGTCAACAA	28670_35571	0.3–2.5	11	202	0.317	0.369	52
CL59-F	TGTCAGACAGTACAGGTGCT	1169_4536	0.2–2.7	14	204	0.251	0.329	55
CL61-F	ATGACATAGGGCACACCAGA	680968_684115	0.37–2.0	14	260	0.320	0.275	55
CL62-F	TAAATAGGGAACGAGGAGCC	178936_183841	0.5–2.7	15	190	0.218	0.306	56
CL63-R	ACATTGTTTGATTCGGGGGG	248490_260632	0.45–2.7	16	166	0.179	0.273	55
Mean	-	-	-	13.37	208.75	0.272	0.327	-

Note: LTR, long terminal repeat; PIC, polymorphic information content.

The marker CL59-F produced the widest range of amplicon sizes, from 200 bp to 2700 bp, while the amplicons generated by the marker CL61-F had the shortest size range, 370–2000 bp. Furthermore, the marker CL34-R produced the highest number of total amplicons (273) in the population followed by CL61-F (260), CL37-R (234), CL3-F (231) and CL42-R (230). All the primers produced 100% polymorphic alleles, except the marker CL22-F, which produced one monomorphic allele and 13 polymorphic alleles showing a polymorphism of 92.8%. The average allele frequency of the IRAP primers ranged between 0.166 (CL22-F) and 0.450 (CL54-R). The polymorphic information content (PIC) values ranged from 0.256 (CL22-F) to 0.400 (CL15-R), having an average of 0.327.

3.2. IRAP Amplicon Variation within and between the Accessions

Of the total of 3340 amplicons obtained, 3282 (98.3%) were polymorphic, and 58 (1.7%) were monomorphic. Among the 47 *Phyllostachys* species used in this study (Table 1), *Ph. edulis* generated the highest number of IRAP amplicons, having an average of 60 alleles across its nine varieties. *Ph. edulis* generated 72 alleles, followed by *Ph. edulis* cv. viridisulcata with 71 alleles. Nine moso bamboo varieties generated a total of 538 amplicons, of which eight alleles were monomorphic in all the nine moso bamboo varieties. Ninety-four alleles showed no amplification among the *Ph. edulis* varieties.

3.3. Multiple Sequence Alignment of IRAP-PCR Products

The multiple sequence alignment of IRAP-PCR amplicons using designed IRAP primers proved that they were indeed LTR-retrotransposon sequences. The neighbor-joining phylogenetic tree showed that the IRAP product sequences were genetically different (Figure 3), suggesting that all sixteen IRAP primers amplified unique PCR bands. A total of 14 bootstrap values were obtained, with 9 values ranging from 56% to 100%, which further confirms that the alleles were unique. The above results suggest that LTR-retrotransposon have unique amplicons for genome integrity and sizes.

Neighbor-joining phylogenetic tree of IRAP-PCR products

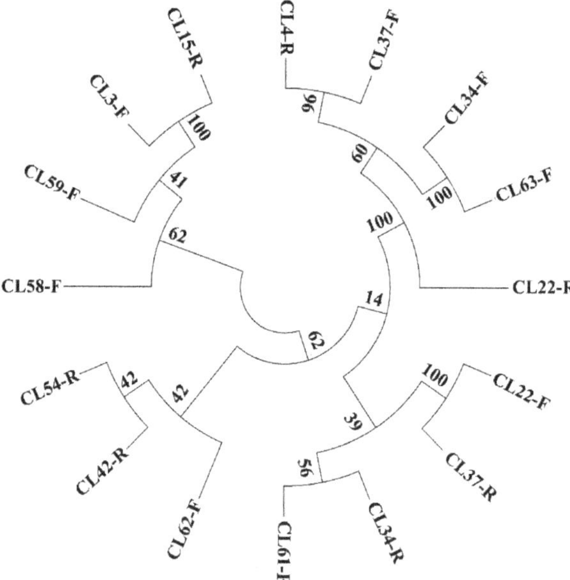

Figure 3. Neighbor-joining phylogenetic tree with bootstrap analysis showing the genetic relationship of 16 inter-retrotransposon amplified polymorphism (IRAP)-PCR amplicon sequences in *Phyllostachys* species (Asian bamboo) by corresponding IRAP markers.

3.4. Diversity Analysis of Asian Bamboo Accessions

The genetic similarity values of the 58 Asian bamboo accessions are shown in Supplementary Table S2. Jaccard similarity coefficient values ranged from 0.09 to 0.98 with an average similarity value of 0.25 among the 58 accessions. A total of 1653 pair-wise similarity coefficients were obtained, among which only seven pairs showed high similarity that ranged between 0.80 and 0.99. Twenty-eight pair wise similarity coefficients were intermediate, with values ranging from 0.60 to 0.79. A large proportion (1618 or 98% of similarity coefficients) were less than 0.59 (Supplementary Table S2). The lowest similarity coefficient of 0.091 was seen between *Ph. longiciliata* and *Ph. edulis* (Carr.) Matsumura followed by that between *Ph. arcana* and *Ph. edulis* (Carr.) Matsumura (0.092). The highest similarity of 0.984 was observed between *Ph. nigra* var. *henonis* and *Ph. edulis* cv. Pachyloen, followed by the similarity between *Ph. edulis* (Carr.) J. Houz and *Ph. edulis* (Carr.) Mitford cv. Gracilis.

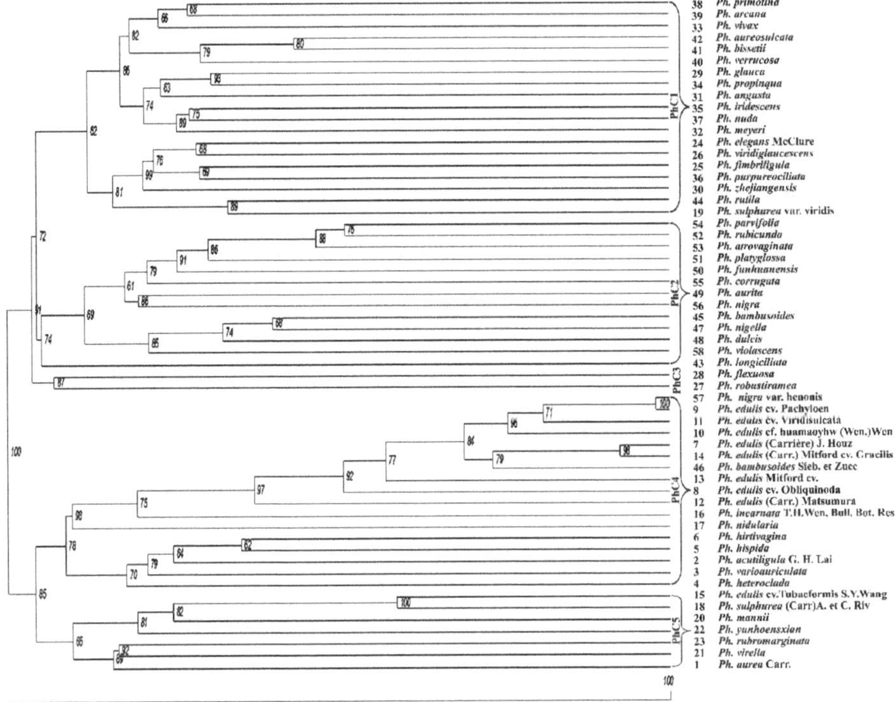

Figure 4. Dendrogram using the genetic distance matrix based on Jaccard's similarity coefficient obtained by hierarchical clustering analysis. The unweighted pair-group method with arithmetic average (UPGMA) with the bootstrap analysis showing the genetic relationship of 58 *Phyllostachys* species (Asian bamboo) based on inter-retrotransposon amplified polymorphism (IRAP) markers. The numbers present inside the clusters represent the bootstrap values. The numbers in bold represent different Asian bamboo species as defined in Table 1.

Based on the UPGMA clustering, the bamboo accessions were grouped into five clusters (Figure 4) viz. *Phyllostachys* clusters, PhC1 to PhC5. The cluster, PhC1 had 19 accessions grouped within, with an average similarity of 0.33. The second cluster, PhC2 which had a mean similarity of 0.34 contained 13 accessions. The cluster PhC3, which had only two accessions had a similarity of 0.26. The highest average similarity of 0.42 was observed in the sub-cluster PhC4, which included 17 accessions. All of the highly similar bamboo accessions were found in this cluster. The remaining sub-cluster, PhC5 encompassed seven accessions, and had average similarity value of 0.33. The clusters themselves showed affinity among themselves, with the first three clusters getting grouped together, as well as the remaining two getting grouped into a separate group. These two groups showed distinct separation with a bootstrap confidence of 100%. In the dendrogram constructed, bootstrap values ranged from 61% to 100% between clusters and the average bootstrap value observed in this study was 81%. For the clusters, PhC1 separated from PhC2 and PhC3 with bootstrap value of 74%, while PhC2 and PhC3 were separate for 91 times out of 100 bootstrap iterations. For the second group, the clusters PhC4 and PhC5 were distinct for 85% of the bootstrap resamplings.

The lowest similarity values observed between several of the accessions studied were due to the highest percentage (98%) of polymorphic bands generated by the IRAP primers. Of the nine varieties of moso bamboo, *Ph. edulis*, eight were found to be included in PhC4, while the remaining, *Ph. edulis* cv. tubaeformis, was found to be placed in sub-cluster PhC5. Most *Ph. edulis* variants were from the Zhejiang, Jiangsu and Jiangxi provinces of China (Table 1). Among 25 accessions collected from the Zhejiang province of China were found placed in different clusters. No specific clusters represented the species collected from the Zhejiang and Anhui province. The species collected from the Jiangsu province of China were grouped in clusters PhC1 and PhC4, and those from Henan provinces in PhC1 and PhC2. Moreover, two varieties of *Ph. nigra* were found to be distributed between PhC2 and PhC4. Similar was the case with two varieties belonging to the species, *Ph. bambusoides*. However, varieties of *Ph. sulphurea* were found to be included in PhC1 and PhC4. The bootstrap value for *Ph. nigra* var. henonis and *Ph. edulis* cv. Pachyloen as well as for *Ph. edulis* cv. Tubaeformis S.Y.Wang and *Ph. sulphurea* (Carr)A. et C. Riv was 100%. A bootstrap confidence of 98% was seen for *Ph. edulis* (Carrière) J. Houz and *Ph. edulis* (Carr.) Mitford cv. Gracilis. The bootstrap values clearly displayed that a good majority (79%) of cluster nodes were well fitted. None of the nodes were found with very low bootstrap value. Based on the resampling method, it was confirmed that the IRAP markers markedly distinguished the Asian bamboo accessions.

3.5. Statistical Fitness Analysis of Clustering Pattern

Two different statistical analyses such as cophenetic correlation coefficient (CCC) and principal component analysis (PCA) were carried out to confirm the grouping pattern of the Asian bamboo species. The CCC value of 0.8848 confirmed that hierarchical clustering was in significant agreement with the similarity matrix obtained from Jaccard similarity coefficients. Based on the IRAP markers the 58 Asian bamboo accessions indicated significant genetic diversity at those loci.

Three-dimensional PCA of the 58 Asian bamboo accessions, based on the variance-covariance matrix, displayed 13%, 7.9%, and 6.5% of the total variance for the first, second, and third component axes, respectively (Figure 5). A total of 57 axes (principal components) were extracted of which five components were retained based on the broken-stick model (Supplementary Figure S1). The selected components accounted for 38.1% of total variation. Three-dimensional plotting of bamboo accessions based on first three components showed four groups (Figure 5). Further resolution of the genotype grouping using hierarchical clustering of Euclidean distances based on the component scores for the five significant PCs, indicated the members under each group. The first group (Group 1) showed grouping of 13 accessions, Group 2 contained 19 accessions, Group 3 had 18 accessions and Group 4 with eight accessions (Figure 6). Comparing the clusters based on the Jaccard's similarity coefficients, the Group 1 had members drawn from PhC4 and PhC5, while Group 4 was almost entirely was drawn from PhC4, and contained seven out of nine *Ph. edulis* accessions used in the study. The Group 2 was exactly similar to PhC1 in both number and membership of accessions. The remaining group, Group 3 contained all the accessions that were members of PhC2 and PhC3, together with three accessions drawn from PhC4 and PhC5. The grouping pattern of the accessions revealed that based on the IRAP polymorphism, Group 2 was more robust and isolated from the subsequent groups of Group 3, Group 1 and Group 4. Group 2 had an accuracy value of 100%, followed by Group 3 (83%), Group 1 (54%) and Group 4 (47%) from the dendrogram based clusters.

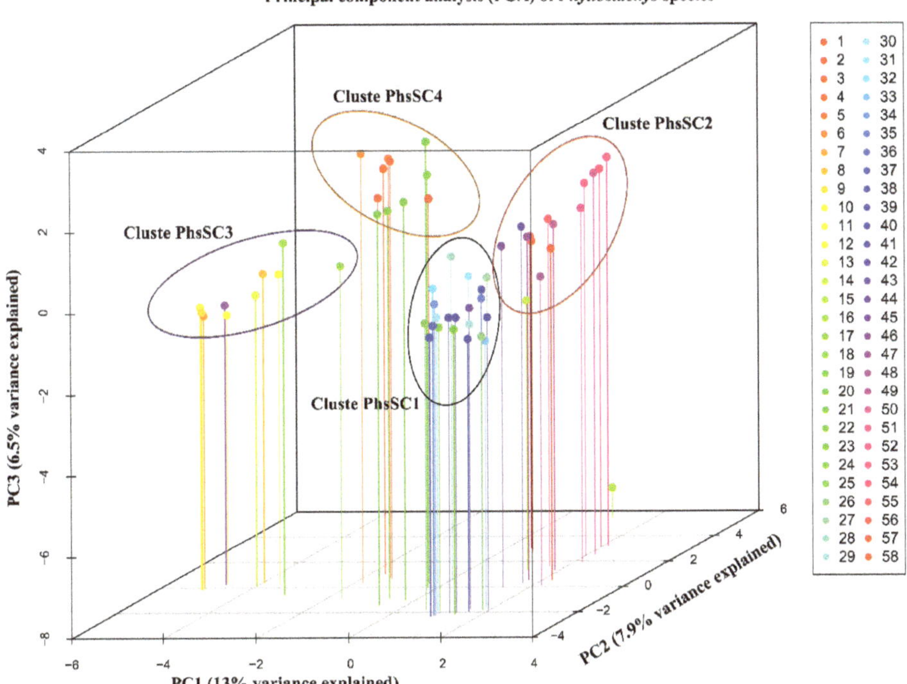

Figure 5. Scatter diagram of three-dimensional principal component analysis (PCA) showing the distribution of 58 *Phyllostachys* species (Asian bamboo) based on inter-retrotransposon amplified polymorphism (IRAP) markers. PC1 (X-axis), PC2 (Y axis), and PC3 (Z-axis) are the first, second, and third principal components, respectively. Numbers 1–58 on the right side represent different Asian bamboo species as defined in Table 1.

3.6. Population Structure of Asian Bamboo Accessions

The population structure analysis performed using IRAP markers to understand the genetic relationship among Asian bamboo accessions, revealed the existence of four apparent sub-populations in the test accessions. The Structure Harvester picked the maximum ΔK value when the inferred number of sub-populations (K) was at four (K = 4), having the maximum delta K value of 265.41 (Figure 7) with the lowest standard deviation (2.34) of the parameter, LnP(K). The resolved sub-populations were designated as PhSP1, PhSP2, PhSP3 and PhSP4 (Figure 8). The first sub-population, PhSP1 accommodated 13 accessions, whereas the remaining sub-populations PhSP2, PhSP3 and PhSP4 carried 19, 17 and 8 accessions, respectively. One of the accessions, *Ph. flexuosa* remained out of the sub-populations being an admixture of all the four sub-populations. Respective proportions of memberships were 23.0% for PhSP1, 35.7% for PhSP2, 27.1% for PhSP3 and 14.2% for PhSP4. The allele frequency divergence between the sub-populations was maximum between PhSP3 and PhSP4 (0.20), followed by between PhSP2 and PhSP4 (0.19). The lowest divergence was observed between PhSP1, PhSP2 and PhSP3. The estimated values of expected heterozygosity that can be construed as the average distance between members within each sub-population were 0.30, 0.27, 0.29 and 0.08 for PhSP1 to PhSP4 respectively. The proportion of the total genetic variance (F_{ST}) explained by the sub-populations ranged from 0.21 (PhSP1) to 0.80 (PhSP4), with the remaining PhSP2 and PhSP3 having F_{ST} values of 0.24 and 0.26 respectively. The inferred ancestry coefficients (Q values) for individual accessions carried by the PhSP1 ranged from 0.63 to 0.99, while that of PhSP2 was between 0.73 and 0.99 (Figure 8). Similarly, PhSP3 had a range of 0.50 to 0.99 for the Q values, and the range for PhSP4 was between 0.74 and 1.00. The average Q values for each sub-population were 0.90 for PhSP1, 0.94 for PhSP2, 0.86 for PhSP3 and 0.91c for PhSP4.

Heatmap of principal component analysis (PCA) of *Phyllostachys* species

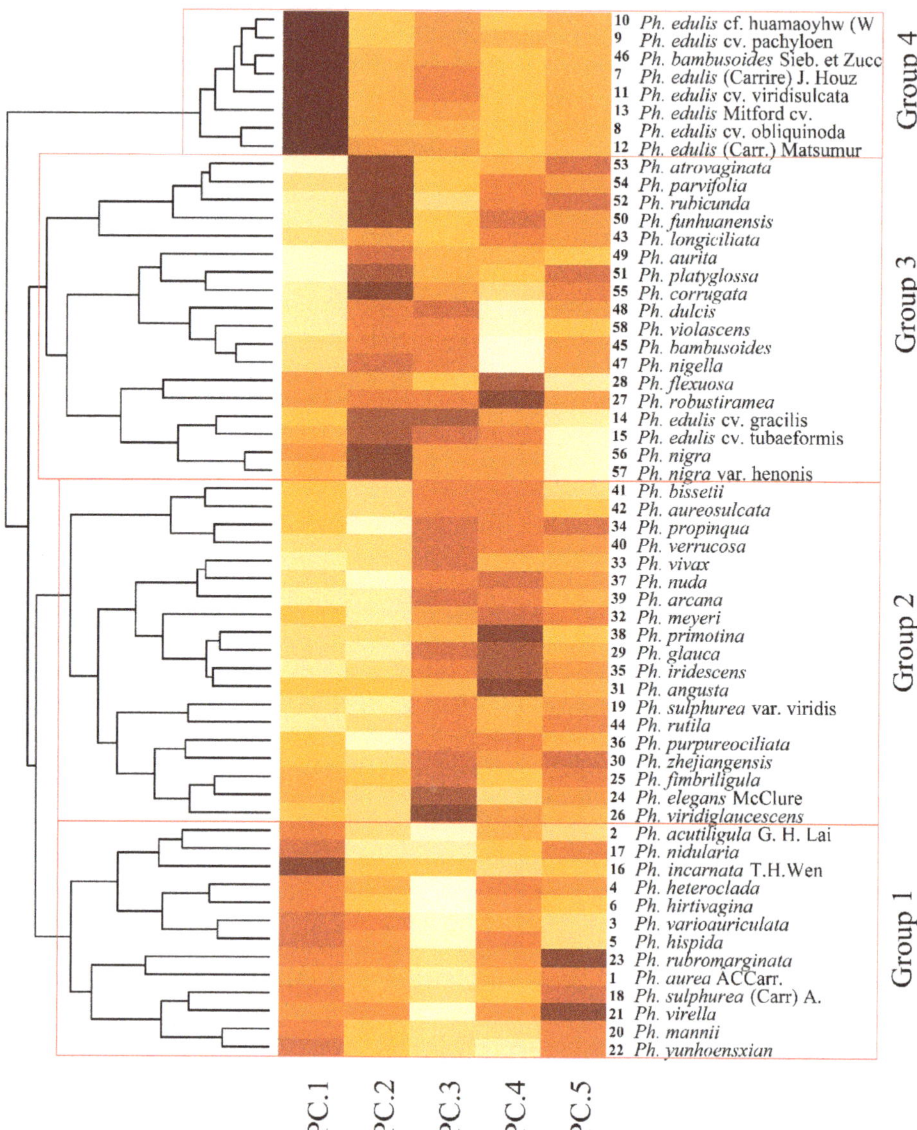

Figure 6. Heatmap of principal component analysis (PCA) showing the distribution of 58 Phyllostachys species (Asian bamboo) based on inter-retrotransposon amplified polymorphism (IRAP) markers. Five significant principal components (PCs) were identified based on the broken-stick model. Hierarchical clustering was done using Euclidean distances. Numbers 1–58 on the right side represent different Asian bamboo species as defined in Table 1.

Individual members of the sub-populations included *Phyllostachys* species such as *Ph. hispida,
Ph. hirtivagina, Ph. virella, Ph. varioauriculata, Ph. heteroclada,* and *Ph. mannii* having placed in PhSP1,
that possessed maximum allele frequency of that sub-population. Similarly, fourteen accessions were
found to possess high inferred ancestry coefficients for PhSP2, eight accessions for PhSP3 and five for
PhSP4. Seven of the moso bamboo varieties (*Ph. edulis*) were placed in PhSP4, and four of them had the
maximum inferred membership coefficient together with one accession, *Ph. bambusoides* (Figure 8B).
All the four sub-populations contained accessions with admixtures of alleles from all, with PhSP3
carrying maximum of admixed accessions, followed by PhSP1 and PhSP4.

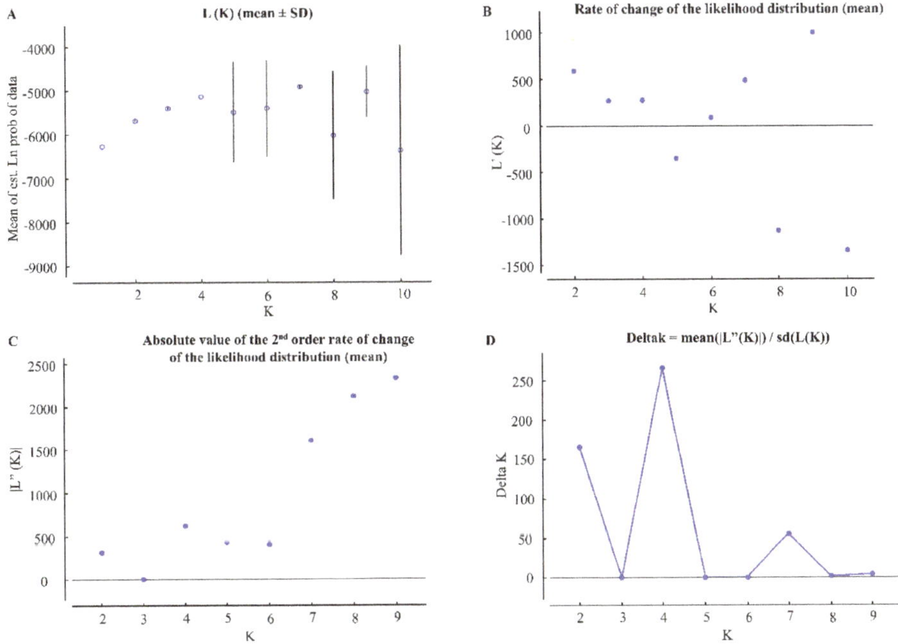

Figure 7. Structure Harvester analysis showing the ΔK value of 58 *Phyllostachys* species (Asian bamboo)
based on inter-retrotransposon amplified polymorphism (IRAP) markers. (**A**), mean of estimated Ln
probability; (**B**), rate of change of the likelihood distribution (mean); (**C**), absolute value of the 2nd
order rate of change of the likelihood distribution (mean); (**D**), ΔK = mean(|L"(K)|)/sd(L(K)). ΔK = 4
indicates the maximum K value.

We have drawn a consensus of distribution of accessions across the sub-populations by comparing
membership of respective groups obtained from hierarchical clustering using Jaccard coefficients and
PCA (Supplementary Table S3). It was identified that PCA showed a very close accuracy with respect to
the population structure than that obtained by hierarchical clustering. Of the total of 58 accessions, only
46 accessions were commonly shared between the distinct groups obtained among population structure,
principal component, and hierarchical clustering analyses (Supplementary Figure S2). Whereas, both
population structure and principal component analyses commonly shared a total of 54 accessions
from four sub-clusters, with an accuracy value of 93%. The sub-population, PhSP1 had 11 consensual
accessions, with an average similarity of 0.28. The remaining sub-populations, PhSP2, PhSP3, and
PhSP4 shared 19, 16, and 8 accessions, respectively sharing average similarity values of 0.33, 0.29 and
0.65 (Table 3).

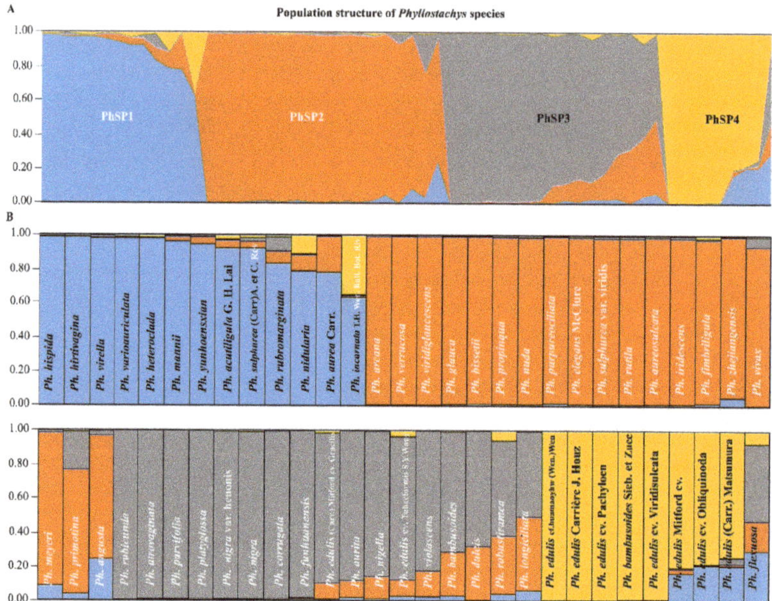

Figure 8. The population structure of 58 *Phyllostachys* species (Asian bamboo)-based inter-retrotransposon amplified polymorphism (IRAP) markers based on the Bayesian model. (**A**), Bar plot of sub-populations; (**B**), Accession based bar-plots indicating the members of different sub-populations showing the proportion of admixture of alleles. The different colors represent different sub-populations. The scale shows the inferred ancestry coefficients.

Table 3. Description of consensual Asian bamboo accessions picked out from the groupings obtained from dendrogram and principal component analysis (PCA) with respect to deduced population structure based on inter-retrotransposon amplified polymorphism (IRAP). There were a total of 54 accessions falling under different sub-populations, PhSP1 to PhSP4.

Sub-Population Name	*Phyllostachys* Species Name	Membership No.	Average Similarity
PhSP1	*Ph. acutiligula* G. H. Lai; *Ph. heteroclada*; *Ph. hirtivagina*; *Ph. varioauriculata*; *Ph. hispida*; *Ph. rubromarginata*; *Ph. aurea* Carr.; *Ph. sulphurea* (Carr)A. et C. Riv; *Ph. virella*; *Ph. mannii*; *Ph. yunhoensxian*.	11	0.28
PhSP2	*Ph. bissetii*; *Ph. aureosulcata*; *Ph. propinqua*; *Ph. verrucosa*; *Ph. vivax*; *Ph. nuda*; *Ph. arcana*; *Ph. meyeri*; *Ph. primotina*; *Ph. glauca*; *Ph. iridescens*; *Ph. angusta*; *Ph. sulphurea var. viridis*; *Ph. rutila*; *Ph. purpureociliata*; *Ph. zhejiangensis*; *Ph. fimbriligula*; *Ph. elegans* McClure; *Ph. viridiglaucescens*.	19	0.33
PhSP3	*Ph. atrovaginata*; *Ph. parvifolia*; *Ph. rubicunda*; *Ph. funhuanensis*; *Ph. longiciliata*; *Ph. aurita*; *Ph. platyglossa*; *Ph. corrugata*; *Ph. dulcis*; *Ph. violascens*; *Ph. bambusoides*; *Ph. nigella*; *Ph. edulis* (Carr.) Mitford cv. Gracilis; *Ph. edulis* cv. Tubaeformis S.Y. Wang; *Ph. nigra*; *Ph. nigra* var. henonis.	16	0.29
PhSP4	*Ph. edulis* cf. huamaoyhw (Wen.) Wen; *Ph. edulis* cv. Pachyloen; *Ph. bambusoides* Sieb. et Zucc; *Ph. edulis* (Carrière) J. Houz; *Ph. edulis* cv. Viridisulcata; *Ph. edulis* Mitford cv.; *Ph. edulis* cv. Obliquinoda; *Ph. edulis* (Carr.) Matsumura.	8	0.65

3.7. Analysis of Molecular Variance

The genetic diversity of the four sub-populations of the Asian bamboo species divulged from the genome wide IRAP polymorphisms indicated significant level of genetic variation within and among populations. The sums of squared deviation (SS) values within populations and among populations were 1518.27 and 492.12, respectively with the variance (MS) values of 28.12 and 164.04. The percentage of total molecular variance within populations was 75%, and, among populations, it was 25%. A molecular variance of 75% indicates that a strong genetic differentiation occurred within populations. These values confirm the size of the four sub-populations and the ratio of the admixture of alleles in the four sub-populations. The genetic diversity value (P) of the Asian bamboo species was highly significant ($p < 0.001$) at two hierarchical levels (among populations and within populations).

4. Discussions

To support genetic bamboo improvement, it is essential to understand the genome-wide variation and diversity among the breeding material. Since TEs form 63% of the bamboo genome [40], of which more than 46% are LTR-retrotransposons, retrotransposon-based DNA-fingerprinting could be an ideal technique to study the genome wide diversity of closely related species or breeding lines [82,83]. The use of retrotransposon-based molecular markers to study plant diversity is a cheap and rapid technique which can provide potentially useful molecular information augmenting other DNA-based markers [82]. The availability of the complete bamboo genome (BambooGDB, http://www.bamboogdb.org) [84] enabled a detailed exploration of the LTR-retrotransposons and their copy numbers based on their structure and subsequently for the development of IRAP primers. Since the development of IRAP primers is a one-time investment, potential primers can be continually be applied [85–87] for studying genetic diversity of bamboo species and their corresponding congeners.

In the current study, we confirmed that LTRs of bamboo species had different copy numbers, and their sequences contained full complement of LTR retrotransposons. The structure analysis revealed that they are transcriptionally active, and could be functional. The IRAP primers developed from the high copy number clusters although could amplify multiple-sites of the genome, their number seems to be low when considering the potential diversity of the *Phyllostachys* accessions used. The designed IRAP primers produced both polymorphic and monomorphic alleles that enabled the use of IRAP DNA fingerprinting to assess the diversity among Asian bamboo accessions. Based on this genome-wide analysis, we concluded that only 29% of IRAP markers showed polymorphism implying that inter-LTR regions in the studied genomes of bamboo species were significantly conserved. This implies that the bamboo genome is still under evolution and LTRs are not very active in contributing to the genome wide variations. This is, to the best of our knowledge, the first report of IRAP primer development and the first report of genetic diversity in bamboo using IRAP based fingerprinting.

4.1. Genetic Diversity of Asian Bamboo Species

The IRAP primers designed and used in the current study offered valuable genetic information about economically important Asian bamboo species and their evolutionary pattern. From the pattern of IRAP polymorphism, we conclude that in spite of their abundance in the bamboo genome, LTR retrotransposons significantly lack polymorphism among the 58 accessions investigated in this study. Although, these accessions belonged to different *Phyllostachys* species as recognized from their collection sites, the lack of IRAP polymorphism suggested significant level of conservation within the bamboo genome. Possible reason for this genome integrity could be the low frequency of sexual reproduction in bamboos that is monocarpic and occur at long breeding cycles [88]. Moreover, bamboos propagate predominantly clonally, leading to less variation on the genomes. However, over a longer period of time, geographic isolation can conserve localized speciation in the bamboo clones leading to identification of different bamboo species, with some level of morphologic variations. Possibly, active LTR-retrotransposons could be playing a major role in this type of speciation, as they are capable of bringing in spontaneous changes in the genome at random loci. We have observed a few polymorphic loci in the present study, but they showed high level of polymorphism between different sub-populations. This could be attributed to the changes infused by active retrotransposons across the genomes. In an earlier study, analysis of 78 accessions of the Asian bamboo species using 23 microsatellite markers had revealed an average of 2.78 alleles per primer [89]. In contrast, the IRAP markers could produce as many as 13 alleles per marker locus, indicating their effectiveness in revealing the genetic diversity of bamboo genomes in the current study. It may be emphasized that, 98% of average polymorphism exhibited by the IRAP amplicons is the highest reported for bamboo. IRAP occurs due to random insertion of retroelements on the genome, resulting in a length variation of the interspersed regions flanked by two elements. If such insertions affect functional genes, the resultant variations are further clonally propagated in the population, resulting in a transient speciation until sexual reproduction occurs. Such transient speciation can be either transmitted to further sexual generations or can be lost or can subsequently crate novel variations. Therefore, species level genetic diversity of the Asian bamboo accessions can be considered predominantly intermediary and is in the course of evolution that may last for several thousands of years to come. High level of genome wide polymorphism using IRAP markers was also reported in other crop species such as among octoploid triticale plants, where 85% polymorphism was observed [90]. Similarly, 94%, 79% and 74% polymorphism have been reported in *Bletilla striata* [62], *Lallemantia iberica* [91] and *Schistosoma japonicum* [92], respectively, using IRAP markers. Like earlier instances, we could employ IRAP markers reliably in assessing the genetic diversity in bamboo.

Based on the inter-retrotransposon distances, 58 Asian bamboo accessions could be separated into four sub-populations. Considering the fact that the study included 47 species of *Phyllostachys*, the four sub-populations could be considered very low if the bamboo populations are in Hardy-Weinberg equilibrium with typical random mating behavior. This further consolidated our inferences that the study accessions used in the study were transient species adapted to local niches and propagated clonally. There was a significant lack of molecular allelic pattern representing different species, except for variations at few loci across the genome that could have been accumulated more recently in the evolutionary time scale due to transposition activity. These observations were similar to previous report on 78 Asian bamboo accessions using SSR markers, which grouped them into three classes [89]. Similarly, two major clusters have been reported in 50 varieties of *Bletilla striata* using IRAP markers [62]. Further, Zhao et al. [89] have reported that the genetic variation between *Ph. nuda* and *Ph. propinqua* was 0.2143 using SSR markers. In the current study, we obtained similar variation (0.371) between *Ph. nuda* and *Ph. propinqua*. Both these species were characterized with having or not having bristles on the back of sheath. *Ph. vivax* and *Ph. aureosulcata* have purple-green or yellow-green with purple culm sheaths and clustered in PhSP2, similar to Zhao et al. [89] study. The current results are consistent with current bamboo's taxonomic classification and agreed with the morphological classification [93]. Our study further proved that, model based and PCA based approaches were significantly better for resolving the population structure of bamboo, in the event of having a few polymorphic IRAP markers, that has produced significantly good number for highly polymorphic alleles. The CCC of 0.88 obtained from the dendrogram based on Jaccard coefficient is suggestive of this. In a previous report on 200 tree accessions in the 20 groups of *Olea europaea* using IRAP markers, CCC value was 0.96, indicating a good fit between the similarity matrix and the dendrogram [94].

4.2. Population Structure of Asian Bamboo

Structure analysis used in our study implements a model-based approach for inferring population structure using unlinked genotype marker data, identifying genetically distinct populations, admixtures of alleles in populations and assign individuals to specific sub-populations [77]. Using this approach, Jiang et al. [95] identified three sub-populations in 803 accessions of moso bamboo using 20 SSRs markers. Further, Nachimuthu et al. [96] have identified two sub-populations in 192 accessions of rice using 61 SSR markers. In the current study too, the initial analyses of genetic diversity using the hierarchical clustering as well as the PCA were suggestive of a low-level population differentiation within the study panel of Asian bamboo accessions. Therefore, we have fixed a sub-population range of 1 to 10 for the model, assuming admixed populations and correlated allele frequencies. Analysis revealed an optimum population structure consisting of four sub-populations.

The inferred ancestry coefficients of Asian bamboo accessions provided the genetic relationship and gene flow pattern between the sub-populations. Few members of all sub-populations had admixture of alleles, while some members were specifically grouped with maximum frequency of sub-population specific alleles. From the molecular diversity pattern of the Asian bamboo accessions, we concluded that the PhSP2 was the sub-population with particularly lower number of admixed clones. Conversely, the PhSP3 had significantly high level of admixture especially from PhSP2. Similarly, the admixture of alleles was found between PhSP1 and PhSP4. However, we could observe that genetic variations among populations was significantly lower than that within populations indicating several subtle allelic variations among the members of each sub-populations. This was similar to previous studies for bamboo species such as *Melocanna baccifera* and *Bashania fangiana* (Dwarf bamboo) using ISSR and AFLP markers, respectively [97,98].

Among the sub-populations, lowest within population variation was observed with PhSP4. This sub-population contained most of the Moso bamboo accessions of the species, *Ph. edulis*. This genetically very close group, however, contained one accession of *Ph. bambusoides*, which could be suspected as a mistaken nomenclature. The significant similarity within this group, could be well explained because all the members were different commercial varieties of Moso bamboo. This category of bamboo is the most commercially exploited bamboos and are chiefly propagated clonally. Moso bamboos also offer long flowering intervals, and long breeding cycle [99]. Further, Moso bamboo displayed very low admixture of alleles, which suggests that LTR-elements had a little role in defining its genetic structure.

5. Conclusions

The aim of the current study was to explore the genome wide abundance of LTR-retrotransposons in the Asian bamboo accessions and to development IRAP based markers for investigating genetic diversity and population structure. Although the variation among the interspaces between the LTR retrotransposons in bamboo species was low, few loci showed apparently high polymorphism aiding the analyses. Since transposon activity is related to environmental factors, geographic speciation could be one of the reasons for high IRAP based diversity at certain loci. This is the first report of population structure using IRAP markers in the Asian bamboo species. From the observed pattern of genetic diversity, it is reasonable to assume that the ancestors of Asian bamboo could be few in number with limited variability, which on evolution, adaptably speciated into different species, with subtle genetic change compared to other rapidly multiplying cross-pollinated species. Each of the IRAP primer had unique differentiation, and this marker system offered highly efficient and reproducible alleles for studying inter-retrotransposon-based genetic diversity.

Supplementary Materials: Supplementary Materials can be found at http://www.mdpi.com/1999-4907/11/1/31/s1. Supplementary Table S1. Description of moso bamboo clusters, with LTR copy numbers, length of LTR sequences and similarity ranges. The cluster sequences were used to generate IRAP amplicons to study the genetic diversity and population structure of different *Phyllostachys* species (Asian bamboo). The positions of the IRAP primer are highlighted by underlined bold letters. A, B, C and D represent cluster name, LTR copy number, length of LTR sequences and similarity (%) (Ranges), respectively. Supplementary Table S2. Jaccard similarity coefficient values of 58 (Asian bamboo) species generated by 16 IRAP markers. Supplementary Table S3. Description of total of the number of accessions commonly shared within each sub-population (PhSP1-PhSP4) of *Phyllostachys* species (Asian bamboo) by population structure, principal component (PCA), and hierarchical clustering analyses. The genetic diversity was assessed from the allele pattern produced by 16 inter-retrotransposon amplified polymorphism (IRAP) markers. Numbers on the table represent different Asian bamboo species as defined in the Table 1. Supplementary Figure S1. Broken-stick model showing the number of significant principal components (PCs) of 58 *Phyllostachys* species (Asian bamboo) based on inter-retrotransposon amplified polymorphism (IRAP) markers. Supplementary Figure S2. Venn diagram showing the number of accessions commonly shared within each sub-population (PhSP1-PhSP4) of *Phyllostachys* accessions (Asian bamboo). The consensus is obtained from population structure, principal component (PCA), and hierarchical clustering analyses. Sub-clusters, PhC2 and PhC3 were combined with PhC2 in the analysis since PhC3 had only two genotypes.

Author Contributions: Conceptualization, M.Z. and M.R.; Methodology, M.R., M.Z. and S.L.; Software, M.R., M.Z. and S.L.; Validation, M.R., M.Z. and S.L.; Formal analysis, M.R., K.K.V. and R.K.; Investigation, M.R. and M.Z.; Resources, M.Z., S.L. and M.R.; Data curation, M.R., M.Z. and S.L.; Writing—original draft preparation, M.R.; Writing—review and editing, M.R., K.K.V., R.K., K.Y. and M.Z.; Visualization, M.R., M.Z., K.K.V. and R.K.; Supervision, M.R. and M.Z.; Project administration, M.Z.; Funding acquisition, M.Z. All authors have read and agreed to the published version of the manuscript.

Funding: This work was funded by the grant from the National Natural Science Foundation of China (grant No 31870656 and 31470615), and the Zhejiang Provincial Natural Science Foundation of China (grant No. LZ19C160001 and 2016C02056-8).

Acknowledgments: The authors would like to extend their sincere appreciation to the Directors of bamboo garden and forests of Fujian, Zhejiang, Anhui, Sichuang, Jiangxi, Guangdong, Hunan, Henan, Jiangsu and Taiwan for supplying the plant materials. We thank all three reviewers for their valuable comments.

Conflicts of Interest: The authors declare that the research was conducted in the absence of any commercial or financial relationships that could be construed as a potential conflict of interest.

Abbreviations

AFLP	Amplified fragment length polymorphism
AMOVA	Analysis of the molecular variance
CCC	Cophenetic correlation coefficient
CTAB	Cetyltrimethylammonium bromide
df	Degree of freedom
EST-SSR	Expressed sequence tags—SSR
IPTG	Isopropyl β-D-1-thiogalactopyranoside
IRAP	Inter-retrotransposon amplified polymorphism
ISSRs	Inter-simple sequence repeats
LTR	Long terminal repeat
MCMC	Markov Chain Monte Carlo
MS	Mean of squared deviations
PCA	Principal component analysis
PhC	Phyllostachys cluster
PhSP	*Phyllostachys* sub-population
PIC	Polymorphic information content
RAPD	Randomly amplified polymorphic DNA
RFLP	Restriction fragment length polymorphism
SCoT	Start codon targeted
siRNAs	Small interfering RNAs
SNP	Single nucleotide polymorphism
SS	Sums of squared deviations
SSRs	Simple sequence repeats
TEs	Transposable elements
UPGMA	Unweighted pair-group method with arithmetic average

References

1. Yeasmin, L.; Ali, M.N.; Gantait, S.; Chakraborty, S. Bamboo: An overview on its genetic diversity and characterization. *3 Biotech* **2015**, *5*, 1–11. [CrossRef] [PubMed]

2. Biswas, P.; Chakraborty, S.; Dutta, S.; Pal, A.; Das, M. Bamboo flowering from the perspective of comparative genomics and transcriptomics. *Front. Plant Sci.* **2016**, *7*, 1900. [CrossRef] [PubMed]

3. Yuan, J.-L.; Yue, J.-J.; Gu, X.-P.; Lin, C.-S. Flowering of woody bamboo in tissue culture systems. *Front. Plant Sci.* **2017**, *8*, 1589. [CrossRef] [PubMed]

4. Zhang, Y.; Tang, D.; Lin, X.; Ding, M.; Tong, Z. Genome-wide identification of MADS-box family genes in moso bamboo (*Phyllostachys edulis*) and a functional analysis of PeMADS5 in flowering. *BMC Plant Biol.* **2018**, *18*, 176. [CrossRef] [PubMed]

5. Tao, G.; Fu, Y.; Zhou, M. Advances in studies on molecular mechanisms of rapid growth of bamboo species. *J. Agric. Biotechnol.* **2018**, *26*, 871–887.

6.   Zheng-ping, W.; Chris, S. *Phyllostachys Siebold* & Zuccarini, Abh. Math.-Phys. Cl. Königl. Bayer. Akad. Wiss. 3: 745. 1843, nom. cons., not Torrey (1836), nom. rej. *Flora China* **2006**, *22*, 163–180.

7.   Panee, J. Potential medicinal application and toxicity evaluation of extracts from bamboo plants. *J. Med. Plant Res.* **2015**, *9*, 681–692.

8.   Friar, E.; Kochert, G. Bamboo germplasm screening with nuclear restriction fragment length polymorphisms. *Theor. Appl. Genet.* **1991**, *82*, 697–703. [CrossRef]

9.   Friar, E.; Kochert, G. A study of genetic variation and evolution of *Phyllostachys* (Bambusoideae: Poaceae) using nuclear restriction fragment length polymorphisms. *Theor. Appl. Genet.* **1994**, *89*, 265–270. [CrossRef]

10.   Gielis, J.; Everaert, I.; De Loose, M. Genetic variability and relationships in *Phyllostachys* using random amplified polymorphic DNA. In *The Bamboos*; Chapman, G.P., Ed.; Linnean Society Symposium Series Academic: London, UK, 1997; Volume 19, pp. 107–124.

11.   Zhang, S.; Ma, Q.; Ding, Y. RAPD analysis for the genetic diversity of *Phyllostachys edulis* China forestry. *Sci. Technol.* **2007**, *21*.

12.   Hodkinson, T.R.; Renvoize, S.A.; Chonghaile, G.N.; Stapleton, C.M.; Chase, M.W. A comparison of ITS nuclear rDNA sequence data and AFLP markers for phylogenetic studies in *Phyllostachys* (Bambusoideae, Poaceae). *J. Plant Res.* **2000**, *113*, 259–269. [CrossRef]

13.   Suyama, Y.; Obayashi, K.; Hayashi, I. Clonal structure in a dwarf bamboo (*Sasa senanensis*) population inferred from amplified fragment length polymorphism (AFLP) fingerprints. *Mol. Ecol.* **2000**, *9*, 901–906. [CrossRef] [PubMed]

14.   Isagi, Y.; Shimada, K.; Kushima, H.; Tanaka, N.; Nagao, A.; Ishikawa, T.; Onodera, H.; Watanabe, S. Clonal structure and flowering traits of a bamboo [*Phyllostachys pubescens* (Mazel) Ohwi] stand grown from a simultaneous flowering as revealed by AFLP analysis. *Mol. Ecol.* **2004**, *13*, 2017–2021. [CrossRef] [PubMed]

15.   Lin, X.; Lou, Y.; Zhang, Y.; Yuan, X.; He, J.; Fang, W. Identification of genetic diversity among cultivars of *Phyllostachys violascens* using ISSR, SRAP and AFLP markers. *Bot. Rev.* **2011**, *77*, 223–232. [CrossRef]

16.   Isagi, Y.; Oda, T.; Fukushima, K.; Lian, C.; Yokogawa, M.; Kaneko, S. Predominance of a single clone of the most widely distributed bamboo species *Phyllostachys edulis* in East Asia. *J. Plant Res.* **2016**, *129*, 21–27. [CrossRef]

17.   Lai, C.; Hsiao, J. Genetic variation of *Phyllostachys pubescens* (Bambusoideae, Poaceae) in Taiwan based on DNA polymorphisms. *Bot. Bull. Acad. Sin.* **1997**, *38*, 145–152.

18.   Tang, D.-Q.; Lu, J.-J.; Fang, W.; Zhang, S.; Zhou, M.-B. Development, characterization and utilization of GenBank microsatellite markers in *Phyllostachys pubescens* and related species. *Mol. Breed.* **2010**, *25*, 299–311. [CrossRef]

19.   Peng, Z.; Lu, T.; Li, L.; Liu, X.; Gao, Z.; Hu, T.; Yang, X.; Feng, Q.; Guan, J.; Weng, Q. Genome-wide characterization of the biggest grass, bamboo, based on 10,608 putative full-length cDNA sequences. *BMC Plant Biol.* **2010**, *10*, 116. [CrossRef]

20.   Zhang, Z.; Guan, Y.; Yang, L.; Yu, L.; Luo, S. Analysis of SSRs information and development of SSR markers from Moso bamboo (*Phyllostachys edulis*) ESTs. *Acta Hortic. Sin.* **2011**, *38*, 989–996.

21.   Lin, X.C.; Ruan, X.S.; Lou, Y.F.; Guo, X.Q.; Fang, W. Genetic similarity among cultivars of *Phyllostachys pubescens*. *Plant Syst. Evol.* **2009**, *277*, 67–73. [CrossRef]

22.   Zhou, M.-B.; Wu, J.-J.; Ramakrishnan, M.; Meng, X.-W.; Vinod, K.K. Prospects for the study of genetic variation among Moso bamboo wild-type and variants through genome resequencing. *Trees* **2018**, *33*, 371–381. [CrossRef]

23.   Chen, C.; Wang, W.; Wang, X.; Shen, D.; Wang, S.; Wang, Y.; Gao, B.; Wimmers, K.; Mao, J.; Li, K.; et al. Retrotransposons evolution and impact on lncRNA and protein coding genes in pigs. *Mob. DNA* **2019**, *10*, 19. [CrossRef] [PubMed]

24.   Li, S.-F.; Guo, Y.-J.; Li, J.-R.; Zhang, D.-X.; Wang, B.-X.; Li, N.; Deng, C.-L.; Gao, W.-J. The landscape of transposable elements and satellite DNAs in the genome of a dioecious plant spinach (*Spinacia oleracea* L.). *Mob. DNA* **2019**, *10*, 3. [CrossRef] [PubMed]

25.   McCue, A.D.; Slotkin, R.K. Transposable element small RNAs as regulators of gene expression. *Trends Genet.* **2012**, *28*, 616–623. [CrossRef]

26.   Gao, D.; Abernathy, B.; Rohksar, D.; Schmutz, J.; Jackson, S.A. Annotation and sequence diversity of transposable elements in common bean (*Phaseolus vulgaris*). *Front. Plant Sci.* **2014**, *5*, 339. [CrossRef]

27. Gao, D.; Jiang, N.; Wing, R.A.; Jiang, J.; Jackson, S.A. Transposons play an important role in the evolution and diversification of centromeres among closely related species. *Front. Plant Sci.* **2015**, *6*, 216. [CrossRef]

28. Bourque, G.; Burns, K.H.; Gehring, M.; Gorbunova, V.; Seluanov, A.; Hammell, M.; Imbeault, M.; Izsvák, Z.; Levin, H.L.; Macfarlan, T.S.; et al. Ten things you should know about transposable elements. *Genome Biol.* **2018**, *19*, 199. [CrossRef]

29. Drongitis, D.; Aniello, F.; Fucci, L.; Donizetti, A. Roles of transposable elements in the different layers of gene expression regulation. *IJMS* **2019**, *20*, 5755. [CrossRef]

30. Orozco-Arias, S.; Isaza, G.; Guyot, R. Retrotransposons in plant genomes: Structure, identification, and classification through bioinformatics and machine learning. *IJMS* **2019**, *20*, 3837. [CrossRef]

31. Feschotte, C.; Pritham, E.J. DNA transposons and the evolution of eukaryotic genomes. *Annu. Rev. Genet.* **2007**, *41*, 331–368. [CrossRef]

32. Finnegan, D.J. Eukaryotic transposable elements and genome evolution. *Trends Genet.* **1989**, *5*, 103–107. [CrossRef]

33. Goodier, J.L. Restricting retrotransposons: A review. *Mob. DNA* **2016**, *7*, 16. [CrossRef] [PubMed]

34. Wicker, T.; Gundlach, H.; Spannagl, M.; Uauy, C.; Borrill, P.; Ramirez-Gonzalez, R.H.; De Oliveira, R.; Mayer, K.F.X.; Paux, E.; Choulet, F. Impact of transposable elements on genome structure and evolution in bread wheat. *Genome Biol.* **2018**, *19*, 103. [CrossRef] [PubMed]

35. Bennetzen, J.L.; Sanmiguel, P. Evidence that a recent increase in maize genome size was caused by the massive amplification of intergene retrotransposons. *Ann. Bot.* **1998**, *82*, 37–44. [CrossRef]

36. Schnable, P.S.; Ware, D.; Fulton, R.S.; Stein, J.C.; Wei, F.; Pasternak, S.; Liang, C.; Zhang, J.; Fulton, L.; Graves, T.A.; et al. The B73 maize genome: Complexity, diversity, and dynamics. *Science* **2009**, *326*, 1112–1115. [CrossRef] [PubMed]

37. Zhou, M.-B.; Lu, J.-J.; Zhong, H.; Tang, K.-X.; Tang, D.-Q. Distribution and polymorphism of mariner-like elements in the Bambusoideae subfamily. *Plant Syst. Evol.* **2010**, *289*, 1–11. [CrossRef]

38. Peng, Z.; Lu, Y.; Li, L.; Zhao, Q.; Feng, Q.; Gao, Z.; Lu, H.; Hu, T.; Yao, N.; Liu, K.; et al. The draft genome of the fast-growing non-timber forest species moso bamboo (*Phyllostachys heterocycla*). *Nat. Genet.* **2013**, *45*, 456–461. [CrossRef]

39. Zhou, M.; Hu, B.; Zhu, Y. Genome-wide characterization and evolution analysis of long terminal repeat retroelements in moso bamboo (*Phyllostachys edulis*). *Tree Genet. Genomes* **2017**, *13*, 43. [CrossRef]

40. Zhao, H.; Gao, Z.; Wang, L.; Wang, J.; Wang, S.; Fei, B.; Chen, C.; Shi, C.; Liu, X.; Zhang, H.; et al. Chromosome-level reference genome and alternative splicing atlas of moso bamboo (*Phyllostachys edulis*). *GigaScience* **2018**, *7*, giy115. [CrossRef]

41. Bennetzen, J.L. Transposable element contributions to plant gene and genome evolution. *Plant Mol. Biol.* **2000**, *42*, 251–269. [CrossRef]

42. Llorens, C.; Munoz-Pomer, A.; Bernad, L.; Botella, H.; Moya, A. Network dynamics of eukaryotic LTR retroelements beyond phylogenetic trees. *Biol. Direct* **2009**, *4*, 41. [CrossRef] [PubMed]

43. Neumann, P.; Novák, P.; Hoštáková, N.; Macas, J. Systematic survey of plant LTR-retrotransposons elucidates phylogenetic relationships of their polyprotein domains and provides a reference for element classification. *Mob. DNA* **2019**, *10*, 1. [CrossRef] [PubMed]

44. Bennetzen, J.L.; Wang, H. The contributions of transposable elements to the structure, function, and evolution of plant genomes. *Annu. Rev. Plant Biol.* **2014**, *65*, 505–530. [CrossRef] [PubMed]

45. Cho, J.; Benoit, M.; Catoni, M.; Drost, H.-G.; Brestovitsky, A.; Oosterbeek, M.; Paszkowski, J. Sensitive detection of pre-integration intermediates of long terminal repeat retrotransposons in crop plants. *Nat. Plants* **2019**, *5*, 26–33. [CrossRef]

46. Griffiths, J.; Catoni, M.; Iwasaki, M.; Paszkowski, J. Sequence-independent identification of active LTR retrotransposons in *Arabidopsis*. *Mol. Plant* **2018**, *11*, 508–511. [CrossRef]

47. Zhou, M.; Zhu, Y.; Bai, Y.; Hänninen, H.; Meng, X. Transcriptionally active LTR retroelement-related sequences and their relationship with small RNA in moso bamboo (*Phyllostachys edulis*). *Mol. Breed.* **2017**, *37*, 132. [CrossRef]

48. Kaeppler, S.M.; Kaeppler, H.F.; Rhee, Y. Epigenetic aspects of somaclonal variation in plants. *Plant Mol. Biol.* **2000**, *43*, 179–188. [CrossRef]

49. Miyao, A.; Nakagome, M.; Ohnuma, T.; Yamagata, H.; Kanamori, H.; Katayose, Y.; Takahashi, A.; Matsumoto, T.; Hirochika, H. Molecular spectrum of somaclonal variation in regenerated rice revealed by whole-genome sequencing. *Plant Cell Physiol.* **2012**, *53*, 256–264. [CrossRef]

50. Paszkowski, J. Controlled activation of retrotransposition for plant breeding. *Curr. Opin. Biotechnol.* **2015**, *32*, 200–206. [CrossRef]

51. Kalendar, R.; Grob, T.; Regina, M.; Suoniemi, A.; Schulman, A. IRAP and REMAP: Two new retrotransposon-based DNA fingerprinting techniques. *Theor. Appl. Genet.* **1999**, *98*, 704–711. [CrossRef]

52. Kalendar, R.; Schulman, A.H. IRAP and REMAP for retrotransposon-based genotyping and fingerprinting. *Nat. Protoc.* **2007**, *1*, 2478. [CrossRef] [PubMed]

53. Schulman, A.H.; Flavell, A.J.; Paux, E.; Ellis, T.H. The application of LTR retrotransposons as molecular markers in plants. *Methods Mol. Biol.* **2012**, *859*, 115–153. [CrossRef] [PubMed]

54. Kalendar, R.; Amenov, A.; Daniyarov, A. Use of retrotransposon-derived genetic markers to analyse genomic variability in plants. *Funct. Plant Biol.* **2019**, *46*, 15–29. [CrossRef] [PubMed]

55. Rahmani, M.-S.; Pijut, P.M.; Shabanian, N.; Nasri, M. Genetic fidelity assessment of in vitro-regenerated plants of *Albizia julibrissin* using SCoT and IRAP fingerprinting. *Vitr. Cell. Dev. Biol. Plant* **2015**, *51*, 407–419. [CrossRef]

56. Vuorinen, L.A.; Kalendar, R.; Fahima, T.; Korpelainen, H.; Nevo, E.; Schulman, H.A. Retrotransposon-based genetic diversity assessment in wild emmer wheat (*Triticum turgidum* ssp. *dicoccoides*). *Agronomy* **2018**, *8*, 107. [CrossRef]

57. Sun, J.; Yin, H.; Li, L.; Song, Y.; Fan, L.; Zhang, S.; Wu, J. Evaluation of new IRAP markers of pear and their potential application in differentiating bud sports and other Rosaceae species. *Tree Genet. Genomes* **2015**, *11*, 25. [CrossRef]

58. Branco, C.J.; Vieira, E.A.; Malone, G.; Kopp, M.M.; Malone, E.; Bernardes, A.; Mistura, C.C.; Carvalho, F.I.; Oliveira, C.A. IRAP and REMAP assessments of genetic similarity in rice. *J. Appl. Genet.* **2007**, *48*, 107–113. [CrossRef]

59. Vukich, M.; Schulman, A.H.; Giordani, T.; Natali, L.; Kalendar, R.; Cavallini, A. Genetic variability in sunflower (*Helianthus annuus* L.) and in the Helianthus genus as assessed by retrotransposon-based molecular markers. *Theor. Appl. Genet.* **2009**, *119*, 1027–1038. [CrossRef]

60. Fan, F.; Cui, B.; Zhang, T.; Ding, G.; Wen, X. LTR-retrotransposon activation, IRAP marker development and its potential in genetic diversity assessment of masson pine (*Pinus massoniana*). *Tree Genet. Genomes* **2014**, *10*, 213–222. [CrossRef]

61. Lee, S.-I.; Kim, J.-H.; Park, K.-C.; Kim, N.-S. LTR-retrotransposons and inter-retrotransposon amplified polymorphism (IRAP) analysis in *Lilium* species. *Genetica* **2015**, *143*, 343–352. [CrossRef]

62. Guo, Y.; Zhai, L.; Long, H.; Chen, N.; Gao, C.; Ding, Z.; Jin, B. Genetic diversity of Bletilla striata assessed by SCoT and IRAP markers. *Hereditas* **2018**, *155*, 35. [CrossRef] [PubMed]

63. Hosid, E.; Brodsky, L.; Kalendar, R.; Raskina, O.; Belyayev, A. Diversity of LTR retrotransposon genome distribution in natural populations of the wild diploid wheat *Aegilops speltoides*. *Genetics* **2012**, *190*, 263–274. [CrossRef] [PubMed]

64. Doyle, J.J.; Doyle, J.L. Isolation of plant DNA from fresh tissue. *Focus* **1990**, *12*, 39–40.

65. Steinbiss, S.; Willhoeft, U.; Gremme, G.; Kurtz, S. Fine-grained annotation and classification of de novo predicted LTR retrotransposons. *Nucleic Acids Res.* **2009**, *37*, 7002–7013. [CrossRef] [PubMed]

66. Li, W.; Godzik, A. Cd-hit: A fast program for clustering and comparing large sets of protein or nucleotide sequences. *Bioinformatics* **2006**, *22*, 1658–1659. [CrossRef] [PubMed]

67. Kumar, S.; Stecher, G.; Tamura, K. MEGA7: Molecular evolutionary genetics analysis version 7.0 for bigger datasets. *Mol. Biol. Evol.* **2016**, *33*, 1870–1874. [CrossRef]

68. Saitou, N.; Nei, M. The neighbor-joining method: A new method for reconstructing phylogenetic trees. *Mol. Biol. Evol.* **1987**, *4*, 406–425. [CrossRef]

69. De Riek, J.; Calsyn, E.; Everaert, I.; Van Bockstaele, E.; De Loose, M. AFLP based alternatives for the assessment of Distinctness, Uniformity and Stability of sugar beet varieties. *Theor. Appl. Genet.* **2001**, *103*, 1254–1265. [CrossRef]

70. Jaccard, P. Nouvelles researches sur la distribution florale. *Bull. Soc. Vaud. Sci. Natl.* **1908**, 223–270.

71. Pavlicek, A.; Hrda, S.; Flegr, J. Free-Tree–freeware program for construction of phylogenetic trees on the basis of distance data and bootstrap/jackknife analysis of the tree robustness. Application in the RAPD analysis of genus *Frenkelia*. *Folia. Biol.* **1999**, *45*, 97–99.

72. Page, R.D. TreeView: An application to display phylogenetic trees on personal computers. *Comput. Appl. Biosci.* **1996**, *12*, 357–358. [PubMed]

73. Hammer, Ø.; Harper, D.; Ryan, P. PAST: Paleontological statistics software package for education and data analysis. *Palaeontol. Electron.* **2001**, *4*, 9.

74. Legendre, P.; Legendre, L. *Numerical Ecology*, 2nd English edition; Elsevier: Amsterdam, The Netherlands, 1998.

75. Jolliffe, I.T. *Principal Component Analysis*, 2nd ed.; Springer: New York, NY, USA, 2002.

76. McGarigal, K.; Cushman, S.A.; Stafford, S. *Multivariate Statistics for Wildlife and Ecology Research*; Springer: New York, NY, USA, 2000.

77. Pritchard, J.K.; Stephens, M.; Donnelly, P. Inference of population structure using multilocus genotype data. *Genetics* **2000**, *155*, 945–959. [PubMed]

78. Fraley, C.; Raftery, A.E. Model-based clustering, discriminant analysis, and density estimation. *J. Am. Stat. Assoc.* **2002**, *97*, 611–631. [CrossRef]

79. Karandikar, R. On the markov chain monte carlo (MCMC) method. *Sadhana* **2006**, *31*, 81–84. [CrossRef]

80. Evanno, G.; Regnaut, S.; Goudet, J. Detecting the number of clusters of individuals using the software STRUCTURE: A simulation study. *Mol. Ecol.* **2005**, *14*, 2611–2620. [CrossRef] [PubMed]

81. Peakall, R.; Smouse, P.E. GenAlEx 6.5: Genetic analysis in Excel. Population genetic software for teaching and research—An update. *Bioinformatics* **2012**, *6*, 2537–2589. [CrossRef] [PubMed]

82. Kalendar, R.; Flavell, A.J.; Ellis, T.H.N.; Sjakste, T.; Moisy, C.; Schulman, A.H. Analysis of plant diversity with retrotransposon-based molecular markers. *Heredity* **2010**, *106*, 520. [CrossRef]

83. Kalendar, R.; Schulman, A.H. Transposon based tagging: IRAP, REMAP, and iPBS. *Methods Mol. Biol.* **2014**, *1115*, 233–255. [CrossRef]

84. Zhao, H.; Peng, Z.; Fei, B.; Li, L.; Hu, T.; Gao, Z.; Jiang, Z. BambooGDB: A bamboo genome database with functional annotation and an analysis platform. *Database* **2014**, *2014*, bau006. [CrossRef]

85. Paux, E.; Faure, S.; Choulet, F.; Roger, D.; Gauthier, V.; Martinant, J.P.; Sourdille, P.; Balfourier, F.; Le Paslier, M.C.; Chauveau, A.; et al. Insertion site-based polymorphism markers open new perspectives for genome saturation and marker-assisted selection in wheat. *Plant Biotechnol. J.* **2010**, *8*, 196–210. [CrossRef] [PubMed]

86. Galindo-Gonzalez, L.; Mhiri, C.; Deyholos, M.K.; Grandbastien, M.A. LTR-retrotransposons in plants: Engines of evolution. *Gene* **2017**, *626*, 14–25. [CrossRef] [PubMed]

87. Belyayev, A.; Kalendar, R.; Brodsky, L.; Nevo, E.; Schulman, A.H.; Raskina, O. Transposable elements in a marginal plant population: Temporal fluctuations provide new insights into genome evolution of wild diploid wheat. *Mob. DNA* **2010**, *1*, 6. [CrossRef]

88. Simmonds, N.W. Monocarpy, calendars and flowering cycles in angiosperms. *Kew Bulletin* **1980**, *35*, 235–245. [CrossRef]

89. Zhao, H.; Yang, L.; Peng, Z.; Sun, H.; Yue, X.; Lou, Y.; Dong, L.; Wang, L.; Gao, Z. Developing genome-wide microsatellite markers of bamboo and their applications on molecular marker assisted taxonomy for accessions in the genus *Phyllostachys*. *Sci. Rep.* **2015**, *5*, 8018. [CrossRef]

90. Szućko, I.; Rogalska, S.M. Application of ISSR-PCR, IRAP-PCR, REMAP-PCR, and ITAP-PCR in the assessment of genomic changes in the early generation of triticale. *Biol. Plant* **2015**, *59*, 708–714. [CrossRef]

91. Cheraghi, A.; Rahmani, F.; Hassanzadeh-Ghorttapeh, A. IRAP and REMAP based genetic diversity among varieties of *Lallemantia iberica*. *Mol. Biol. Res. Commun.* **2018**, *7*, 125–132. [CrossRef]

92. Li, J.; Zhao, G.H.; Li, X.Y.; Chen, F.; Chen, J.B.; Zou, F.C.; Yang, J.F.; Lin, R.Q.; Weng, Y.B.; Zhu, X.Q. IRAP: An efficient retrotransposon-based electrophoretic technique for studying genetic variability among geographical isolates of *Schistosoma japonicum*. *Electrophoresis* **2011**, *32*, 1473–1479. [CrossRef]

93. Li, J. Flora of China. *Harv. Pap. Bot.* **2007**, *13*, 301–303. [CrossRef]

94. Khaleghi, E.; Sorkheh, K.; Chaleshtori, M.H.; Ercisli, S. Elucidate genetic diversity and population structure of *Olea europaea* L. germplasm in Iran using AFLP and IRAP molecular markers. *3 Biotech* **2017**, *7*, 71. [CrossRef]

95. Jiang, W.; Bai, T.; Dai, H.; Wei, Q.; Zhang, W.; Ding, Y. Microsatellite markers revealed moderate genetic diversity and population differentiation of moso bamboo (*Phyllostachys edulis*)—A primarily asexual reproduction species in China. *Tree Genet. Genomes* **2017**, *13*, 130. [CrossRef]

96. Nachimuthu, V.V.; Muthurajan, R.; Duraialaguraja, S.; Sivakami, R.; Pandian, B.A.; Ponniah, G.; Gunasekaran, K.; Swaminathan, M.; Suji, K.K.; Sabariappan, R. Analysis of population structure and genetic diversity in rice germplasm using SSR markers: An initiative towards association mapping of agronomic traits in *Oryza Sativa*. *Rice* **2015**, *8*, 30. [CrossRef] [PubMed]
97. Nilkanta, H.; Amom, T.; Tikendra, L.; Rahaman, H.; Nongdam, P. ISSR marker based population genetic study of *Melocanna baccifera* (Roxb.) Kurz: A commercially important bamboo of Manipur, North-East India. *Scientifica* **2017**, *2017*, 9. [CrossRef]
98. Ma, Q.-Q.; Song, H.-X.; Zhou, S.-Q.; Yang, W.-Q.; Li, D.-S.; Chen, J.-S. Genetic structure in dwarf bamboo (*Bashania fangiana*) clonal populations with different genet ages. *PLoS ONE* **2013**, *8*, e78784. [CrossRef] [PubMed]
99. Gao, J.; Zhang, Y.; Zhang, C.; Qi, F.; Li, X.; Mu, S.; Peng, Z. Characterization of the floral transcriptome of Moso bamboo (*Phyllostachys edulis*) at different flowering developmental stages by transcriptome sequencing and RNA-seq analysis. *PLoS ONE* **2014**, *9*, e98910. [CrossRef] [PubMed]

 **forests**

Article

Geographical Gradients of Genetic Diversity and Differentiation among the Southernmost Marginal Populations of *Abies sachalinensis* Revealed by EST-SSR Polymorphism

Keiko Kitamura [1], Kentaro Uchiyama [2], Saneyoshi Ueno [2], Wataru Ishizuka [3], Ikutaro Tsuyama [1] and Susumu Goto [4],*

[1] Hokkaido Research Center, Forestry and Forest Products Research Institute, 7 Hitsujigaoka, Toyohira, Sapporo, Hokkaido 062-8516, Japan; kitamq@ffpri.affrc.go.jp (K.K.); itsuyama@affrc.go.jp (T.I.)

[2] Department of Forest Molecular Genetics and Biotechnology, Forestry and Forest Products Research Institute, 1 Matsunosato, Tsukuba, Ibaraki 305-8687, Japan; kruchiyama@affrc.go.jp (U.K.); saueno@ffpri.affrc.go.jp (U.S.)

[3] Forestry Research Institute, Hokkaido Research Organization, Koushunai, Bibai, Hokkaido 079-0166, Japan; wataru.ishi@gmail.com

[4] Education and Research Center, The University of Tokyo Forests, Graduate School of Agricultural and Life Sciences, The University of Tokyo, 1-1-1 Yayoi, Bunkyo-ku, Tokyo 113-8657, Japan

* Correspondence: gotos@uf.a.u-tokyo.ac.jp

Received: 17 December 2019; Accepted: 17 February 2020; Published: 20 February 2020

Abstract: *Research Highlights*: We detected the longitudinal gradients of genetic diversity parameters, such as the number of alleles, effective number of alleles, heterozygosity, and inbreeding coefficient, and found that these might be attributable to climatic conditions, such as temperature and snow depth. *Background and Objectives*: Genetic diversity among local populations of a plant species at its distributional margin has long been of interest in ecological genetics. Populations at the distribution center grow well in favorable conditions, but those at the range margins are exposed to unfavorable environments, and the environmental conditions at establishment sites might reflect the genetic diversity of local populations. This is known as the central-marginal hypothesis in which marginal populations show lower genetic variation and higher differentiation than in central populations. In addition, genetic variation in a local population is influenced by phylogenetic constraints and the population history of selection under environmental constraints. In this study, we investigated this hypothesis in relation to *Abies sachalinensis*, a major conifer species in Hokkaido. *Materials and Methods*: A total of 1189 trees from 25 natural populations were analyzed using 19 EST-SSR loci. *Results*: The eastern populations, namely, those in the species distribution center, showed greater genetic diversity than did the western peripheral populations. Another important finding is that the southwestern marginal populations were genetically differentiated from the other populations. *Conclusions*: These differences might be due to genetic drift in the small and isolated populations at the range margin. Therefore, our results indicated that the central-marginal hypothesis held true for the southernmost *A. sachalinensis* populations in Hokkaido.

Keywords: central-marginal hypothesis; cline; Pinaceae; trailing edge population; Sakhalin fir; sub-boreal forest

---

1. Introduction

The spatial distribution of genetic variation provides essential information for conservation programs and management of forest tree species [1]. The geographic distribution center provides favorable biotic and abiotic environments for species persistence. On the contrary, peripheral populations are exposed to unfavorable environmental conditions and the populations at the range margin are smaller and more isolated from each other than those at the range core [2–4]. The central-marginal hypothesis states that marginal populations show lower genetic variation and higher differentiation than do central populations (see review [5,6]). The above evidence was obtained from a study of Scots pine [7] and silver fir (*Abies alba*), which showed a decline in gene diversity among the western margin populations [8].

Long-lived plant species are sessile organisms and their survival is attributed to their adaptation to the local environment. Such adaptation processes inevitably involve genetic changes in local populations [9]. However, genetic variation in a local population is influenced by phylogenetic constraints and the population history of selection under environmental constraints. Genetic markers are, therefore, useful for inferring the genetic structure of local populations. They are also useful for inferring adaptability to similar environments according to similarities in genetic characteristics, as geographic genetic variations reflect the results of selection and adaptation to the local environment [10].

Sub-boreal conifer species play an important role in the sustainability of its forest ecosystems across the northern hemisphere. *Abies sachalinensis* is a major sub-boreal conifer in East Asia. The geographical distribution of *A. sachalinensis* ranges from Sakhalin in the north, to the southern Kuril Islands in the east, and Hokkaido, the northernmost island of the Japanese Archipelago, in the south [11,12]. The species often occurs across a broad altitudinal range; namely, from sea level to 1500 m a.s.l. It is one of the major conifers of the montane forest in Hokkaido and often forms mixed forests with other conifers, such as *Picea jezoensis, P. glehnii,* and broadleaf species, such as *Quercus mongolica* var. *crispula, Betula ermanii, Fagus crenata,* and *Tilia japonica.* At its southern and western distribution, *A. sachalinensis* is scarce, whereas it is abundant in the northern and eastern regions [13,14]. However, the distribution center of *A. sachalinensis,* including genetic core populations, has not yet been determined to date.

The species has been known to show both regional geographic and altitudinal variations in morphological traits [15–17]. Kurahashi and Hamaya [15] earlier noticed a wide variety of altitudinal differences in growth traits. This phenomenon was later revealed to be the result of adaptation to the local environment of the establishment sites [18]. Ishizuka et al. [19] confirmed that the altitudinal gradient of the autumn phenology related to cold tolerance was genetically controlled. Further, Goto et al. [20] found natural selection at the QTL (quantitative trait loci) of phenological traits across altitudinal differences. Hatakeyama [16] reported that regional differences in morphological traits were closely related to the snow-related climatic conditions based on a common garden experiment. Further, Eiga [17] reported regional genetic clines in both the altitudinal differences and longitudinal range differences in cold tolerance in Hokkaido, which are attributable to the level of snow acclimation at the establishment sites. A longitudinal gradient was also observed in allozyme variations [21]. Moreover, Suyama et al. [22] suggested a single lineage based on the cpDNA sequence of *A. sachalinensis* among Hokkaido populations. However, due to the limited numbers of populations and loci, the regional genetic relationships among *A. sachalinensis* populations in Hokkaido have not yet been resolved.

In this study, we investigated regional geographic variations in *A. sachalinensis* in Hokkaido by EST-SSR polymorphism with a special emphasis on the southern range marginal population. We sought to answer two questions; is the central-marginal hypothesis applicable to *A. sachalinensis*? And does the environmental gradient affect genetic clines as indicated by Nagasaka et al. [21]?

2. Materials and Methods

We chose 25 natural populations of *A. sachalinensis* from across the island of Hokkaido and a total of 1189 mature individuals as sample trees (Figure 1, Table 1). Total DNA was extracted from 100 mg of fresh needle leaves by the DNeasy Plant Mini Kit (QIAGEN K. K., Tokyo, Japan).

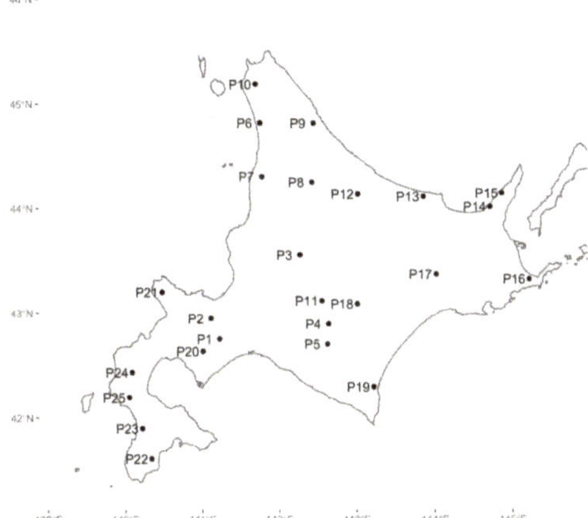

Figure 1. Locations of the 25 natural populations of *Abies sachalinensis* used in this study.

Table 1. Study sites of *Abies sachalinensis* populations in Hokkaido, northern Japan.

Population ID	Longitude East (°)	Latitude North (°)	Alt [1] (m)	N [2]	Nttl_a [3]	Na [4]	Nef [5]	H_O	H_E	G_{IS}
P1	141.2365	42.7691	800	44	68	3.579	1.815	0.385	0.383	−0.006
P2	141.1201	42.9663	580	48	66	3.474	1.975	0.379	0.410	0.074
P3	142.3221	43.5685	300	50	75	3.947	1.871	0.383	0.401	0.044
P4	142.6851	42.9003	930	47	75	3.947	1.928	0.408	0.417	0.021
P5	142.6656	42.7072	700	48	73	3.842	1.962	0.419	0.418	−0.003
P6	141.8080	44.8361	5	46	70	3.684	1.862	0.356	0.382	0.067
P7	141.8255	44.3187	90	48	71	3.737	1.978	0.394	0.409	0.038
P8	142.4986	44.2581	285	48	72	3.789	1.964	0.419	0.409	−0.025
P9	142.5356	44.8255	260	48	67	3.526	1.908	0.392	0.386	−0.016
P10	141.7521	45.2064	10	48	66	3.474	1.916	0.389	0.395	0.014
P11	142.6033	43.1250	360	43	68	3.579	1.930	0.387	0.396	0.024
P12	143.1218	44.1365	250	48	76	4.000	2.000	0.399	0.422	0.054
P13	144.0025	44.0930	200	48	73	3.842	1.903	0.395	0.400	0.014
P14	144.8912	43.9676	73	48	70	3.684	1.965	0.376	0.411	0.085
P15	145.0595	44.0907	100	47	77	4.053	1.916	0.355	0.399	0.110
P16	145.3731	43.2588	20	48	70	3.684	1.888	0.382	0.407	0.062
P17	144.1321	43.3482	950	48	77	4.053	1.975	0.377	0.416	0.093
P18	143.0752	43.0874	750	48	75	3.947	1.904	0.392	0.409	0.041
P19	143.2620	42.2881	73	48	75	3.947	2.022	0.380	0.424	0.102
P20	141.0124	42.6518	450	48	69	3.632	1.887	0.393	0.402	0.023
P21	140.4678	43.2183	200	48	62	3.263	1.934	0.425	0.385	−0.105
P22	140.3397	41.6239	90	48	47	2.474	1.768	0.312	0.313	0.001
P23	140.2135	41.9132	470	48	62	3.263	1.803	0.379	0.374	−0.014
P24	140.0727	42.4468	620	48	67	3.526	1.897	0.389	0.393	0.010
P25	140.0379	42.2073	100	48	66	3.474	1.939	0.378	0.384	0.017
Overall				1189				0.386	0.398	0.030

[1] Altitude, [2] number of individuals analyzed, [3] total number of alleles, [4] number of alleles, [5] effective number of alleles.

We analyzed the previously reported 11 EST-SSR loci: Aat01, Aat02, Aat04, Aat05, Aat06, Aat08, Aat09, Aat10, Aat11, Aat13, Aat15 [23]. In addition, we developed 8 new EST-SSR markers from *A. sachalinensis* transcriptome data. All the transcript sequences (158,542) from TodoFirGene [24] were used as input for the CMIB (CD-HIT-EST, MISA, ipcress and BlastCLUST) pipeline [25] to obtain PCR primers for amplifying unique microsatellite sequences with the number of repeat units ≥ 6, 5, 4, 3, and 3 for di-, tri-, tetra-, penta-, and hexa-simple sequence repeats (SSRs), respectively. For each primer pair, genomic DNA from one individual was used to check PCR amplification. PCR reactions were carried out following the standard protocol included in the QIAGEN Multiplex PCR Kit (QIAGEN, Hilden, Germany). For the primer pairs that exhibited clear microsatellite peaks at the expected fragment length, the extracted DNA of 32 individuals of *A. sachalinensis* representative of the species' range was used to evaluate EST-SSR polymorphism. Among them, 8 polymorphic markers, Egm1005,

Egm14860, Egm16822, Egm26233, Egm4191, Egm4389, Egm5979, and Egm55338, were used for the following analysis. Genotyping data has been deposited in the TreeGenes Database under accession number TGDR252.

PCR was carried out with a Type-it Microsatellite PCR Kit (QIAGEN K. K., Tokyo, Japan). Fragment analyses were performed by ABI 3130-xl Genetic Analyzer, 600 LIZ size standard, and GENESCAN for Windows (Thermo Fischer Scientific, Tokyo, Japan).

We used GenoDive ver. 2.0b17 [26] to calculate the following genetic parameters; observed (H_O), expected heterozygosity (H_E), total heterozygosity (H_T), inbreeding coefficient (G_{IS}), genetic differentiation (G'_{ST}), number of alleles (Na), effective number of alleles (Nef), and Nei's genetic distance (D). Principal component analysis (PCA) was conducted to clarify the genetic relationship among populations using the same software. Four genetic diversity parameters; number of alleles, effective number of alleles, H_E, and G_{IS}, were then spatially interpolated by kriging [27] using R [28] and laid out on the contour map. A neighbornet phylogenetic tree based on Nei's genetic distance was drawn using SplitsTree4 ver. 4.14.6 [29].

The climatic conditions for each population were estimated by the Mesh Climate Data 2000 [30]. We calculated the following environmental factors; WI, warmth index; CI, cold index; TMC, mean minimum temperature of the coldest month; PRS, precipitation in summer (May to September); PRW, precipitation in winter (October to April); MSD, maximum snow depth; WinSR, solar radiation in winter (October to April); SprSR, solar radiation in spring (May); SumSR, solar radiation in summer (June to August); and AutSR, solar radiation in autumn (September). Classification of four seasons for solar radiation was determined based on the regression coefficient ($r > 0.7$) between months (Table S1).

3. Results

The 19 EST-SSR loci used in this study were polymorphic, with the number of alleles ranging from 2 to 13 (Table S2). The genetic diversity (H_E) of the 25 populations ranged from 0.313 to 0.424, and the overall H_E was 0.398. The G'_{ST} ranged from −0.105 to 0.110, and the overall value was 0.016 (Table S2). Each genetic diversity parameter, along with the 95% confidence interval for the 25 populations, is shown in Figure S1. The effective number of alleles and H_E did not differ significantly among the populations. The number of alleles of P22 was significantly lower than those of P3, P4, P5, P12, P15, P17, P18, and P19, and the G_{IS} of P21was significantly lower than that of P15 (Figure S1). The geographic patterns of genetic diversity were represented by contour diagrams (Figure 2). All four parameters showed longitudinal gradients; that is, the eastern populations showed higher values for the genetic diversity parameters (Figure 3). In addition, the four genetic parameters for each locus are shown in Figures S2–S5. Most of the loci showed eastward increases in genetic parameters, but several of them, for example, Aat09, Aat10, and Egm1005, showed opposite results for the effective number of alleles and H_E (Figures S2 and S4).

Pairwise differentiation matrices between populations by G'_{ST} (Table S3) revealed significant differentiation between the isolated populations (P19, P22) and southwestern populations (P20–21, P23–25) and the rest of the populations.

Principle component analysis by covariance matrix revealed that the plots of the geographically peripheral populations (P19 to P25) were in outlying positions on the first and second axis plane (Figure 4). The plots of the southern populations, in particular, showed smaller Co1 and larger Co2 scores (grouped within the dotted line in Figure 4). P19 and P21 showed the highest Co1 and the lowest Co2 scores, respectively. The other populations did not show any geographical clustering and were located at the center of the axis plane.

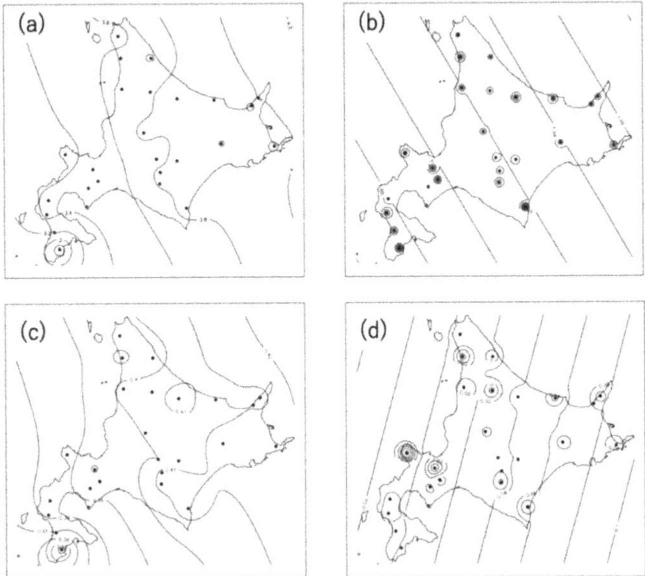

Figure 2. Spatial contour maps of four genetic parameters, the number of alleles (**a**), effective number of alleles (**b**), H_E (**c**), and G_{IS} (**d**), measured for the 25 natural populations. Dots indicate study sites (cf., Table 1, Figure 1).

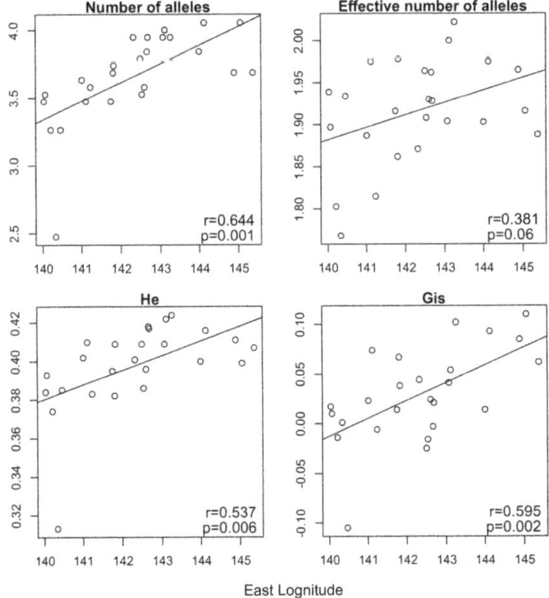

Figure 3. Relationships among genetic parameters and longitude east. Lines indicate the least squares.

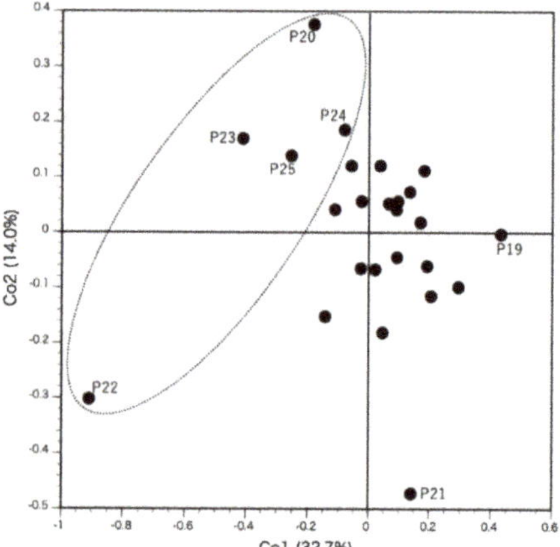

Figure 4. Principle component analysis based on the covariance matrix of the 25 populations. The first (Co1) and second (Co2) axes' % of variances are in parentheses. Southern range populations are grouped by the dotted line.

The phylogenetic relationships of the 25 populations are shown by neighbornet tree based on Nei's genetic distances (Figure 5). The tree shows that the plots of the geographically peripheral populations (P19 to P25) were in outlying positions. Moreover, the southern populations were placed on the same branch (grouped within the dotted line in Figure 5). The southernmost population (P22) was located at the furthest position in the cluster made up of the southern populations. Other geographically peripheral populations, namely, P19 and P25, were clustered to individual branches and were differentiated from the central populations.

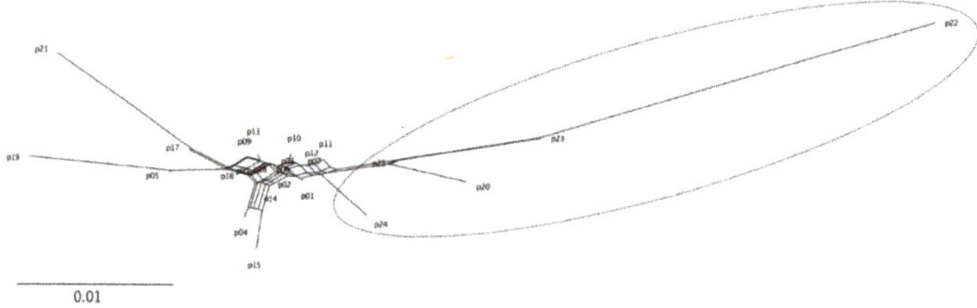

Figure 5. Neighbornet tree among the 25 populations based on Nei's genetic distances. Southern range populations are grouped by the dotted line.

An environmental gradient along longitude was found among the climatic conditions for *A. sachalinensis* populations. The results from the correlation test (Table S4) and principal component analysis (Figure 6) between longitude and climatic conditions indicated negative correlations between longitude and winter precipitation (PRW), maximum snow depth (MSD), mean minimum temperature of the coldest month (TMC), and solar radiation in autumn (AurSR), as well as a positive correlation between longitude and solar radiation in winter (WinSR). Although PRW showed a significant *p*-value alone after Bonferroni correction (Table S4), the climatic conditions for the 25 populations studied along longitude can be assumed; thus, the eastern populations are characterized by less snow, colder winters, and more solar radiation in winter. On the contrary, the western populations have more snow, warmer winters, and less solar radiation in winter. We did not detect any relationship between solar radiation and precipitation in summer.

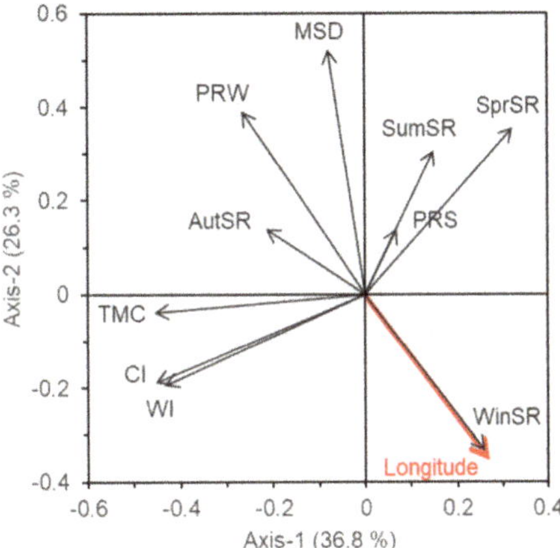

Figure 6. Principal component analysis using longitude and selected environmental factors for 25 populations of *A. sachalinensis*.

In addition, we detected significant relationships between Na and climatic environmental factors (Table S5). We observed that the number of alleles at 10 loci showed significant relationships with climatic factors, with seven of these loci showing significant relationships with more than two factors.

4. Discussion

Central populations are characterized by high genetic diversity due to a large effective population size. In contrast, peripheral populations are often characterized by a small effective population size. Our present results regarding genetic diversity parameters; namely, Na, Nef, and H_E, were greater in the northeastern than in the southwestern populations (Figures 2a–c and 3). This indicated that the distribution center of *A. sachalinenesis* in eastern Hokkaido has greater genetic diversity than do the southwestern peripheral populations. The species distribution model indicated that the suitable habitat for *A. sachalinenesis* was from central to eastern Hokkaido, and the marginal habitat was on the southwestern Oshima Peninsula [31]. Thus, the northern and eastern parts of Hokkaido can be regarded as the distribution center of this species, as corroborated by the cumulative volume of *A. sachalinensis* in previous studies [13,14]. The island of Hokkaido includes the southernmost distribution range of *A. sachalinensis* on its southwestern peninsula. Similar results were obtained for European silver fir, *A. alba*, which has a long pollen flow distance similar to *A. sachalinensis* and shows a decline in gene diversity in its western marginal populations [8]. Thus, the central-marginal hypothesis is supported in *A. sachalinensis* in this study.

On the one hand, the inbreeding coefficient, G_{IS}, also showed the same gradient (Figures 2d and 3). The reason for this tendency has not yet been clarified, but the same tendency has been reported in previous studies of marginal populations of the anemogamous species, *Fagus crenata* [4]; that is, the smaller the population, the lower the inbreeding coefficient.

Previous studies of *A. sachalinensis* revealed east-to-west longitudinal genetic clines in morphological traits, resistance against disease, tolerance against environmental conditions by provenance tests [16,17] and allozymes polymorphisms among natural populations [21]. The present study also revealed the same gradient in genetic parameters determined by EST-SSRs. Nagasaka et al. [21] detected significant east–west directional variation patterns in H_E, effective number of alleles, and number of alleles based on 4 allozyme loci, and they regarded this variation to be due to temperature and precipitation. A longitudinal gradient in gene diversity can be attributable to the climatic factors at the establishment site [32]. A longitudinal gradient in climatic factors in Hokkaido might involve temperature and snow depth [17,33], with eastern Hokkaido being colder but having less accumulation of snow than southwestern Hokkaido. Okada et al. [34] revealed that the number of winter bud scales of *A. sachalinensis* differed between the eastern and western populations, and they inferred that this indicated a response to drought hardiness in the winter season. Eiga [17] indicated a longitudinal gradient in freezing resistance in seedlings. These morphological traits might result from the natural selection and adaptation to the local climate, which was one of the major causes of the longitudinal gradient in genetic diversity.

Our results showed that some environmental factors for the 25 populations showed a longitudinal gradient (Table S4, Figure 6). The relationships between the number of alleles and climatic factors were revealed to be significant for 10 EST-SSR loci (Table S5). Among them, the most highly correlated climatic factor was TMC, with the second most being PRW, while PRS and SumSR did not show any significance. This result indicated that the climate conditions in winter were more correlated to genetic traits than were those in summer. Specifically, colder temperatures, less snow, and more solar radiation in the winter and spring were indicative of a greater number of alleles within a population. Several loci were correlated with more than one climatic factor, such as Aat02, Egm26233, and Egm4389 with WI, TMC, and PRW. These climatic factors also showed longitudinal clines (Table S5). Thus, the specific relationships between loci and climate factors could be reflected in the regional differences in gene diversity observed in *A. sachalinensis* in this study.

Previous studies have revealed relationships between climatic conditions and gene diversity in natural populations of forest tree species. For example, associations between AFLP loci and temperature were observed in *Fagus sylvatica* [35] and *Betula pendula* [36]. Grivet et al. [37] found two SNPs that were correlated with temperature in *Pinus pinaster* and *P. halepensis*. Annotated EST-SSR loci in *Eucalyptus gomphocephala* showed clines in allele frequencies for climatic factors, such as solar radiation, potential evaporation, summer precipitation, and aridity [38]. Former studies of allozyme variation also revealed adaptive differentiation in conifer species (e.g., [39]). Not only allele frequencies but also the heterozygosity of an individual is often associated with fitness among a changing environment (e.g., [40]). Correlations have been reported between growth rate and heterozygosity in *Populus tremuloides* [41] and survivorship and heterozygosity in *Picea jezoensis* [42]. It has been observed that the heterozygosity of a population influences its productivity, fitness, and stability [40,43,44]. Indeed, genetic diversity was affected by many other factors, such as demographic history and genetic drift, and we have to wait for association analyses, such as the outlier test, to reveal the natural selection of certain alleles against the environment for *A. sachalinensis*.

Another interesting issue regarding the longitudinal gradient was the observed counteraction among individual loci (Figures S2–S4). Most of the loci showed higher genetic diversity in the eastern than in the western populations, but several loci showed an opposite trend. This might offer evidence of opposite selection or adaptation forces to the environmental gradient between different loci. Currently, we are not able to clarify this discrepancy, but it is likely that these loci were affected by the selective sweep of adaptive genes [45,46] as the loci used in this study were EST-SSR, whose primers were developed among the sequences in close proximity to expressed gene sequences.

STRUCTURE analysis [47] by F-model with 70,000 burn-in and 30,000 MCMC detected differences in the admixture coefficient between the southwestern and the other populations (Figure S6) but did not indicate any regionally specific ancestral clusters among the 25 populations. This might be attributable to the long pollen flow distance of anemogamous species such as *A. sachalinensis*. A long distance gene flow may result in gene mixture and lead to homogeneous gene pools among populations [48]. It is also known that two major spruce species in Hokkaido, *P. jezoensis* and *P. glehnii*, demonstrate homogeneous gene pools in Hokkaido [49,50]. In this regard, anemogamous conifer species can be said to generally show homogeneous gene pools among local populations in Hokkaido.

However, there is evidence of genetic differentiation between the southwestern populations and the rest of the populations. The results from PCA revealed that the southern peripheral populations were differentiated from the other central populations (Figure 4). Geographic isolation of some populations was reflected in the PCA. P20 is located in a volcanic region so that the forest soil is composed of volcanic ash or pumice as the base materials. Due to recent volcanic activity, the continuity of the species distribution might also be interrupted. P21 is located in a deep ravine at the tip of a peninsula and P19 in a coastal area, with both populations being isolated from other *A. sachalinensis* populations. Therefore, P19, P20, and P21 were plotted at the periphery of the PCA plane due to their geographic isolation (Figure 4).

The phylogenetic tree indicated that the six southernmost populations, P20 to P25, were differentiated from the other populations (Figure 5). These results indicated that the populations of the southernmost distribution were highly differentiated from the other populations. This finding is consistent with the previous studies based on morphological traits [34], allozyme variations [21], and organelle DNA haplotypes [51].

The high genetic divergence of southernmost populations may be also explained by the fossil pollen record in southwestern Hokkaido [52], which revealed that *Abies*, supposedly *A. sachalinensis*, that had dominated sub-boreal forests was completely replaced by cool temperate forest at the end of the last glacial period (10,000 yrs. BP), indicating that the southwestern populations became smaller and more isolated from each other at that time. Pairwise differentiation by G'_{ST} indicated significantly high values in the southwestern populations (Table S3). These relic populations might have lost gene exchange with the distributional center. This would cause a genetic drift that affected the genetic structure of the small, isolated populations on the distribution periphery. In addition, southwestern Hokkaido is made up of a long, narrow peninsula, which can be a topographical barrier against the frequent exchange of individuals.

In conclusion, the central-marginal hypothesis that marginal populations show less genetic variation and higher differentiation than do central populations [4–6] was found to be relevant to natural populations of *A. sachalinensis* in Hokkaido, with the southwestern populations being highly differentiated from the other populations. In addition, the longitudinal genetic cline revealed by Nagasaka et al. [21] was supported by the 19 EST-SSR markers in this study. This cline may be related to adaptation to the environmental gradient in Hokkaido.

5. Conclusions

We analyzed the genetic diversity of 25 natural populations of a major sub-boreal conifer, *Abies sachalinensis*, including the species distribution range from its center to its southern margin. Nineteen EST-SSR loci were applied and revealed that the genetic diversity parameters were higher among the eastern populations and lower among the southwestern ones. This result supported the central-marginal hypothesis that the distribution center possesses higher gene diversity because the eastern populations are located in the core of species distribution. Phylogenetic analysis revealed that the marginal populations at the southern range limit showed further genetic distances from the central populations. The eastern to southwestern gradient of genetic diversity indicated a relationship to the species' adaptation to certain environmental factors.

Supplementary Materials: The following are available online at http://www.mdpi.com/1999-4907/11/2/233/s1, Figure S1: Genetic diversity parameters for the 25 populations. Bars indicate the 95% confidence intervals, Figure S2: Relationships among number of alleles and longitude east. Lines indicate the least squares, Figure S3: Relationships among effective number of alleles and longitude east. Lines indicate the least squares, Figure S4: Relationships among H_E and longitude east. Lines indicate the least squares, Figure S5: Relationships among G_{IS} and longitude east. Lines indicate the least squares, Figure S6: STRUCTURE results from K = 2 to 4. Populations are arranged from SW (left) to NE (right), Table S1: Climatic conditions for the 25 populations of *A. sachalinensis* used in this study, Table S2: Gene diversity among 19 EST-SSR loci used in this study, Table S3: Pairwise differentiation matrices by G'_{ST} (lower triangle) and *p*-values (upper triangle) between populations, Table S4: Correlation analysis between longitude and selected environmental factors for the 25 populations of *A. sachalinensis*, Table S5: Regression coefficients between environmental factors and the number of alleles for each locus.

Author Contributions: Conceptualization, G.S., K.K., and I.W.; methodology, K.K., U.K., and U.S.; formal analysis, K.K.; investigation, K.K., U.K., I.W., T.I., and G.S.; writing—original draft preparation, K.K.; writing—review and editing, G.S., K.K., U.K., I.W., T.I., and U.S. All authors have read and agree to the published version of the manuscript.

Funding: This research was funded by a Grant-in-Aid for Scientific Research from the Japan Society for the Promotion of Science, grant number 16H02554 and 16H06279 (PAGS).

Acknowledgments: We thank T. Kawahara and A. Takazawa for their technical support.

Conflicts of Interest: The authors declare no conflict of interest.

References

1. Petit, R.J.; El-Mousadik, A.; Pons, O. Identifying populations for conservation on the basis of genetic markers. *Conserv. Biol.* **1998**, *12*, 844–855. [CrossRef]

2. Brown, J.H. On the relationship between abundance and distribution of species. *Am. Nat.* **1984**, *124*, 255–279. [CrossRef]

3. Sagarin, R.D.; Gaines, S.D. The 'abundant centre' distribution: To what extent is it a biogeographical rule? *Ecol. Lett.* **2002**, *5*, 137–147. [CrossRef]

4. Kitamura, K.; Matsui, T.; Kobayashi, M.; Saitou, H.; Namikawa, K.; Tsuda, Y. Decline in gene diversity and strong genetic drift in the northward-expanding marginal populations of *Fagus crenata*. *Tree Genet. Genomes.* **2015**, *11*, 36. [CrossRef]

5. Eckert, C.G.; Samis, K.E.; Lougheed, S.C. Genetic variation across species' geographical ranges: The central-marginal hypothesis and beyond. *Mol. Ecol.* **2008**, *17*, 1170–1188. [CrossRef]

6. Arnaud-Haond, S.; Teixeira, S.; Massa, S.I.; Billot, C.; Saenger, P.; Coupland, G.; Duarte, C.M.; Serrao, E.A. Genetic structure at range edge: Low diversity and high inbreeding in Southeast Asian mangrove (*Avicennia marina*) populations. *Mol. Ecol.* **2006**, *15*, 3515–3525. [CrossRef]

7. Tóth, E.G.; Vendramin, G.G.; Bagnoli, F.; Cseke, K.; Höhn, M. High genetic diversity and distinct origin of recently fragmented Scots pine (*Pinus sylvestris* L.) populations along the Carpathians and the Pannonian Basin. *Tree Genet. Genomes* **2017**, *13*, 47.

8. Dobrowolska, D.; Bončina, A.; Klumpp, R. Ecology and silviculture of silver fir (*Abies alba* Mill.): A review. *J. For. Res.* **2017**, *22*, 326–335. [CrossRef]

9. Young, A.; Boyle, T.; Brown, T. The population genetic consequences of habitat fragmentation for plants. *Trends Ecol. Evol.* **1996**, *11*, 413–418. [CrossRef]

10. Petit, R.J.; Bialozyt, R.; Garnier-Géré, P.; Hampe, A. Ecology and genetics of tree invasions: From recent introductions to Quaternary migrations. *For. Ecol. Manag.* **2004**, *197*, 117–137. [CrossRef]

11. Tatewaki, M. Forest ecology of the islands of the north Pacific ocean. *J. Fac. Agric. Hokkaido Univ.* **1958**, *50*, 341–486.

12. Kitamura, S.; Murata, G. *Colored Illustrations of Woody Plants of Japan Volume II*; Hoikusya Co., Ltd.: Osaka, Japan, 1979; p. 545.

13. Matsuda, S. Distribution status of *Abies sachalinensis* and *Picea jezoensis* in the Hokkaido national forest. *Hokkaido Ringyo Kaiho* **1936**, *34*, 351–361.

14. Matsui, T.; Kitamura, K.; Saito, H.; Namikawa, K.; Terazawa, K.; Haruki, M.; Itaya, A.; Honma, Y.; Miyoshi, Y.; Uchida, K.; et al. Habitat and vegetation of an outlying *Fagus crenata* population in its northern range limit at Rebunge Pass, Hokkaido. *J. Phyto. Taxon.* **2012**, *59*, 113–123.

15. Kurahashi, A.; Hamaya, T. Variation of morphological characters and growth response of Saghalin fir (*Abies sachalinensis*) in different altitude. *Bull. Tokyo Univ. For.* **1981**, *71*, 101–151.

16. Hatakeyama, S. Genetical and breeding studies on the regional differences of interprovenance variation in *Abies sachalinensis* Mast. *Bull. Hokkaido For. Exp. Stn.* **1981**, *19*, 1–91.

17. Eiga, S. Ecological study on the freezing resisntace of Saghalin fir (*Abies sachalinensis* Mast.) in Hokkaido. *Bull. For. Tree Breed. Inst.* **1984**, *2*, 61–107.

18. Ishizuka, W.; Goto, S. Modeling intraspecific adaptation of *Abies sachalinensis* to local altitude and responses to global warming, based on a 36-year reciprocal transplant experiment. *Evol. Appl.* **2012**, *5*, 229–244. [CrossRef]

19. Ishizuka, W.; Ono, K.; Hara, T.; Goto, S. Use of intraspecific variation in thermal responses for estimating an elevational cline in the timing of cold hardening in a sub-boreal conifer. *Plant Biol.* **2015**, *17*, 177–185. [CrossRef]

20. Goto, S.; Kajiya-Kanegae, H.; Ishizuka, W.; Kitamura, K.; Ueno, S.; Hisamoto, Y.; Kudoh, H.; Yasugi, M.; Nagano, A.J.; Iwata, H. Genetic mapping of local adaptation along the altitudinal gradient in *Abies sachalinensis*. *Tree Genet. Genomes* **2017**, *13*, 104. [CrossRef]

21. Nagasaka, K.; Wang, Z.M.; Tanaka, K. Genetic variation among natural *Abies sachalinensis* population in relation to environmental gradients in Hokkaido, Japan. *For. Genet.* **1997**, *4*, 43–50.

22. Suyama, Y.; Yoshimaru, H.; Tsumura, Y. Molecular phylogenetic position of Japanese *Abies* (Pinaceae) based on chloroplast DNA sequences. *Mol. Phylogenetics Evol.* **2000**, *16*, 271–277. [CrossRef] [PubMed]

23. Postolache, D.; Leonarduzzi, C.; Piotti, A.; Spanu, I.; Roig, A.; Fady, B.; Roschanski, A.; Liepelt, S.; Vendramin, G.G. Transcriptome versus genomic microsatellite markers: Highly informative multiplexes for genotyping *Abies alba* Mill. and congeneric species. *Plant Mol. Biol. Rep.* **2014**, *32*, 750–760. [CrossRef]

24. Ueno, S.; Nakamura, Y.; Kobayashi, M.; Terashima, S.; Ishizuka, W.; Uchiyama, K.; Tsumura, Y.; Yano, K.; Goto, S. TodoFirGene: Developing transcriptome resources for genetic analysis of *Abies sachalinensis*. *Plant Cell Physiol.* **2018**, *59*, 1276–1284. [CrossRef] [PubMed]

25. Ueno, S.; Moriguchi, Y.; Uchiyama, K.; Ujino-Ihara, T.; Futamura, N.; Sakurai, T.; Shinohara, K.; Tsumura, Y. A second generation framework for the analysis of microsatellites in expressed sequence tags and the development of EST-SSR markers for a conifer, *Cryptomeria japonica*. *BMC Genom.* **2012**, *13*, 136. [CrossRef]

26. Meirmans, P.G.; Van Tienderen, P.H. GENOTYPE and GENODIVE: Two programs for the analysis of genetic diversity of asexual organisms. *Mol. Ecol. Notes.* **2004**, *4*, 792–794. [CrossRef]

27. Mase, S. *Geostatistics and Kriging Method*; Ohmsha: Tokyo, Japan, 2010.

28. R Core Team. *A Language and Environment for Statistical Computing*; R Foundation for Statistical Computing: Vienna, Austria, 2016; Available online: https://www.R-project.org/ (accessed on 18 July 2018).

29. Huson, D.H.; Bryant, D. Application of phylogenetic networks in evolutionary studies. *Mol. Biol. Evol.* **2006**, *23*, 254–267. [CrossRef]

30. The Japan Meterological Agency. *Mesh Climate Data 2000, CD-ROM*; The Japan Meterological Agency: Tokyo, Japan, 2002.

31. Tanaka, N.; Nakao, K.; Tsuyama, I.; Matsui, T. Assessing impact of climate warming on potential habitats of ten conifer species in Japan. *AIRIES* **2009**, *14*, 153–164.

32. Conord, C.; Gurevitch, J.; Fady, B. Large-scale longitudinal gradients of genetic diversity: A meta-analysis across six phyla in the Mediterranean basin. *Ecol. Evol.* **2012**, *2*, 2600–2614. [CrossRef]

33. Yokota, S. Scleroderris canker of Todo-fir in Hokkaido, Northern Japan. *Eur. J. For. Path.* **1975**, *5*, 13–21. [CrossRef]

34. Okada, S.; Mukaide, H.; Sakai, A. Genetic variation in Saghalien fir from different areas of Hokkaido. *Silvae Genet.* **1973**, *22*, 1–2.

35. Jump, A.S.; Hunt, J.M.; Martínez-Izquierdo, J.A.; Peñuelas, J. Natural selection and climate change: Temperature-linked spatial and temporal trends in gene frequency in *Fagus sylvatica*. *Mol. Ecol.* **2006**, *15*, 3469–3480. [CrossRef] [PubMed]

36. Kelly, C.K.; Chase, M.W.; De Bruijn, A.; Fay, M.F.; Woodward, F.I. Temperature-based population segregation in birch. *Ecol. Lett.* **2003**, *6*, 87–89. [CrossRef]

37. Grivet, D.; Sebastiani, F.; Alía, R.; Bataillon, T.; Torre, S.; Zabal-Aguirre, M.; Vendramin, G.G.; González-Martínez, S.C. Molecular footprints of local adaptation in two Mediterranean conifers. *Mol. Biol. Evol.* **2010**, *28*, 101–116. [CrossRef] [PubMed]

38. Bradbury, D.; Smithson, A.; Krauss, S.L. Signatures of diversifying selection at EST-SSR loci and association with climate in natural *Eucalyptus* populations. *Mol. Ecol.* **2013**, *22*, 5112–5129. [CrossRef] [PubMed]

39. Jump, A.S.; Peñuelas, J. Running to stand still: Adaptation and the response of plants to rapid climate change. *Ecol. Lett.* **2005**, *8*, 1010–1020. [CrossRef]

40. Jump, A.S.; Marchant, R.; Peñuelas, J. Environmental change and the option value of genetic diversity. *Trends Plant Sci.* **2009**, *14*, 51–58. [CrossRef] [PubMed]

41. Mitton, J.B.; Grant, M.C. Observations on the ecology and evolution of quaking aspen, *Populus tremuloides*, in the Colorado front range. *Am. J. Bot.* **1980**, *67*, 202–209. [CrossRef]

42. Okada, M.; Kitamura, K.; Lian, C.; Goto, S. The effects of multilocus heterozygosity on the longevity of seedlings established on fallen logs in *Picea jezoensis* and *Abies sachalinensis*. *Open J. For.* **2015**, *5*, 422–430.

43. Leimu, R.; Mutikainen, P.I.A.; Koricheva, J.; Fischer, M. How general are positive relationships between plant population size, fitness and genetic variation? *J. Ecol.* **2006**, *94*, 942–952. [CrossRef]

44. Mitton, J.B.; Grant, M.C. Associations among protein heterozygosity, growth rate, and developmental homeostasis. *Annu. Rev. Ecol. Syst.* **1984**, *15*, 479–499. [CrossRef]

45. Smith, J.M.; Haigh, J. The hitch-hiking effect of a favourable gene. *Genet. Res.* **1974**, *23*, 23–35. [CrossRef] [PubMed]

46. Kado, T.; Yoshimaru, H.; Tsumura, Y.; Tachida, H. DNA variation in a conifer, *Cryptomeria japonica* (*Cupressaceae sensu lato*). *Genetics* **2003**, *164*, 1547–1559. [PubMed]

47. Pritchard, J.K.; Stephens, M.; Donnelly, P. Inference of Population Structure Using Multilocus Genotype Data. *Genetics* **2000**, *155*, 945. [PubMed]

48. Leonarduzzi, C.; Piotti, A.; Spanu, I.; Vendramin, G.G. Effective gene flow in a historically fragmented area at the southern edge of silver fir (*Abies alba* Mill.) distribution. *Tree Genet. Genomes* **2016**, *12*, 95. [CrossRef]

49. Iwaizumi, M.G.; Aizawa, M.; Watanabe, A.; Goto, S. Highly polymorphic nuclear microsatellite markers reveal detailed patterns of genetic variation in natural populations of Yezo spruce in Hokkaido. *J. For. Res.* **2015**, *20*, 301–307. [CrossRef]

50. Aizawa, M.; Yoshimaru, H.; Takahashi, M.; Kawahara, T.; Sugita, H.; Saito, H.; Sabirov, R.N. Genetic structure of Sakhalin spruce (*Picea glehnii*) in northern Japan and adjacent regions revealed by nuclear microsatellites and mitochondrial gene sequences. *J. Plant Res.* **2015**, *128*, 91–102. [CrossRef]

51. Hayashi, E.; Ubukata, M.; Iizuka, K.; Itahana, N. Genetic differentiation of organelle DNA polymorphisms in Sakhalin fir from Hokkaido. *For. Genet.* **2000**, *7*, 31–38.

52. Takiya, M.; Hagiwara, N. Vegetational history of Mt. Yokotsudake, southwestern Hokkaido, since the Last Glacial. *Quat. Res.* **1997**, *36*, 217–234. [CrossRef]

Article

Complete Chloroplast Genome of Japanese Larch (*Larix kaempferi*): Insights into Intraspecific Variation with an Isolated Northern Limit Population

Shufen Chen [1], Wataru Ishizuka [2], Toshihiko Hara [3] and Susumu Goto [1,*

[1] Education and Research Center, The University of Tokyo Forests, Graduate School of Agricultural and Life Sciences, The University of Tokyo, 1-1-1 Yayoi, Bunkyo-ku, Tokyo 113-8657, Japan; suky0414@hotmail.com
[2] Forestry Research Institute, Hokkaido Research Organization, Koushunai, Bibai, Hokkaido 079-0166, Japan; wataru.ishi@gmail.com
[3] Institute of Low Temperature Science, Hokkaido University, Sapporo-city, Hokkaido 060-0819, Japan; t-hara@pop.lowtem.hokudai.ac.jp
* Correspondence: gotos@uf.a.u-tokyo.ac.jp; Tel.: +81-3-5841-5493

Received: 25 July 2020; Accepted: 11 August 2020; Published: 14 August 2020

Abstract: *Research Highlights:* The complete chloroplast genome for eight individuals of Japanese larch, including from the isolated population at the northern limit of the range (Manokami larch), revealed that Japanese larch forms a monophyletic group, within which Manokami larch can be phylogenetically placed in Japanese larch. We detected intraspecific variation for possible candidate cpDNA markers in Japanese larch. *Background and Objectives:* The natural distribution of Japanese larch is limited to the mountainous range in the central part of Honshu Island, Japan, with an isolated northern limit population (Manokami larch). In this study, we determined the phylogenetic position of Manokami larch within Japanese larch, characterized the chloroplast genome of Japanese larch, detected intraspecific variation, and determined candidate cpDNA markers. *Materials and Methods:* The complete genome sequence was determined for eight individuals, including Manokami larch, in this study. The genetic position of the northern limit population was evaluated using phylogenetic analysis. The chloroplast genome of Japanese larch was characterized by comparison with eight individuals. Furthermore, intraspecific variations were extracted to find candidate cpDNA markers. *Results:* The phylogenetic tree showed that Japanese larch forms a monophyletic group, within which Manokami larch can be phylogenetically placed, based on the complete chloroplast genome, with a bootstrap value of 100%. The value of nucleotide diversity (π) was calculated at 0.00004, based on SNP sites for Japanese larch, suggesting that sequences had low variation. However, we found three hyper-polymorphic regions within the cpDNA. Finally, we detected 31 intraspecific variations, including 19 single nucleotide polymorphisms, 8 simple sequence repeats, and 4 insertions or deletions. *Conclusions:* Using a distant genotype in a northern limit population (Manokami larch), we detected sufficient intraspecific variation for the possible candidates of cpDNA markers in Japanese larch.

Keywords: cpDNA; next generation sequencing; northern limit; nucleotide diversity; phylogeny; In/Del; SNP; SSR; Pinaceae

---

1. Introduction

The chloroplast genome is highly conserved and has a much lower mutation rate than the nuclear genome [1]. Chloroplast DNA (cpDNA) has been widely used to clarify interspecific relationships, and to evaluate the magnitude of intraspecific variation [2,3]. The cpDNA of gymnosperms, particularly of the conifers, is characterized by high levels of intraspecific variation [4,5] and paternal inheritance [6].

A high-resolution chloroplast-specific polymorphic assay would facilitate the analysis of population differentiation and gene flow in gymnosperms [7].

The next-generation sequencing (NGS) technique enables the sequencing of whole chloroplast genomes. The chloroplast genome has a circular molecular structure, with a length ranging from 120 to 160 kbp in most plants. The cpDNA contains a pair of inverted repeats (IRs), a large single-copy region (LSC), and a small single-copy region (SSC) [8]. IRs are a crucial feature of the chloroplast genome in most plants, likely contributing to the maintenance of a conserved arrangement of cpDNA sequences. Previous studies have reported that the length of observed short IRs is roughly consistent among gymnosperm species [9]. The whole chloroplast genome is of significant use for phylogenetic studies [2,3]; Parks et al. [10] presented complete chloroplast genomes for 37 pine species and documented a notable degree of variation at several loci (particularly at *ycf*1 and *ycf*2). Intraspecific variation in whole chloroplast genomes derived from multiple individuals can clarify the phylogenic lineage of target individuals [11]. In particular, single nucleotide polymorphisms (SNPs) have been efficiently used in the fields of phylogeography and conservation biology [12].

Japanese larch (*Larix kaempferi* (Lamb.) Carr.) is a deciduous coniferous tree species endemic to Japan, and integral to the country's forestry efforts. The natural distribution of Japanese larch is limited to the mountainous range in the central part of Honshu Island, Japan [13]. An isolated population with ten mature trees was discovered at Manokami (hereafter Manokami larch), in the Zao Mountains in 1932 [14], extending the known northern limit of the species. Manokami larch was initially believed to be *Larix gmelinii* var. *japonica*, based on the morphological traits. An analysis of partial cpDNA sequences, and random amplified polymorphic DNA (RAPD) analysis, indicated that the Manokami larch population was actually Japanese larch [15]. However, the phylogenic position of Manokami larch, with relation to Japanese larch, has not yet been sufficiently defined [16,17].

The complete chloroplast genome of the genus *Larix* has been reported for several species [11,18,19], and the complete chloroplast genome of Japanese larch introduced in Korea, was reported by Kim et al. [20]. However, the intraspecific variations of Japanese larch have not yet been examined based on the complete chloroplast genome.

In this study, we identified the complete chloroplast genome for eight individuals of Japanese larch, including from the isolated population at the northern limit of the range (Manokami larch), to (1) determine the phylogenetic position of Manokami larch within Japanese larch, (2) characterize the chloroplast genome of Japanese larch, with included chloroplast data from Manokami larch, and (3) detect intraspecific variation and determine candidate cpDNA markers.

2. Materials and Methods

Eight individuals of Japanese larch were used in this study. Five individuals (*Lk_Ho1*, *Lk_Ho2*, *Lk_Ho3*, *Lk_Ho4*, and *Sorachi 3*) were collected from test plantations or the Arboretum Garden at the Forestry Research Institute, Hokkaido Research Organization (HRO). *Sorachi 3* was selected specifically due to its superior growth in a larch-breeding program in Hokkaido. Two individuals (*Lk_Ka1* and *Lk_Ka2*), were collected from the open-pollinated progeny of artificial plantations of Japanese larch, at the southern edge of Sakhalin Island, Russia. The detailed location of the seed collection of *Lk_Ka2* was described in a previous study [11]. These seven individuals were originally derived from the eastern region of Nagano, in the central part of Honshu Island in Japan, near the center of distribution for this species. We added one grafted tree (*Manokami 15*), as an isolated germplasm of Manokami larch. The ortet of this tree was conserved in situ with the label "No.15" on Mt. Manokami in the Zao Mountains.

Fresh leaves from all eight individuals were collected in 2016 March for *Lk_Ka2*, June for *Lk_Ho2*, *Lk_Ho3*, *Lk_Ho4*, *Sorachi3*, 2017 March for *Lk_Ka1*, June for *Lk_Ho1*, and 2018 July for *Manokami15*. Leaf sampling, isolation of purified intact chloroplasts, and extraction of high-concentrate chloroplast DNA were performed as previously described [11]. Briefly, we used a saline Percoll (GE Healthcare,

Uppsala, Sweden) gradient for chloroplast isolation and the DNeasy Plant Mini kit (QIAGEN, Hilden, Germany) for DNA extraction.

The cpDNA sequence reads were obtained using the Illumina platform. CLC Genomics Workbench 9.5.3 software (CLC bio, Aarhus, Denmark) was used for genetic analysis. After trimming low-quality sequences from the reads, bulked reads for all eight individuals were used to determine the draft consensus sequence for *L. kaempferi*. Reference mapping to *L. gmelinii* var. *japonica* (LC228570; [11]) was performed with parameter settings of mismatch cost 3, In/Del cost 3, length fraction 0.9, and similarity fraction 0.9. The complete chloroplast genome for each sample was then determined by mapping reads for each sample to our consensus sequence, using the same parameter settings as described above. The initial annotation of the chloroplast genome was performed using DOGMA [21]. Prediction of tRNA genes was performed using tRNAscan (http://lowelab.ucsc.edu/tRNAscan-SE). The annotation was finalized, with reference to that of *L. gmelinii* var. *japonica* (LC228570). To estimate the pseudogenes of *ndh* (subunits of an NADH dehydrogenase), we referred to the *Pinus thunbergii* chloroplast genome (NC_001631; [9]). REPuter [22] was used to confirm repetitive sequences in the chloroplast genome, (i.e., tandem repeats, duplicated genes, and IR regions). Finally, the gene map of the circular chloroplast genome of Japanese larch was drawn using OrganellarGenomeDRAW [23].

A phylogenetic tree was constructed based on the chloroplast genome sequences of the eight individuals identified in this study, and of MF990369 in the NCBI database (https://www.ncbi.nlm.nih.gov/) for Japanese larch, as well as five reference sequences of related *Larix* species derived from the database: LC228570 (*L. gmelinii* var. *japonica*), MF990370 (*L. gmelinii* var. *olgensis*), NC_016058 (*L. decidua*), KX880508 (*L. potaninii*), and NC_036811 (*L. sibirica*). The alignment of these fourteen chloroplast genome sequences was performed in MAFFT [24], and the final alignment was checked using CLC Genomics Workbench 9.5.3. A phylogenetic tree was constructed by MEGA X [25], based on maximum likelihood (ML) methods. A total of 1000 bootstrap replicates were applied to evaluate the branch supports.

The SNP data from the eight Japanese larch individuals was used for subsequent analyses. Haplotype networks have been demonstrated to show alternative genealogical relationships at the intraspecific population level, with low divergence [26]. We estimated haplotype networks for the chloroplast data using the software Network 10 (https://www.fluxus-engineering.com/sharenet.htm). Nucleotide diversity can be used as an inference parameter for evolutionary and demographic forces [12]; here, nucleotide diversity was calculated using DnaSP v6 software [27] and estimated as π. Divergent regions of the chloroplast genomes were identified according to the variation in π, by sliding window analysis, with a 500 bp step size and 10,000 bp window length.

Tandem repeats were identified using the Tandem Repeats Finder website [28]. In addition, simple sequence repeats (SSRs) were identified by MISA-web (https://webblast.ipk-gatersleben.de/misa/) [29], with minimum repetition numbers of 10 for mononucleotides, 6 for dinucleotides, and 5 for trinucleotides, tetranucleotides, pentanucleotides, and hexanucleotides each.

3. Results

3.1. Phylogenetic Analysis

The phylogenetic tree showed that Japanese larch forms a monophyletic group, within which Manokami larch can be phylogenetically placed based on the complete chloroplast genome, with a bootstrap value of 100% (Figure 1). Japanese larch is genetically close to *L. decidua* and *L. gmelinii*, but distant from *L. sibirica* and *L. potaninii* (Figure 1). The haplotype network among the eight sampled individuals revealed that Manokami larch was genetically distinct from other Japanese larches (Figure 2).

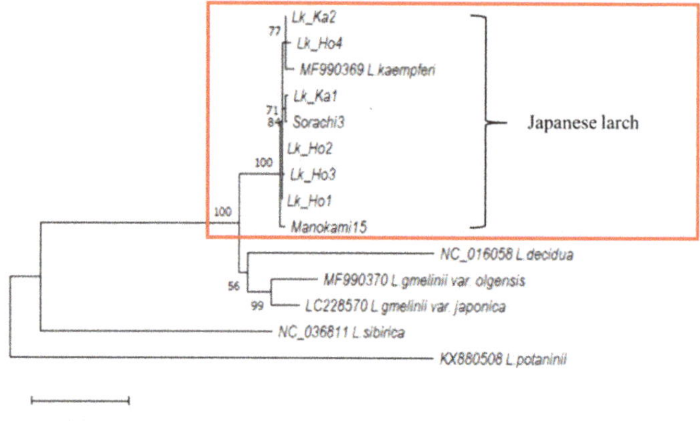

Figure 1. The maximum likelihood (ML) phylogenetic tree based on 14 chloroplast genomes of *Larix* species. The red square represents Japanese larch.

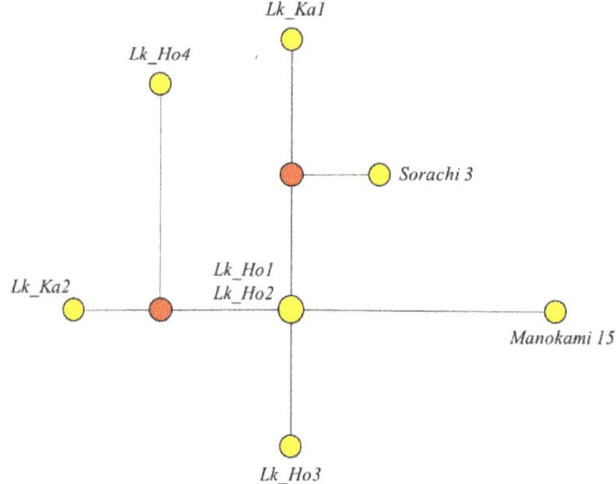

Figure 2. Haplotype network based on SNP sites within eight individuals of Japanese larch. The red dot represents missing haplotypes.

3.2. Characteristics of the Japanese Larch Chloroplast Genome

Japanese larch circular chloroplast genomes were characterized in the range of 122,394–122,409 bp with accession numbers from the DNA Data Bank of Japan (DDBJ) from LC574969 to LC574976. The gene type, number, and order were identical among the Japanese larch chloroplast genomes used in this study. *Lk_Ho1* (LC574969) was used as a representative of Japanese larch; Figure 3 illustrates the physical mapping of its chloroplast genome, which contained a pair of IRs (436 bp each) separated by large single copy (LSC), and small single copy (SSC) regions, of 65,398 bp and 56,136 bp, respectively. The *trnI-CAU* gene was duplicated within inverted repeats, the *trnS-GCU* and *psbI* genes were duplicated as another inverted repeat of 457 bp in the LSC region (Figure 3), and two *trnT-GGU* genes were dispersed in the LSC and SSC regions, respectively. A total of 119 genes were identified (Table S1), including 72 protein genes, 35 transfer RNA genes, 4 ribosomal RNA genes, and 8 pseudogenes. Thirteen genes contained an intron, including *trnA-UGC, trnG-UCC, trnI-GAU, trnK-UUU, trnL-UAA, trnV-UAC, rpoC1, rps12, rpl2, rpl16, petB, petD,* and *atpF*. In addition, *ycf3* contained two introns. Furthermore, *rps12* was a trans-splicing gene with 5′ end and 3′ end exons, located in the LSC region and the SSC region, respectively. The G+C content of the complete chloroplast genome of Japanese larch was 38.7%.

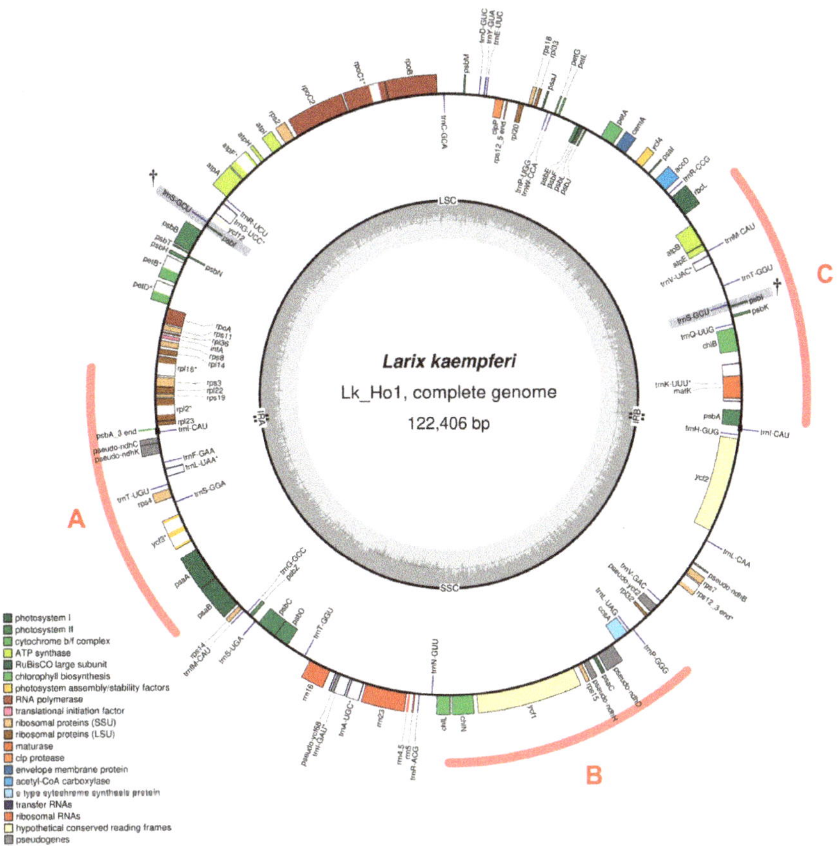

Figure 3. A gene map of Japanese larch chloroplast genome (accession number: LC574969). Genes shown outside and inside the circle are transcribed clockwise, and transcribed counterclockwise, respectively. Genes were colour-coded to distinguish different functional groups. The dark and light gray inner circle indicates the GC and AT content of the chloroplast genome, respectively. "✝" represents the location of a longer inverted repeat. A, B and C represent hotspot of variation.

3.3. Nucleotide Diversity Analysis

The value of nucleotide diversity (π) was calculated at 0.00004, based on SNP sites for Japanese larch, suggesting that sequences had low variation. As shown in Figure 4, there were three divergent regions (A, B, and C) in Japanese larch. Two regions (A, C), which were roughly in the range of the *rpl16* gene *psaB* and the *rbcL* gene *psbA*, respectively, were classified as moderately variable ($\pi > 0.00004$); these regions contained variant sites in *psaB*, *rpl16*, *ψndhK*, *atpB*, *psbK*, and *matK*, and three intergenic spacers (between the *rpl23* and *psbA*-partial genes, between the *trnS-GGA* and *ycf3* genes, and between the *trnS-GCU* and *trnT-GGU* genes). The B region (roughly from *chlL* to *rpl32*, $\pi > 0.0001$) was identified as a hypervariable region, in which mutation occurred twice in the *ψndhD* and the *ycf1*.

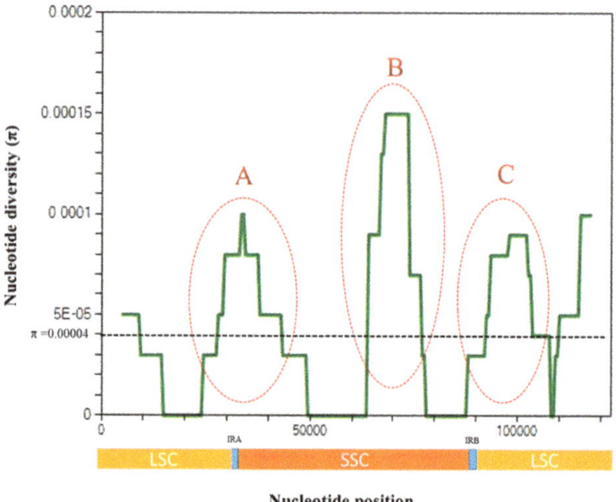

Nucleotide position

Figure 4. The nucleotide diversity (π) of Japanese larch chloroplast genomes, based on sliding window analysis.

3.4. Repeat Sequence Analysis of the Japanese Larch Chloroplast Genome

Tandem repeats were detected in approximately 25 sites in the Japanese larch chloroplast genomes. Repeated lengths of tandem repeats varied from 12 to 117 bp, and 64% of all tandem repeats occurred in *ycf1*, which belongs to a protein-coding region containing 76% of all detected tandem repeats. Nineteen SSR motifs were detected in the Japanese larch chloroplast genome. The majority of the detected SSR motifs were mononucleotide motifs, of which the SSR motif of mononucleotide T was the most frequent, followed by mononucleotide A and mononucleotide G. With the exception of three SSR motifs of dinucleotide AT, no other multiple nucleotide motifs were detected in the Japanese larch chloroplast genome. Furthermore, most (77.8%) of the detected SSR motifs were found in the intergenic region, followed by introns (16.7%), and protein-coding genes (5.5%). Eight cpSSR variants out of nineteen cpSSR motifs were detected in the intergenic region of the chloroplast genome of Japanese larch.

3.5. Genetic Variation among Japanese Larch Chloroplast Genomes

Among the eight individuals sequenced in this study, 31 variants (including 19 SNPs, 8 SSRs, and 4 In/Dels) were detected. For SNP variants, six and thirteen SNPs were identified in the intergenic spacer (IGS) and coding sequence (CDS) regions, respectively. These were detected in the *ψndhK* (one SNP), *ψndhD* (two SNPs), and protein-coding *ycf1* (two SNPs), all of which belong to the SSC region. Six SSR variants were identified in the IGS, whereas two SSR variants were detected in the CDS region. Four In/Del variants were identified in the *ψndhK* gene (one In/Del variant) and the *ycf1* gene (three In/Del variants), belonging to CDS region. (Table 1)

Table 1. Genetic variation among chloroplast genomes of Japanese larch. CDS: coding sequence; IGS: intergenic spacer; In/Del: insertion or deletion; SNP: single nucleotide polymorphism; SSR: simple sequence repeat.

Sequence	N	In/Del		SNP		SSR	
		CDS	IGS	CDS	IGS	CDS	IGS
Lk_Ho1(Reference)	—	—	—	—	—	—	—
Lk_Ho2	2						2
Lk_Ho3	3	1		1	1		
Lk_Ho4	8			4	2		2
Lk_Ka1	4			3	1		
Lk_Ka2	5	1		3			1
Sorachi3	3			3			
Manokami15	12	2		2	3	1	4
Whole sequences	31	4		19		8	

4. Discussion

The phylogenetic position of Manokami larch, has been discussed by several researchers [14,16,17]. This study clearly indicates that Manokami larch should be phylogenetically categorized into Japanese larch, with a bootstrap value of 100% (Figure 1). Our findings support the assertion by Shiraishi et al. [15] that Manokami larch must be a Japanese larch. Manokami larch is located far from other Japanese larches (Figure 2); genetically divergent genotypes, such as that of Manokami larch, could be used to efficiently detect intraspecific variation in Japanese larch.

The chloroplast genomes of Japanese larch obtained from this study were similar in size and gene order to those of *L. gmelinii* [11], *L. sibirica* [19], *L. decidua* [30], and *L. potaninii* [18]. The chloroplast structure types were classified in Pinaceae according to their alignment order and the orientation of the F1 (fragment flanked by *trnG-UCC* and *trnE-UUC*), F2 (fragment flanked by *clpP* and *trnT-GGU*), T1 (type 1 Pinaceae-specific repeat containing *trnS-GCU* and *psbI*), and T2 (type 2 Pinaceae-specific repeats in intergenic spacers) fragments in the LSC region, which can produce eight different cpDNA forms, including A, B, C, D, E, F, and G forms [30]. The chloroplast DNA form used in this study was classified into the C form, the same form identified for *L. gmelinii*, *L. decidua*, *L. griffithiana*, and *Pinus elliottii* [11,30] based on the alignment order and orientation of T1, T2, −F1 (reverse strand), T2, +F2 (forward strand), and T1. Due to this T1 repeat, there were longer inverted repeats (457 bp) in the LSC region than two IRs (436 bp) in Japanese larch. Extremely shortened IRs, with another pair of inverted repeats, is regarded as a common feature in Pinaceae [30,31].

In this study, three hotspots of variation were detected throughout the entire chloroplast genome (Figures 3 and 4). The *ycf1* and *ψndhD* were included in the hypervariable region (region B), and the *ψndhK* was included in the moderately variable region (region A). Three In/Del variants occurred in the *ycf1*, and previous research has reported insertions or deletions in the *ycf1* of *L. gmelinii* [11]. Although it was considered a possibility that the *ycf1* might be a nonfunctional pseudogene, another study [32] indicated that *ycf1* is a functional gene, and encodes a product essential for cell survival. Dong et al. [33] revealed that the divergence of the *ycf1* was obvious in gymnosperms. Additionally, Firetti et al. [34] indicated that the *ycf1* was more divergent than the non-coding regions in the genus *Anemopaegma*. Regarding pseudogenes, eleven *ndh* genes (*ndhA—K*) have been identified in the cpDNA sequences of photosynthetic land plants [9,35,36]. In our study, five *ψndh* genes were found only in Japanese larch, of which *ψndhD* (two SNPs) and *ψndhK* (one SNP, one SSR, one In/Del variant) belonged to the region of frequent variation; these genes did not, however, exhibit a function consistent with other *Pinus* and *Larix* species [9,11].

Repeat sequences may play an important role in chloroplast genome arrangement and sequence divergence. In particular, tandem repeats may induce In/Dels [37,38]. In this study, tandem repeats were primarily identified in the *ycf1*. Tandem repeats were also located in the *ycf1* of other conifers, such as *Cryptomeria japonica* [39] and *L. gmelinii* [11].

Among the eight Japanese larch individuals, we detected 31 variants (19 SNPs, 8 SSRs, and 4 In/Dels) located in *psaB*, *ψndhD*, *ψndhK*, *psbE*, *psbK*, *rpoC1*, *rpoC2*, the intron of *rpl16*, *matK*, *atpB*, *ycf1*, six intergenic spacers (between the *rpl23* and *psbA*-partial genes, between the *trnS-GGA* and *ycf3* genes, between the *trnS-GCU* and *trnT-GGU* genes, between the *clpP* and *trnE-UUC* genes twice, between the *psbE* and *petL* genes) with SNPs, the intron of *atpF*, *ψndhK*, six intergenic spacers (between the *trnC-GCA* and *rpoB* genes, between the *ycf1* and *rps15* genes, between the *trnL-CAA* and *ycf2* genes, between the *ψycf2* and *trnV-GAC* genes, between the *trnT-GGU* and *trnV-UAC* genes twice) with SSRs, *ψndhK* and *ycf1* with In/Dels that could prove useful for providing candidate cpDNA markers. Chloroplast simple sequence repeat (cpSSR) markers often contain highly polymorphic variations within a population of conifers (see [7]), although Zhang et al. [40] found only three polymorphic cpSSR markers among 11 candidate markers in Japanese larch. We identified 19 SSR motifs within the chloroplast genome of Japanese larch, preferentially within the intergenic space, and only 8 SSR motifs occurred among 19 SSR motifs in the intergenic region of the chloroplast genome. These results lay a foundation for the development of cpDNA markers for Japanese larch.

5. Conclusions

The complete chloroplast genome of Japanese larch (122,398–122,409 bp) was obtained using next-generation sequencing technology. The comparison of whole chloroplast genomes clearly indicated that the isolated population, forming the northern limit of the species' range (Manokami larch), should be placed phylogenetically within Japanese larch. The Manokami larch was found to be genetically different from other Japanese larches, indicating that sufficient genetic variation should be detected within the samples used in this study. Based on an analysis of intraspecific variation, 31 variants were detected, including 19 SNPs, 8 SSRs, and 4 In/Dels, all of which can be applied for the development of cpDNA markers. These variations should be useful for paternity analysis and population genetics analysis of Japanese larch in future studies.

Supplementary Materials: The following are available online at http://www.mdpi.com/1999-4907/11/8/884/s1, Table S1: List of estimated chloroplast genes of Japanese larch. Genes with * have intron(s). Duplicated genes are shown in parenthesis.

Author Contributions: Conceptualization, W.I. and T.H.; Founding acquisition, T.H.; Software, W.I. and S.C.; visualization, W.I; Validation, W.I, S.C., T.H. and S.G.; Formal analysis, S.C. and W.I.; data curation, S.C. and W.I.; Writing—original draft preparation, S.C., S.G.; Writing—review and editing, S.C., W.I., T.H. and S.G. All authors have read and agreed to the published version of the manuscript.

Funding: This study was partly supported by the Japan Society for the Promotion of Science [15K18715] and the Grant for Joint Research Program of the Institute of Low Temperature Science, Hokkaido University.

Acknowledgments: We would like to thank K. Ono for the help of laboratory work.

Conflicts of Interest: The authors declare no conflict of interest.

References

1. Wolfe, K.H.; Li, W.-H.; Sharp, P.M. Rates of nucleotide substitution vary greatly among plant mitochondrial, chloroplast, and nuclear DNAs. *Proc. Natl. Acad. Sci. USA* **1987**, *84*, 9054–9058. [CrossRef]

2. Wu, F.H.; Kan, D.P.; Lee, S.B.; Daniell, H.; Lee, Y.W.; Lin, C.C.; Lin, N.S.; Lin, C.S. Complete nucleotide sequence of *Dendrocalamus latiflorus* and *Bambusa oldhamii* chloroplast genomes. *Tree Physiol.* **2009**, *29*, 847–856. [CrossRef]

3. Nock, C.J.; Waters, D.L.E.; Edwards, M.A.; Bowen, S.G.; Rice, N.; Cordeiro, G.M.; Henry, R.J. Chloroplast genome sequences from total DNA for plant identification. *Plant Biotechnol. J.* **2011**, *9*, 328–333. [CrossRef]

4. Wagner, D.B.; Furnier, G.R.; Saghai-Maroof, M.A.; Williams, S.M.; Dancik, B.P.; Allard, R.W. Chloroplast DNA polymorphisms in lodgepole and jack pines and their hybrids. *Proc. Natl. Acad. Sci. USA* **1987**, *84*, 2097. [CrossRef]

5. Tsumura, Y.; Suyama, Y.; Taguchi, H.; Ohba, K. Geographical cline of chloroplast DNA variation in *Abies mariesii*. *Theor. Appl. Genet.* **1994**, *89*, 922–926. [CrossRef]

6. Neale, D.B.; Sederoff, R.R. Paternal inheritance of chloroplast DNA and maternal inheritance of mitochondrial DNA in loblolly pine. *Theor. Appl. Genet.* **1989**, *77*, 212–216. [CrossRef]

7. Vendramin, G.G.; Lelli, L.; Rossi, P.; Morgante, M. A set of primers for the amplification of 20 chloroplast microsatellites in Pinaceae. *Mol. Ecol.* **1996**, *5*, 595–598. [CrossRef]

8. Sugiura, M. The chloroplast chromosomes in land plants. *Annu. Rev. Cell Biol.* **1989**, *5*, 51–70. [CrossRef]

9. Wakasugi, T.; Tsudzuki, J.; Ito, S.; Nakashima, K.; Tsudzuki, T.; Sugiura, M. Loss of all *ndh* genes as determined by sequencing the entire chloroplast genome of the black pine *Pinus thunbergii*. *Proc. Natl. Acad. Sci. USA* **1994**, *91*, 9794–9798. [CrossRef]

10. Parks, M.; Cronn, R.; Liston, A. Increasing phylogenetic resolution at low taxonomic levels using massively parallel sequencing of chloroplast genomes. *BMC Biol.* **2009**, *7*, 84. [CrossRef]

11. Ishizuka, W.; Tabata, A.; Ono, K.; Fukuda, Y.; Hara, T. Draft chloroplast genome of *Larix gmelinii* var. *japonica*: Insight into intraspecific divergence. *J. For. Res.* **2017**, *22*, 393–398. [CrossRef]

12. Branca, A.; Paape, T.D.; Zhou, P.; Briskine, R.; Farmer, A.D.; Mudge, J.; Bharti, A.K.; Woodward, J.E.; May, G.D.; Gentzbittel, L.; et al. Whole-genome nucleotide diversity, recombination, and linkage disequilibrium in the model legume. *Proc. Natl. Acad. Sci. USA* **2011**, *108*, E864. [CrossRef] [PubMed]

13. Ngamitsu, T.; Nagasaka, K.; Yoshimaru, H.; Tsumura, Y. Provenance tests for survival and growth of 50-year-old Japanese larch (*Larix kaempferi*) trees related to climatic conditions in central Japan. *Tree Genet. Genomes* **2014**, *10*, 87–99. [CrossRef]

14. Japan Forest Tree Breeding Association. Characteristics and conservation of isolated and northernmost natural population of larch in Japan-observing the population and trying to reveal what it really is like. In Proceedings of the 1st Forest Genetics Seminar, Miyagi, Japan, July 1994; p. 97, NAID: 10015777891. (In Japanese).

15. Shiraishi, S.; Isoda, K.; Watanabe, A.; Kawasaki, H. DNA systematical study on the *Larix* relicted at Mt.Manokami, the Zao Mountains. *J. Jpn. For. Soc.* **1996**, *78*, 175–182.

16. Oda, H. The northern natural Japanese larch forest at Manokami, the Zao Mountaions. *For. Sci.* **2013**, *38*, 52–58. (In Japanese)

17. Tozawa, T. Ecological studies on the natural forest of the Japanese larch in Tohoku districts (Northeast of Honshu Island, Japan), 1: Natural forest of the now as its northern limit in Japan. *Univ. For. Rep. Iwate Univ.* **1975**, *6*, 1–18.

18. Han, K.; Li, J.; Zeng, S.; Liu, Z.-L. Complete chloroplast genome sequence of Chinese larch (*Larix potaninii* var. *chinensis), an endangered conifer endemic to China*. *Conserv. Genet. Resour.* **2017**, *9*, 111–113.

19. Bondar, E.I.; Putintseva, Y.A.; Oreshkova, N.V.; Krutovsky, K.V. Siberian larch (*Larix sibirica* Ledeb.) chloroplast genome and development of polymorphic chloroplast markers. *BMC Bioinform.* **2019**, *20*, 38. [CrossRef]

20. Kim, S.-C.; Lee, J.-W.; Lee, M.-W.; Baek, S.-H.; Hong, K.-N. The complete chloroplast genome sequences of *Larix kaempferi* and *Larix olgensis* var. *koreana (Pinaceae)*. *Mitochondrial DNA Part B* **2018**, *3*, 36–37. [CrossRef]

21. Wyman, S.K.; Jansen, R.K.; Boore, J.L. Automatic annotation of organellar genomes with DOGMA. *Bioinformatics* **2004**, *20*, 3252–3255. [CrossRef]

22. Kurtz, S.; Choudhuri, J.V.; Ohlebusch, E.; Schleiermacher, C.; Stoye, J.; Giegerich, R. REPuter: The manifold applications of repeat analysis on a genomic scale. *Nucleic Acids Res.* **2001**, *29*, 4633–4642. [CrossRef] [PubMed]

23. Lohse, M.; Drechsel, O.; Kahlau, S.; Bock, R. OrganellarGenomeDRAW: A suite of tools for generating physical maps of plastid and mitochondrial genomes and visualizing expression data sets. *Nucleic Acids Res.* **2013**, *41*, W575–W581. [CrossRef] [PubMed]

24. Kuraku, S.; Zmasek, C.M.; Nishimura, O.; Katoh, K. aLeaves facilitates on-demand exploration of metazoan gene family trees on MAFFT sequence alignment server with enhanced interactivity. *Nucleic Acids Res.* **2013**, *41*, W22–W28. [CrossRef] [PubMed]

25. Kumar, S.; Stecher, G.; Li, M.; Knyaz, C.; Tamura, K. MEGA X: Molecular evolutionary genetics analysis across computing platforms. *Mol. Biol. Evol.* **2018**, *35*, 1547–1549. [CrossRef]

26. Templeton, A.R.; Crandall, K.A.; Sing, C.F. A cladistic analysis of phenotypic associations with haplotypes inferred from restriction endonuclease mapping and DNA sequence data. III. *Cladogram Estim. Genet.* **1992**, *132*, 619.

27. Rozas, J.; Ferrer-Mata, A.; Sánchez-DelBarrio, J.C.; Guirao-Rico, S.; Librado, P.; Ramos-Onsins, S.E.; Sánchez-Gracia, A. DnaSP 6: DNA Sequence Polymorphism Analysis of Large Data Sets. *Mol. Biol. Evol.* **2017**, *34*, 3299–3302. [CrossRef]

28. Benson, G. Tandem repeats finder: A program to analyze DNA sequences. *Nucleic Acids Res.* **1999**, *27*, 573–580. [CrossRef]

29. Beier, S.; Thiel, T.; Münch, T.; Scholz, U.; Mascher, M. MISA-web: A web server for microsatellite prediction. *Bioinformatics* **2017**, *33*, 2583–2585. [CrossRef]

30. Wu, C.-S.; Lin, C.-P.; Hsu, C.-Y.; Wang, R.-J.; Chaw, S.-M. Comparative chloroplast genomes of Pinaceae: Insights into the mechanism of diversified genomic organizations. *Genome Biol. Evol.* **2011**, *3*, 309–319. [CrossRef]

31. Kang, H.-I.; Lee, H.O.; Lee, I.H.; Kim, I.S.; Lee, S.-W.; Yang, T.J.; Shim, D. Complete chloroplast genome of *Pinus densiflora* Siebold & Zucc. and comparative analysis with five pine trees. *Forests* **2019**, *10*, 600.

32. Drescher, A.; Ruf, S.; Calsa, T., Jr.; Carrer, H.; Bock, R. The two largest chloroplast genome-encoded open reading frames of higher plants are essential genes. *Plant J.* **2000**, *22*, 97–104. [CrossRef]

33. Dong, W.; Xu, C.; Li, C.; Sun, J.; Zuo, Y.; Shi, S.; Cheng, T.; Guo, J.; Zhou, S. ycf1, the most promising plastid DNA barcode of land plants. *Sci. Rep.* **2015**, *5*, 8348. [CrossRef] [PubMed]

34. Firetti, F.; Zuntini, A.R.; Gaiarsa, J.W.; Oliveira, R.S.; Lohmann, L.G.; Van Sluys, M.-A. Complete chloroplast genome sequences contribute to plant species delimitation: A case study of the *Anemopaegma* species complex. *Am. J. Bot.* **2017**, *104*, 1493–1509. [CrossRef] [PubMed]

35. Ohyama, K.; Fukuzawa, H.; Kohchi, T.; Shirai, H.; Sano, T.; Sano, S.; Umesono, K.; Shiki, Y.; Takeuchi, M.; Chang, Z. Chloroplast gene organization deduced from complete sequence of liverwort *Marchantia polymorpha* chloroplast DNA. *Nature* **1986**, *322*, 572–574. [CrossRef]

36. Hiratsuka, J.; Shimada, H.; Whittier, R.; Ishibashi, T.; Sakamoto, M.; Mori, M.; Kondo, C.; Honji, Y.; Sun, C.-R.; Meng, B.-Y.; et al. The complete sequence of the rice (*Oryza sativa*) chloroplast genome: Intermolecular recombination between distinct tRNA genes accounts for a major plastid DNA inversion during the evolution of the cereals. *Mol. Gen. Genet.* **1989**, *217*, 185–194. [CrossRef] [PubMed]

37. Guisinger, M.M.; Kuehl, J.V.; Boore, J.L.; Jansen, R.K. Extreme reconfiguration of plastid genomes in the Angiosperm family Geraniaceae: Rearrangements, repeats, and codon usage. *Mol. Biol. Evol.* **2011**, *28*, 583–600. [CrossRef]

38. Weng, M.-L.; Blazier, J.C.; Govindu, M.; Jansen, R.K. Reconstruction of the ancestral plastid genome in Geraniaceae reveals a correlation between genome rearrangements, repeats, and nucleotide substitution rates. *Mol. Biol. Evol.* **1994**, *31*, 645–659. [CrossRef]

39. Hirao, T.; Watanabe, A.; Kurita, M.; Kondo, T.; Takata, K. A frameshift mutation of the chloroplast matK coding region is associated with chlorophyll deficiency in the *Cryptomeria japonica* virescent mutant Wogon-Sugi. *Curr. Genet.* **2009**, *55*, 311–321. [CrossRef]

40. Zhang, X.Y.; Shiraishi, S.; Huang, M.R. Analysis of genetic structure in population of *Larix Kaempferi* by chloroplast SSR markers. *Hereditas* **2004**, *26*, 486–490.

Article

Genetic Diversity and Structure of Japanese Endemic Genus *Thujopsis* (Cupressaceae) Using EST-SSR Markers

Michiko Inanaga [1], Yoichi Hasegawa [2], Kentaro Mishima [1] and Katsuhiko Takata [3,*

[1] Forest Tree Breeding Center, Forestry and Forest Products Research Institute,
Forest Research and Management Organization, 3809-1 Ishi, Juo, Hitachi, Ibaraki 319-1301, Japan;
inasuzume@affrc.go.jp (M.I.); mishimak@affrc.go.jp (K.M.)

[2] Department of Forest Molecular Genetics and Biotechnology, Forestry and Forest Products Research
Institute, Forest Research and Management Organization, 1 Matsunosato, Tsukuba, Ibaraki 305-8687, Japan;
yohase@gmail.com

[3] Institute of Wood Technology, Akita Prefectural University, 11-1 Kaieizaka, Noshiro, Akita 016-0876, Japan

* Correspondence: katsu@iwt.akita-pu.ac.jp; Tel.: +81-185-52-6900

Received: 24 July 2020; Accepted: 24 August 2020; Published: 27 August 2020

Abstract: The genus *Thujopsis* (Cupressaceae) comprises monoecious coniferous trees endemic to Japan. This genus includes two varieties: *Thujopsis dolabrata* (L.f.) Siebold et Zucc. var. *dolabrata* (southern variety, Td) and *Thujopsis dolabrata* (L.f.) Siebold et Zucc. var. *hondae* Makino (northern variety, Th). The aim of this study is to understand the phylogeographic and genetic population relationships of the genus *Thujopsis* for the conservation of genetic resources and future breeding. A total of 609 trees from 22 populations were sampled, including six populations from the Td distribution range and 16 populations from the Th distribution range. The genotyping results for 19 expressed sequence tag (EST)-based simple sequence repeat (SSR) markers, followed by a structure analysis, neighbor-joining tree creation, an analysis of molecular variance (AMOVA), and hierarchical F statistics, supported the existence of two genetic clusters related to the distribution regions of the Td and Th varieties. The two variants, Td and Th, could be defined by their provenance, in spite of the ambiguous morphological differences between the varieties. The distribution ranges of both variants, which have been defined from their morphology, was confirmed by genetic analysis. The Th populations exhibited relatively uniform genetic diversity, most likely because Th refugia in the glacial period were scattered throughout their current distribution area. On the other hand, there was a tendency for Td's genetic diversity to decrease from central to southern Honshu island. Notably, the structure analysis and neighbor-joining tree suggest the hybridization of the two varieties in the contact zone. More detailed studies of the genetic structure of Td are required in future analyses.

Keywords: *Thujopsis dolabrata*; EST-SSR markers; varieties; population structure

---

1. Introduction

Conifers are a dominant plant type found in the vast boreal forests of the North American and Eurasian continents. They represent an important forest resource in many countries because of their superior wood properties, including straighter trunks and stronger yet lighter wood compared to those of most angiosperms [1]. Generally, wild populations of trees are an important genetic resource and are required for breeding programs that aid in the selection of new plant varieties suitable for withstanding a range of environmental conditions, diseases, and future climate change [2–4]. A principal requirement for conserving forest genetic resources is maintaining the genetic diversity within and among the populations of a species [5]. Phylogeographic and population genetic studies using neutral molecular markers provide a means of identifying in situ units and can be used to determine diversity level and distribution, gene flow routes, and major genetic disjunctions within the species [6]. Therefore, it is important to examine the genetic diversity and phylogeography of a wide range of wild populations, including in the context of conifer breeding.

Thujopsis (Cupressaceae) is a genus of monoecious coniferous trees with wind-mediated pollen- and seed-dispersal systems. It is endemic to Japan and is one of the basal lineages of Cupressoideae in the Cupressaceae phylogenetic tree [7]. This genus includes two varieties: *Thujopsis dolabrata* (L.f.) Siebold et Zucc. var. *dolabrata* (Td) and *Thujopsis dolabrata* (L.f.) Siebold et Zucc. var. *hondae* Makino (Th). Taxonomically, the former is regarded as the southern variety and the latter as the northern variety. Td, which features somewhat horned cones, is widely distributed in the southern region of the Japanese Archipelago, whereas Th, which is characterized by denser needles and rounder cones, is a race distributed in northern Honshu and the southern region of Hokkaido [8,9] (Figures 1a and A1). It is difficult to distinguish between these varieties using morphology alone, as their morphology tends to vary continuously. Therefore, in this study, both varieties will be classified according to their regional distribution. Because of the valuable properties of the wood, the genus *Thujopsis* is one of the most important tree species in Japanese forestry, and plantations of Th have been actively established in Aomori, Niigata, and Ishikawa Prefectures (see Figure 1b and Table 1) [10]. It is therefore essential to understand the phylogeographic and population genetic relationships of the genus *Thujopsis* for the conservation of genetic resources and future breeding.

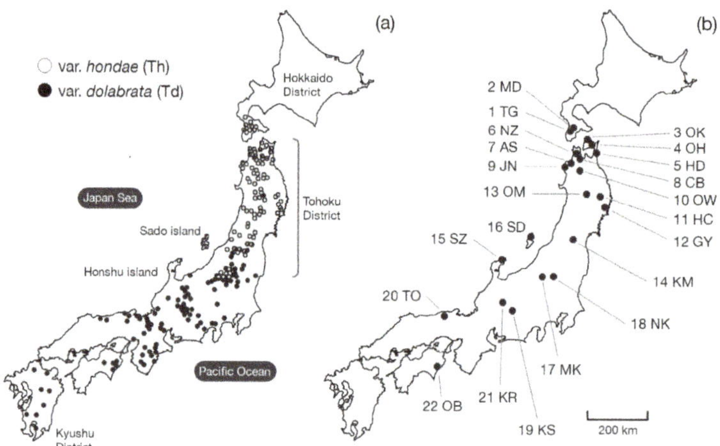

Figure 1. (**a**) Classical distribution range of genus *Thujopsis* defined by morphological differences between varieties shown in Kurata [9], var. *dolabrata* (Td, black dots) and var. *hondae* (Th, white dots). (**b**) Locations of the 22 sampled populations. Numbers correspond to the population numbers in Table 1.

Higuchi et al. [11] suggested that seven natural populations from the Th distribution region, as well as other Japanese conifers distributed over a wide area, showed a relatively low F_{ST} (0.046). The two populations that were located in the marginal part of the Th distribution region and the five populations in the northern part of the distribution region were genetically different in the neighbor-joining tree and structure analysis [11]. Ikeda et al. [12] also analyzed the genetic structure of natural forests using populations distributed over almost the same geographical area as studied by Higuchi et al. [11]. They suggested that the distribution area of Th included four regional groups (Hokkaido and Aomori prefectures, Iwate and Yamagata prefectures, Niigata, and Ishikawa prefectures) of natural populations [12]. These findings are important for illustrating the genetic structure of Th;

however, they did not include Td populations. Therefore, a study examining both Td and Th is needed to determine the extent to which these two varieties are genetically distinct.

Table 1. Locations of *Thujopsis dolabrata*. Populations 1 to 16 fall within the range of var. *hondae* and 17 to 22 fall within var. *dolabrata*.

No	Pop Name	Pop Code	Prefecture	Distribution Region	Latitude/Longitude	N
1	Minamidate	MD	Hokkaido	var. *hondae*	41.870072,140.283394	15
2	Todogawa	TG	Hokkaido	var. *hondae*	41.823216,140.208334	23
3	Okoppe	OK	Aomori	var. *hondae*	41.478333,140.952778	27
4	Ohata	OH	Aomori	var. *hondae*	41.386911,141.048666	30
5	Higashidori	HD	Aomori	var. *hondae*	41.085490,141.314990	30
6	Nakazato	NZ	Aomori	var. *hondae*	40.987595,140.456850	31
7	Ajigasawa	AS	Aomori	var. *hondae*	40.677765,140.184354	12
8	Choubousan	CB	Aomori	var. *hondae*	40.903131,140.605565	15
9	Juniko	JN	Aomori	var. *hondae*	40.558209,139.964216	30
10	Owani	OW	Aomori	var. *hondae*	40.448351,140.590625	22
11	Omyoujin	OM	Iwate	var. *hondae*	39.654310,140.896561	29
12	Hayachine	HC	Iwate	var. *hondae*	39.582572,141.481268	24
13	Goyousan	GY	Iwate	var. *hondae*	39.207849,141.716187	24
14	Kaminoyama	KM	Yamagata	var. *hondae*	38.083196,140.304773	19
15	Sado	SD	Niigata	var. *hondae*	38.215756,138.451880	31
16	Suzu	SZ	Ishikawa	var. *hondae*	37.402306,137.165028	17
17	Minakami	MK	Gunma	var. *dolabrata*	36.839137,138.972695	35
18	Nikko	NK	Tochigi	var. *dolabrata*	36.822614,139.440342	48
19	Kiso	KS	Nagano	var. *dolabrata*	35.727433,137.620934	32
20	Kuraiyama	KR	Gifu	var. *dolabrata*	35.985774,137.216638	35
21	Toyo-oka	TO	Hyougo	var. *dolabrata*	35.509750,134.637190	44
22	Obitani	OB	Tokushima	var. *dolabrata*	33.844595,134.327933	36

Population numbers (No); sample size of each population (N).

The aim of this study was to examine the genetic structure of the genus *Thujopsis* and to determine the relative contributions of the genetic structure between varieties. In the present study, simple sequence repeat (SSR) markers, which were representative neutral and co-dominant genetic markers, were used in order to provide basic information that would be useful in the conservation of *Thujopsis* genetic resources in natural forests.

2. Materials and Methods

2.1. Sampling and Study Sites

A total of 609 trees from 22 populations, including 6 populations from the Td distribution range and 16 populations from the Th distribution range, were sampled (Table 1; Figure 1b). Fresh needles were collected and stored at −30 °C. The sampled trees were each separated by more than 30 m. Sample sizes varied among the populations (Table 1). This variation reflects the sampling of large individuals (older trees) from some populations. Moreover, sampling was not attempted in dangerous areas where the topography was too steep.

2.2. DNA Extraction and Genotyping

Total genomic DNA was extracted from 100-mg needles using the hexadecyltrimethylammonium bromide (CTAB) method [13] with minor modifications. The genotypes of the sample trees were determined for 19 markers that were developed from expressed sequence tag (EST)-based simple sequence repeat (SSR) markers for the genus *Thujopsis* [14] (Table 2). In addition, universal primers were attached to each forward primer to efficiently incorporate fluorescent dyes during PCR for multiplexing [15]. The PCR was performed in a volume of 10.0 µL containing 10–120 ng of template DNA, 0.15 µM of forward primer, 0.5 µM reverse primer, 0.2 µM of either one of the tail primers fluorescently labeled by FAM, VIC, or NED, and 5.0 µL of Go Taq Master mix (Promega Corporation, Madison, MI, USA). A GeneAmp PCR System 9700 thermal cycler (Applied Biosystems, Foster City, CA,

USA) was used (Applied Biosystems, Foster City, CA, USA) with the following thermal profile: initial denaturation at 94 °C for 2 min, followed by 30 cycles of denaturation at 94 °C for 30 s; an annealing temperature of 60 °C for 30 s, and an extension at 72 °C for 30 s; then a final extension at 72 °C for 5 min. The amplified PCR products of each individual were classified into five groups according to the fragment size and the type of fluorescent marker (Table A1). Each group of PCR products was separated by capillary electrophoresis using a 3100 Genetic Analyzer (Applied Biosystems), and genotypes were scored with Geneious 7.0.4 software (Biomatters Ltd., Auckland, New Zealand).

Table 2. Genetic diversity measures estimated at 19 microsatellite loci.

Variety	No	Pop Code	N	Allele Number	Allelic Richness	Private Allele	H_o	H_e	F_{IS}
Th	1	MD	15	6.30	5.86	1	0.646	0.639	0.025
	2	TG	23	6.70	5.54	1	0.593	0.623	0.071
	3	OK	27	7.50	5.97	2	0.673	0.662	0.002
	4	OH	30	8.00	6.02	1	0.628	0.664	0.071
	5	HD	30	7.80	5.93	2	0.646	0.641	0.010
	6	NZ	31	8.70	6.55	0	0.689	0.682	0.005
	7	AS	12	5.80	5.84	3	0.746	0.646	−0.112
	8	CB	15	6.50	6.12	0	0.662	0.660	0.032
	9	JN	30	7.80	6.08	3	0.681	0.664	−0.009
	10	OW	22	6.80	5.76	1	0.624	0.636	0.042
	11	OM	29	7.90	6.13	4	0.679	0.663	−0.006
	12	HC	24	6.80	5.56	1	0.618	0.603	−0.004
	13	GY	24	7.60	6.19	0	0.672	0.651	−0.012
	14	KM	19	6.70	5.87	3	0.637	0.624	0.006
	15	SD	31	7.30	5.67	0	0.637	0.623	−0.005
	16	SZ	17	6.50	5.74	1	0.573	0.587	0.054
	Average			7.17	5.93	1.4	0.650	0.642	
Td	17	MK	35	7.20	5.61	1	0.609	0.611	0.018
	18	NK	48	7.30	5.24	5	0.554	0.584	0.062
	19	KS	32	6.40	5.15	1	0.554	0.561	0.027
	20	KR	35	6.20	4.70	2	0.496	0.500	0.022
	21	TO	44	5.90	4.68	1	0.542	0.548	0.023
	22	OB	36	5.50	4.20	1	0.513	0.504	−0.005
	Average			6.42	4.93	1.8	0.545	0.551	

Thujopsis dolabrata (L.f.) Siebold et Zucc. var. *hondae* Makino (Th); *T. dolabrata* (L.f.) Siebold et Zucc. var. *dolabrata* (Td); population numbers (No); sample size of each population (N); number of alleles per locus (allele number); allelic richness for standardized samples of 24 gene copies (allelic richness); number of private alleles per population (private allele); average observed heterozygosity (H_o); average expected heterozygosity (H_e); fixation index (F_{IS}).

2.3. Data Analysis

2.3.1. Genetic Diversity within Populations

The total number of detected alleles (TA), the observed heterozygosity (H_o), the total gene diversity (H_T), and Wright's inbreeding coefficient (F_{IS}) were calculated at each locus using FSTAT software version 2.9.4 [16].

The total number of detected alleles (allele number), the observed (H_o) and expected heterozygosity (H_e), and the number of private alleles were calculated for each population using GenAlEx 6.51b2 [17].

The allelic richness was standardized based on 12 individuals (24 gene copies), which was the smallest sample size among the populations. Fixation index values estimating inbreeding within individuals in a population (F_{IS}) were calculated. The significance of positive or negative values of F_{IS} was tested based on 8360 randomizations with a Bonferroni correction. These calculations were performed using FSTAT [16]. The statistical independence of loci (linkage disequilibrium for all pairs of loci across populations) was also evaluated using FSTAT.

To compare genetic differentiation among populations and between loci, the relative genetic differentiation among populations defined under the infinite allele model (IAM; F_{ST}) [18] and the stepwise mutation model (SMM; R_{ST}) [19] were calculated using SPAGeDi version 1.5 [20]. The genetic differentiation coefficients (G_{ST}; analogous to F_{ST}) and a standardized measure, which had a range of 0–1 for all levels of genetic diversity (G'_{ST}), were calculated based on the allele frequencies for each locus using GenAlEx [21,22]. The significance of the deviations of F_{ST}, R_{ST}, G_{ST} and G'_{ST} (from zero) was evaluated by permutation tests.

The likelihood of a bottleneck within each population was examined using the two-phase model with 95% single-step mutations and 5% multi-step mutations with 1000 iterations, implemented in the program BOTTLENECK version 1.2.02 [23]. The one-tailed Wilcoxon test was used to detect an excess of expected heterozygosity (H_e) compared to that expected under mutation–drift equilibrium (H_{EQ}).

2.3.2. Genetic Structures among Populations and Distribution Regions

The presence of isolation-by-distance patterns in population differentiation was investigated by applying the Mantel test to the pairwise relationship between the geographic distances (transformed to natural logarithms) and genetic distances, $F_{ST}/(1 - F_{ST})$, between the populations according to Rousset's method [24]. Comparison analyses for Td population vs. Td population, Th population vs. Th population, and the species as a whole were performed, in order to separately examine the effect of isolation-by-distance within each distribution region and within the species. These calculations were performed using SPAGeDi version 1.5 [20].

For the analysis of molecular variance (AMOVA), we used Arlequin version 3.5.2.2 [25].

Hierarchical F statistics were estimated using the R hierfstat package [26] ("varcomp.glob" and "boot.vc" functions), with the individuals ("Ind") nested within populations ("Pop"), nested within distribution regions ("Region"), and with 95% confidence intervals (CIs) from 1000 bootstraps. The significance of the hierarchical F statistics was assessed using 10,000 permutations ("test.between" and "test.within" functions) [27]. These calculations were performed using R version 3.6.1 [28].

The genetic relationships among the populations were evaluated by constructing a neighbor-joining tree [29] using Poptree2 web [30,31]. Nei's chord distance (D_a) [32] was used to estimate the degree of genetic divergence of the populations. The node significances of the trees were evaluated using bootstrap probabilities based on 1000 replicates.

The Bayesian approach, which infers population structure and assigns individuals into clusters, was used, implemented in STRUCTURE version 2.3.4 software [33]. We performed 10 runs for each value of K (number of putative populations) from 1 to 10, and employed the Markov chain method with 100,000 iterations (burn-in) and 100,000 Markov chain Monte Carlo repetitions. The simulation was performed under the admixture model with correlated allele frequencies (default parameters). Then the most appropriate cluster number (K) was selected using the criterion from Evanno et al. [34], which is based on ΔK. To choose K and identify sets of highly similar runs in multiple independent runs at a single K value, we used CLUMPAK software [35].

3. Results

3.1. Genetic Diversity across All Populations

The total number of detected alleles for all populations at each locus ranged from five to 32, with an average value of 14.3 (Table A2). On average, over all loci, the observed heterozygosity (H_o) and gene diversity in the total population (H_T) were 0.621 and 0.691, respectively. The inbreeding coefficient (F_{IS}) ranged from −0.056 to 0.096. At all loci, the F_{IS} values deviated significantly from zero.

The ranges of genetic diversity within the populations were 5.5 to 8.7 for allele number, 4.20 to 6.55 for allelic richness, 0.496 to 0.746 for H_o, and 0.500 to 0.682 for H_e (Table 2). The average genetic diversities for Td and Th were 6.42 and 7.17 for allele number, 4.93 and 5.93 for allelic richness, 0.545 and 0.650 for H_o, and 0.551 and 0.642 for H_e, respectively. Private alleles were found in 18 populations, with a maximum value of five. There were four populations without private alleles. In all populations, observed heterozygosity was not significantly different from that expected for Hardy–Weinberg equilibrium. No evidence of significant linkage disequilibrium was detected in any of the total of 3420 tests for linkage disequilibrium between loci in the populations.

The genetic differentiation among population parameters F_{ST}, R_{ST}, G_{ST} and G'_{ST} varied among loci, ranging from 0.037 to 0.238, from 0.021 to 0.289, from 0.031 to 0.204, and from 0.098 to 0.502, respectively (Table A2). The genetic differentiation between populations over all loci (F_{ST}, R_{ST}, G_{ST} and G'_{ST}) was 0.105, 0.096, 0.088, and 0.246, respectively. All these measures were significantly different from zero at all loci.

No evidence of a recent bottleneck was found (i.e., no H_e excess compared to H_{EQ}) in the genotypes of all the populations.

3.2. Genetic Structure among Populations

All of the $F_{ST}/(1 - F_{ST})$ values between pairs of populations calculated for the three categories were significantly related to the natural logarithms of the geographic distances between them (Figure A2). For the Td vs. Td category, the intercept of the regression line (a) = −0.073 and the slope of the regression line (b) = 0.038, $R^2 = 0.567$ and $p < 0.01$. The Th vs. Th category showed a = −0.075, b = 0.022, $R^2 = 0.479$, $p < 0.000$. The species as a whole category showed a = −0.292, b = 0.071, $R^2 = 0.564$, $p < 0.000$.

The AMOVA suggested that the majority of the variation existed within population (Within Pop, 5.889, $p < 0.001$, Table 3). The results of the AMOVA and hierarchical F statistics revealed a highly significant genetic divergence between the two distribution regions (Among Region, 0.606; $F_{Region/Total} = 0.088$; $p < 0.001$). The genetic differentiation was highly significant among populations, whereas the contribution of variety was relatively low between the two distribution regions (Among Pop, 0.393; $F_{Pop/Region} = 0.062$; $p < 0.001$).

Table 3. Hierarchical analysis of genetic structure using analysis of molecular variance (AMOVA) and hierarchical F statistics.

	AMOVA			Hierarchical F Statistics			
Source of Variation	**Variance Components**	**Percentage of Variation**				**F**	**95% CI**
Among Region	0.606	8.80	*	$F_{Region/Total}$	0.088	*	0.050–0.136
Among Pop	0.393	5.71	*	$F_{Pop/Region}$	0.062	*	0.049–0.079
Within Pop	5.889	85.49	*	$F_{Ind/Pop}$	0.018	ns	0.004–0.035

$* p < 0.001$.

The neighbor-joining tree based on D_a distance reflected the geographical locations of the populations for the most part (Figure 2). However, genetic relationships between two distribution regions (Td and Th) and two intermediate populations (Minakami (MK) and Nikko (NK)) were not reliably supported because of low bootstrap values.

According to clustering analysis in STRUCTURE, the cluster number (*K*) was 2 (Figure A3). The distribution of populations for the two clusters was found to be geographically structured between the populations along the distribution regions (Figure 3). The proportion of cluster 1 was much higher in the Td distribution region, while that of cluster 2 was much higher in the Th region. Although the MK and NK populations were located in the Td region, individuals of these two populations belonged to both clusters 1 and 2.

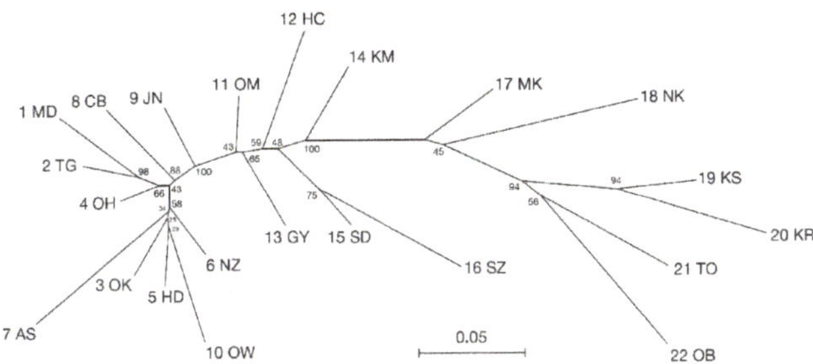

Figure 2. Neighbor-joining tree based on the D_a distance of the 22 populations. The populations from No. 1 to 16 represent var. *hondae* (Th), whereas the populations from No. 17 to 22 represent var. *dolabrata* (Td).

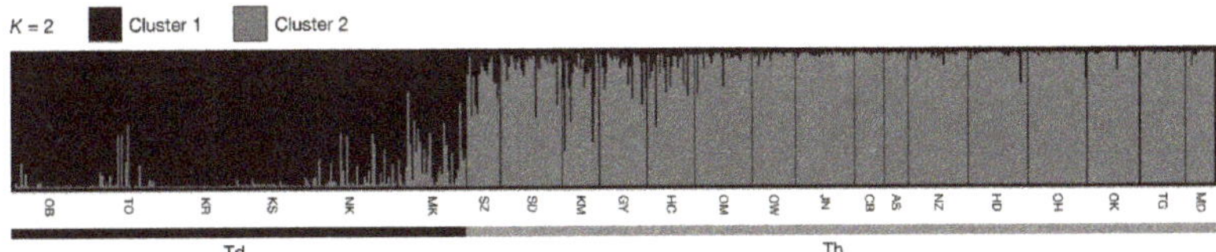

Figure 3. The proportions of cluster memberships at the individual level in the 22 genus *Thujopsis* populations based on STRUCTURE software analysis. Populations are represented from the left (southern populations) to the right (northern populations).

4. Discussion

4.1. Genetic Diversity at EST-SSR in Thujopsis

In the present study, 19 EST-SSR loci were used to estimate population genetic diversity and to investigate the genetic structure in 22 natural populations of the genus *Thujopsis*. The EST-SSR polymorphisms of the genus *Thujopsis* retained a nearly equal or slightly higher diversity compared to other Cupressaceae species (Table 2) [36–38]. In general, EST-SSR markers have a lower polymorphism than nuclear SSR markers [39,40]. Therefore, the EST-SSR marker loci used in the present study showed lower levels of polymorphism than the previous research on Th [11] and other conifers that are widely distributed in Japan [4,41,42].

A significant and relatively high value of overall population differentiation was found among the populations we examined (Table A2; F_{ST} = 0.105, $p < 0.001$; R_{ST} = 0.096, $p < 0.001$; G_{ST} = 0.088, $p < 0.001$; G'_{ST} = 0.246, $p < 0.001$). These values were clearly greater than those obtained in Ikeda et al. [12] (F_{ST} = 0.039; G'_{ST} = 0.114; EST-SSR). Katsuki et al. [43] reanalyzed and summarized the measures of population differentiation for major conifers distributed in the Japanese Archipelago, including *Cryptomeria japonica* (F_{ST}, 0.028; R_{ST}, 0.032; G'_{ST}, 0.125; SSR; [41]), *Chamaecyparis obtusa* (F_{ST} = 0.039; G_{ST} = 0.040; G'_{ST} = 0.188; SSR; [4]), *Picea alcoquiana* (F_{ST} = 0.071; G'_{ST} = 0.164; SSR; [44]), and *Picea jezoensis* (F_{ST} = 0.101; SSR; [45]). Additionally, Iwaizumi et al. [46] performed these measurements for *Pinus densiflora* populations distributed in Japan (F_{ST} = 0.013; G_{ST} = 0.013; G'_{ST} = 0.122; SSR). Compared with these species, we found a relatively moderate value for F_{ST} and a high value for G'_{ST} in the genus *Thujopsis*. In contrast to previous studies, which were conducted on a single species, measures of population differentiation in the genus *Thujopsis* may be higher because the study populations included two varieties (Td and Th). On the other hand, higher values for population differentiation measures have been observed in species, including isolated populations, of *Picea koyamae* (F_{ST} = 0.209; R_{ST} = 0.173; G'_{ST} = 0.410; SSR; [43]), *Sciadopitys verticillata* (F_{ST} = 0.142; SSR; [47]), *Abies mariesii* (G_{ST} = 0.144; allozyme; [48]), and *Pinus pumila* (G_{ST} = 0.170; allozyme; [49]). These species have narrow, isolated distributions that could reflect restricted gene flow between populations because of habitat discontinuity [50]. Therefore, it is likely the case that the two populations were not completely genetically isolated, although we identified relatively high values for the population differentiation measures in the genus *Thujopsis*.

4.2. Comparison of Genetic Structure between Td and Th

The population relationships and genetic structures for the genus *Thujopsis* were analyzed, focusing on the relationship between the two variants. Structure analysis supported the existence of two genetic clusters related to the distribution regions (Figure 1a), i.e., the Td and Th varieties (Figure 3). These clusters were significantly differentiated based on AMOVA and hierarchical *F* statistics (Table 3). The neighbor-joining tree also supported these results, according to the high (100%) bootstrap probability of branches between the KM population (No. 14, Th) and the MK population (No. 17, Td) (Figure 2). Therefore, the two variants, Td and Th, could be defined by their provenance, in spite of the ambiguous morphological differences between these varieties.

The average values of allelic richness, H_o and H_e, were relatively lower in Td than Th. Two factors may have contributed to the decline of genetic diversity in Td. First, demographic factors, such as postglacial colonization and a history of human overexploitation, could have played a role. If the refugia of the species were restricted to the southern region, a postglacial rapid expansion to northern regions would be expected to cause a series of founding events that would lead to a loss of alleles and homozygosity [51]. Similarly, tree species that have experienced population declines due to human overexploitation may show low genetic diversity and genetic bottlenecks [52–54]. However, no significant bottlenecks were detected in the population in the present study. This suggests that the low genetic diversity exhibited by each of the Td populations was probably caused by the natural characteristics of this variety, or other factors.

Since the Japanese Archipelago extends in a narrow arc from northeast to southwest, with the various mountain ranges probably acting as physical barriers, temperate plant species have generally migrated along the Pacific side, the Sea of Japan side, or the mountain slopes of the Archipelago. Thus, plants species migrated either southwards along the coasts or to lower altitudes into refugia during glacial periods, and expanded either northwards or to higher altitudes during interglacial periods [55]. In fact, many plant species distributed in Japan exhibit genetic divergence between the Pacific side and the Sea of Japan side (e.g., *Fagus crenata*, [56]; *Kalopanax septemlobus*, [36]). Additionally, as many tree species exhibit a long generation time, it is likely that not many generations have elapsed since the initial postglacial colonization. As a result, there has been less opportunity for genetic drift, and the large size of many plant populations could fossilize the genetic structure established at the time of colonization [57]. In the case of conifers, Kimura et al. [58] identified clear genetic divergence between two and four gene pools in *Cryptomeria japonica*. Two gene pools were distributed along the Sea of Japan side and along the Pacific Ocean side, while four gene pools suggested the potential of northern cryptic refugia and/or the potential of admixture events from several refugia between populations in the northern Tohoku district and an isolated gene pool on Yakushima Island (south of Kyushu district). As an example of fossilized genetic structures, Tsuda and Ide [59] suggested that the populations of *Betula maximowicziana* could be divided into a southern group (Central Honshu island) and a northern group (Hokkaido and Tohoku district) that originated from different refugia. They detected significant bottlenecks that may have been caused by processes of postglacial colonization and the species' characteristics and/or life history as a long-lived pioneer tree species. However, the present study indicates that the relatively high genetic differentiation of the genus *Thujopsis* did not fit these frequent patterns of genetic differentiation. The Td and Th varieties were distributed in the southern and northern parts of the Japanese Archipelago. No evidence of a genetic structure between the Sea of Japan side and Pacific Ocean side was found in this study. Th has similar diversity among populations, with relatively uniform values of allelic richness, H_o and H_e. On the other hand, there was a tendency for allelic richness, H_o and H_e, to decrease from the Minakami and Nikko groups to the Obitani group in Td. Structure analysis, and the locations of the Minakami and Nikko populations in the neighbor-joining tree with low node significances, suggested that these two populations may contain Td and Th hybridization. In this case, the reason for the high genetic diversity in the Minakami and Nikko populations could be due to hybridization. The remaining four populations (Kiso, Kuraiyama, Toyo-oka, and Obitani) most likely represented pure Td populations, but we found no evidence of refugia in any of these.

Comparing the current distribution of Th with that of *Cr. japonica*, Aoyama [60] indicated that Th is more drought tolerant and can survive in colder climates. These physiological characteristics may have allowed Th to form refugia in southern Hokkaido and the areas below 500 m elevation in the Tohoku district during the last glacial period [60]. In the case of Th, refugia may have been scattered throughout the Tohoku district, which is the approximate current distribution range of Th. The relatively uniform values of allelic richness, H_o and H_e, among the Th populations could be explained by this hypothesis. However, additional populations must be examined to fully understand the characteristics of Td.

4.3. Contributions of the Breeding Program for the Genus Thujopsis

Both varieties, Td and Th, are essential elements of natural forests in Japan, and logs from natural forests have historically been used for timber and other wooden materials. However, at present, Th is more important than Td in forestry. Plantations of Th are active mainly in Aomori, Niigata, and Ishikawa Prefectures (see Table 1) [10]. In Aomori, the demand for Th has recently increased. As a result, seed orchards were established in the beginning of 2003 [61]. In Niigata, plus trees on Sado Island were selected in 1989, and a seed orchard was established on Honshu island in 2009 [62]. Plantations in Ishikawa were established using rooted cuttings or saplings created by layering, and the majority of trees in these plantations are clones of 14 Th cutting/saplings [10]. In the Niigata and Ishikawa plantations, particular attention to interactions between natural populations through pollen flow and hybridization with Td is required. Th is found on Sado Island, and both Td and Th are distributed in the closest area on Honshu island (Figure 1). Similarly, the Suzu (SZ) population in Ishikawa, the present study site, is classified as Th; however, Td is distributed in neighboring areas. This study showed that these varieties are genetically different, and there is a risk that the seed orchard in Niigata may unintentionally produce seeds derived from hybridization between the variants. Furthermore, Th plantations in Ishikawa might risk introducing pollen of different variants into the surrounding natural Td forests. Since the breeding program of Th in Niigata and Ishikawa started relatively recently, the results of the present study could be useful for ongoing and future breeding programs, especially for the proper development and maintenance of seed orchards.

The results of the present study suggest that Td and Th can be distinguished by EST-SSR. As mentioned previously, distinguishing among varieties may be an important task in the future for breeding within the genus, *Thujopsis*. Several loci containing alleles with characteristic frequencies for Td and Th, respectively, were identified (Table A3). A combination of four loci, Tdest24, 39, 42, and 56, maintain a 100% bootstrap probability between varieties when constructing a neighbor-joining tree. Therefore, an analysis of these four loci is sufficient for the identification of the varieties.

5. Conclusions

Evidence from EST-SSR markers suggested that the two variants of the genus *Thujopsis*, *Thujopsis dolabrata* (L.f.) Siebold et Zucc. var. *dolabrata* (Td) and *Thujopsis dolabrata* (L.f.) Siebold et Zucc. var. *hondae* Makino (Th), were clearly distinct, as assessed using structure analysis and a neighbor-joining tree. Using these techniques, the two varieties could be defined by their provenance, in spite of the ambiguous morphological differences between them. The relatively uniform values of genetic diversity among the Th populations suggest that refugia of Th may have been scattered throughout the Tohoku district. On the other hand, there was a tendency for genetic diversity to decrease from central to southern Honshu island in Td. Structure analysis and the neighbor-joining tree suggested hybridization in the contact zone between the two varieties. More detailed studies of the genetic structure of Td will be needed in the future.

Author Contributions: Study design and funding acquisition, K.M. and K.T.; needle sampling and fieldwork, M.I., Y.H., K.M., and K.T.; molecular and data analyses, M.I., and Y.H.; writing—original draft preparation, M.I.; writing—review and editing, M.I., Y.H., K.M., and K.T.; supervision and project administration, K.T. All authors have read and agreed to the published version of the manuscript.

Funding: This research received no external funding.

Acknowledgments: We would like to express our deepest gratitude to Miyako Sato, who was the first to summarize the research for this paper and provide draft results. Our heartfelt appreciation goes to Satomi Akiyama, whose enormous experimental support and encouraging words were invaluable. We would like to thank Seishiro Taki for his assistance in the fieldwork and for compiling information on sampling locations. Permission to use photographs of Td and Th used in Figure A1 in Appendix A was provided by the Forestry and Forest Products Research Institute (FFPRI) Database of Japanese Woods (http://db.ffpri.affrc.go.jp/WoodDB/JWDB-E/home.php). We are also indebted to the anonymous reviewers, who gave us invaluable comments.

Conflicts of Interest: The authors indicated no conflicts of interest.

Appendix A

Figure A1. Typical morphology of (**a**) *Thujopsis dolabrata* (L.f.) Siebold et Zucc. var. *dolabrata* (Td) and (**b**) *Thujopsis dolabrata* (L.f.) Siebold et Zucc. var. *hondae* Makino (Th). Td (**a**) have horned cones, and Th (**b**) have denser needles and rounder cones. Photographs are cited from Forestry and Forest Products Research Institute (FFPRI) Database of Japanese Woods ((**a**) http://db.ffpri.affrc.go.jp/WoodDB/JWDB-E/detailA_coll.php?-action=browse&-recid=262420, and (**b**) http://db.ffpri.affrc.go.jp/WoodDB/JWDB-E/detailA_coll.php?-action=browse&-recid=265188).

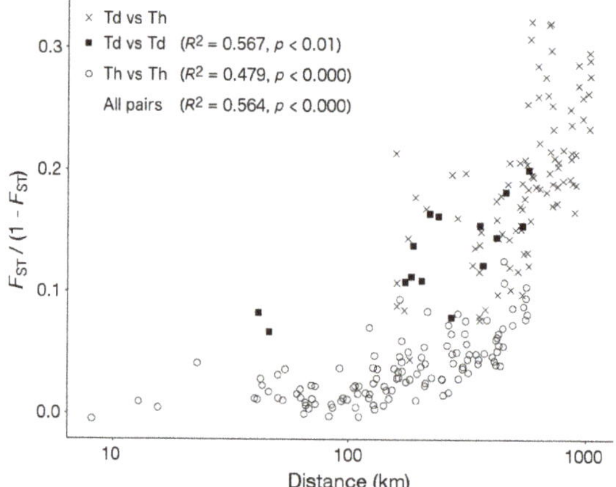

Figure A2. Relationships between pairwise genetic distances, $F_{ST}/(1 - F_{ST})$, and the geographic distance separating 22 *Thujopsis* populations.

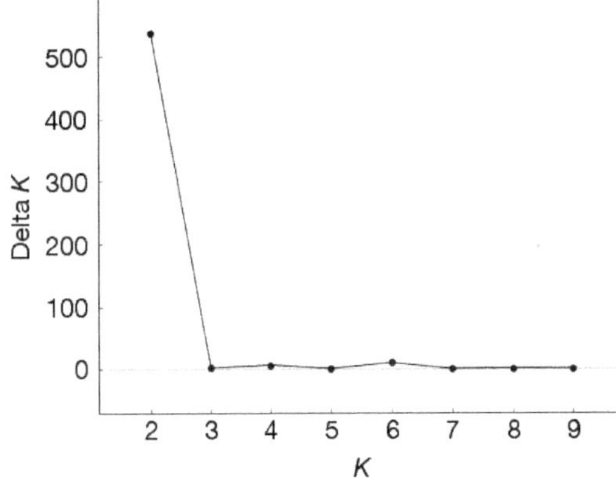

Figure A3. Relationships between the number of clusters (K) and the rate of the change in lnP(X|K) (Delta *K*), based on STRUCTURE analysis.

Table A1. Characteristics of 19 EST-SSR loci for use in the genus *Thujopsis*.

Locus	Group	Repeat Motif		Primer Sequence (5′–3′)	Allele Size Range (bp)
Tdest1	5	(CT)11	F:	GCCTCCCTCGCGCCATCAG GATTTTCTGACAGGCTTTGTTCTC	137–173
			R:	GTTTCTTAATTCCCAAGAGTGCTTATGAGTTC	
Tdest3	1	(AT)11	F:	GCCTCCCTCGCGCCATCAG CGGCCCAGGTTTCTGTACTC	155–184
			R:	GTTTCTTGCCCATTAAAGTCGGGTATTG	
Tdest11	3	(AT)12	F:	GCCTCCCTCGCGCCATCAGTGGGATACATACTGCATTTGTTAGG	136–161
			R:	GTTTCTTCTCCCCAAGCAAGTCACCAC	
Tdest14	4	(AG)12	F:	GCCTCCCTCGCGCCATCAGCAGTAGACAATTTCTGCAAATCACC	152–190
			R:	GTTTCTTTCCCTTTTGTTGGCATTATAGG	
Tdest17	3	(AG)12	F:	GCCTCCCTCGCGCCATCAGGCTTTTGATGTCCGCTATATCCTC	160–176
			R:	GTTTCTTGGAGATTCCAATGTTTGTCATGC	
Tdest21	3	(AG)13	F:	GCCTCCCTCGCGCCATCAGGTCCATCCATTCTCACTCCAAAG	228–292
			R:	GTTTCTTAGCAGACCCTATTTCACAGCATC	
Tdest24	4	(AT)15	F:	GCCTCCCTCGCGCCATCAGATACCATACAGCTTTCAGCCAG	239–266
			R:	GTTTCTTGCAGAACAAACGAATCAATGAGAG	
Tdest29	3	(AC)16	F:	GCCTCCCTCGCGCCATCAGAAACGACTCTGCTGGATTTCAC	215–243
			R:	GTTTCTTTTCCGCTCTTGATTTTCTCTCC	
Tdest35	2	(CT)15	F:	GCCTCCCTCGCGCCATCAGAAGCTATTGACCCTTCTCAGGATAC	191–227
			R:	GTTTCTTCCATGTTGAATTGTTCCCTTTC	
Tdest37	5	(ATC)9	F:	GCCTCCCTCGCGCCATCAGCCAAGCGACAGAAAACCATTC	158–175
			R:	GTTTCTTTCAGTCTCTTCCTCCTCCTCCTC	
Tdest38	1	(ACC)9	F:	GCCTCCCTCGCGCCATCAGTGACCATTCCTCCTCCTCCTC	114–137
			R:	GTTTCTTCATGTTTGCAGTTGAGAGAAGACC	
Tdest39	1	(GCT)9	F:	GCCTCCCTCGCGCCATCAGGCAGCACAGGAGAAGAAAGATG	153–186
			R:	GTTTCTTACAACAGCCACAACGTGTCC	
Tdest42	1	(ACC)9	F:	GCCTCCCTCGCGCCATCAGCTCCCTATCCCAACACCAACAC	225–258
			R:	GTTTCTTTGCCTACCTATCCTTCTTCTTCTCC	
Tdest43	2	(CGG)9	F:	GCCTCCCTCGCGCCATCAGGGTCCAATGCAGGTAATACAAGAAG	134–167
			R:	GTTTCTTTCCCCGCCAAGATACTCAAC	
Tdest45	5	(GGT)12	F:	GCCTCCCTCGCGCCATCAGTGAGGGTGGTGAGACAATTC	208–235
			R:	GTTTCTTCAAGATTTGGAACTCCTGCAAC	
Tdest49	2	(GAT)10	F:	GCCTCCCTCGCGCCATCAGGTGCCCTCAAAGTTACAGCAGTC	221–248
			R:	GTTTCTTGCAATCACCTCATCCTCACTTC	
Tdest53	4	(CTT)13	F:	GCCTCCCTCGCGCCATCAGCCAAAGCCCTTCCAGTAACATC	241–305
			R:	GTTTCTTGATGGAATGAGTGAATCTCAGGAAC	
Tdset56	2	(AAG)9	F:	GCCTCCCTCGCGCCATCAGCATTGCCCTTTGGAATATAGGATC	142–167
			R:	GTTTCTTGTTGCCCATCTGCTCTTCTTC	
Tdest58	4	(AAG)13	F:	GCCTCCCTCGCGCCATCAGCTGAACGGCGCCCTAATCTC	151–188
			R:	GTTTCTTGCCCACTCCTCAAATCCAAC	

Groups of loci that were mixed when amplified PCR products of individuals were separated by capillary electrophoresis (Group).

Table A2. Genetic diversity measures estimated at 19 microsatellite loci.

Locus	TA	H_o	H_T	F_{IS}		F_{ST}		R_{ST}		G_{ST}		G'_{ST}	
Tdest1	19	0.889	0.904	−0.037	*	0.065	**	0.097	**	0.052	**	0.380	**
Tdest3	13	0.503	0.606	0.064	**	0.135	**	0.058	**	0.113	**	0.251	**
Tdest11	14	0.789	0.873	0.032	**	0.078	**	0.289	**	0.066	**	0.371	**
Tdest14	20	0.864	0.904	0.014	*	0.037	**	0.024	*	0.031	**	0.260	**
Tdest17	10	0.624	0.781	0.054	**	0.175	**	0.069	**	0.157	**	0.474	**
Tdest21	32	0.891	0.927	0.004	**	0.041	**	0.021	*	0.035	**	0.349	**
Tdest24	16	0.710	0.767	0.023	**	0.054	**	0.205	**	0.053	**	0.199	**
Tdest29	8	0.358	0.409	0.025	**	0.118	**	0.153	**	0.101	**	0.163	**
Tdest35	32	0.852	0.935	0.040	**	0.059	**	0.095	**	0.051	**	0.470	**
Tdest37	5	0.398	0.484	0.018	**	0.207	**	0.205	**	0.164	**	0.281	**
Tdest38	10	0.655	0.692	−0.034	**	0.108	**	0.104	**	0.085	**	0.239	**
Tdest39	7	0.456	0.490	−0.056	**	0.143	**	0.173	**	0.118	**	0.212	**
Tdest42	13	0.565	0.732	0.031	**	0.238	**	0.054	**	0.204	**	0.502	**
Tdest43	14	0.714	0.753	−0.011	**	0.077	**	0.121	**	0.062	**	0.218	**
Tdest45	9	0.462	0.581	0.096	**	0.144	**	0.136	**	0.120	**	0.252	**
Tdest49	8	0.228	0.272	0.051	**	0.121	**	0.125	**	0.116	**	0.155	**
Tdest53	21	0.863	0.888	−0.025	**	0.064	**	0.099	**	0.052	**	0.342	**
Tdest56	8	0.626	0.730	−0.010	**	0.183	**	0.082	**	0.152	**	0.411	**
Tdest58	12	0.361	0.402	0.044	**	0.065	**	0.068	**	0.060	**	0.098	**
average	14.3	0.621	0.691	0.014	**	0.105	**	0.096	**	0.088	**	0.246	**

Total number of detected alleles (TA); average observed heterozygosity (H_o); total gene diversity (H_T); Wright's inbreeding coefficient (F_{IS}) and significant deviations from Hardy–Weinberg expectations were tested; Weir and Cockerham's F_{ST} (F_{ST}); relative genetic differentiation among populations defined under the stepwise mutation model (R_{ST}); genetic differentiation coefficient (G_{ST}); standardized measure of relative genetic differentiation among populations (G'_{ST}); *, $p < 0.01$; **, $p < 0.001$; p values indicate the significance of deviations of F_{IS}, F_{ST}, R_{ST}, G_{ST}, and G'_{ST} from zero, evaluated by permutation tests.

Table A3. Recommended set of markers for distinguishing Td and Th and their allele characteristics.

		Allele Frequencies across All Populations			
Locus	Allele Size	Td	Th	Td/Th	Th/Td
Tdest24	N	230	379		
	239	0.152	0.003	57.6	0.0
	242	0.000	0.001		
	243	0.030	0.000		
	245	0.217	0.090	2.4	0.4
	247	0.424	0.429	1.0	1.0
	249	0.072	0.110	0.7	1.5
	251	0.057	0.173	0.3	3.1
	253	0.004	0.074	0.1	17.0
	255	0.002	0.051	0.0	23.7
	256	0.002	0.008	0.3	3.6
	257	0.026	0.028	0.9	1.1
	259	0.002	0.003	0.8	1.2
	261	0.009	0.004	2.2	0.5
	263	0.002	0.022	0.1	10.3
	265	0.000	0.004		
	266	0.000	0.001		
Tdest39	N	230	379		
	153	0.952	0.575	1.7	0.6
	156	0.000	0.001		
	159	0.039	0.243	0.2	6.2
	162	0.000	0.003		
	165	0.009	0.170	0.1	19.6
	168	0.000	0.003		
	186	0.000	0.005		

Table A3. *Cont.*

Allele Frequencies across All Populations					
Locus	Allele Size	Td	Th	Td/Th	Th/Td
Tdest42	N	230	379		
	225	0.028	0.001	21.4	0.0
	228	0.000	0.003		
	231	0.004	0.018	0.2	4.2
	234	0.022	0.000		
	237	0.041	0.215	0.2	5.2
	240	0.015	0.001	11.5	0.1
	243	0.498	0.042	11.8	0.1
	244	0.102	0.000		
	246	0.048	0.600	0.1	12.5
	249	0.185	0.062	3.0	0.3
	252	0.007	0.041	0.2	6.3
	255	0.050	0.007	7.6	0.1
	258	0.000	0.009		
Tdest56	N	230	379		
	142	0.002	0.000		
	147	0.000	0.001		
	152	0.370	0.080	4.6	0.2
	153	0.011	0.000		
	155	0.022	0.463	0.0	21.3
	161	0.478	0.230	2.1	0.5
	164	0.117	0.223	0.5	1.9
	167	0.000	0.003		

Sample size (N); *T. dolabrata* (L.f.) Siebold et Zucc. var. *dolabrata* (Td); *Thujopsis dolabrata* (L.f.) Siebold et Zucc. var. *hondae* Makino (Th); ratio of allele frequency, Td divided by Th (Td/Th); ratio of allele frequency, Th divided by Td (Th/Td).

References

1. Neale, D.B.; Wheeler, N.C. *The Conifers: Genomes, Variation and Evolution*; Springer Science+Business Media: New York, NY, USA, 2019; ISBN 978-3-319-46806-8.
2. Rice, K.J.; Emery, N.C. Managing microevolution: Restoration in the face of global change. *Front. Ecol. Environ.* **2003**, *1*, 469–478. [CrossRef]
3. Grassi, F.; Labra, M.; Imazio, S.; Rubio, R.O.; Failla, O.; Scienza, A.; Sala, F. Phylogeographical structure and conservation genetics of wild grapevine. *Conserv. Genet.* **2006**, *7*, 837–845. [CrossRef]
4. Matsumoto, A.; Uchida, K.; Taguchi, Y.; Tani, N.; Tsumura, Y. Genetic diversity and structure of natural fragmented *Chamaecyparis obtusa* populations as revealed by microsatellite markers. *J. Plant Res.* **2010**, *123*, 689–699. [CrossRef]
5. Rajora, O.P.; Mosseler, A. Challenges and opportunities for conservation of forest genetic resources. *Euphytica* **2001**, *118*, 197–212. [CrossRef]
6. Cavers, S.; Navarro, C.; Lowe, A.J. Targeting genetic resource conservation in widespread species: A case study of *Cedrela odorata* L. *For. Ecol. Manag.* **2004**, *197*, 285–294. [CrossRef]
7. Yang, Z.Y.; Ran, J.H.; Wang, X.Q. Three genome-based phylogeny of *Cupressaceae* s.l.: Further evidence for the evolution of gymnosperms and southern hemisphere biogeography. *Mol. Phylogenet. Evol.* **2012**, *64*, 452–470. [CrossRef]
8. Hayashi, Y. *Taxonomical and Phytogeographical Study of Japanese Conifers*; Norin Shuppan: Tokyo, Japan, 1960. (In Japanese)
9. Kurata, S. *Illustrated Important Forest Trees of Japan*; Chikyu Shuppan Co., LTD: Tokyo, Japan, 1964. (In Japanese)
10. Ikeda, T.; Tokuda, N.; Tomaru, N. Identification of *Thujopsis dolabrata* var. *hondae* clones and their distribution across plantations in Ishikawa Prefecture. *J. For. Res.* **2018**, *23*, 105–111. [CrossRef]
11. Higuchi, Y.; Matsumoto, A.; Moriguchi, Y.; Mishima, K.; Tanaka, K.; Yada, Y.; Takata, K.; Watanabe, A.; Hirao, T.; Tsumura, Y.; et al. Genetic diversity and structure using microsatellite markers in natural and breeding populations of *Thujopsis dolabrata* var. *hondae*. *J. Jpn. For. Soc.* **2012**, *94*, 247–251. (In Japanese) [CrossRef]

12. Ikeda, T.; Mishima, K.; Takata, K.; Tomaru, N. The origin and genetic variability of vegetatively propagated clones identified from old planted trees and plantations of *Thujopsis dolabrata* var. *hondae* in Ishikawa Prefecture, Japan. *Tree Genet. Genomes* **2019**, *15*, 80. [CrossRef]

13. Murray, M.G.; Thompson, W.F. Rapid isolation of high molecular weight plant DNA. *Nucleic Acids Res.* **1980**, *8*, 4321–4326. [CrossRef]

14. Sato, M.; Hasegawa, Y.; Mishima, K.; Takata, K. Isolation and characterization of 22 EST-SSR markers for the Genus *Thujopsis* (*Cupressaceae*). *Appl. Plant Sci.* **2015**, *3*, 1400101. [CrossRef]

15. Blacket, M.J.; Robin, C.; Good, R.T.; Lee, S.F.; Miller, A.D. Universal primers for fluorescent labelling of PCR fragments—An efficient and cost-effective approach to genotyping by fluorescence. *Mol. Ecol. Resour.* **2012**, *12*, 456–463. [CrossRef]

16. Goudet, J. A Program to Estimate and Test Population Genetics Parameters. Available online: https://www2.unil.ch/popgen/softwares/fstat.htm (accessed on 25 May 2020).

17. Peakall, R.; Smouse, P.E. GenAlEx 6.5: Genetic analysis in Excel. Population genetic software for teaching and research—An update. *Bioinformatics* **2012**, *28*, 2537–2539. [CrossRef]

18. Weir, B.S.; Cockerham, C.C. Estimating F-statistics for the analysis of population structure. *Evolution* **1984**, *38*, 1358–1370.

19. Slatkin, M. A measure of population subdivision based on microsatellite allele frequencies. *Genetics* **1995**, *139*, 457–462.

20. Hardy, O.J.; Vekemans, X. SPAGeDi: A versatile computer program to analyse spatial genetic structure at the individual or population levels. *Mol. Ecol. Notes* **2002**, *2*, 618–620. [CrossRef]

21. Hedrick, P.W. A standardized genetic differentiation measure. *Evolution* **2005**, *59*, 1633–1638. [CrossRef]

22. Meirmans, P.G.; Hedrick, P.W. Assessing population structure: F_{ST} and related measures. *Mol. Ecol. Resour.* **2011**, *11*, 5–18. [CrossRef]

23. Piry, S.; Luikart, G.; Cornuet, J.M. BOTTLENECK: A computer program for detecting recent reductions in the effective size using allele frequency data. *J. Hered.* **1999**, *90*, 502–503. [CrossRef]

24. Rousset, F. Genetic differentiation and estimation of gene flow from F-statistics under isolation by distance. *Genetics* **1997**, *145*, 1219–1228.

25. Excoffier, L.; Lischer, H.E.L. Arlequin suite ver 3.5: A new series of programs to perform population genetics analyses under Linux and Windows. *Mol. Ecol. Resour.* **2010**, *10*, 564–567. [CrossRef] [PubMed]

26. Goudet, J. HierFstat, a package for r to compute and test hierarchical F-statistics. *Mol. Ecol. Notes* **2005**, *5*, 184–186. [CrossRef]

27. De Meeûs, T.; Goudet, J. A step-by-step tutorial to use HierFstat to analyse populations hierarchically structured at multiple levels. *Infect. Genet. Evol.* **2007**, *7*, 731–735. [CrossRef] [PubMed]

28. R Core Team. *R: A Language and Environment for Statistical Computing*; R Foundation for Statistical Computing: Vienna, Austria, 2019.

29. Saitou, N.; Nei, M. The neighbor-joining method: A new method for reconstructing phylogenetic trees. *Mol. Biol. Evol.* **1987**, *4*, 406–425. [PubMed]

30. Takezaki, N.; Nei, M.; Tamura, K. POPTREE2: Software for constructing population trees from allele frequency data and computing other population statistics with windows interface. *Mol. Biol. Evol.* **2010**, *27*, 747–752. [CrossRef]

31. Takezaki, N.; Nei, M.; Tamura, K. POPTREE2_web. Available online: http://www.kms.ac.jp/~{}genomelb/takezaki/poptree2/poptree2_web.html (accessed on 12 June 2020).

32. Nei, M.; Tajima, F.; Tateno, Y. Accuracy of estimated phylogenetic trees from molecular data. II. Gene frequency data. *J. Mol. Evol.* **1983**, *19*, 153–170. [CrossRef]

33. Pritchard, J.K.; Stephens, M.; Donnelly, P. Inference of population structure using multilocus genotype data. *Genetics* **2000**, *155*, 945–959.

34. Evanno, G.; Regnaut, S.; Goudet, J. Detecting the number of clusters of individuals using the software structure: A simulation study. *Mol. Ecol.* **2005**, *14*, 2611–2620. [CrossRef]

35. Kopelman, N.M.; Mayzel, J.; Jakobsson, M.; Rosenberg, N.A.; Mayrose, I. CLUMPAK: A program for identifying clustering modes and packaging population structure inferences across K. *Mol. Ecol. Resour.* **2015**, *15*, 1179–1191. [CrossRef]

36. Sakaguchi, S.; Takeuchi, Y.; Yamasaki, M.; Sakurai, S.; Isagi, Y. Lineage admixture during postglacial range expansion is responsible for the increased gene diversity of *Kalopanax septemlobus* in a recently colonised territory. *Heredity* **2011**, *107*, 338–348. [CrossRef]

37. Ding, M.; Meng, K.; Fan, Q.; Tan, W.; Liao, W.; Chen, S. Development and validation of EST-SSR markers for *Fokienia hodginsii* (*Cupressaceae*). *Appl. Plant Sci.* **2017**, *5*, 1600152. [CrossRef] [PubMed]

38. Huang, C.J.; Chu, F.H.; Liu, S.C.; Tseng, Y.H.; Huang, Y.S.; Ma, L.T.; Wang, C.T.; You, Y.T.; Hsu, S.Y.; Hsieh, H.C.; et al. Isolation and characterization of SSR and EST-SSR loci in *Chamaecyparis formosensis* (*Cupressaceae*). *Appl. Plant Sci.* **2018**, *6*, e01175. [CrossRef] [PubMed]

39. Leigh, F.; Lea, V.; Law, J.; Wolters, P.; Powell, W.; Donini, P. Assessment of EST- and genomic microsatellite markers for variety discrimination and genetic diversity studies in wheat. *Euphytica* **2003**, *133*, 359–366. [CrossRef]

40. Kalia, R.K.; Rai, M.K.; Kalia, S.; Singh, R.; Dhawan, A.K. Microsatellite markers: An overview of the recent progress in plants. *Euphytica* **2011**, *177*, 309–334. [CrossRef]

41. Takahashi, T.; Tani, N.; Taira, H.; Tsumura, Y. Microsatellite markers reveal high allelic variation in natural populations of *Cryptomeria japonica* near refugial areas of the last glacial period. *J. Plant Res.* **2005**, *118*, 83–90. [CrossRef] [PubMed]

42. Iwaizumi, M.G.; Takahashi, M.; Isoda, K.; Austerlitz, F. Consecutive five-year analysis of paternal and maternal gene flow and contributions of gametic heterogeneities to overall genetic composition of dispersed seeds of *Pinus densiflora* (*Pinaceae*). *Am. J. Bot.* **2013**, *100*, 1896–1904. [CrossRef]

43. Katsuki, T.; Shimada, K.; Yoshimaru, H. Process to extinction and genetic structure of a threatened Japanese conifer species, *Picea koyamae*. *J. For. Res.* **2011**, *16*, 292–301. [CrossRef]

44. Aizawa, M.; Yoshimaru, H.; Katsuki, T.; Kaji, M. Imprint of post-glacial history in a narrowly distributed endemic spruce, *Picea alcoquiana*, in central Japan observed in nuclear microsatellites and organelle DNA markers. *J. Biogeogr.* **2008**, *35*, 1295–1307. [CrossRef]

45. Aizawa, M.; Yoshimaru, H.; Saito, H.; Katsuki, T.; Kawahara, T.; Kitamura, K.; Shi, F.; Sabirov, R.; Kaji, M. Range-wide genetic structure in a north-east Asian spruce (*Picea jezoensis*) determined using nuclear microsatellite markers. *J. Biogeogr.* **2009**, *36*, 996–1007. [CrossRef]

46. Iwaizumi, M.G.; Tsuda, Y.; Ohtani, M.; Tsumura, Y.; Takahashi, M. Recent distribution changes affect geographic clines in genetic diversity and structure of *Pinus densiflora* natural populations in Japan. *For. Ecol. Manag.* **2013**, *304*, 407–416. [CrossRef]

47. Kawase, D.; Tsumura, Y.; Tomaru, N.; Seo, A.; Yumoto, T. Genetic structure of an endemic Japanese conifer, *Sciadopitys verticillata* (*Sciadopityaceae*), by using microsatellite markers. *J. Hered.* **2010**, *101*, 292–297. [CrossRef]

48. Suyama, Y.; Tsumura, Y.; Ohba, K. A cline of allozyme variation in *Abies mariesii*. *J. Plant Res.* **1997**, *110*, 219–226. [CrossRef]

49. Tani, N.; Tomaru, N.; Araki, M.; Ohba, K. Genetic diversity and differentiation in populations of Japanese stone pine (*Pinus pumila*) in Japan. *Can. J. For. Res.* **1996**, *26*, 1454–1462. [CrossRef]

50. Hamrick, J.L.; Godt, M.J.W.; Sherman-Broyles, S.L. Factors influencing levels of genetic diversity in woody plant species. *New For.* **1992**, *6*, 95–124. [CrossRef]

51. Hewitt, G.M. Post-glacial re-colonization of European biota. *Biol. J. Linn. Soc.* **1999**, *68*, 87–112. [CrossRef]

52. Baucom, R.S.; Estill, J.C.; Cruzan, M.B. The effect of deforestation on the genetic diversity and structure in *Acer saccharum* (Marsh): Evidence for the loss and restructuring of genetic variation in a natural system. *Conserv. Genet.* **2005**, *6*, 39–50. [CrossRef]

53. O'Connell, L.M.; Mosseler, A.; Rajora, O.P. Impacts of forest fragmentation on the mating system and genetic diversity of white spruce (*Picea glauca*) at the landscape level. *Heredity* **2006**, *97*, 418–426. [CrossRef]

54. Fady, B.; Lefèvre, F.; Vendramin, G.G.; Ambert, A.; Régnier, C.; Bariteau, M. Genetic consequences of past climate and human impact on eastern Mediterranean *Cedrus libani* forests. Implications for their conservation. *Conserv. Genet.* **2008**, *9*, 85–95. [CrossRef]

55. Tsukada, M. The history of *Cryptomeria japonica* (sugi): The last 15000 years. *Kagaku Sci.* **1980**, *50*, 538–546. (In Japanese)

56. Hiraoka, K.; Tomaru, N. Genetic divergence in nuclear genomes between populations of *Fagus crenata* along the Japan Sea and Pacific sides of Japan. *J. Plant Res.* **2009**, *122*, 269–282. [CrossRef]

57. Hu, F.S.; Hampe, A.; Petit, R.J. Paleoecology meets genetics: Deciphering past vegetational dynamics. *Front. Ecol. Environ.* **2009**, *7*, 371–379. [CrossRef]
58. Kimura, M.K.; Uchiyama, K.; Nakao, K.; Moriguchi, Y.; Jose-Maldia, L.S.; Tsumura, Y. Evidence for cryptic northern refugia in the last glacial period in *Criptomeria japonica*. *Ann. Bot.* **2014**, *114*, 1687–1700. [CrossRef]
59. Tsuda, Y.; Ide, Y. Wide-range analysis of genetic structure of *Betula maximowicziana*, a long-lived pioneer tree species and noble hardwood in the cool temperate zone of Japan. *Mol. Ecol.* **2005**, *14*, 3929–3941. [CrossRef]
60. Aoyama, T. The properties of cold environment for *Thujopsis dolabrata* var. *hondai* and *Criptomeria japonica* in the northeastern Japan. *Q. J. Geogr.* **1995**, *47*, 91–102. (In Japanese) [CrossRef]
61. Tanaka, K. Forest tree breeding project in Aomori Prefecture to promote excellent seeds and seedlings. *For. Genet. Tree Breed.* **2013**, *2*, 139–141. (In Japanese)
62. Higuchi, Y.; Itou, S. Test on breeding of Hiba plus trees (I) seed production characteristics of Hiba plus trees selected from Sado Island. *Bull. Niigata Prefect. For. Res. Inst.* **2012**, *53*, 1–6. (In Japanese)

Article

Population Genetic Diversity and Structure of Ancient Tree Populations of *Cryptomeria japonica* var. *sinensis* Based on RAD-seq Data

Mengying Cai [1], Yafeng Wen [2], Kentaro Uchiyama [3], Yunosuke Onuma [1] and Yoshihiko Tsumura [4,*]

1 Graduate School of Life and Environmental Science, University of Tsukuba, Tsukuba 305-8572, Japan; mengyingcai92@gmail.com (M.C.); okna8.sa@gmail.com (Y.O.)
2 College of Landscape and Architecture, Central South University of Forestry and Technology, Changsha 410004, China; t19930899@csuft.edu.cn
3 Department of Forest Molecular Genetics and Biotechnology, Forestry and Forest Products Research Institute, Tsukuba 305-8687, Japan; kruchiyama@affrc.go.jp
4 Faculty of Life and Environmental Science, University of Tsukuba, Tsukuba 305-8572, Japan
* Correspondence: tsumura.yoshihiko.ke@u.tsukuba.ac.jp; Tel.: +81-29-853-4629

Received: 4 October 2020; Accepted: 6 November 2020; Published: 12 November 2020

Abstract: Research highlights: Our study is the first to explore the genetic composition of ancient *Cryptomeria* trees across a distribution range in China. Background and objectives: *Cryptomeria japonica* var. *sinensis* is a native forest species of China; it is widely planted in the south of the country to create forests and for wood production. Unlike *Cryptomeria* in Japan, genetic Chinese *Cryptomeria* has seldom been studied, although there is ample evidence of its great ecological and economic value. Materials and methods: Because of overcutting, natural populations are rare in the wild. In this study, we investigated seven ancient tree populations to explore the genetic composition of Chinese *Cryptomeria* through ddRAD-seq technology. Results: The results reveal a lower genetic variation but higher genetic differentiation ($Ho = 0.143$, $F_{ST} = 0.1204$) than Japanese *Cryptomeria* ($Ho = 0.245$, $F_{ST} = 0.0455$). The 86% within-population variation is based on an analysis of molecular variance (AMOVA). Significant excess heterozygosity was detected in three populations and some outlier loci were found; these were considered to be the consequence of selection or chance. Structure analysis and dendrogram construction divided the seven ancient tree populations into four groups corresponding to the geographical provinces in which the populations are located, but there was no obvious correlation between genetic distance and geographic distance. A demographic history analysis conducted by a Stairway Plot showed that the effective population size of Chinese *Cryptomeria* had experienced a continuing decline from the mid-Pleistocene to the present. Our findings suggest that the strong genetic drift caused by climate fluctuation and intense anthropogenic disturbance together contributed to the current low diversity and structure. Considering the species' unfavorable conservation status, strategies are urgently required to preserve the remaining genetic resources.

Keywords: *Cryptomeria japonica* var. *sinensis*; genetic diversity; population structure; demographic history; SNP; RAD-seq; ancient tree; conservation

1. Introduction

Cryptomeria (Cupressaceae) is a relic genus that was widely distributed throughout Eurasia during the Cenozoic era [1]. Today, there is only one extant species, *Cryptomeria japonica* (Linn. f.) D. Don, which has three recognized varieties. Two Japanese varieties, var. *japonica* and var. *radicans*, are found in the moist temperate region from Aomori Prefecture to Yakushima Island on the Japanese archipelago [2]; var. *japonica* mainly occurs on the Pacific Ocean side and var. *radicans* mainly on the Sea of Japan side. These two varieties are present in 44% of all Japanese planted forests [3], and the species is known as "the national tree" of Japan. The third variety, var. *sinensis*, is limited to southern China, with a few natural occurrences in Fujian (Nanping), Jiangxi (Lushan mountain), and Zhejiang (Tianmu mountain) provinces. However, these wild forests are hard to distinguish from an enormous number of artificial stands [4]. Planted forests in China are not as common as in Japan, but still play an important role in forestry in the south of the country. This species has been long cultivated in China, and the earliest historical document can be traced back to 1279, referring to the Tianmu mountains area. Some ancient trees are still well preserved in villages such as Fengshui forest, and they are supposed to bring fortune and happiness. Since the founding of modern China, the government has launched a series of greening campaigns. Because var. *sinensis* is a common and popular species, it has been widely planted throughout southern China for afforestation and to exploit as timber in the future. Recently, studies have revealed that Chinese *Cryptomeria* (*Cryptomeria japonica* var. *sinensis* Miquel) forests have great ecological benefits with respect to soil properties, water infiltration, and biodiversity, and also have substantial economic benefits in terms of wood production. However, we know little about its population genetics.

Knowledge of population genetic diversity and structure is of fundamental importance for conifer conservation and breeding programs [5]. Chen et al. [6] investigated the demographic structure of Chinese *Cryptomeria* on Tianmu mountain using microsatellites markers. This area contains the most famous and largest ancient tree population. Luo et al. [7] examined the genetic diversity of 96 clones from 12 provenances in seed orchards, using 26 polymorphic microsatellite loci. However, a large-scale study on natural or artificial resources of Chinese *Cryptomeria* has yet to be conducted.

Single-nucleotide polymorphisms (SNP) have proved to be the most abundant form of variation within a species at the genome level and can provide detailed insight into the genetic basis of a population [8]. Combined with Next-Generation Sequencing (NGS) technology, SNP markers are having substantial impacts on population genetics as well as plant breeding [9,10]. Among large-scale sequencing-based approaches, Restriction-site Associated DNA sequencing (RAD-seq) technology has been shown to be cost-effective for generating genome-wide markers for a large number of samples simultaneously [11–14]. This approach has great advantages, including generating a large quantity of data across the genome, having reasonable costs, needing simpler procedures for library construction, needing a short duration of experiment, having no requirements for a reference genome, and having a well-developed pipeline for data treatment and analysis [15,16].

In the present study, we employed ddRAD-seq (double digest RAD-seq) technology, based on 122 samples from seven ancient tree populations in China, to (1) evaluate the level of genetic diversity, (2) explore the genetic structure among current ancient tree populations, and (3) estimate the demographic history. We also used six natural Japanese populations to (4) compare the genetic diversity between Japanese and Chinese *Cryptomeria*. From this analysis, we hope to gain insight into the status of genetic resources of *C. japonica* var. *sinensis,* and put forward some suggestions for conservation strategies.

2. Materials and Methods

2.1. Population Sampling

We investigated the sites where natural forests supposedly occurred in China, according to previous studies [17]. Unfortunately, most forests have been exposed to severe disturbance as a result of human activities, and the species is now found in patches in villages and national forest parks. To avoid materials from unknown sources, only ancient trees with a DBH (diameter at breast height) greater than 100 cm were selected for this study. A total of 122 individuals from seven populations were collected, covering all the recorded natural forest sites. For each population, needles were collected from 10 to 22 mature trees from each population (Figure 1). The name, geographic location, altitude, and sample size for each population are listed in Table 1. We also used six natural populations of *C. japonica* from Japan, covering the whole natural distribution range and the most important forests, in order to compare the genetic diversity with Chinese populations [18] (Table S1).

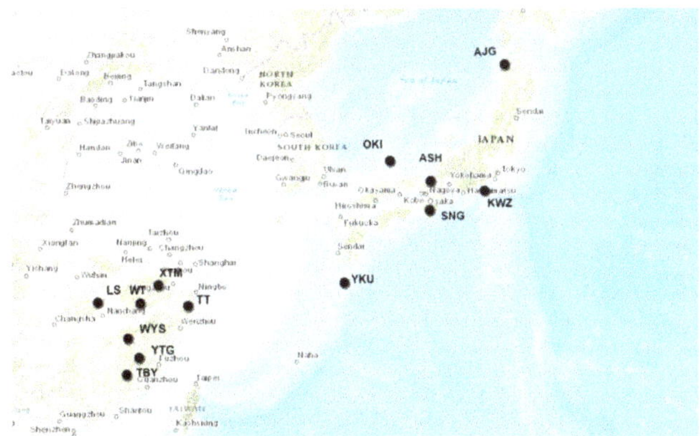

Figure 1. Sampling locations of seven populations of *C. japonica* var. *chinensis* in China and six populations of *C. japonica* in Japan.

Table 1. Geographic locations and sample size (*n*) of the seven Chinese *Cryptomeria* populations.

Populations		Longitude (E)	Latitude (N)	Altitude (m)	*n*
Province	location				
Jiangxi	Lushan (LS)	115.5801	29.3259	895 m	22
Anhui	Huangshan (WT)	118.1600	29.2743	436 m	17
Fujian	Tianbaoyan (TBY)	117.3044	25.5756	1132 m	17
	Nanping (YTG)	118.0048	26.4418	846 m	10
	Wuyishan (WYS)	117.4136	27.4535	892 m	18
Zhejiang	Tiantaishan (TT)	121.0541	29.1458	897 m	17
	Tianmushan (XTM)	119.2608	30.2017	1089 m	21
Total					122

2.2. DNA Extraction and RAD Sequencing

The total genomic DNA was extracted from fresh needles using a modified CTAB (cetyltrimethylammonium bromide) method [19]. Purified DNA was digested with *Pst*I and *Sph*I, ligated with Y-shaped adaptors, and amplified by PCR with KAPA HiFi polymerase (Kapa Biosystems, Wilmington, USA). After PCR amplification with adapter-specific primer pairs (Access Array Barcode Library for Illumina, Fluidigm, South San Francisco, USA), an equal amount of DNA from each sample was mixed and size-selected with the BluePippin agarose gel (Sage Science, Beverly, MA, USA). Approximately 450 bp library fragments were retrieved. Further details of the library preparation method are given by Ueno et al. [20]. The quality of the library was checked using a 2100 Bioanalyzer with a high-sensitivity DNA chip (Agilent Technologies, Waldbronn, Germany) and finally sequenced using an Illumina Hi-Seq X to generate paired-end reads 150 bp long.

2.3. SNP Calling and Filtering

SNPs were called by dDocent (version 2.17) [21], which is a pipeline containing a series of statistical tools. Because no reference genome was available for *Cryptomeria japonica*, a reference was constructed using the dDocent de novo assembly and optimized utilizing the reference optimization steps provided on the dDocent assembly tutorial. We followed the default settings of dDocent for mapping and SNP calling, and the resulting vcf-file was used for filtering by VCFtools [22] in the dDocent environment.

Specifically, for the first filtering sites with >50% missing data across all individuals and sites with a minor allele count <3 and quality value <30 were excluded. Secondly, we removed individuals with >10% missing data, and further filtered SNPs with the following criteria: mean depth ≥ 20, the proportion of missing data >95%, a Minor Allele Frequency (MAF) ≥ 0.05. In addition, we removed sites that deviated greatly from the Hardy–Weinberg equilibrium within populations and thinned sites that were tightly linked at <1 kb intervals using VCFtools. For the stairway plot analysis, we included all SNPs located less than 1 kb apart with no MAF filtering.

2.4. Genetic Diversity and Genetic Differentiation

Neutral loci tests of the genotype data for all populations and markers were conducted using BayeScan [23] and the "Fsthet" package [24]. Genetic indices, such as the number of alleles (Na), number of effective alleles (Ne), observed heterozygosity (Ho), expected heterozygosity (He), and Fixation index (F_{IS}) were estimated within each population using GenALEx 6.502 [25]. HP-Rare v.1.1 [26] was employed to calculate the allelic richness (Ar) and private allelic richness (pAr) with a minimum sample size of eight.

To examine differences between populations, genetic differentiation coefficients were calculated following Meirmans's method [27] in GenALEx 6.502 and Weir and Cockerham's method [28] in the R "hierfstat" package. Hierarchical analysis of molecular Variance (AMOVA) [29] was performed using GenALEx 6.502. Gene flows (Nm) based on F_{ST} and private alleles were calculated using GenALEx 6.502 and GENEPOP v4.3 [30]. Genetic distance matrices of pairwise population F_{ST} and pairwise population gene flow were also calculated in GenALEx 6.502.

2.5. Population Structure

We inferred the most likely number of genetic clusters using STRUCTURE v.2.3.4 [31]: 10 independent runs were performed at $K = 1\text{--}10$ with a burn-in period of 50,000 iterations and 100,000 MCMC repetitions, using no prior information, under the admixture and correlated allele frequencies models. The outputs of STRUCTURE were analyzed in Structure Harvester [32] to determine the most likely number of clusters according to ΔK [33] and mean LnP(K) [31]. CLUMPP v.1.1 [34] was then used to calculate the average pairwise similarity of runs based on the Greedy method, and finally the outputs of CLUMPP were visualized in Distruct v.1.1 [35].

The pairwise F_{ST} distance matrix was used to generate a dendrogram in MEGA v.7.0.26 [36] with the neighbor-joining method [37] and a network in SplitsTree v.4.14.8 [38] with the neighbor-net method [39]. In addition, we tested the correlations between genetic distance and geographic distance by correlating $F_{ST}/(1 - F_{ST})$ with geographic distance (km) in a Mantel test with 9999 permutations, as implemented in GenAIEx.

In order to assess the relationship structure within each population, we employed the COANCESTRY software to calculate the pairwise relatedness for all individuals with Wang estimator [40]. These relatedness coefficients(r) vary from 0 to 1, and a value of 0.5 indicates that individuals are first-order relatives, such as parents–offspring or full siblings. A value of 0.25 indicates second-order relatedness, such as half sibling, grandparents–grandchildren, avuncular, or double first cousins.

2.6. Demographic History

The variation in effective population size (Ne) over time was inferred using the composite likelihood approach with a multi-epoch model implemented in the Stairway plot software [41]. This method evaluates the difference between the observed site frequency spectrum (SFS) and its expectation under a specific demographic history [42]. The software was run using the two-epoch method, following the recommended 67% of sites for training and 200 bootstraps on the folded SFS. We excluded singletons from the estimation to minimize errors due to genotype calling. We assumed a mutation rate per generation of 1.50×10^{-9} based on a previous study by Moriguchi et al. [43]. *Cryptomeria japonica* is a long-lived species, and there are many ancient trees older than 1000 years in the wild. Suzuki and Susukida [44] estimated that 100 to 300 years were necessary for the regeneration of the natural forest on Yakushima Island, thus we set the generation time to 150, 200, or 300 years in different runs.

3. Results

3.1. Genetic Diversity and Differentiation

A total of 922 SNPs were obtained and used to assess the genetic diversity of seven populations of Chinese *Cryptomeria*. The SNP data were deposited in Dryad (DOI: https://doi.org/10.5061/dryad.nk98sf7rf); the loci did not depart from neutrality according to Bayescan. The number of alleles in each population ranged from 1.550 to 1.939, with an average of 1.789. The observed heterozygosity and expected heterozygosity were in the ranges Ho = 0.187 to 0.307 and He = 0.174 to 0.316, with an average of Ho = 0.269 and He = 0.253, respectively. The fixation index (F_{IS}) for populations LS and WT indicated significant inbreeding, while populations YTG, WYS, and XTM exhibited significant excess heterozygosity. The allelic richness varied from 1.42 in population WT to 1.77 in population LS. Notably, population LS had a relatively higher private allelic richness of 0.03 and the other populations were lower (pAr = 0.01 or 0). Overall, the highest diversity within a population was in LS and the lowest was in WT (Table 2).

Table 2. Genetic diversity indices of the seven populations of *C. japonica* var. *sinensis* based on 922 loci.

Pop		N	Na	Ne	Ho	He	F	Ar	pAr
LS	Mean	16	1.939	1.532	0.307	0.316	0.023 *	1.77	0.03
	SE		0.008	0.011	0.006	0.005	0.009		
WT	Mean	11	1.550	1.309	0.187	0.174	0.023 *	1.42	0.00
	SE		0.016	0.013	0.009	0.007	0.015		
TBY	Mean	15	1.900	1.458	0.286	0.277	−0.023	1.70	0.01
	SE		0.010	0.011	0.006	0.005	0.009		
YTG	Mean	8	1.614	1.351	0.239	0.209	−0.123 *	1.53	0.00
	SE		0.016	0.012	0.008	0.006	0.011		
WYS	Mean	11	1.871	1.466	0.306	0.279	−0.087 *	1.71	0.00
	SE		0.011	0.011	0.007	0.006	0.009		
TT	Mean	17	1.838	1.429	0.266	0.257	−0.033	1.64	0.01
	SE		0.012	0.011	0.007	0.006	0.08		
XTM	Mean	15	1.807	1.427	0.279	0.257	−0.082 *	1.64	0.01
	SE		0.013	0.011	0.007	0.006	0.007		
Total	Mean	13.3	1.789	1.424	0.267	0.253	−0.042	1.63	0.0086
	SE		0.005	0.004	0.003	0.002	0.004		

N: sample size; Na: No. alleles; Ne: No. effective alleles; Ho: observed Heterozygosity; He: expected Heterozygosity; F_{IS}: Fixation index; Ar: Allelic richness; pAr: private Allelic richness. * Significance (>confidence interval 99%).

The overall population differentiation coefficient (F_{ST}) among all Chinese populations for the 922 loci was 0.119 (Meirmans's method) and 0.134 (Weir and Cockerham's method). Correspondingly, the gene flows (Nm) based on F_{ST} were 1.858 and 1.618 respectively, while the Nm based on private allele frequency was only 0.0811. The AMOVA results (Table 3) showed that the proportion of variation among populations was 12%; among individuals, it was 12%' and within individual, it was 76%. The majority of variation occurred within individuals. The pairwise F_{ST} and pairwise Nm for each population in this study suggested significant differentiation in every pair of populations, and the gene flow varied widely between different pairs. The greatest gene flow occurred between populations TBY and WYS (7.846), and the lowest occurred between populations YTG and WT (0.696) (Table 4).

Table 3. Analysis of Molecular Variance (AMOVA) for the seven populations of *C. japonica* var. *sinensis*.

Source	df	SS	MS	Est. Var.	%
Among Pops	6	4033.190	672.198	18.885	12%
Among Indiv	89	14,142.633	158.906	18.552	12%
Within Indiv	96	11,693.000	121.802	121.802	76%
Total	191	29,868.823		159.239	100%
F_{ST}	0.119				
Nm	1.858				

Table 4. Pairwise Nm (above the diagonal) and pairwise genetic differentiation (F_{ST}) (below the diagonal) of the seven populations of *C. japonica* var. *sinensis*.

	LS	WT	TBY	YTG	WYS	TT	XTM
LS	-	1.014	2.460	1.251	2.630	1.533	1.963
WT	0.198 ***	-	1.296	0.696	1.293	0.995	1.047
TBY	0.092 ***	0.162 ***	-	1.927	7.846	2.902	3.259
YTG	0.167 ***	0.264 ***	0.115 ***	-	2.287	1.282	1.490
WYS	0.087 ***	0.162 ***	0.031 **	0.099 ***	-	2.780	4.953
TT	0.140 ***	0.201 ***	0.079 ***	0.163 ***	0.083 ***	-	4.788
XTM	0.113 ***	0.193 ***	0.071 ***	0.144 ***	0.048 **	0.050 ***	-

Significance levels: ** $p < 0.01$, *** $p < 0.001$.

Six Japanese populations were also sequenced and merged with the Chinese populations into a CHN-JPN dataset. We obtained 183 SNPs from this dataset to compare the genetic diversity and structure between the Chinese and Japanese groups. Japanese populations ($Na = 1.842$, $Ne = 1.393$, $He = 0.267$, $Ho = 0.245$) showed higher genetic diversity than the Chinese populations ($Na = 1.511$, $Ne = 1.232$, $He = 0.150$, $Ho = 0.143$). Interestingly, the highest diversity population in China, LS, harbors a very similar level of diversity to the Japanese populations (Table S2). In addition, a higher genetic differentiation among Chinese populations ($F_{ST} = 0.1204$) was detected than among Japanese populations ($F_{ST} = 0.0455$).

3.2. Genetic Structure

We explored the genetic structure of the Chinese and Japanese populations based on 183 SNPs. Two groups were clearly identified, but the population LS was placed with the Japanese group in the network (Figure S1).

We subsequently analyzed the genetic structure within Chinese populations using the 922 SNPs. The Bayesian cluster analysis assigned the seven populations into four distinct clusters (Figure 2). The results based on ΔK and mean LnP(K) indicated optimal values of 4 and 5, respectively (Figure S2). The presence of four clusters is consistent with division according to the four geographical provinces in which the populations are located, but note that population WYS from cluster 3 (Fujian prov.) shows a certain amount of mixing with cluster 4 (Zhejiang prov.). When $K = 5$, population YTG from Fujian province is allocated to a separate cluster. The other values of K also provided some additional information. Population LS (Jiangxi prov.) was the first to split from the other populations when $K = 2$, followed by population WT (Anhui prov.) when $K = 3$. Separation occurred within cluster 3 (Fujian prov.) when $K = 5$–7. Cluster 4, two populations from Zhejiang province, were always closely related. Similar results were obtained from the dendrogram based on the pairwise F_{ST} matrix (Figure 3) (Figure S3).

In summary, we partitioned the seven populations of *C. japonica* var. *sinensis* into four clusters (LS from Jiangxi prov.; WT from Anhui prov.; WYS, TBY, and YTG from Fujian prov.; and TTS and XTM from Zhejiang prov.) which seems to be a reasonable classification. However, population LS should be regarded as Japanese *Cryptomeria* (Figure S1).

We did not detect a significant correlation between geographic distance and genetic distance based on a Mantel test (R^2 = 0.007, $p = 0.382$) (Figure 4). No isolation by distance (IBD) was found.

Here, we considered six populations, excluding LS because this population appears to be an old plantation derived from Japanese stock.

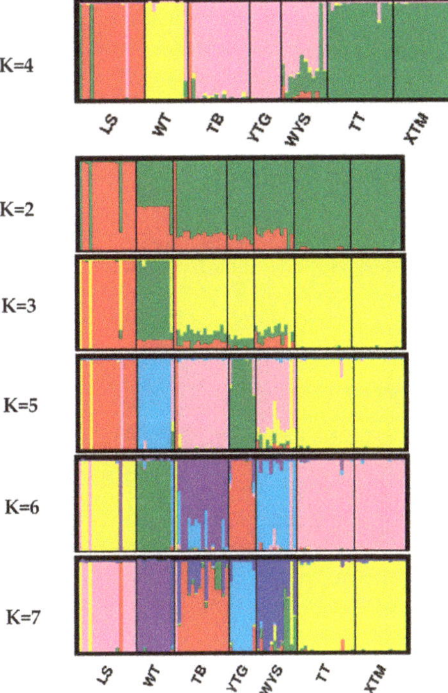

Figure 2. Population genetic structure of seven populations of *C. japonica* var. *sinensis* by STRUCTURE.

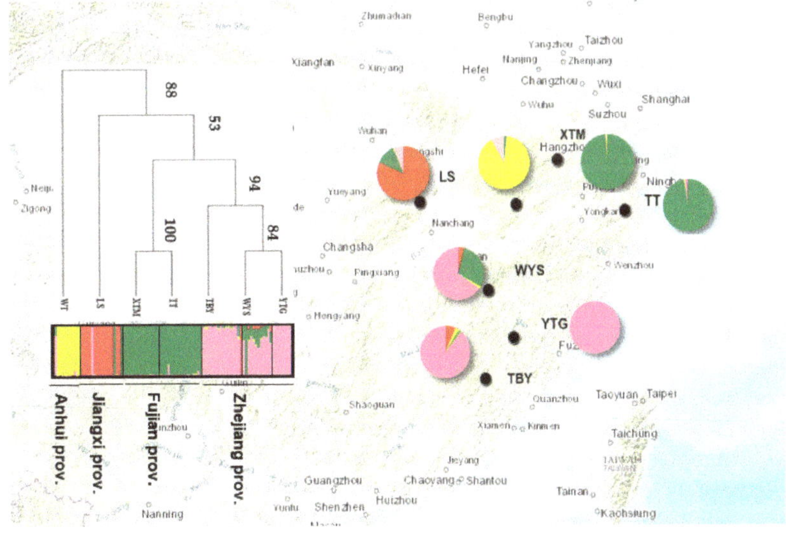

Figure 3. Genetic structure of *C. japonica* var. *sinensis* based on 922 SNPs. The pie diagrams on the map represents the membership coefficients to the four clusters inferred in Structure software. The neighbor-joining dendrogram based on pairwise F_{ST} values with 1000 bootstrap replicates.

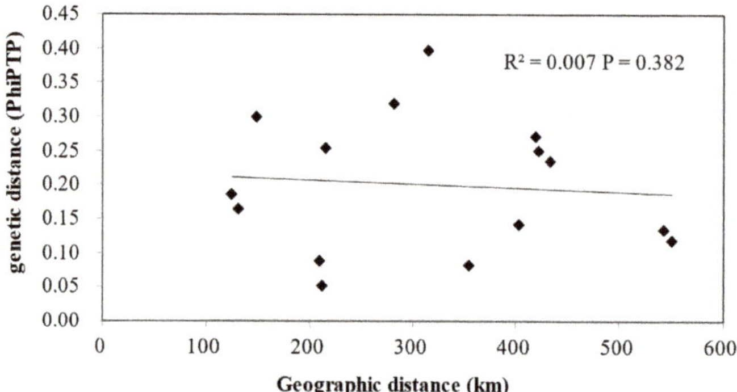

Figure 4. Mental test between geographical distance (km) and genetic distance (PhiPTP) for only Chinese populations where suspected Japanese individuals in LS were removed.

3.3. Demographic History

We obtained very similar trends in the three different scenarios (generation time = 150, 200, and 300 years)—namely, that the effective population size of Chinese *Cryptomeria* has experienced a continuous decline from the mid-Pleistocene to the present. The first decline occurred from 1 Mya to 0.4 Mya BP, coinciding with the onset of the Naynayxungla Glaciation (0.8–0.5 Mya) in China. The second decline began ca. 0.1 Mya to 0.06 Mya BP, when the Last Glacial Period (LGP) commenced. However, the range of *Cryptomeria* in China did not increase but continued to decline throughout the Holocene (Figure 5).

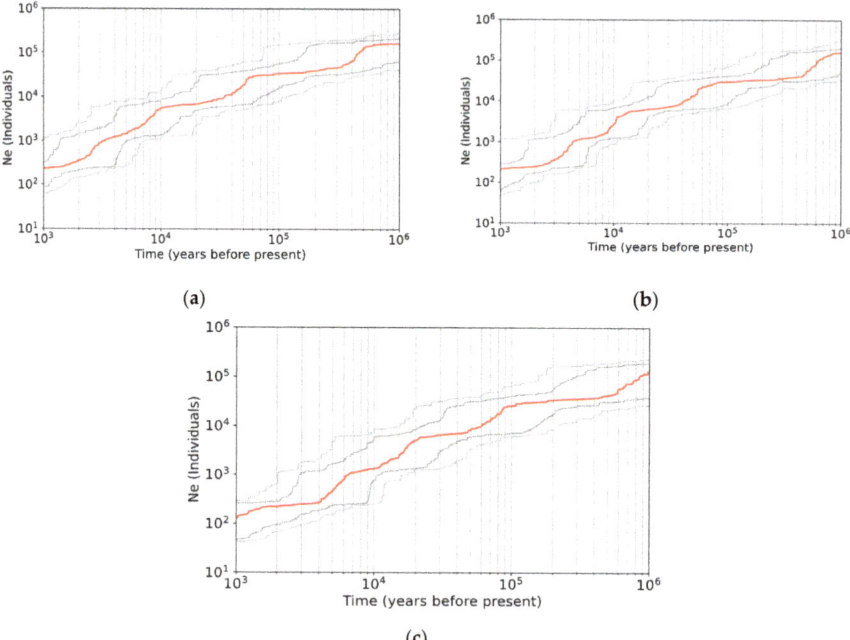

Figure 5. Effective population size (Ne) estimated based on folded SFS (Site Frequency Spectrum) with a generation time of (**a**) 150 years, (**b**) 200 years, and (**c**) 300 years, according to the Stairway Plot software. Red and grey lines represent the medians and the 2.5 and 97.5 percentiles, respectively.

4. Discussion

4.1. An Old Plantation Derived from Japanese Origin

The result of the NeighborNet based on F_{ST} revealed a clear separation between the Chinese and Japanese *Cryptomeria* (Figure S1), with population LS from Lushan mountain in Jiangxi province clustered with the Japanese group. Moreover, the genetic diversity of this population was apparently higher than that of the other Chinese populations and showed a very similar level of genetic diversity to the Japanese populations. There are historical records showing that, from ancient times, certainly as early as the 1st century BC, there was trade and the movement of people between Japan and China. After that, this kind of relationship continued until now. As a result, many items were exchanged between the two countries. Useful and important goods and ideas were shared, with rice cultivation being one of the best examples, having originated in China and then been exported to Japan. It is likely that a visitor to Japan in ancient times saw a huge *Cryptomeria* tree, brought back the seeds, and planted them in China. More recently, large-scale introductions of C. japonica from Japan occurred in the early 20th century. We, therefore, consider the population LS to be a Japanese *Cryptomeria* population based on our genetic data and historical evidence. However, some Chinese *Cryptomeria* individuals are present in this population, because individuals LS001 and LS017 belong to the Chinese population according to the NeighborNet result (Figure S4). Bayesian cluster analysis also showed a similar result to that generated by NeighborNet. With regard to individual LS001, this famous tree is recorded in an ancient book, "Travel Notes of Xu Xiake", written by a geographer in 1618, and it is estimated to be more than 600 years old (Figure S5). The *Cryptomeria* forest in Lushan mountain is, therefore, a mix of Chinese and Japanese *Cryptomeria*, and we found no significant phenotypic difference between them during our field investigation.

4.2. Low Genetic Diversity and High Genetic Differentiation in Chinese Cryptomeria

Ancient tree populations of C. *japonica* var. *sinensis* in China harbor a very low genetic diversity, but there is high genetic differentiation between populations compared to its congener, C. *japonica* in Japan. Here, we excluded LS population when discussing the genetic diversity and differentiation of Chinese *Cryptomeria*. Previous studies such as those by Chen et al. [6] and Tsumura et al. [18] also indicate a lower genetic variation in Chinese populations. Theoretically, inbreeding, genetic drift, restricted gene flow, and small population size all contribute to a reduction in genetic diversity [45]. In this study, we found evidence of a continuous decline in effective population size since the mid-Pleistocene. Genetic drift caused by climate fluctuation probably played an important role during the evolutionary process. Moreover, there was no expansion after the retreat of glaciers. We speculate that habitat loss or degradation and artificial selection caused by intense human activities have further accelerated the decline of the already low diversity, also leading to the great differentiation between different regions. Although gene flow in *Cryptomeria*, an allogamous, wind-pollinated conifer species, is expected to be high, we detected restricted gene flow ($Nm = 1.618$) between all the ancient tree populations, lower than the normal value of $Nm > 3$ in conifers [46]. Given that 76% of the variation occurred within populations, we consider that limited gene flow is also a factor accounting for the low genetic diversity. However, we did not find any sign of inbreeding in any population except WT. The result of relatedness analysis also presented a low proportion of kinship in most cases except the population of WT and YTG (Figure S6). The low diversity in WT may be caused by family relationship, while in YTG is even in the small sample size (only 8) there is no inbreeding ($F_{IS} = -0.123$) found in it. As far as other Chinese *Cryptomeia* populations go, we believe that the low level of genetic diversity can be attributed to climate change and strong human activity. Interesting, three populations (WYS, YTG, and XTM) showed signs of significant excess heterozygosity, which we consider probably to be the consequence of selection. However, the F_{IS} values of the three populations are not particularly high (WYS = −0.087, YTG = −0.123, and XTM = −0.082, Table 2) and thus this result may be related to the relatively small number of individuals investigated or may just be a chance occurrence. Some outlier loci under selection were detected by "Fsthet", and these might also be related to this result (Figure S7).

Similar situations with low genetic variation and high genetic differentiation have been found in some isolated and threatened species, such as *Tsuga caroliniana* Englem. [47], *Podocarpus sellowii* Klotzsch ex Endl. [48], and *Lupinus alopecuroides* Desr. [49]. Chinese *Cryptomeria*, as an important timber species, has been widely planted throughout southern China, and the unexpectedly low genetic diversity is probably associated with population size reduction because of global climate change, while the high genetic differentiation is probably the result of long isolation and human disturbance.

4.3. Specific Genetic Structure with an Absence of IBD

We separated the seven ancient tree populations into four groups coinciding with the four different administrative provinces in which the trees are located. Population LS from Lushan province and population WT from Anhui province were clearly separated, while the divergence between the Zhejiang and Fujian groups had relatively lower support (Figure 3). A certain mixing of the two groups occurred in the contact population, WYS. Ln(K) is also provided an alternative optimal structure that showed differentiation within the Fujian province group when $K = 5$ (Figure S2) (Figure 2). We noticed that it was YTG, not the contact population of WYS, that was separated. As we mentioned before, this may related to the small sample size of that population.

Even though a clear genetic structure was found, we did not detect a significant correlation between geographical distance and genetic distance, but Japanese *Cryptomeria* does exhibit such a correlation [50]. Generally, most tree species exhibit clear isolation by distance (IBD) if there is no strong human disturbance and selection [50,51]. In this case, geographical isolation is not the major factor responsible for the current structure. Since *Cryptomeria* is a wind-pollinated monoecious species, its mating system cannot explain the large differentiation either. As discussed above, the genetic drift associated with climate oscillations greatly reduced genetic diversity; one key factor may have been drought stress, which *Cryptomeria* is particularly sensitive to, as reported by Tsumura et al. [18] and Mori et al. [52]. On the other hand, the resulting habitat fragmentation also led to great genetic differentiation. We think that the reason for the unexpected absence of IBD was probably human disturbance. The topography of the range investigated, southern China, is characterized by numerous plains and basins between low hills [53]. In ancient times, the relatively flat terrain, with abundant grain cultivation, along with the development of handcraft industries resulted in the viability of commercial activities based on the timber and silk trade in this area, especially creating links between Fujian and Zhejiang province. A study of the genetic structure of horses suggested an important role for trade routes in facilitating exchange over topographically, ecologically, culturally, and politically diverse landscapes and large geographical distances [54]. Thus, we speculate that ancient trade routes in southern China may have provided opportunities for the transfer of material between different regions, resulting in ambiguities in the genetic origin of trees across the whole distribution range. However, given that we still found a pattern of genetic structure, the transfer of material must have been restricted.

Because of the very limited number of ancient trees currently in existence—we discovered only seven populations—the genetic structure presented in this study may deviate more or less from the original pattern without human interference. Despite this, we found a similar structure pattern—namely, an absence of IBD—to some species in southern China, including *Miscanthus lutarioriparius* L. Liou ex Renvoize & S.L. Chen [55], *Houpoea officinalis* (Rehder & E.H. Wilson) N.H. Xia & C.Y. Wu [56], and *Brasenia schreberi* J.F.Gmel. [57]. This may indicate a similar evolutionary history under anthropogenic pressure.

4.4. Continuous Decline of Population Size without Postglacial Recolonization

Climate oscillations throughout the Late Quaternary had a dramatic effect on the species ranges of both plants and animals in subtropical mainland Asia and the Japanese Archipelago [58]. The same goes for Chinese *Cryptomeria*: a remarkable decline in effective population size has been detected since the mid-Pleistocene. However, unlike many widely spread temperate plant species in Japan and East China [59], there was no recolonization after contractions during the LGM, but the population kept declining throughout the Holocene. Tsumura et al. [18] also did not find an obvious range expansion in the mid-Holocene and during the present using Species Distribution Modelling (SDM). There are two possible explanations for the continuous decline in population size after the LGM. First, the Ne of Chinese *Cryptomeria* may have decreased to a threshold size that constrained recovery. The low genetic diversity may have undermined any adaptive potential of the population during migrations. Second, humans in the Holocene directly reduced population size by cutting trees and clearing land [60]. Similar cases can be seen in some plant species in eastern China, including the genus *Croomia* [61], *Davidia involucrate* Baill. [62], *Ostrya rehderiana* Chun [60], and *Kalopanax septemlobus* (Thunb.) Koidz. [63].

In Japan, the range of *Cryptomeria* contracted to several refugia, mainly concentrated in the southwestern part of Japan during the last glaciation [64], and some natural stands have been retained up to the present. Japan may have had a favorable environment, with sufficient precipitation and fertile soil [18], and less anthropogenic disturbance. Thus, *Cryptomeria* in Japan maintained a higher level of genetic diversity and also presumably a larger effective population size than in China.

4.5. Conversation Considerations

Our investigated sites included the most famous and well-conserved forest stand of Chinese *Cryptomeria*, population XTM, which is located in Tianmu national nature reserve, Zhejiang province. This area contains many relict species of the Paleogene glaciation and is known for various ancient trees, including *Ginkgo biloba* L., *Liquidambar formosana* Hance, and *Pseudolarix amabilis* (J. Nelson) Rehder; among them, *C. japonica* var. *sinensis* is the dominant species. Previous studies on *Ginkgo biloba* [65], *Liriodendron chinensis* (Hemsl.) Sarg. [66], and *Quercus acutissima* Carruth [67] presented some evidence of glacial refugia in this region. In our case, we found a moderate level of genetic diversity and some private alleles, but no evidence to show that this area is the origin of Chinese *Cryptomeria*, even though it may seem to be [18]. Chen et al. [6] also suggested that the ancient population of *C. japonica* var. *sinensis* on Tianmu mountain was introduced originally, with subsequent natural regeneration.

Before the 1950s, most of China's forests were naturally regenerated. Since then, demand for timber has resulted in the extensive cutting of forests, and timber harvests increased from 20 million m^3/year in the 1950s to 63 million m^3/year in the 1990s [68]. Large-diameter trees are the first targets of timber extraction [69]. Government policy did not require that native tree species be planted after logging, but promoted the planting of fast-growing tree species, such as larch (*Larix* sp.), poplar (*Populus* sp.), and Chinese fir (*Cunninghamia lanceolata* (Lamb.) Hook). As a consequence, forest coverage has increased substantially, while natural forest has declined to 30% of the total forest area in China [68]. Moreover, genetic diversity loss is irreversible. In 1998, the Chinese government established the National Forest Conservation program (NPCP) to protect existing natural forest from excessive cutting. However, natural stands of Chinese *Cryptomeria* are now very rare, and the ancient populations which probably contain the ancestral genetic signature exhibit very low diversity, highlighting the precarious status of genetic resources of Chinese *Cryptomeria*.

During our field investigation, we found that some ancient trees occur sporadically in secondary forest and they are vulnerable to habitat fragmentation, atmospheric drought, and other competing species such as moso bamboo (*Phyllostachys edulis* (Carrière) J.Houz) [70]. However, some ancient trees in human-dominated landscapes are known for their high cultural and socioeconomic value, so citizens are willing to pay for their conservation [69]. Overall, the fate of ancient trees of *Cryptomeria* in China still hangs in the balance.

Under these circumstances, a more profound understanding and effective measurement are urgently needed. Here we propose both the in situ and ex situ conservation of this species. First, we appeal for stricter implementation of regulations to protect existing ancient trees in the wild. We suggest a complete ban on cutting and grazing; in addition, the thinning of dense, competitive bamboos or other problem species in the habitat is required. Even those trees that are worshipped as "sacred" still face some threats from industrial pollution, urbanization, and other forms of economic development. Sightseeing activities should be properly restricted. Compared to other populations, measures for the conservation of population WT, which is located on Huangshan mountain in Anhui province, are urgently required because it has the lowest diversity and diverges most from the other populations. Secondly, ex situ collections, which represent the highest genetic variation in the wild, need to be established. So far, two large germplasm gardens of Chinese *Cryptomeria* have been constructed and abundant sources have been collected from Fujian province and Zhejiang province, but we recommend a wider range of collection, targeting every single ancient tree in China. In view of afforestation, even though this species has been planted widely across southern China, many plantations lack clear provenance information, thus obscuring the genetic composition of planted forests; better data is required in order to develop a better afforestation strategy. As revealed in our study, populations from Tianmu mountain (XTM) in Zhejiang province and Tianbaoyan nature reserve (TBY) in Fujian province have a relatively high diversity and harbor some private alleles; these populations should be considered core resources.

5. Conclusions

Although the Chinese *Cryptomeria* population in Tianmu mountain has been studied and its low genetic diversity has been reported before, our study is the first study to explore the genetic composition of ancient *Cryptomeria* trees across a distribution range in China. The findings confirm that the ancient *Cryptomeria* population in China contains a low level of genetic diversity and high differentiation. Populations in different provinces were genetically differentiated, but no IBD was detected. We infer that the genetic drift caused by climate oscillations during the Last Glacial Period greatly reduced the population size of Chinese *Cryptomeria*, and this was followed by intense anthropogenic disturbance, which accelerated the loss of diversity and also led to a clear differentiation between regions. In addition, we present some theoretical guidance for conservation work in the future. Our study is the first to explore the genetic composition of ancient *Cryptomeria* trees in China and lays a firm foundation for further molecular research.

Supplementary Materials: The following are available online at http://www.mdpi.com/1999-4907/11/11/1192/s1: Figure S1: NeighborNet based on pairwise F_{ST} matrix of Chinese and Japanese populations; Figure S2: The number of inferred cluster K based on ΔK and mean LnP(K) obtained from Structure Harvester; Figure S3 Neighbor-joining dendrogram based on F_{ST} pairwise matrix of *C. japonica* var. *sinensis* after removing the suspected Japanese population of LS; Figure S4: NeighborNet based on pairwise F_{ST} matrix of all individuals. Individuals in red color indicated two trees in population of LS that belong to Chinese group; Figure S5: A ancient tree growing in Lushan mountain, coded LS001 in this study, was estimated more than 600 years old; Figure S6: The result of relatedness analysis of 7 populations, the darker blue indicated the higher relatedness coefficiency. Figure S7: Distribution of F_{ST}–H_T (expected heterozygosity) relationship based on 922 loci in *C. japonica* var *sinensis*. Two red line indicated the confidence interval of high and low value. A total of 20 outlier loci detected among 922 loci. Table S1: Geographic locations and sample size (N) of the 6 Japanese *Cryptomeria* populations; Table S2: Genetic diversity indices of six populations of *C. japonica* and seven populations of *C. japonica* var. *sinensis* based on 183 loci.

Author Contributions: Conceptualization, Y.T. and Y.W.; methodology, K.U.; software, K.U.; validation, Y.T. and K.U.; formal analysis, K.U., M.C., and Y.O.; investigation, Y.W. and M.C.; sources, Y.W.; data curation, K.U.; writing—original draft preparation, M.C.; writing—review and editing, Y.T., K.U., and M.C.; visualization, M.C.; supervision, Y.T., K.U., and Y.W.; project administration, Y.T. and Y.W.; funding acquisition, Y.T., K.U., and Y.W. All authors have read and agreed to the published version of the manuscript.

Funding: This study was partly supported by the Sumitomo Foundation Grant for Environmental Research Projects, JSPS KAKENHI (grant no. JP18H02248) and the National Key Research and Development Program of China (grant no. 2016YFE0127200).

Acknowledgments: We sincerely offer thanks for the assistance provided by Minqui Wang, Xingtong Wu, Xingyu Li, and Liang Wang during the material sampling and DNA extracting process. We also thank Eko Prasetyo and Ye Chen for offering help on the data analysis.

Conflicts of Interest: The authors declare no conflict of interest.

References

1. Ding, W.-N.; Kunzmann, L.; Su, T.; Huang, J.; Zhou, Z.-K. A new fossil species of *Cryptomeria* (Cupressaceae) from the Rupelian of the Lühe Basin, Yunnan, East Asia: Implications for palaeobiogeography and palaeoecology. *Rev. Palaeobot. Palynol.* **2018**, *248*, 41–51. [CrossRef]
2. Hayashi, Y. *Taxonomical and Phytogeographical Study of Japanese Conifers*; Norin-Shuppan: Tokyo, Japan, 1960.
3. Tsumura, Y. Cryptomeria. In *Wild Crop Relatives: Genomic and Breeding Resources: Forest Trees*; Kole, C., Ed.; Springer: Berlin, Germany, 2011; pp. 49–63.
4. Fu, L.G.; Yu, Y.F.; Mill, R.R. Taxodiaceae. In *Flora of China*; Wu, Z.Y., Raven, P.H., Eds.; Science Press: Beijing, China, 1999.
5. Zheng, H.; Hu, D.; Wei, R.; Yan, S.; Wang, R. Chinese Fir Breeding in the High-Throughput Sequencing Era: Insights from SNPs. *Forests* **2019**, *10*, 681. [CrossRef]
6. Chen, Y.; Yang, S.Z.; Zhao, M.S.; Ni, B.Y.; Liu, L.; Chen, X.Y. Demographic genetic structure of *Cryptomeria japonica* var. *sinensis in Tianmushan Nature Reserve, China. J. Integr. Plant Biol.* **2008**, *50*, 1171–1177. [CrossRef]
7. Luo, P.; Cao, Y.; Mo, J.; Wong, H.; Shi, J.; Xu, J. Analysis of genetic diversity and construction of DNA fingerprinting of clones in *Cryptomeria fortune. J. Nanjing For. Univ. (Nat. Sci. Ed.)* **2017**, *41*, 191–196.
8. Howe, G.T.; Yu, J.; Knaus, B.; Cronn, R.; Kolpak, S.; Dolan, P.; Lorenz, W.W.; Dean, J.F. A SNP resource for Douglas-fir: De novotranscriptome assembly and SNP detection and validation. *BMC Genom.* **2013**, *14*, 137. [CrossRef]
9. Davey, J.W.; Blaxter, M.L. RADSeq: Next-generation population genetics. *Brief. Funct. Genom.* **2010**, *9*, 416–423. [CrossRef]
10. Iwata, H.; Hayashi, T.; Tsumura, Y. Prospects for genomic selection in conifer breeding: A simulation study of *Cryptomeria japonica. Tree Genet. Genom.* **2011**, *7*, 747–758. [CrossRef]
11. Zhou, W.; Ji, X.; Obata, S.; Pais, A.; Dong, Y.; Peet, R.; Xiang, Q.-Y.J. Resolving relationships and phylogeographic history of the *Nyssa sylvatica* complex using data from RAD-seq and species distribution modeling. *Mol. Phylogen. Evol.* **2018**, *126*, 1–16. [CrossRef]
12. Brandrud, M.K.; Paun, O.; Lorenz, R.; Baar, J.; Hedrén, M. Restriction-site associated DNA sequencing supports a sister group relationship of *Nigritella* and *Gymnadenia* (Orchidaceae). *Mol. Phylogenetics Evol.* **2019**, *136*, 21–28. [CrossRef]
13. Miller, M.R.; Dunham, J.P.; Amores, A.; Cresko, W.A.; Johnson, E.A. Rapid and cost-effective polymorphism identification and genotyping using restriction site associated DNA (RAD) markers. *Genome Res.* **2007**, *17*, 240–248. [CrossRef]
14. Peterson, B.K.; Weber, J.N.; Kay, E.H.; Fisher, H.S.; Hoekstra, H.E. Double Digest RADseq: An Inexpensive Method for De Novo SNP Discovery and Genotyping in Model and Non-Model Species. *PLoS ONE* **2012**, *7*, e37135. [CrossRef]
15. Eaton, D.A.; Ree, R.H. Inferring phylogeny and introgression using RADseq data: An example from flowering plants (*Pedicularis*: Orobanchaceae). *Syst. Biol.* **2013**, *62*, 689–706. [CrossRef]
16. McCormack, J.E.; Hird, S.M.; Zellmer, A.J.; Carstens, B.C.; Brumfield, R.T. Applications of next-generation sequencing to phylogeography and phylogenetics. *Mol. Phylogen. Evol.* **2013**, *66*, 526–538. [CrossRef]
17. Wang, J.; Liu, J.; Huang, Y.; Yang, J. The origin and natural distribution of *Cryptomeria. J. Sichuan For. Sci. Technol.* **2007**, *28*, 92–94.
18. Tsumura, Y.; Kimura, M.; Nakao, K.; Uchiyama, K.; Ujino-Ihara, T.; Wen, Y.; Tong, Z.; Han, W. Effects of the last glacial period on genetic diversity and genetic differentiation in *Cryptomeria japonica* in East Asia. *Tree Genet. Genom.* **2020**, *16*, 19. [CrossRef]
19. Zhang, J. The Analysis on Genetic Diversity of Superior *Cryptomehia Fortunei* Resources and Screening of Hybrid Parent. Master's Thesis, Zhejiang Agriculture and Forestry University, Zhejiang, China, 2014.
20. Ueno, S.; Uchiyama, K.; Moriguchi, Y.; Ujino-Ihara, T.; Matsumoto, A.; Wei, F.-J.; Saito, M.; Higuchi, Y.; Futamura, N.; Kanamori, H. Scanning RNA-Seq and RAD-Seq approach to develop SNP markers closely linked to MALE STERILITY 1 (MS1) in *Cryptomeria japonica* D. Don. *Breed. Sci.* **2019**, *69*, 19–29. [CrossRef]
21. Puritz, J.B.; Hollenbeck, C.M.; Gold, J.R. dDocent: A RADseq, variant-calling pipeline designed for population genomics of non-model organisms. *PeerJ* **2014**, *2*, e431. [CrossRef]
22. Danecek, P.; Auton, A.; Abecasis, G.; Albers, C.A.; Banks, E.; DePristo, M.A.; Handsaker, R.E.; Lunter, G.; Marth, G.T.; Sherry, S.T. The variant call format and VCFtools. *Bioinformatics* **2011**, *27*, 2156–2158. [CrossRef]

23. Foll, M.; Gaggiotti, O. A genome-scan method to identify selected loci appropriate for both dominant and codominant markers: A Bayesian perspective. *Genetics* **2008**, *180*, 977–993. [CrossRef]

24. Flanagan, S.P.; Jones, A.G. Constraints on the F_{ST}–heterozygosity outlier approach. *J. Hered.* **2017**, *108*, 561–573. [CrossRef]

25. Peakall, R.; Smouse, P.E. GenAlEx 6.5: Genetic analysis in Excel. Population genetic software for teaching and research–an update. *Bioinformatics* **2012**, *28*, 2537–2539. [CrossRef] [PubMed]

26. Kalinowski, S.T. hp-rare 1.0: A computer program for performing rarefaction on measures of allelic richness. *Mol. Ecol. Notes* **2005**, *5*, 187–189. [CrossRef]

27. Meirmans, P.G. Using the AMOVA framework to estimate a standardized genetic differentiation measure. *Evolution* **2006**, *60*, 2399–2402. [CrossRef] [PubMed]

28. Weir, B.S.; Cockerham, C.C. Estimating F-statistics for the analysis of population structure. *Evolution* **1984**, *38*, 1358–1370. [PubMed]

29. Excoffier, L.; Smouse, P.; Quattro, J. Analysis of Molecular Variance Inferred From Metric Distances among DNA Haplotypes: Application to Human Mitochondrial DNA Restriction Data. *Genetics* **1992**, *131*, 479–491.

30. Rousset, F. genepop'007: A complete re-implementation of the genepop software for Windows and Linux. *Mol. Ecol. Resour.* **2008**, *8*, 103–106. [CrossRef]

31. Pritchard, J.K.; Stephens, M.; Donnelly, P. Inference of Population Structure Using Multilocus Genotype Data. *Genetics* **2000**, *155*, 945.

32. Earl, D.A. STRUCTURE HARVESTER: A website and program for visualizing STRUCTURE output and implementing the Evanno method. *Conserv. Genet. Resour.* **2012**, *4*, 359–361. [CrossRef]

33. Evanno, G.; Regnaut, S.; Goudet, J. Detecting the number of clusters of individuals using the software STRUCTURE: A simulation study. *Mol. Ecol.* **2005**, *14*, 2611–2620. [CrossRef]

34. Jakobsson, M.; Rosenberg, N.A. CLUMPP: A cluster matching and permutation program for dealing with label switching and multimodality in analysis of population structure. *Bioinformatics* **2007**, *23*, 1801–1806. [CrossRef]

35. Rosenberg, N.A. distruct: A program for the graphical display of population structure. *Mol. Ecol. Notes* **2004**, *4*, 137–138. [CrossRef]

36. Kumar, S.; Stecher, G.; Tamura, K. MEGA7: Molecular evolutionary genetics analysis version 7.0 for bigger datasets. *Mol. Biol. Evol.* **2016**, *33*, 1870–1874. [CrossRef] [PubMed]

37. Saitou, N.; Nei, M. The neighbor-joining method: A new method for reconstructing phylogenetic trees. *Mol. Biol. Evol.* **1987**, *4*, 406–425.

38. Huson, D.H.; Bryant, D. Application of phylogenetic networks in evolutionary studies. *Mol. Biol. Evol.* **2006**, *23*, 254–267. [CrossRef] [PubMed]

39. Bryant, D.; Moulton, V. Neighbor-Net: An Agglomerative Method for the Construction of Phylogenetic Networks. *Mol. Biol. Evol.* **2004**, *21*, 255–265. [CrossRef]

40. Wang, J. Estimating pairwise relatedness in a small sample of individuals. *Heredity* **2017**, *119*, 302–313. [CrossRef] [PubMed]

41. Liu, X.; Fu, Y.-X. Exploring population size changes using SNP frequency spectra. *Nat. Genet.* **2015**, *47*, 555–559. [CrossRef]

42. Khimoun, A.; Doums, C.; Molet, M.; Kaufmann, B.; Peronnet, R.; Eyer, P.; Mona, S. Urbanization without isolation: The absence of genetic structure among cities and forests in the tiny acorn ant *Temnothorax nylanderi*. *Biol. Lett.* **2020**, *16*, 20190741. [CrossRef]

43. Moriguchi, N.; Uchiyama, K.; Miyagi, R.; Moritsuka, E.; Takahashi, A.; Tamura, K.; Tsumura, Y.; Teshima, K.M.; Tachida, H.; Kusumi, J. Inferring the demographic history of Japanese cedar, *Cryptomeria japonica*, using amplicon sequencing. *Heredity* **2019**, *123*, 371–383. [CrossRef]

44. Suzuki, E.; Susukida, J. Age Structure and Regeration Process of Temperate Coniferous Stands in the Segire River Basin, Yakushima Island. *Jpn. J. Ecol.* **1989**, *39*, 45–51.

45. Furlan, E.; Stoklosa, J.; Griffiths, J.; Gust, N.; Ellis, R.; Huggins, R.; Weeks, A. Small population size and extremely low levels of genetic diversity in island populations of the platypus, *Ornithorhynchus anatinus*. *Ecol. Evol.* **2012**, *2*, 844–857. [CrossRef] [PubMed]

46. Ledig, F.T. Genetic Variation in Pinus. In *Ecology and Biogeography of Pinus*; Richardson, D.M., Ed.; Cambridge University Press: Cambridge, UK, 1998; pp. 251–280.

47. Potter, K.M.; Campbell, A.R.; Josserand, S.A.; Nelson, C.D.; Jetton, R.M. Population isolation results in unexpectedly high differentiation in Carolina hemlock (*Tsuga caroliniana*), an imperiled southern Appalachian endemic conifer. *Tree Genet. Genom.* **2017**, *13*, 105. [CrossRef]

48. Dantas, L.G.; Esposito, T.; de Sousa, A.C.B.; Félix, L.; Amorim, L.L.; Benko-Iseppon, A.M.; Batalha-Filho, H.; Pedrosa-Harand, A. Low genetic diversity and high differentiation among relict populations of the neotropical gymnosperm *Podocarpus sellowii* (Klotz.) in the Atlantic Forest. *Genetica* **2015**, *143*, 21–30. [CrossRef] [PubMed]

49. Vásquez, D.L.; Balslev, H.; Hansen, M.M.; Sklenář, P.; Romoleroux, K. Low genetic variation and high differentiation across sky island populations of *Lupinus alopecuroides* (Fabaceae) in the northern Andes. *Alp. Bot.* **2016**, *126*, 135–142. [CrossRef]

50. Tsumura, Y.; Kado, T.; Takahashi, T.; Tani, N.; Ujino-Ihara, T.; Iwata, H. Genome scan to detect genetic structure and adaptive genes of natural populations of *Cryptomeria japonica*. *Genetics* **2007**, *176*, 2393–2403. [CrossRef]

51. Aoki, K.; Ueno, S.; Kamijo, T.; Setoguchi, H.; Murakami, N.; Kato, M.; Tsumura, Y. Genetic differentiation and genetic diversity of *Castanopsis* (Fagaceae), the dominant tree species in Japanese broadleaved evergreen forests, revealed by analysis of EST-associated microsatellites. *PLoS ONE* **2014**, *9*, e87429. [CrossRef]

52. Mori, H.; Yamashita, K.; Saiki, S.-T.; Matsumoto, A.; Ujino-Ihara, T. Climate sensitivity of *Cryptomeria japonica* in two contrasting environments: Perspectives from QTL mapping. *PLoS ONE* **2020**, *15*, e0228278. [CrossRef]

53. Cao, X.; Flament, N.; Müller, D.; Li, S. The dynamic topography of eastern China since the latest Jurassic Period. *Tectonics* **2018**, *37*, 1274–1291. [CrossRef]

54. Warmuth, V.M.; Campana, M.G.; Eriksson, A.; Bower, M.; Barker, G.; Manica, A. Ancient trade routes shaped the genetic structure of horses in eastern Eurasia. *Mol. Ecol.* **2013**, *22*, 5340–5351. [CrossRef]

55. Yang, S.; Xue, S.; Kang, W.; Qian, Z.; Yi, Z. Genetic diversity and population structure of *Miscanthus lutarioriparius*, an endemic plant of China. *PLoS ONE* **2019**, *14*, e0211471. [CrossRef]

56. Yang, X.; Yang, Z.; Li, H. Genetic diversity, population genetic structure and protection strategies for *Houpoëa officinalis* (Magnoliaceae), an endangered Chinese medical plant. *J. Plant Biol.* **2018**, *61*, 159–168. [CrossRef]

57. Li, Z.-Z.; Gichira, A.W.; Wang, Q.-F.; Chen, J.-M. Genetic diversity and population structure of the endangered basal angiosperm *Brasenia schreberi* (Cabombaceae) in China. *PeerJ* **2018**, *6*, e5296. [CrossRef]

58. Harrison, S.; Yu, G.; Takahara, H.; Prentice, I. Diversity of temperate plants in east Asia. *Nature* **2001**, *413*, 129–130. [CrossRef]

59. Qiu, Y.; Sun, Y.; Zhang, X.; Lee, J.; Fu, C.-X.; Comes, H.P. Molecular phylogeography of East Asian *Kirengeshoma* (Hydrangeaceae) in relation to quaternary climate change and landbridge configurations. *New Phytol.* **2009**, *183*, 480–495. [CrossRef]

60. Yang, Y.; Ma, T.; Wang, Z.; Lu, Z.; Li, Y.; Fu, C.; Chen, X.; Zhao, M.; Olson, M.S.; Liu, J. Genomic effects of population collapse in a critically endangered ironwood tree *Ostrya rehderiana*. *Nat. Commun.* **2018**, *9*, 5449. [CrossRef]

61. Li, E.-X.; Yi, S.; Qiu, Y.-X.; Guo, J.-T.; Comes, H.P.; Fu, C.-X. Phylogeography of two East Asian species in *Croomia* (Stemonaceae) inferred from chloroplast DNA and ISSR fingerprinting variation. *Mol. Phylogen. Evol.* **2008**, *49*, 702–714. [CrossRef]

62. Chen, Y.; Ma, T.; Zhang, L.; Kang, M.; Zhang, Z.; Zheng, Z.; Sun, P.; Shrestha, N.; Liu, J.; Yang, Y. Genomic analyses of a "living fossil": The endangered dove-tree. *Mol. Ecol. Resour.* **2020**, *20*, 756–769. [CrossRef]

63. Sakaguchi, S.; Qiu, Y.-X.; Liu, Y.-H.; Qi, X.-S.; Kim, S.-H.; Han, J.; Takeuchi, Y.; Worth, J.R.P.; Yamasaki, M.; Sakurai, S.; et al. Climate oscillation during the Quaternary associated with landscape heterogeneity promoted allopatric lineage divergence of a temperate tree *Kalopanax septemlobus* (Araliaceae) in East Asia. *Mol. Ecol.* **2012**, *21*, 3823–3838. [CrossRef]

64. Tsukada, M. Altitudinal and latitudinal migration of *Cryptomeria japonica* for the past 20,000 years in Japan. *Quatern. Res.* **1986**, *26*, 135–152. [CrossRef]

65. Gong, W.; Chen, C.; Dobeš, C.; Fu, C.-X.; Koch, M.A. Phylogeography of a living fossil: Pleistocene glaciations forced *Ginkgo biloba* L. (Ginkgoaceae) into two refuge areas in China with limited subsequent postglacial expansion. *Mol. Phylogenetics Evol.* **2008**, *48*, 1094–1105. [CrossRef]

66. Shen, Y.; Cheng, Y.; Li, K.; Li, H. Integrating Phylogeographic Analysis and Geospatial Methods to Infer Historical Dispersal Routes and Glacial Refugia of *Liriodendron chinense*. *Forests* **2019**, *10*, 565. [CrossRef]

67. Zhang, X.-W.; Li, Y.; Zhang, Q.; Fang, Y.-M. Ancient east-west divergence, recent admixture, and multiple marginal refugia shape genetic structure of a widespread oak species (*Quercus acutissima*) in China. *Tree Genet. Genom.* **2018**, *14*, 88. [CrossRef]
68. Zhang, P.; Shao, G.; Zhao, G.; Le Master, D.C.; Parker, G.R.; Dunning, J.B.; Li, Q. China's forest policy for the 21st century. *Science* **2000**, *288*, 2135–2136. [CrossRef]
69. Wu, C.; Jiang, B.; Yuan, W.; Shen, A.; Yang, S.; Yao, S.; Liu, J. On the Management of Large-Diameter Trees in China's Forests. *Forests* **2020**, *11*, 111. [CrossRef]
70. Wu, C.; Mo, Q.; Wang, H.; Zhang, Z.; Huang, G.; Ye, Q.; Zou, Q.; Kong, F.; Liu, Y.; Wang, G.G. Moso bamboo (*Phyllostachys edulis* (Carriere) J. Houzeau) invasion affects soil phosphorus dynamics in adjacent coniferous forests in subtropical China. *Ann. For. Sci.* **2018**, *75*, 24. [CrossRef]

Publisher's Note: MDPI stays neutral with regard to jurisdictional claims in published maps and institutional affiliations.

Article

Reforestation or Genetic Disturbance: A Case Study of *Pinus thunbergii* in the Iki-no-Matsubara Coastal Forest (Japan)

Aziz Akbar Mukasyaf [1,*], Koji Matsunaga [2], Miho Tamura [1], Taiichi Iki [3], Atsushi Watanabe [1] and Masakazu G. Iwaizumi [4,*]

[1] Department of Forest Environmental Science, Faculty of Agriculture, Kyushu University, 744 Motooka, Nishiku, Fukuoka 819-0395, Japan; mtamura@agr.kyushu-u.ac.jp (M.T.); nabeatsu@agr.kyushu-u.ac.jp (A.W.)
[2] Kyushu Regional Breeding Office, Forest Tree Breeding Center, Forestry and Forest Product Research Institute, 2320-5 Suya, Koshi, Kumamoto 861-1102, Japan; makoji@affrc.go.jp
[3] Tohoku Regional Breeding Office, Forest Tree Breeding Center, Forestry and Forest Product Research Institute, 95 Osaki, Takizawa, Iwate 020-0621, Japan; iki@affrc.go.jp
[4] Kansai Regional Breeding Office, Forest Tree Breeding Center, Forestry and Forest Product Research Institute, 1043 Uetsukinaka, Shoo, Katsuta, Okayama 709-4335, Japan
* Correspondence: mukasyaf.aziz.777@s.kyushu-u.ac.jp (A.A.M.); ganchan@ffpri.affrc.go.jp (M.G.I.)

Abstract: In the twentieth century, a substantial decline in *Pinus thunbergii* populations in Japan occurred due to the outbreak of pine wood nematode (PWN), *Burshaphelencus xylophilus*. A PWN-*P. thunbergii* resistant trees-breeding project was developed in the 1980s to provide reforestation materials to minimalize the pest damage within the population. Since climate change can also contribute to PWN outbreaks, an intensive reforestation plan instated without much consideration can impact on the genetic diversity of *P. thunbergii* populations. The usage and deployment of PWN-*P. thunbergii* resistant trees to a given site without genetic management can lead to a genetic disturbance. The Iki-no-Matsubara population was used as a model to design an approach for the deployment management. This research aimed to preserve local genetic diversity, genetic structure, and relatedness by developing a method for deploying Kyushu PWN-*P. thunbergii* resistant trees as reforestation-material plants into Iki-no-Matsubara. The local genotypes of the Iki-no-Matsubara population and the Kyushu PWN-*P. thunbergii* resistant trees were analyzed using six microsatellite markers. Genotype origins, relatedness, diversity, and structure of both were investigated and compared with the genetic results previously obtained for old populations of *P. thunbergii* throughout Japan. A sufficient number of Kyushu PWN-*P. thunbergii* resistant trees, as mother trees, within seed orchards and sufficient status number of the seedlings to deploy are needed when deploying the Kyushu PWN-*P. thunbergii* resistant trees as reforestation material planting into Iki-no-Matsubara population. This approach not only be used to preserve Iki-no-Matsubara population (genetic diversity, genetic structure, relatedness, and resilience of the forests) but can also be applied to minimize PWN damage. These results provide a baseline for further seed sourcing as well as develop genetic management strategies within *P. thunbergii* populations, including Kyushu PWN-*P. thunbergii* resistant trees.

Keywords: genetic conservation; genetic management; pine wood nematode; *Pinus thunbergii*; pine wood nematode-*Pinus thunbergii* resistant trees

1. Introduction

In general, forests can be categorized based on their purpose as conservation forests, protected forests, production forests, and forests with specific functions such as mitigation or tourism [1]. Different management strategies are required to protect forests with multiple functions [2], such as in Indonesia [3], China and Germany [4], and recently genetic approach methods have been proposed for long-term management [5,6]. Forests today face numerous threats, including diseases and pests [7], human interference [8], loss of unique or

---

---

rare species and genetic resources [9,10], and loss of genetic diversity, which provides forest ecosystem resilience [11,12]. In Japan, climate change has led to changes in environmental conditions in, such as increased annual sunshine, temperature, and rainfall (precipitation); rapid sea-level rise at a rate of 3.2 mm/year from 1993–2010; and higher intensity and more frequent storm surges (27 tropical cyclones slightly above normal) [13,14]. Climate change is an unpredictable factor and one of the most serious threats to forest ecosystems [15].

By area, Japanese forests comprise almost 50% conifers; however, *Pinus thunbergii* accounts for only 1% of the conifer composition [16]. In the Kyushu area, the species has been planted in coastal areas since more than 400 years ago [17]. A characteristic of *P. thunbergii* is tolerance to extreme conditions such as high salinity, high temperature, and low precipitation. Moreover, as a pine forest, it provides protection to coastal areas, by reducing wind damage, inhibiting sand movement, and decreasing tsunami wave energy [18,19].

Severe outbreaks of *Burshaphelencus xylophilus* (pine wood nematode; PWN) depleted *P. thunbergii* populations in Japan between the 1900s and 2000s. The spread of PWN in Japan is the most significant occurrences of pest-disease damage than another country. The individuals damaged by PWN reached its peak in 1979, exceeded 2.43 million m^3; as of 2016, the damage was one-fifth of the peak volume [16]. Air temperatures significantly influence the growth of PWN [20,21]. From a forest pest/disease perspective, climate change can directly or indirectly affect forest dynamics, changing the way that host trees and pathogens interact [22]. The warming climate may provide conditions for further PWN outbreaks and damage in the future [23].

In 1978, a breeding project to develop a PWN-*P. thunbergii* resistant trees as a countermeasure against outbreaks [24] was established at Breeding Region Institutions in Japan. This breeding project was initiated to select surviving pine trees from heavily damaged forests in Southwestern Japan. In the case of *P. thunbergii*, 14,620 trees were selected as candidate, and after the artificial inoculation tests, 16 clones were certified as resistant trees [24,25]. Three rounds of breeding program, based on individual performance selection-trial, have been performed throughout Japan until 2018 with gradual changes in the methodology [26]. In each prefecture of Japan, seed orchards were designed based on these resistant trees breeding program and the seedlings were used as reforestation-material plant. To date, 211 PWN-*P. thunbergii* resistant trees from 71 forests had been developed [27]. The purpose of the *P. thunbergii* breeding project was to create PWN-resistant trees for use as reforestation materials to enhance old populations of *P. thunbergii* in Japan for mitigation functions. Before the existence of breeding project, artificial planting with natural seedling recruitment, as reforestation-material plants, had repeatedly performed to maintain the forest. Unfortunately, the mitigation functions have been given priority with little consideration of the seed sources or genetic impacts of artificial planting.

From a forest protection perspective, the deployment of PWN-*P. thunbergii* resistant trees at a given site would indeed protect forests against PWN infection, minimizing damage. However, deployment without proper genetic management could lead to a genetic disturbance within the population, such as genetic diversity loss and modification of the genetic structure, reduced adaptability to local environments, "gene swamping," and increased homogeneity; thus, negatively impacting the population as a gene resource [7,10,28–30]. Therefore, genetic diversity management must be considered when implementing tree improvement-products such as PWN-*P. thunbergii* resistant trees [31]. Genetic management and silviculture are fundamental components of forest management systems that have the potential to affect one another [2]. Both strategies are important for preserving local genetic diversity and maintaining forest resilience against environmental changes [32] even to the ecosystem [33], especially in extreme environments such as coastal areas.

In Japan, genetic diversity as well as genetic management of *P. thunbergii* populations and PWN-*P. thunbergii* resistant trees topics have not been discussed. In this study, we developed a genetic management based on the current genetic informations (genetic diversity, genetic structure, and relatedness) of a local pine forest, Iki-no-Matsubara, that has been repeatedly planted for mitigation functions under the situation of PWN damage is not yet under control. The origin of seedlings in this site were inferred based on their genetic relationships with neighboring *P. thunbergii* populations in Kyushu area and throughout Japan. In addition, we investigated the genetic diversity, genetic structure, and relatedness of Kyushu PWN-*P. thunbergii* resistant trees with *P. thunbergii* populations in Kyushu area. In this way, this study aimed to preserve the Iki-no-Matsubara *P. thunbergii* population (current genetic diversity, genetic structure, and relatedness) as genetic resources throughs the use of Kyushu PWN-*P. thunbergii* resistant trees with the possibility of genetic disturbance when deploying it into the site. The genetic knowledges obtained from this case study are expected to provide a baseline for further seed sourcing as well as develop genetic management strategies within *P. thunbergii* populations, including Kyushu PWN-*P. thunbergii* resistant trees.

2. Materials and Methods

2.1. Study Field

Total individual and diversity of *P. thunbergii* within the populations in Japan has been declined. Most of current *P. thunbergii* populations in Japan, including Iki-no-Matsubara, are an uneven-aged forest, because it has been replanted repeatedly to preserve the forest. There was no historical record of the origin of material-planted and genetic information.

The research area of Iki-no-Matsubara (33°34′52.8″ N 130°17′59.7″ E) was 12.56 hectares. A folktale claims that the forest was established in tribute to Empress Jingu for Silla conquest around 200 or 300 AD (Yamato periods) [34]. Iki-no-Matsubara is one of Japan's top 100 beautiful green pine forests [35]. It is not only an education forest that belongs to Kyushu University since 1922 but also as an urban forest and conservation forest with mitigation functions since Edo Era (1603–1868) or earlier [36]. Iki-no-Matsubara locates within Genkai Quasi-national Park, which under Nature Conservation Law based on Natural Park Act, designated by prefectural government as a conservation forest for mitigation functions [34]. Field survey, tree census, and measurement of the trees' diameter at Iki-no-Matsubara was conducted from January 2017 until June 2019. From tree census data, the diameter was classified into three classes: 1–30 cm DBH (Diameter at breast height) range, 31–60 cm DBH range, and 61–90 cm DBH range (Table 1). Then, measured the stumps to estimate the age based on the DBH class ranges [37]. Based on cross-dating dendrochronology observation of annual ring of the stump in different location within Iki-no-Matsubara [38], the oldest tree was estimated to be around 200 years old (Table 1). *P. thunbergii* was highly regarded by Japan's religion and culture and became Japan's cultural identity. Hence, it possible that domesticated and artificial regeneration has been conducted repeatedly by local people since 1500 BP [39].

Table 1. DBH class range based on stump wood and number of samples within the Iki-no-Matsubara population.

No	DBH Class Range (cm)	DBH Stump (cm)		Replication	Estimation of the Age (Years)			Sample (Trees)
		Min	Max		Min	Max	Mean	
1	1–30	4	29	9	12	36	23	109
2	31–60	32	51	9	32	190	84	108
3	61–90	65	85	7	170	195	178	52
							Total:	269

2.2. DNA Analysis

A total of 269 mature leaves were collected from selected trees at Iki-no-Matsubara experimental research field representing each DBH ranges class (Table 1). Selected trees were chosen randomly which represented each DBH classification and research field. Genomic DNA was extracted from 50 mg of tissue per individual by using the cetyltrimethyl ammonium bromide (CTAB) method [40] with slightly modifications and a DNeasy Plant

Kit (Qiagen Inc., Valencia, CA, USA) following the manufacturer's protocol. Simple Sequence Repeat (SSR) analysis was carried out by six markers, bcpt1075, bcpt1671, bcpt834, bcpt1823, bcpt2532, and bcpt1549 [18]. A total of 12 μL for PCR analysis was carried out by using 2 μL DNA elution, 1 μL of primer mix, DNase/RNase-free water, and 2× multiplex PCR kit by Qiagen (Qiagen Inc., USA). PCR reaction was carried out by Touchdown PCR [41]. PCR protocol began with denaturing 95 °C for 15 min, two step annealing: (1) 10 cycles of denaturation 94 °C (30 s), annealing 60 °C (90 s), annealing temperature was decreased by 0.5 °C per cycle until 55 °C, and extension 72 °C (1 min); (2), 20 cycles of 94 °C (30 s), 55 °C (90 s), and 72 °C (1 min), and final extension 60 °C for 30 min. Then, 10 μL of DNA amplicon mixed with Genescan 500 Liz Size Standard and Hi-Di Formamide (Applied Biosystems Inc., Bedford, MA, USA) was electrophoresed by ABI PRISM 3730 Genetic Analyzer (Applied Biosystems Inc., USA). Genotype data was analyzed using Genemapper 4.0 software (Applied Biosystems Inc., USA).

2.3. Statistical Analysis

Genotype data of 42 old populations of *P. thunbergii* from Iwaizumi et al. (2018) and PWN-*P. thunbergii* resistant trees (Watanabe, unpublished data, see Appendix A Table A2), which have been selected based on three breeding programs, was analyzed with data from Iki-no-Matsubara. Old populations are remaining populations of *P. thunbergii* that had decline due to overbreak of PWN. PWN-*P. thunbergii* resistant trees are tree improvement products that have high PWN resistance, which managed by Japan Tree Breeding Institution office in each region (Tohoku, Kansai, Kanto, and Kyushu) except Hokkaido. GeneAlex version 6.503 [42] was used to measure genetic diversity, Hardy-Weinberg Equilibrium, private alleles, genetic differentiation pattern through by principal coordinates analysis (PCoA) among populations and investigated gene flow (*Nm*) for examining the relationship between genetic differentiation and number of migrants variable per generation at each locus. Allelic richness (*AR*) and F_{IS} (inbreeding coefficient) at each locus was calculated by Fstat version 2.9.3.2 software [43]. Structure 2.3.4 [44] was used to determine individual-based genetic structure assessment by Bayesian method with a simulation run 15 times replicated, K-set 1–6 for 30,000 iterations burn-in period, and 30,000 iterations LOCPRIOR model under admixture ancestral model. The optimum value of each cluster *K* and the ΔK value within the genetic structure was determined by Evanno method [45] then upload the results to structure harvester [46].

3. Results

3.1. Inference of Origin and Genetic Structure in Iki-no-Matsubara Based on DBH

Table 2 shows the genetic diversity in Iki-no-Matsubara. N_a values was ranged from 14 (bcpt1549) to 29 (bcpt2532), N_e value from 3.18 (bcpt1549) to 7.83 (bcpt2532), *AR* value from 5.75 (bcpt1549) to 11.49 (bcpt2532), H_O and H_E from 0.57 (bcpt2532) to 0.85 (bcpt1075), and 0.69 (bcpt1549) to 0.87 (bcpt2532), respectively. Lowest value on F_{IS} was −0.03 (bcpt1075) and highest was 0.35 (bcpt2532). Three markers, bcpt834, bcpt1823, and bcpt2532 showed deviation from Hardy-Weinberg equilibrium ($p < 0.05$, $p < 0.001$, and $p < 0.001$, respectively).

The N_a, *AR*, *Ho*, and F_{IS} values for Iki-no-Matsubara were higher than those reported by Iwaizumi et al. (2018). Iki-no-Matsubara had more private alleles than another population within the Kyushu region and the presence private alleles in the same loci were none to be found in nearby populations in Kyushu area. Among 269 trees, 92 carried a total of 18 private alleles at four out of six loci (Table 3). Four trees were in the 61–90 cm DBH range, six trees were in the 31–60 cm DBH range, and the remaining were in the 1–30 cm DBH range (Appendix A Table A1).

Table 2. Genetic diversity of the Iki-no-Matsubara *P. thunbergii* population using six primer markers.

Locus	Size Range (bp)	N_a	N_e	AR	H_O	H_E	F_{IS}	HWE
bcpt1075	141–201	18	5.62	7.42	0.85	0.82	−0.03	ns
bcpt1671	162–225	22	5.79	8.25	0.82	0.83	0.01	ns
bcpt834	139–183	16	5.3	7.83	0.74	0.81	0.09	*
bcpt1823	128–169	15	5.71	7.51	0.62	0.82	0.25	***
bcpt2532	128–190	29	7.83	11.49	0.57	0.87	0.35	***
bcpt1549	93–130	14	3.18	5.75	0.67	0.69	0.03	ns
Mean		19	5.57	8.04	0.71	0.81	0.12	

N_a: number of allele, N_e: number of effective allele, AR: allelic richness, H_O: observed heterozygosity, H_E: expected heterozygosity, F_{IS}: inbreeding coefficient within the population, HWE: Hardy-Weinberg equilibrium (ns = not significant, * $p < 0.05$, ** $p < 0.01$, *** $p < 0.001$).

Table 3. Private alleles in Iki-no-Matsubara with 42 old populations of *P. thunbergii* throughout Japan using six primers.

Pop	Locus	PA	Pop	Locus	PA	Pop	Locus	PA	Pop	Locus	PA
Fukiage	bcpt1075	1	Miyajima	bcpt1549	1	Oki	bcpt2532	1	Jusan	bcpt2532	1
Miyazaki	bcpt1075	1	Kubokawa	bcpt1075	2	Kaga	bcpt834	1	Wakinosawa	bcpt2532	1
Karatsu	bcpt1075	1	Imabari	bcpt2532	1	Komatsu	bcpt1549	1			
Iki-no-Matsubara (Fukuoka)	bcpt1671	3	Tsuda	bcpt1075	1	Komatsu	bcpt1075	1			
	bcpt834	2	Kaihu	bcpt834	1						
	bcpt1823	1	Kaihu	bcpt1823	1						
	bcpt2532	12	Suzuka	bcpt834	2						

The genetic structure of Iki-no-Matsubara showed two color patterns (Figure 1A), even on K2, K3, or K4 structure result (number 8 in Appendix A Figure A1). Further analysis of the spatial distribution of genetic structure at Iki-no-Matsubara showed that the blue pattern (blue color $\geq$ 55%) was dominantly observed on the east side of the research field (Appendix A Figure A2). In the 1–30 cm DBH range, 20 out of 109 trees showed the blue pattern, while in the 31–60 cm and 61–90 cm DBH range 12 out of 108 trees and no trees, respectively, showed the blue pattern

Figure 1B shows the genetic structure among *P. thunbergii* populations throughout Japan. Iki-no-Matsubara was dominated by the yellow pattern, same with the other populations from the Kyushu area. Principle Coordinate Analysis (PCoA) showed that Iki-no-Matsubara was more similar to the Minami-Shimabara population than the Karatsu population, which is geographically closer to Iki-no-Matsubara. The Minami-Shimabara and Amakusa populations had the highest probability of taking part in the gene flow (*Nm*) into Iki-no-Matsubara, at 47.99% and 35.35%, respectively (Figure 2). Based on DBH class range (Figure 3) specifically, the similarity between the Minami-Shimabara and Iki-no-Mastubara 61–90 cm DBH, 31–60 cm DBH, and 1–30 cm DBH ranges were 14.18%, 36.93%, and 15.99%, respectively. More importantly, the relationship between the DBH ranges indicated that the 61–90 cm DBH range shared 53.44% of genetic similarity with the 31–60 cm range, and 6.51% with the 1–30 cm DBH range. This finding suggests that the origin of the young trees, 1–30 cm DBH range class, were not from the Iki-no-Matsubara population but another area.

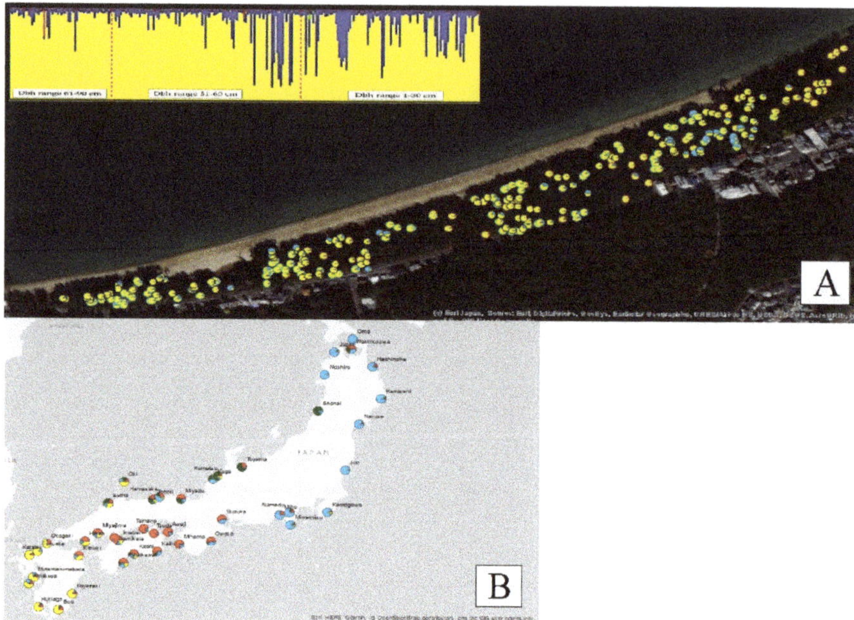

Figure 1. Spatial distribution of genetic structure at Iki-no-Matsubara (**A**); Iki-no-Matsubara with 42 old populations of *P. thunbergii* throughout Japan (**B**).

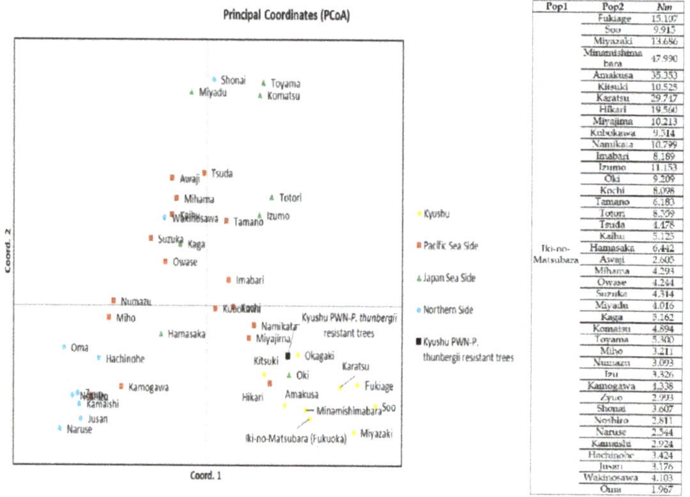

Pop1	Pop2	Nm
Iki-no-Matsubara	Fukiage	15.107
	Soo	9.915
	Miyazaki	13.686
	Minamishimabara	47.990
	Amakusa	35.353
	Kitsuki	10.525
	Karatsu	29.717
	Hikari	19.560
	Miyajima	10.213
	Kubokawa	9.514
	Namikata	10.799
	Imabari	8.189
	Izumo	11.153
	Oki	9.209
	Kochi	8.098
	Tamano	6.183
	Totori	8.559
	Tsuda	4.478
	Kaihu	5.125
	Hamasaka	6.442
	Awaji	2.605
	Mihama	4.293
	Owase	4.244
	Suzuka	4.314
	Miyadu	4.016
	Kaga	3.162
	Komatsu	4.894
	Toyama	5.300
	Miho	3.211
	Numazu	3.093
	Izu	3.326
	Kamogawa	4.338
	Zyuo	2.993
	Shonai	3.607
	Noshiro	2.811
	Naruse	2.544
	Kamaishi	2.924
	Hachinohe	3.424
	Jusan	3.176
	Wakinosawa	4.103
	Oma	1.967

Figure 2. PCoA of Iki-no-Matsubara with other populations of *P. thunbergii* and the Kyushu PWN-*P. thunbergii* resistant trees with gene flow (*Nm*) between Iki-no-Matsubara and the other populations.

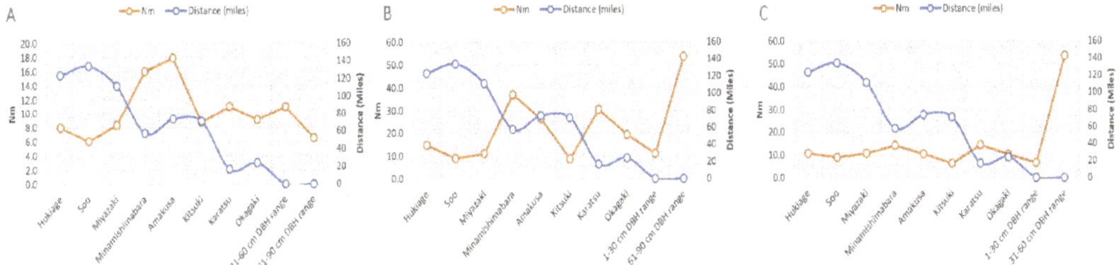

Figure 3. Relationship between gene flow (*Nm*) and distance with other populations in the Kyushu area on the following basis: (**A**) 1–30 cm DBH class range; (**B**) 31–60 cm DBH class range; (**C**) 61–90 cm DBH class range.

3.2. Genetic Diversity and Genetic Structure of PWN-P. thunbergii Resistant Trees

Since the 1990s, PWN-*P. thunbergii* resistant trees have been planted to enhance the old populations of *P. thunbergii*. Therefore, analyzing the local genotype of the Iki-no-Matsubara population (genetic diversity, genetic structure, and relatedness with other populations in Kyushu area) provides a baseline when deploying Kyushu PWN-*P. thunbergii* resistant trees. In general, the genetic structure of PWN-*P. thunbergii* resistant trees within each region (Figure 4) showed Kyushu (yellow color) and Kanto (green) PWN-*P. thunbergii* resistant trees had the most distinct genetic structure (dominated by region's structure pattern). In contrast, Tohoku and Kansai PWN-*P. thunbergii* resistant trees exhibited mixed patterns. The PCoA results show that the Kyushu PWN-*P. thunbergii* resistant trees are similar to the Okagaki populations (Figure 2). Some North Kyushu populations likely had a higher possibility of contributing to the gene flow than populations on the other side of Kyushu (see Appendix A Table A3). The genetic diversity of Kyushu PWN-*P. thunbergii* resistant trees was low compared with the mean genetic diversity of the *P. thunbergii* populations in entire Kyushu area (Table 1).

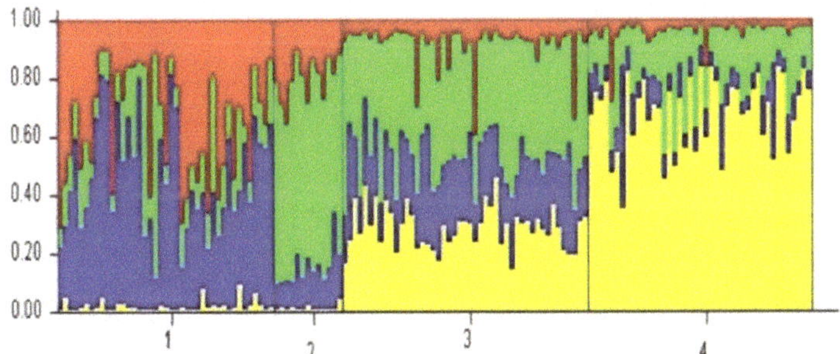

Figure 4. Genetic structure of PWN-*P. thunbergii* resistant trees on K4 (1) Tohoku region, (2) Kanto region, (3) Kansai region, and (4) Kyushu region.

Table 4. Genetic diversity of the Kyushu PWN-*P. thunbergii* resistant trees using six primer markers compared with the overall genetic diversity of populations in the Kyushu area.

Locus	N_a	N_e	*AR*	H_O	H_E	F_{IS}	HWE
bcpt1075	12	6.29	8.23	0.7	0.84	0.18	ns
bcpt1671	14	5.34	8.85	0.81	0.81	0.01	ns
bcpt834	9	3.88	6.05	0.67	0.74	0.1	ns
bcpt1823	9	3.77	6.14	0.51	0.73	0.31	***
bcpt2532	14	7.07	10.06	0.51	0.86	0.41	***
bcpt1549	7	2.86	4.84	0.63	0.65	0.05	ns
Mean	**10.83**	**4.87**	**7.36**	**0.64**	**0.77**	**0.18**	
Overall Populations in Kyushu Area	12.22	5.17	7.57	0.68	0.78	0.12	

N_a: number of allele, N_e: number of effective allele, *AR*: allelic richness, H_O: observed heterozygosity, H_E: expected heterozygosity, F_{IS}: inbreeding coefficient within the population, HWE: Hardy-Weinberg equilibrium (ns = not significant, *** $p < 0.001$).

4. Discussion

4.1. Inference of Origin and Genetic Structure in Iki-no-Matsubara based on DBH

Most *P. thunbergii* forests are located in coastal regions, including the Iki-no-Matsubara population. They have been expected for conservation area, especially to preserve mitigation functions such as reducing wind damage, inhibiting sand movement, and decreasing tsunami wave energy [19]. Before the existence of breeding project, artificial planting with natural seedling recruitment had repeatedly performed to maintain the forest.

In wind-pollinated conifers, the genetic diversity within the population has a tendency to be higher than that among populations. However, the genetic diversity within Iki-no-Matsubara was low in this study. Many *P. thunbergii* in Japan were damaged by the strong impact of PWN. After the 1980s, individuals with pest damage in the Iki-no-Matsubara population were removed and replanting has been continuously performed; however, the origin of seedlings were unknown. The number of private alleles was highest in Iki-no-Matsubara, and the presence of private alleles in the same loci were none to be found in the nearby populations in the Kyushu area. The lack of private alleles in a particular population within the Kyushu area is likely due to the small sample size compared to Iki-no-Matsubara [47]. The presence of private alleles in the Iki-no-Matsubara (Appendix A Table A1), interestingly, showed 31–60 cm DBH class range and 61–90 cm DBH class range shared on the same loci, while 1–30 cm DBH class range on different loci. Based on the structure analysis and PCoA results (Figure 2), we postulate that the Iki-no-Matsubara could be derived from the Kyushu area, especially the Minami-Shimabara or Amakusa population, which was farther from Iki-no-Matsubara than the Karatsu or Okagaki populations. In detail, 1–30 cm DBH class range was highly associated with Minami-shimabara and Amakusa. Meanwhile the 31–60 cm and 61–90 cm DBH class ranges displayed strong associated each other and the closest neighbour, Karatsu population (Figure 3). Such results, showed the recently planted the 1–30 cm DBH class range indicate that they were planted without considering genetic origin.

The genetic structure within the population was clearly divided into two patterns, and younger individual corresponding to DBH was remarkable. Furthermore, the genetic structure deviated to the area in the field. In more detail, some individuals exhibited the same pattern, yet different diameter class range (Figure 1A). The yellow patterns observed in Iki-no-Matsubara were common among populations in the Kyushu area (Figure 1B), while the blue pattern was not recognized in Karatsu nor Okagaki. There two possible explanations for this finding: (1) the materials planted in Iki-no-Matsubara were introduced from a different origin area, especially at DBH range 1–30 cm, which show dominantly blue color patterns; (2) Iki-no-Matsubara had more than two patterns of genetic structure in the past, including the patterns observed in Karatsu and Okagaki, but the population was reduced as a result of a bottleneck [18]. The exact cause is still uncertain due to the lack of historical records regarding the artificially-planted materials and the *P. thunbergii* genetic structure of Iki-no-Matsubara in the past. It would be reasonable to assume that the origin of the seedling was not considered when new planting was performed after removing individuals damaged by pine wilt disease.

4.2. Genetic Management of P. thunbergii in Iki-no-Matsubara with Kyushu PWN-P. thunbergii Resistant Trees

From a forest protection viewpoint, artificially planting Kyushu PWN-*P. thunbergii* resistant trees to enhance Iki-no-Matsubara population and counter PWN infection still has its merits; however, the genetic aspects such as genetic diversity (avoid homogeneity), genetic structure, resilience of the forest, and relatedness with another populations must also be properly considered. Thus, there are two crucial points to consider: (1) how well the PWN-*P. thunbergii* resistant trees as seed-sourcing strategy and (2) genetic management within the population, including the PWN-*P. thunbergii* resistant trees.

In addition, there two aspects should be considered for the genetic management of PWN-*P. thunbergii* resistant trees as seed-source strategy: (1) How well the mother trees represent the genetic diversity and relatedness in the selected area? (2) A sufficient number of resistant trees should be sourced as mother trees? [48–51]. The mother trees will represent the genetic diversity, structure, and gene flow pattern of the population where it was taken [52,53]. The extent of gene flow among populations shows how alleles are shared (similarities) and play an important role in genetic differentiation among populations [54,55].

From the perspective of genetic structure, Kyushu PWN-*P. thunbergii* resistant trees were noticeably displayed a yellow pattern (Kyushu region's structure pattern) (Figure 4). However, from relatedness viewpoint, Kyushu PWN-*P. thunbergii* resistant trees were located in the middle between the Kyushu area and Pacific seaside area and shared similarity with the Okagaki populations (Figure 2). This may have occurred because the selected trees for Kyushu PWN-*P. thunbergii* resistant trees were not sufficiently balanced to represent all Kyushu area populations. In fact, among 43 Kyushu PWN-*P. thunbergii* resistant trees, ten were from Okagaki, and none were from Iki-no-Matsubara (Appendix A Table A2).

The sufficient number of mother trees, act as effective population size in seed orchard, must be examined first to manage the diversity and relatedness within Iki-no-Matsubara with other populations in Kyushu area [56]. The effective population size is a concept used to predict the ideal size of the population, considering that the genes transmitted to seeds will still possess the same level of genetic diversity after many generations [57]. However, this study case only provide the Iki-no-Matsubara, not of the entire *P. thunbergii* in Kyushu area. Thus, in the future, the breeding project of Kyushu PWN-*P. thunbergii* resistant trees need to develop a perspective based on genetic management according to the genetic characteristics in each local pine forest in the Kyushu area.

4.3. Kyushu PWN-P. thunbergii Resistant Trees Deployment Management as Part of Genetic Management

To maintain *P. thunbergii* population in Iki-no-Matsubara, both PWN resistance and genetic diversity must be considered as part of genetic management, which is PWN-*P. thunbergii* resistant trees deployment management. Only using clones (vegetative) or seeds of specific Kyushu PWN-*P. thunbergii* resistant trees as reforestation-material plants on a large scale repeatedly for long-terms would cause a genetic disturbance such as increased homogeneity, inbreeding depression, reduced genetic diversity and adaptability to local environments [30,31,58]; thus, negatively impacting the population as a gene resource. Therefore, it is necessary to determine the status numbers of Kyushu PWN-*P. thunbergii* resistant trees [59] using information from genetic analysis within the population by practice selective seed-cone harvesting to balance genetic gain and diversity [48,59] for the necessary reforestation. When considering genetic diversity in the next generation and the status number of Kyushu PWN-*P. thunbergii* resistant trees, we can first refer to the local seed pool for reference, where at least 24 seedlings (generative) from each of the 30 mother trees will be needed to provide complete coverage for genetic diversity in the Iki-no-Matsubara population in the next generation [60]. Genetic diversity is defined as the genetic variation carried by individuals within a population as a part of their evolutionary path, providing a basis to form responses to environmental changes, as resilience of the forest [61].

Seedlings from a local seed pool or a neighbour population, such as Karatsu (geographically near of Iki-no-Matsubara), should be given priority. A seedling's adaptive potential from the local seed pool will have the optimal genotype because it has undergone many life cycles within the local environment over several generations. Proper seedling selection for planting is necessary to avoid maladaptation and improve the survival rate [62,63]. Furthermore, determining the origin of seedlings according to the Japan Forest Seeds and Seedlings Law 1939 [64] so that, at least, the structure pattern among the four areas shown in Figure 1B could be maintained. Subsequently, PWN-*P. thunbergii* resistant trees should be managed separately in each Japan Tree Breeding Institution office region. Using a non-local seed pool or non-local genetic pattern could lead to uncertain results in terms of adaptation and genetic differentiation among populations [30,65].

5. Conclusions

Declining *P. thunbergii* populations as a result of PWN outbreaks triggered to the consideration of genetic diversity management of the current populations for necessary genetic resources [18]. A forest with high genetic diversity provides a foundation for individuals to survive and adapt through evolution, especially when the forest has undergone human intervention [9,47,66,67]. Nevertheless, understanding the current genetic informations of Iki-no-Matsubara (genetic diversity, genetic structure, and relatedness) are essential for deploy Kyushu PWN-*P. thunbergii* resistant trees into the site, as part of genetic management. Genetic diversity (H_O) in Iki-no-Matsubara was 0.71 and dominated by yellow pattern from structure viewpoint. However, information based on DBH class range showed high relatedness with Minami-Shimabara and Amakusa, and there was a possibility that the origin of the materials that had been planted were not from the local seed pool was proposed, which was especially likely for the 1–30 cm DBH class range.

Additionally, the genetic structure of Kyushu PWN-resistant trees revealed a clear yellow genetic pattern. The Kyushu PWN-*P. thunbergii* resistant trees genetic diversity was lower than that of the overall population in the Kyushu area. An insufficient number of Kyushu PWN-*P. thunbergii* resistant trees unbalanced the gene flow, thus genetically to be found similar to the Okagaki population. A sufficient number of Kyushu PWN-*P. thunbergii* resistant trees, as mother trees, within seed orchards and sufficient status number of the seedlings need to be considered to safely deploy Kyushu PWN-*P. thunbergii* resistant trees as reforestation-material plants into Iki-no-Matsubara population. This approach can be used not only to preserve Iki-no-Matsubara population (genetic diversity, genetic structure, relatedness, and resilience of the forests) but can also be applied to minimize PWN damage. These results provide a baseline for further seed sourcing as well as develop genetic management strategies within *P. thunbergii* populations, including the PWN-*P. thunbergii* resistant trees.

Author Contributions: Methodology, Investigation, Conceptualisation, Formal Analysis, Writing—original draft and editing, Validation, A.A.M.; Methodology, Investigation, Conceptualisation, Resources, M.T.; Investigation, Resources, Supervision, Writing—review, K.M., T.I., M.G.I.; Methodology, Investigation, Conceptualisation, Supervision, Resources, Validation, Writing—review and editing, A.W. All authors have read and agreed to the published version of the manuscript.

Funding: This research was funded by Japan Society for the Promotion of Science, KAKENHI, grant number 17K07853.

Institutional Review Board Statement: This study did not require ethical approval. This study did not involve humans or animals.

Informed Consent Statement: This study did not involve humans or animals.

Data Availability Statement: Data available on request due to restrictions eg privacy or ethical. The data presented in this study are available upon request to the authors. The data are not publicly available due to the data are managed by Instituional authority in Japan.

Conflicts of Interest: The authors declare no conflict of interest.

Appendix A

Table A1. Number of private alleles within Iki-no-Matsubara.

No	DBH Class Range	Number of Trees	Loci
1	1–30 cm	82	bcpt2532
		1	bcpt1823
2	31–60 cm	5	bcpt1671
		1	bcpt834
3	61–90 cm	1	bcpt1671
		4	bcpt834

Table A2. List of PWN-*P. thunbergii* resistant trees based on the region [68].

No	PWN-*P. thunbergii* Resistant Trees			
	Tohoku	Kansai	Kanto	Kyushu
1	Naruse39	Tanabe54	Odaka37	Shima64
2	Naruse72	Bizen143	Odaka203	Tsuyazaki50
3	Naruse6	Mitoyo103	Iwaki27	Karatsu1
4	Watari5	Namikata37	Osuga5	Karatsu4
5	Yamamoto82	Namikata73	Osuga6	Karatsu7
6	Yamamoto84	Misaki90	Osuga12	Karatsu9
7	Yamamoto90	Yoshida2	Osuga15	Karatsu11
8	Yuza27	Yasu37	Osuga23	Karatsu16
9	Yuza72	Tosashimizu63	Utchihara5	Karatsu17
10	Yuza33	Kumihama10	Tomiura7	Obama30
11	Yuza54	Kumihama21	Okazaki25	Oseto12
12	Yuza56	Kumihama109	Okazaki34	Kawaura8
13	Yuza58	Amino31	Okazaki35	Kawaura13
14	Yuza60	Amino43		Amakusa20
15	Yuza57	Tango47		Oita8
16	Yuza59	Tango50		Sadowara8
17	Yuza77	Tango51		Sadowara14
18	Murakami2	Tango58		Sadowara15
19	Murakami5	Tango60		Miyazaki20
20	Murakami11	Tango65		Sendai20
21	Murakami16	Tango69		Ei425
22	Murakami44	Tango71		Hiyoshi1
23	Murakami9	Totori7		Hiyoshi5
24	Murakami15	Totori13		Hukiage25
25	Nigata8	Iwami63		Okagaki1
26	Nigata40	Nisinosima142		Okagaki5

Table A2. *Cont.*

27	Nigata3	Komatsu99	Okagaki6
28	Aikawa27	Ota39	Okagaki8
29	Nagaoka15	Hamada6	Okagaki25
30	Nagaoka8	Hamada12	Okagaki29
31	Ozika151	Hamada24	Okagaki31
32	Sendai35	Hamada28	Okagaki32
33	Ishimaki251	Gotsu29	Okagaki35
34	Ishimaki260	Yunotsu52	Okagaki20
35	Ishimaki259	Hukube51	Munakata2
36	Atsumi43	Hukube54	Munakata4
37	Tsuruoka38	Hukube60	Munakata12
38	Tsuruoka44	Hukube61	Munakata19
39	Tsuruoka46	Hukube71	Shingu2
40	Zyoetsu1	Koryo60	Shingu5
41	Zyoetsu10	Koryo77	Shingu11
42		Kaga387	Shingu14
43		Kaga388	Shingu17
44		Kaga295	
45		Shiga396	
46		Tsuruga14	
47		Tsuruga15	

Table A3. Gene flow (Nm) of Kyushu PWN-*P. thunbergii* resistant trees with the populations within Kyushu area.

Pop1	Pop2	F_{ST}	Nm
Hukiage	Kyushu PWN-*P. thunbergii* resistant trees	0.020	12.308
Soo	Kyushu PWN-*P. thunbergii* resistant trees	0.033	7.349
Miyazaki	Kyushu PWN-*P. thunbergii* resistant trees	0.027	8.956
Minamishimabara	Kyushu PWN-*P. thunbergii* resistant trees	0.007	34.354
Amakusa	Kyushu PWN-*P. thunbergii* resistant trees	0.008	32.530
Kitsuki	Kyushu PWN-*P. thunbergii* resistant trees	0.016	15.698
Karatsu	Kyushu PWN-*P. thunbergii* resistant trees	0.008	32.144
Iki-no-Matsubara (Fukuoka)	Kyushu PWN-*P. thunbergii* resistant trees	0.010	25.961
Okagaki	Kyushu PWN-*P. thunbergii* resistant trees	0.010	24.077

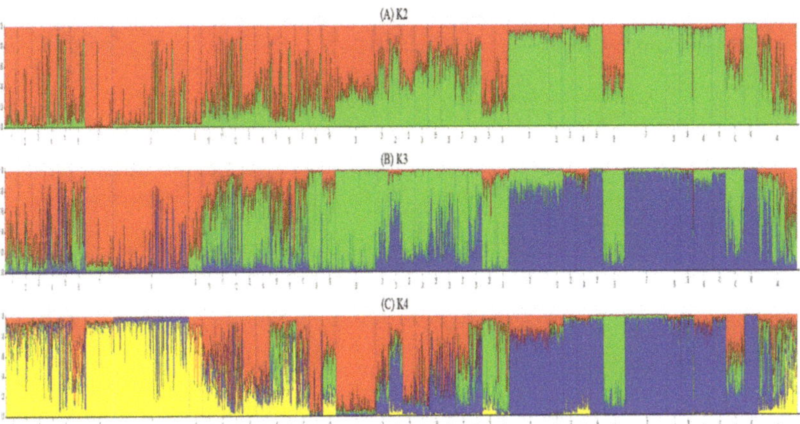

Figure A1. Genetic structure of Iki-no-Matsubara (Fukuoka) with 42 old populations of *P.thunbergii* on *K*2, *K*3, and *K*4 (From South-West (**Left**) to North-East (**Right**)).

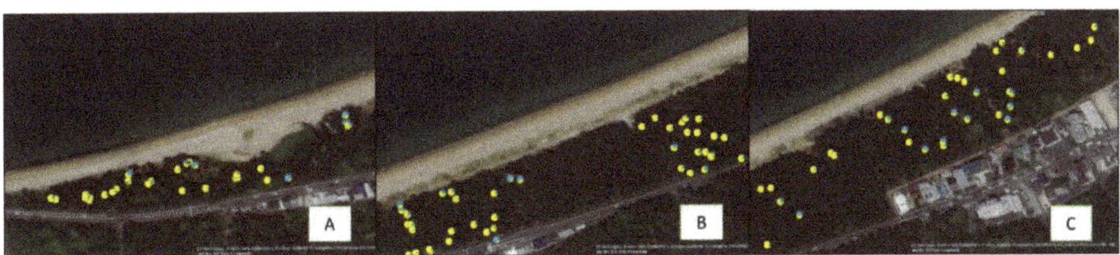

1–30 cm DBH range

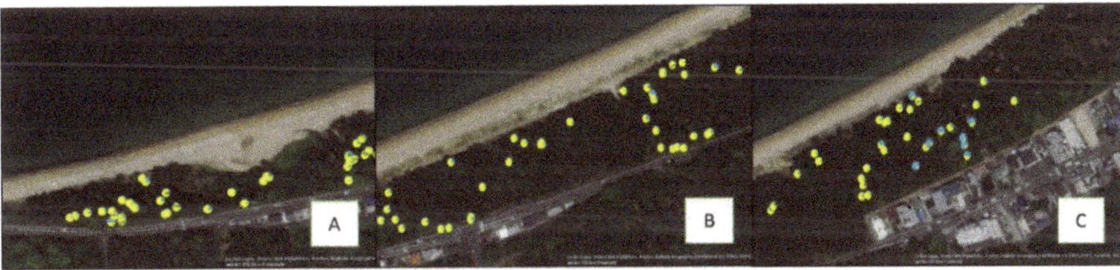

31–60 cm DBH range

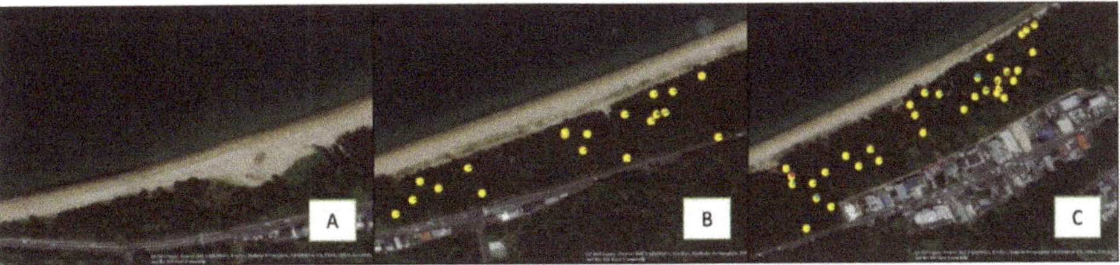

61–90 cm DBH range

Figure A2. Spatial distribution of Iki-no-Matsubara genetic structure per DBH range (**A**) West side, (**B**) Central side, and (**C**) East side.

References

1. Fuhrer, E. Forest function, ecosystem stability and management. *For. Ecol. Manag.* **2000**, *132*, 29–38. [CrossRef]
2. Krott, M. *Forest Policy Analysis*; European Forest Institute; Springer: Amsterdam, The Netherlands, 2005; ISBN 978-1-4020-3485-5.
3. Santika, T.; Meijaard, E.; Wilson, K.A. Designing multifunctional landscapes for forest conservation. *Environ. Res. Lett.* **2015**, *10*. [CrossRef]
4. Benz, J.P.; Chen, S.; Dang, S.; Dieter, M.; Labelle, E.R.; Liu, G.; Hou, L.; Mosandl, R.M.; Pretzsch, H.; Pukall, K.; et al. Multifunctionality of forests: A white paper on challenges and opportunities in China and Germany. *Forests* **2020**, *11*, 266. [CrossRef]
5. Aravanopoulos, F. Do silviculture and forest management affect the genetic diversity and structure of long-impacted forest tree populations? *Forests* **2018**, *9*, 355. [CrossRef]
6. Ratnam, W.; Rajora, O.P.; Finkeldey, R.; Aravanopoulos, F.; Bouvet, J.M.; Vaillancourtf, R.E.; Kanashiro, M.; Fady, B.; Tomita, M.; Vinson, C. Genetic effects of forest management practices: Global synthesis and perspectives. *For. Ecol. Manag.* **2014**, *333*, 52–65. [CrossRef]
7. Tubby, K.V.; Webber, J.F. Pests and diseases threatening urban trees under a changing climate. *For. Int. J. For. Res.* **2010**, *83*, 451–459. [CrossRef]
8. Milner-Gulland, E.J. Interactions between human behaviour and ecological systems. *Philos. Trans. R. Soc. Lond. B Biol. Sci.* **2012**, *367*, 270–278. [CrossRef]
9. Aravanopoulos, F.A. Conservation and monitoring of tree genetic resources in temperate forests. *Curr. For. Rep.* **2016**, *2*, 119–129. [CrossRef]
10. Ledig, F.T. The conservation of diversity in forest trees: Why and how should genes be conserved? *BioScience* **1988**, *38*, 471–479. [CrossRef]
11. Rajora, O.P.; Pluhar, S.A. Genetic diversity impacts of forest fires, forest harvesting, and alternative reforestation practices in black spruce (*Picea mariana*). *Theor. Appl. Genet.* **2003**, *106*, 1203–1212. [CrossRef]
12. White, T.L.; Adams, W.T.; Neale, D.B. *Forest Genetics*; Cabi: Cambridge, MA, USA, 2007; ISBN 978-0-85199-083-5.
13. Japan Meteorological Agency. *Climate Change Monitoring Report 2017*; Japan Meteorological Agency: Tokyo, Japan, 2018; pp. 31–55.
14. Ministry of the Environment; Ministry of the Education Culture, Sports, Science, and Technology; Ministry of the Agriculture, Forestry, and Fisheries; Ministry of the Land, Infastructure, Transport, and Tourism; Japan Meteorological Agency. *Synthesis Report on Observations, Projections, and Impact Assesments of Climate Change, 2018: Climate Change in Japan and It's Impacts*; Ministry of the Environment: Tokyo, Japan, 2018; pp. 1–7.
15. Sturrock, R.N.; Frankel, S.J.; Brown, A.V.; Hennon, P.E.; Kliejunas, J.T.; Lewis, K.J.; Worrall, J.J.; Woods, A.J. Climate change and forest diseases. *Plant Pathol.* **2011**, *60*, 133–149. [CrossRef]
16. Forestry Agency. *State of Japan's Forest and Forest Management: 3rd Country Report of Japan to the Montreal Process*; Forest Agency: Tokyo, Japan, 2019.
17. Ministry of Agriculture, Forestry, and Fisheries. Niji-no-Matsubara (Pine Grove) Recreation Forest. Available online: https://www.rinya.maff.go.jp/e/national_forest/recreation_forest/niji.html (accessed on 20 April 2020).
18. Iwaizumi, M.G.; Miyata, S.; Hirao, T.; Tamura, M.; Watanabe, A. Historical seed use and transfer affect geographic specificity in genetic diversity and structure of old planted *Pinus thunbergii* population. *For. Ecol. Manag.* **2018**, *408*, 211–219. [CrossRef]
19. Suwa, R. Evaluation of the wave attenuation function of a coastal black pine *Pinus thunbergii* forest using the individual-based dynamic vegetation model SEIB-DGVM. *J. For. Res.* **2013**, *18*, 238–245. [CrossRef]
20. Ichihara, Y.; Fukuda, K.; Suzuki, K. Early symptom development and histological change associated with migration of *Bursaphelenchus xylophilus* in seedling tissues of *Pinus thunbergii*. *Plant Dis.* **2000**, *84*, 675–680. [CrossRef]
21. Yamaguchi, R.; Matsunaga, K.; Watanabe, A. Influence of temperature on pine wilt disease progression in *Pinus thunbergii* seedlings. *Eur. J. Plant Pathol.* **2019**, *156*, 1–10. [CrossRef]
22. Linnakoski, R.; Kasanen, R.; Dounavi, A.; Forbes, K.M. Forest health under climate change: Effects on tree resilience, and pest and pathogen dynamics. In *Frontiers in Plant Science*; Frontiers Media SA: Lausanne, Switzerland, 2019; pp. 5–7. [CrossRef]
23. Hirata, A.; Nakamura, K.; Nakao, K.; Kominami, Y.; Tanaka, N.; Ohashi, H.; Takano, K.T.; Takeuchi, W.; Matsui, T. Potential distribution of pine wilt disease under future climate change scenarios. *PLoS ONE* **2017**, *12*, 1–18. [CrossRef]
24. Fujimoto, Y.; Toda, T.; Nishimura, K.; Yamate, H.; Fuyuno, S. Breeding project on resistance to pine-wood nematode-an outline of the research and the achievement of the project for ten years. *Bull. For. Tree Breed. Inst.* **1989**, *3*, 1–84.
25. Kurinobu, S. Current status of resistance breeding of Japanese pine species to pine wilt disease. *For. Sci. Technol.* **2008**, *4*, 51–57. [CrossRef]
26. Matsunaga, K.; Watanabe, A. characterization of pine wood nematode and development of more improved second generation resistant trees [In Japanese]. *For. Genet. Tree Breed.* **2018**, *7*, 115–119. [CrossRef]
27. FFPRI. Forest tree breeding center and forest bio-research center brochure. In *Forest Research and Management Organization National Research and Development Agency*; FFPRI: Tokyo, Japan, 2018.
28. Aitken, S.N.; Whitlock, M.C. Assisted gene flow to facilitate local adaptation to climate change. *Annu. Rev. Ecol. Evol. Syst.* **2013**, *44*, 367–388. [CrossRef]
29. Içgen, Y.; Kaya, Z.; Çengel, B.; Velioğlu, E.; Öztürk, H.; Önde, S. Potential impact of forest management and tree improvement on genetic diversity of Turkish red pine (*Pinus brutia* Ten.) plantations in Turkey. *For. Ecol. Manag.* **2006**, *225*, 328–336. [CrossRef]

30. Konnert, M.; Fady, B.; Gomory, D.; A'Hara, S.; Wolter, F.; Ducci, F.; Koskela, J.; Bozzano, M.; MaKowalczyk, J. Use and transfer of forest reproductive material in europe in the context of climate change. In *European Forest Genetic Resources Programme (EUFORGEN), Bioversity International*; Euforgen: Rome, Italy, 2015; ISBN 978-92-9255-031-8.

31. Ingvarsson, P.K.; Dahlberg, H. The effect of clonal forestry on genetic diversity in wild and domesticated stands of forest trees. *Scandinavian J. For. Res.* **2019**, *34*, 370–379. [CrossRef]

32. Kavaliauskas, D.; Fussi, B.; Westergren, M.; Aravanopoulos, F.; Finzgar, D.; Baier, R.; Alizoti, P.; Bozic, G.; Avramidou, E.; Konnert, M.; et al. The interplay between forest management practices, genetic monitoring, and other long-term monitoring systems. *Forests* **2018**, *9*, 133. [CrossRef]

33. Bailey, J.K.; Schweitzer, J.A.; Ubeda, F.; Koricheva, J.; LeRoy, C.J.; Madritch, M.D.; Rehill, B.J.; Bangert, R.K.; Fischer, D.G.; Allan, G.J.; et al. From genes to ecosystems: A synthesis of the effects of plant genetic factors across levels of organization. *Philos. Trans. Royal Soc. B Biol. Sci.* **2009**, *364*, 1607–1616. [CrossRef] [PubMed]

34. Fukuoka City. The Iki-no-Matsubara pine forest and genko borui. *Hakata Cult.* **2016**, *133*. Available online: http://www.fukuoka-now.com (accessed on 23 December 2019).

35. Awano, T. An evaluation of the intrinsic value of the takada pine forest (Takakada-no-Matsubara) as a scenic beauty spot in japan. *Urban Reg. Plan. Rev.* **2015**, *2*, 18–30. [CrossRef]

36. Park, K.W.; Sokh, H.; Inoue, S. Local inhabitants consciousness of using and managing urban forest Aaea-a case study of iki no matsubara in fukuoka city. *J. Fac. Agric. Kyushu Univ.* **2003**, *47*, 267–275.

37. Studhalter, R.A.; Glock, W.S.; Agerter, S.R. Tree gowth: Some historical chapters in the study of diameter growth. *Bot. Rev.* **1963**, *29*, 245–365. [CrossRef]

38. Pilcher, J.R. Sample preparation, cross-dating, and measurement. In *Method of Dendrochronology: Application in the Environmental Sciences*; Editor Cook, E., Kairiukstis, L., Eds.; Springer Science & Business Media Dordrecht: Berlin, Germany, 1992; pp. 40–50. ISBN 978-94-015-7879-0.

39. Omura, H. Tree, forests, and religion in japan. *Mt. Res. Dev.* **2004**, *24*, 179–182. [CrossRef]

40. Fukatsu, E.; Isoda, K.; Hirao, T.; Watanabe, A. Development and characterization of simple sequence repeat dna markers for *Zelkova serrata. Mol. Ecol. Notes* **2005**, *5*, 378–380. [CrossRef]

41. Korbie, D.J.; Mattick, J.S. Protocol: Touchdown pcr for increased specifiy and sensitivity in pcr amplification. *Nat. Protoc.* **2008**, *3*, 1452–1456. [CrossRef] [PubMed]

42. Peakall, R.; Smouse, P.E. GENALEX 6.5: Genetic analysis in excel. Population genetic software for teaching and research-an update. *Bioinformatics* **2012**, *28*, 2537–2539. [CrossRef] [PubMed]

43. Goudet, J. FSTAT, a Program to Estimate and Test Gene Diversities and Fixation Indices (Version 2.9.3.2). 2001. Available online: http://www2.unil.ch/popgen/softwares/fstat.html (accessed on 20 December 2019).

44. Pritchard, J.K.; Stephens, M.; Donnelly, P. Inference of population structure using multilocus genotype data. *Genetics* **2000**, *155*, 945–959. [PubMed]

45. Evanno, G.; Regnaut, S.; Goudet, J. Detecting the number of clusters of individuals using the software structure: A simulation study. *Mol. Ecol.* **2005**, *14*, 2611–2620. [CrossRef]

46. Earl, D.A.; von Holdt, B.M. Stcuture harvester: A website and program for visualizing structure output and implementing the evanno method. *Conserv. Genet. Resour.* **2012**, *4*, 359–361. [CrossRef]

47. Welt, R.S.; Litt, A.; Franks, S.J. Analysis of population genetic structure and gene fow in an annual plant before and after a rapid evolutionary response to drought. *AoB Plants* **2015**, *7*, 1–11. [CrossRef]

48. Funda, T.; Istibůrek, M.; Lachout, P.; Klápště, J.; El-Kassaby, Y.A. Optimization of combine genetic gain and diversity for collection and deployment of seed orchard crops. *Tree Genet. Genomes* **2009**, *5*, 583–593. [CrossRef]

49. Kang, K.S.; Lindgren, D.; Mullin, T.J. Prediction of genetic gain and genetic diversity in seed orchard crops under alternative management strategies. *Theor. Appl. Genet.* **2001**, *103*, 1099–1107. [CrossRef]

50. David, A.; Pike, C.; Stine, R. Comparison of selection methods for optimizing genetic gain and gene diversity in a red pine (*Pinus resinosa* Ait.). *Theor. Appl. Genet.* **2003**, *107*, 843–849. [CrossRef]

51. Funda, T. Population Genetics of Conifer Seed Orchard. Ph.D. Thesis, The Faculty of Graduate Studies (Forestry), The University of British Columbia, Vancouver, BC, Canada, 2012.

52. Kitzmiller, J.H. Managing genetic diversity in a tree improvement program. *For. Ecol. Manag.* **1990**, *35*, 131–149. [CrossRef]

53. Wheeler, N.C.; Jech, K.S. The use of electrophoretic markers in seed orchard research. *New For.* **1992**, *6*, 311–328. [CrossRef]

54. Slatkin, M. Gene flow and the geographic structure of natural population. *Science* **1987**, *236*, 787–792. [CrossRef] [PubMed]

55. Stefenon, V.M.; Gailing, O.; Finkeldey, R. The role of gene flow in shaping genetic structures of the subtropical conifer species *Araucaria angustifolia. Plant Biol.* **2008**, *10*, 356–364. [CrossRef]

56. Lindgren, D.; Mullin, T.J. Relatedness and status number in seed orchard crops. *Can. J. For. Res.* **1998**, *28*, 276–283. [CrossRef]

57. Crow, J.F.; Kimura, M. *An Introduction to Population Genetic Theory*; Blackburn Press: Caldwell, NJ, USA, 2009; pp. 345–364, ISBN 978-1-932846-12-6.

58. Wu, H.X. Benefits and risks of using clones in forestry-a review. *Scandinavian J. For. Res.* **2018**, *34*, 352–359. [CrossRef]

59. Lindgren, D.; El-Kassaby, Y. Genetic consequences of combining selective cone harvesting and genetic thinning in clonal seed orchards. *Silvae Genet.* **1989**, *38*, 65–70.

60. Iwaizumi, M.; Kawai, Y.; Miyamoto, N.; Nasu, J.; Kubota, M.; Mukasyaf, A.A.; Tamura, M.; Watanabe, A. Evaluation of Genetic Diversity of Mother Trees and Seed Pool for Gene Conservation of A *Pinus thunbergii* Population. *Jpn. For. Soc. Congr.* **2020**, *131*. (In Japanese) [CrossRef]
61. Jost, L.; Archer, F.; Flanagan, S.; Gaggiotti, O.; Hoban, S.; Latch, E. Differentiation measures for conservation genetics. *Evol. Appl.* **2018**, *11*, 1139–1148. [CrossRef]
62. Johnson, G.R.; Sorensen, F.C.; St Clair, J.B.; Cronn, R.C. Pacific northwest forest tree seed zones-a template for native plants? *Nativ. Plants J.* **2004**, *5*, 131–140. [CrossRef]
63. O'Brien, E.K.; Krauss, S.L. Testing the home-site advantage in forest trees on disturbed and undisturbed sites. *Restor. Ecol.* **2010**, *18*, 359–372. [CrossRef]
64. Nagamitsu, T.; Shimada, K.; Kanazashi, A. A reciprocal transplant trial suggests a disadvantage of northward seed transfer in survival and growth of japanese red pine (*Pinus densiflora*) trees. *Tree Genet. Genomes* **2015**, *11*, 1–10. [CrossRef]
65. Adams, W.T.; Burczyk, J. Magnitude and implications of gene flow in gene conservation reserves. In *Forest Conservation Genetics: Principles and Practice*; CABI: Wallingford, UK, 2000; pp. 215–224. ISBN 0-643-06260-2.
66. Jump, A.S.; Marchant, R.; Peñuelas, J. Environmental change and the option value of genetic diversity. *Trends Plant Sci.* **2009**, *14*, 51–58. [CrossRef] [PubMed]
67. Hughes, A.R.; Inouye, B.D.; Johnson, M.T.J.; Underwood, N.; Vellen, M. Ecological consequences of genetic diversity. *Ecol. Lett.* **2008**, *11*, 609–623. [CrossRef] [PubMed]
68. FFPRI. *Variety of Pine Wood Nematode Resistant Tree Brochure*; Forest Tree Breeding Center and Forest Bio-Research Center; Forest Research and Management Organization National Research and Development Agency: Tokyo, Japan, 2019. Available online: https://www.ffpri.affrc.go.jp/ftbc/business/sinhijnnsyu/teikousei.html (accessed on 21 April 2020). (In Japanese)

Article

Variability and Plasticity in Cuticular Transpiration and Leaf Permeability Allow Differentiation of *Eucalyptus* Clones at an Early Age

André Carignato [1], Javier Vázquez-Piqué [1], Raúl Tapias [1], Federico Ruiz [2] and Manuel Fernández [1,*]

[1] Department of Agroforestry Sciences, School of Engineering, University of Huelva, 21071 Huelva, Spain; andrecarignatoflorestal@hotmail.com (A.C.); jpique@dcaf.uhu.es (J.V.-P.); rtapias@dcaf.uhu.es (R.T.)

[2] ENCE, Energía y Celulosa, S.A. (ENCE)-Ctra A-5000, km. 7, 5. Apdo. 223, 21007 Huelva, Spain; fruiz@ence.es

* Correspondence: manuel.fernandez@dcaf.uhu.es

Received: 5 November 2019; Accepted: 14 December 2019; Published: 18 December 2019

Abstract: *Background and Objectives.* Water stress is a major constraining factor of *Eucalyptus* plantations' growth. Within a genetic improvement program, the selection of genotypes that improve drought resistance would help to improve productivity and to expand plantations. Leaf characteristics, among others, are important factors to consider when evaluating drought resistance evaluation, as well as the clone's ability to modify leaf properties (e.g., stomatal density (d) and size, relative water content at the time of stomatal closure (RWC_c), cuticular transpiration (E_c), specific leaf area (SLA)) according to growing conditions. Therefore, this study aimed at analyzing these properties in nursery plants of nine high-productivity *Eucalyptus* clones. *Material and Methods*: Five *Eucalyptus globulus* Labill. clones and four hybrids clones (*Eucalyptus urophylla* S.T. Blake × *Eucalyptus grandis* W. Hill ex Maiden, 12€; *Eucalyptus urograndis* × *E. globulus*, HE; *Eucalyptus dunnii* Maiden–*E. grandis* × *E. globulus*, HG; *Eucalyptus saligna* Sm. × *Eucalyptus maidenii* F. Muell., HI) were studied. Several parameters relating to the aforementioned leaf traits were evaluated for 2.5 years. *Results:* Significant differences in stomatal d and size, RWC_c, E_c, and SLA among clones ($p < 0.001$) and according to the dates ($p < 0.001$) were obtained. Each clone varied seasonally the characteristics of its new developing leaves to acclimatize to the growth conditions. The pore opening surface potential (i.e., the stomatal d × size) did not affect transpiration rates with full open stomata, so the water transpired under these conditions might depend on other leaf factors. The clones HE, HG, and 12€ were the ones that differed the most from the drought resistant *E. globulus* control clone (C14). Those three clones showed lower leaf epidermis impermeability (HE, HG, 12€), higher SLA (12€, HG), and lower stomatal control under moderate water stress (HE, HG) not being, therefore, good candidates to be selected for drought resistance, at least for these measured traits. *Conclusions:* These parameters can be incorporated into genetic selection and breeding programs, especially E_c, SLA, RWC_c, and stomatal control under moderate water stress.

Keywords: early selection; stomatal characteristics; water stress; water relations; specific leaf area; *Eucalyptus* clones

---

1. Introduction

Leaf morphology (e.g., specific leaf area, SLA), stomatal characteristics (e.g., stomatal size and density (d)) and stomatal opening are closely linked to physiological activity and transpiration control, in turn related to growth and survival. Responding to the available resources, plants can adjust these characteristics and acclimatize to changing environmental conditions [1,2], and this acclimatization process is genetically influenced [3,4]. Nevertheless, the relationships between stomatal size, *d* and stomatal conductance (gs) should be treated with caution, because the speed of variation of gs is not necessarily related to *d* or stomatal size in different species, or individuals within a species [5].

The regulation of stomatal opening is multigenic, resulting in a multiple control mechanism (water status, illuminance, vapor-pressure deficit (VPD), photosynthetic activity, CO_2 concentration, etc.). However, how simultaneous stomatal signals interact and influence stomatal behavior is relatively unexplored [6]. The tightness of stomatal closure also constitutes an important component of stomatal control, particularly during a drought when it is necessary to restrict transpiration water-loss [7]. When exposed to drought, stomata can become increasingly sensitive to CO_2 concentration and leaf abscisic acid concentration (ABA) in comparison to photosynthetically active radiation (PAR), leaf to air VPD, and leaf water potential [6]. Consequently, water loss through the stomata is based on guard cells opening variations [2], which are in turn produced by fluxes of potassium ions (K^+) in or out of the guard cell [8]. Such mechanisms could include physiological de-/activation of ion transport in the stomatal guard cells, or a genetic control of the expression of ion transport channels [9]. When an area of leaf is placed under conditions of high leaf to air VPD stomata usually close to prevent desiccation, while low leaf to air VPD is conducive to higher rates of gs as the potential of excessive water-loss is diminished [10,11]. Species that have more effective stomatal control are therefore expected to withstand water deficit situations more successfully. However, not all of them have equally effective stomatal control, whether in terms of number of stomata during leaf development or of their regulation of stomatal opening [12]. However, stomatal behavior is not universal, with some species altering the number of stomata on newly developed leaves in response to (CO_2) rather than utilizing physiological regulation of stomatal aperture [13]. Stomatal density and size are usually highly sensitive to environmental abiotic stress such as drought because of stomatal resistance to transpiration [14,15]. Stomatal density (*d*) is often negatively related to stomatal size [16]. A signaling mechanism from the mature leaves to the developing leaves seems to exist, leading to the optimization of *d* and of stomatal size to face the changes to come in future environmental conditions [17].

The *SLA*, for its part, indicates how the leaf biomass is distributed, seeking a balance between carbon gain and water loss, since at equal mass, a broader and thinner leaf blade favors not only photosynthesis but also water loss by transpiration. This property plays an important role in allowing plants to adapt to environmental conditions and this plasticity is often understood as a way of optimizing light absorption as well as water use efficiency (WUE) [3,18]. Thus, as they develop their leaves, plants can resort to certain morphological alterations such as palisade parenchyma thickness [19] or epidermis and cuticle water-tightness [20], the latter being essential to control water losses when stomata are closed, especially during drought periods, by means of so-called cuticular transpiration (E_c).

In areas with marked seasons, especially in regions that have dry seasons, such as the Mediterranean, evaporative demands vary considerably throughout the year and plants must constantly acclimatize: it is thus interesting in these cases to understand how plants transpire over a complete annual cycle [21]. The *Eucalyptus* genus stands out as one of the most widely planted exotic genera in tropical and Mediterranean climate regions and, together with *Pinus*, represents 98% of the world's forestry production [22,23]. Within the genetic improvement programmes of this genus, it is possible to associate desirable characteristics of different species by synthesizing interspecific hybrids [24] and, together with the cloning technique, to generate homogeneous plantations that are highly productive and resistant to pests and diseases [25,26]. Therefore, it is necessary to incorporate simple tools in the evaluation methods of a genetic improvement program in order to evaluate the genotypes and that these tools be applicable on a large scale (i.e., that they measure easily and quickly). For instance, it is essential today to select taxa of *Eucalyptus* spp. that are resistant to water deficit, mainly in regions subject to irregular and scarce annual rainfall regimes [27], and at an early age in order to shorten any improvement program [22]. Among the most widely planted species of eucalypts, it is known that *Eucalyptus camaldulensis* Dehnh. is a drought-tolerant species [28], *Eucalyptus globulus* offers certain genotypes that potentially tolerate environments with low water availability [29], and the hybrid *Eucalyptus × urograndis* is usually sensitive to water deficit [30,31].

WUE relates photosynthetic rate or plant growth to water consumption and has been shown to be a useful physiological parameter to assess plant drought adaptation [22,32,33], and to differentiate the behavior of different taxa [28]. However, WUE is not always a constant trait of a given taxon: it varies according to a specific combination of conditions of the site, weather, and tree age [34]. While the physiological characteristics may vary within a very short interval, the morphological characteristics do not. They maintain the same structure while the organ is functional despite possible environmental changes. Hence the importance of developing organs with an appropriate structure to withstand coming environmental conditions and of studying the effect of some anatomical leaf characteristics on water loss due to plant transpiration, with either fully open or totally closed stomata. For example, anatomical structures of leaves such as palisade parenchyma and stomatal density, and leaf morphology such as leaf thickness and specific leaf weight, regulate the physiological functions (i.e., photosynthesis and transpiration), which vary in different cultivars or clones [35]. Moreover, during drought conditions, gs is directly affected, and the consequent stomatal closure is a way of reducing the water loss due to leaf transpiration and the susceptibility of xylem vessels to cavitation (i.e., embolism or dysfunction) that results in a reduction in hydraulic conductance [36,37]. Therefore, since drought-resistance is a multiple control mechanism, it is the conjunction of several factors, not just one factor, that represents the true degree of each taxon's water consumption and drought-resistance [38,39]. In addition, because annual plant growth and water consumption depend not only the dry season but on the whole year, we hypothesized that differences would exist among genotypes regarding stomatal characteristics and leaf structure, which would vary throughout the year depending on environmental conditions and would be detectable at an early age. The present study deals with nursery plants of nine high productivity *Eucalyptus* clones belonging to a breeding program and that could be used in commercial plantations. It focused on comparing these clones and the seasonal development of (1) leaf stomatal size and density (*d*); (2) cuticular transpiration (E_c); (3) and specific leaf area (*SLA*). The objective was to detect traits that could be incorporated into the improvement programs of this plant genus.

2. Materials and Methods

2.1. Plant Material and Growing Conditions

The starting plant material consisted of 10-month-old plants of five clones belonging to *E. globulus* (reference codes: C14, 225, 227, 358, 437) and four hybrid clones (12€, *Eucalyptus urophylla* S.T. Blake × *Eucalyptus grandis* W. Hill ex Maiden; HE, *E. urograndis* × *E. globulus*; HG, *Eucalyptus dunnii* Maiden–*E. grandis* × *E. globulus*; HI, *Eucalyptus saligna* Sm. × *Eucalyptus maideni* F. Muell.), obtained by rooting cuttings in a commercial nursery, in 150 cm³ containers, provided by the plant selection and breeding program of the ENCE, energía y celulosa, Inc. (Madrid, Spain) The first clone, C14, belonged to the first generation (F_0). It offers high productivity and plasticity and is widely used by the company in commercial plantations, even in areas subject to dry summer seasons. The others corresponded to clones of later generations of improvement and field trials have shown that they can increase productivity by up to 25% with respect to C14 under favorable growing conditions. However, no significant differences were obtained between clones for plant growth in the three assays performed in this study under nursery conditions (data not shown). Generally, within the *Eucalyptus* genus, *E. grandis*, *E. dunnii*, and *E. saligna* are described as low drought-resistant species, while *E. globulus* and *E. urophylla* are moderately resistant [22,34,40,41]. According to field trials with 4–10 year-old plants, the ENCE company classifies seven clones for their drought resistance following this approximate ranking: C14, 437 ≥ 227, 358, 225 ≥ HE, 12€. Further information, however, is not available.

For three consecutive years, they were transplanted in December into 10 L containers allowing for a full vegetative period. Each year, the experimental design consisted of four randomly distributed plants per clone (36 plants per year, 108 plants in total). The substrate consisted of a mixture of peat, coconut fiber and perlite (2:2:1 by volume), that was well watered to field capacity and fertilized using Ferticote 16-7-8 + 2 MgO + Micros (Burés profesional S.A., Girona, Spain) applying a dose of 1.5 kg m^{-3}. Four additional plants per clone and year were grown under the same conditions in order to replace any of the tested plants in case of need. No plants, however, died during the experimental period. The plants were placed outdoors and fully exposed to sunlight in an experimental plot at the University of Huelva (37°12′03″ N, 6°54′53″ W, 5 m a.s.l.).

2.2. Stomatal Characteristics

Across the four seasons of the year, at 10 different dates, and for 2.5 years in a row, 3–4 fully developed leaves were collected per clone and season (i.e., the leaves developed during spring, summer, autumn, or winter were collected at every measurement date), from the 3rd to the 5th whorl of the main stem (34 leaves per clone, 306 leaves in total during the period of study). In the case of *E. globulus* clones, which present foliar dimorphism between mature and juvenile leaves, the harvested leaves were always juvenile due to the plants' size and age. To select the sample leaves, we considered hardness to the touch, the growth stoppage and the constant value of the *SLA*. For this, previous tests were carried out and the additional plants were used to verify these characteristics. Each year, on the dates of the first measurement, the plants averaged 6 mm in stem diameter and 60 cm in height, while on date of the last measurements, after a vegetative period, they averaged 17 mm and 150 cm, respectively. The leaf samples were considered to have developed during the 90-day period prior to each measurement date. This foliar development is highly influenced by environmental conditions (light radiation, humidity and VPD, temperature, etc.), which in turn affect the physiological state of the plant, showing phenotypic plasticity [1,4,18–20,42]. The leaf's morphology and internal structure are formed during its growth and development and, once fully developed, its anatomy does not vary substantially during the rest of its life. Therefore, growth conditions during the leaf's development are more interesting to consider than the conditions of the rest of the year or of subsequent months. Table 1 shows the values of the relevant climatic variables in the area during the study.

Table 1. Temperature, relative humidity, and solar radiation in the nursery 90 days prior to each measurement date: the measurements of February, May, and November were made at the beginning of each month, and the July measurements were made at the end of the month.

Variables	Measurement Dates									
	Feb 2015	May 2015	July 2015	Nov 2015	Feb 2016	May 2016	July 2016	Nov 2016	Feb 2017	May 2017
T_{90} [a] (°C)	16.3	19.9	30.0	24.9	19.8	19.3	28.6	27.3	17.9	20.9
t_{90} [b] (°C)	3.3	6.9	15.1	14.0	8.5	7.5	14.4	13.8	5.7	8.0
RH_{90} [a] (%)	102.1	100.1	89.6	95.8	95.2	93.6	86.7	86.9	89.8	91.6
rh_{90} [b] (%)	59.0	52.3	33.1	51.0	52.7	45.9	37.0	42.8	50.5	46.3
R_{90} [c] (MJ m^{-2} d^{-1})	9.3	16.2	26.3	16.0	9.1	16.8	26.2	17.1	9.5	16.2

[a] T_{90}/RH_{90}: average maximum daily temperature/relative humidity over the 90-day period. [b] t_{90}/rh_{90}: average minimum daily temperature/relative humidity over the 90-day period. [c] R_{90}: cumulative daily solar radiation over the 90-day period.

Leaf prints of these leaves were collected using nail varnish, and the stomata could be observed. Having detected the absence or extreme scarcity of the stomata on the leaves' adaxial side, prints were made of three zones (basal, B; central, C; and apical, A) on the abaxial side (Figure 1a). The image recorded in the nail varnish prints was mounted on a slide that allowed viewing under optical microscope (Leica DM/LS, Leica Microsystems) using an image capture software (Leica LAS EZ, Leica Microsystems). Two images magnified 100 times were randomly captured of each zone (six images per leaf) to determine *d* (number of stomata per mm^2). Subsequently, the number of stomata present in five randomly selected squares measuring 250×250 μm (0.0625 mm^2) on each image were counted. Since each image was 1.05 mm^2 (1.200×0.875 mm), the area measured in the five grids accounted for

29.8% of the image taken. A total of 30 grids per leaf (10 grids per zone) were measured. Five images magnified 400 times were also randomly taken to define the width and length both of the stomatal cells (*SW* and *SL*, respectively) and of the epidermis opening where the ostiole was located (*OW*, *OL*) (Figures 1b and S1). To determine stomatal size, 93 randomly chosen stomata on each leaf (31 stomata per zone) were measured.

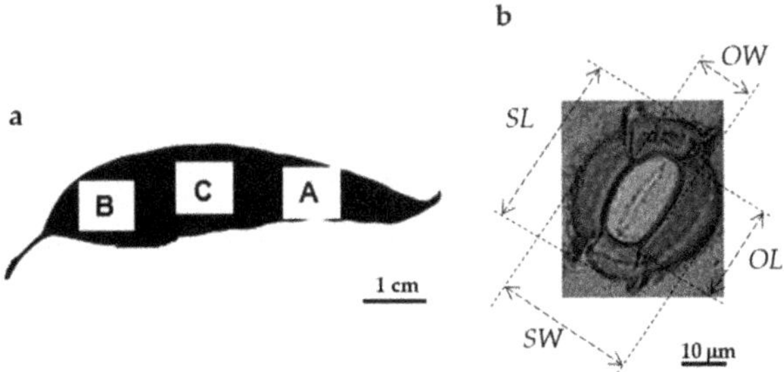

Figure 1. (**a**) Leaf zones (B; C; A) of which leaf prints were made to observe the stomata, and (**b**) the width and length of the stomatal cell (*SW*, *SL*) and the epidermis opening where the ostiole is located (*OW*, *OL*).

2.3. Cuticular Transpiration

On the same dates the stomata were characterized, three more leaves with similar characteristics to those used for the stomata were taken per clone in order to measure E_c under laboratory conditions. The leaves were sampled 1–2 h after dawn, presenting an apparently good water status. Just after cutting, they were placed in sealed plastic bags and taken to the laboratory chilled in a portable cooler. Once in the laboratory (less than 20 min after they were sampled), and following a standard methodology [43], the leaves were hydrated until saturation (in the dark, at 4 °C for 16 h). The next day, before starting the measurements, the leaves were left in the laboratory in the dark until they reached room temperature. They were then weighed in succession using precision scales (±0.1 mg) at short time intervals (every 5 min during the first hour, every 10 min during the following 2 h and every 20 min thereafter, until weight loss was constant over time). During the measurements, the leaves were exposed to light (430 μmol m^{-2} s^{-1} of PAR, using LED lamps), with their abaxial facing downwards on a grid, to allow free circulation of air on both sides, in an environment conditioned at 20–23 °C and 45%–60% relative humidity. E_c was measured under laboratory conditions, maintaining homogeneous environmental conditions for all leaves and at all measurement dates, so that the data could be compared.

Once the leaf area (*LA*) was measured and using the previously collected data (fresh weight, *FW*; and time), it was possible to determine the time elapsed until stomata closure (t_c), estimated by the cut-off point between the curve generated by all value pairs (fresh weight-time) and the regression line generated by the points marking a leaf's constant weight drop (Figure S2); E_c as the slope of the regression line, since the stomata are supposed to be closed when the weight drop is constant; and the relative water content and the moisture content at the time of stomatal closure (RWC_c and M_c, respectively). Initially, transpiration occurred through the stomata (stomatal transpiration, E_s) and leaf epidermis (E_c), but after stomatal closure, only the E_c remained. The leaves were then heated in an oven at 70 °C until they reached a constant weight to determine their dry weight (*DW*). With all these data, it was possible to determine the E_c based on *DW* and *LA*, as well as *SLA* ($SLA = LA/DW$, m^2 kg^{-1}). Additionally, to know the relationship between gas exchange and leaf water potential (Ψ), a test was carried out during the summer of the last year in which the plants were subjected to progressive water stress. These data are shown as supplementary material (Figures S3–S6; Table S1) because they are not the main objective of this study and were measured only at one moment in the year.

2.4. Data Analysis

Regarding d on the abaxial leaf surface, the data were analyzed following a Generalized Mixed Model with Poisson distribution and a logarithmic link function. Regarding stomatal size (SW, SL), the data were analyzed by means of a Mixed Linear Model with Gaussian distribution and identity link function. We took into account the fixed Clone, Date, and Clone × Date interaction effects, and the plant nested within the clone as a random effect. The Akaike information criterion (AIC) was used to choose the selected models [44]. The differences between the groups of the different factors were analyzed by conducting a Scheffé test. Regarding E_c and associated parameters, as well as SLA, the data were analyzed by means of a General Linear two-factor Model (clone, date), which were considered fixed, and the differences between the groups of the different factors were analyzed using Dunnett's T3 test. The fixed Clone, Date, and Clone × Date interaction effects were taken into account. The statistical package SAS® 9.2 was used. The differences were deemed significant obtaining a significance level of $p \leq 0.05$.

3. Results

3.1. Stomatal Characteristics

Significant differences relating to d, stomatal size (SL, SW) and the SW/SL ratio among clones ($p < 0.001$) and among dates ($p < 0.001$) were detected (Tables 2 and 3). The interaction between the two factors was also significant ($p < 0.001$), indicating a different seasonal development pattern among clones (Figure 2). The results obtained for SW, OW, and OL are not presented in more detail due to the high and significant correlations obtained: SL vs. SW ($r = 0.746$), SL vs. OL ($r = 0.795$), OL vs. OW ($r = 0.683$), 14,043 being the sample size and $p < 0.001$ for all of them. These three parameters showed seasonal development and differentiation among clones and dates similar to SL. The mean values (±SE) obtained for the series of clones and measurement dates were $SW = 16.6 \pm 0.3$ μm; $OL = 14.4 \pm 0.2$ μm; $OW = 10.1 \pm 0.2$ μm. The SW/SL ratio indicated that the stomata tended to be elliptical, and only the three clones with the lowest average value (HE, HG, and HI) were significantly differentiated from the clone with the highest value (437) (Table 2).

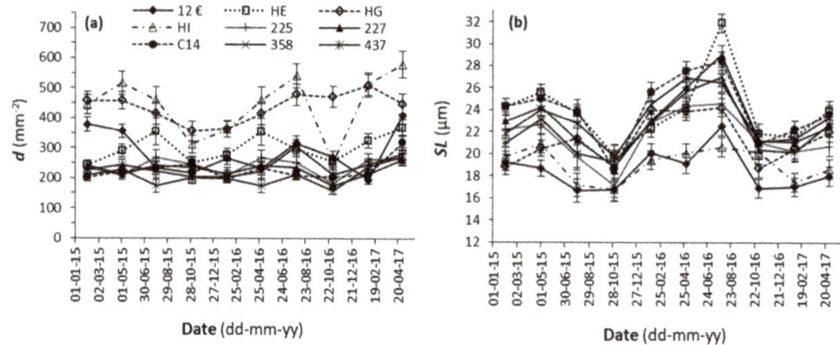

Figure 2. Seasonal evolution (average ±SE) of (**a**) stomatal density (d), and (**b**) stomatal length (SL) of the nine *Eucalyptus* clones studied on the 10 measurement dates.

Taking the maximum possible stomatal opening into account—that is, the share of LA that all open ostioles would entirely occupy if they covered the full window left free by the epidermis, calculated via the expression $d \times (\pi \times OW \times OL)/4$—the *E. globulus* clones as well as the 12€ hybrid were in the lowest range (from 2.3% for 12€ to 3.2% for C14), while the other three hybrids (HI, HE, and HG) were within a range of 3.7% (HI) to 4.4% (HG).

Table 2. Average values (±SE) of stomatal density (d) and length (SL) of the occlusive cells, ratio between length and width of the occlusive cells (SW/SL ratio), relative water content at stomatal closure (RWC_c), time elapsed (t_c) before stomatal closure, moisture content at stomatal closure (M_c), cuticular transpiration rate (E_c), and specific leaf area (SLA) of the different *Eucalyptus* genotypes under study. Different letters in each column indicate significant differences between clones.

Clone	d (mm^{-2})	SL (μm)	SW/SL	RWC_c (%)	t_c (min)	M_c (%)	E_c (mmol m^{-2} s^{-1})	SLA (m^2 kg^{-1})
12€	279 ± 7 b	18.5 ± 0.3 a	0.77 ± 0.01 ab	78.2 ± 1.3 ab	70.2 ± 3.0 b	61.8 ± 1.0 a	0.12 ± 0.02 a	13.4 ± 0.6 c
HE	297 ± 9 b	23.9 ± 0.3 d	0.75 ± 0.01 a	74.8 ± 1.4 a	80.6 ± 3.7 b	59.9 ± 0.9 a	0.27 ± 0.02 b	9.9 ± 0.3 ab
HG	435 ± 11 c	21.3 ± 0.3 b	0.75 ± 0.01 a	80.9 ± 1.3 ab	64.5 ± 3.8 ab	61.7 ± 0.8 a	0.16 ± 0.02 a	12.2 ± 0.5 c
HI	431 ± 11 c	19.1 ± 0.3 a	0.75 ± 0.01 a	83.4 ± 1.0 b	55.5 ± 3.2 a	61.4 ± 0.8 a	0.12 ± 0.01 a	11.0 ± 0.4 bc
225	238 ± 6 a	21.6 ± 0.3 bc	0.77 ± 0.01 ab	79.7 ± 1.1 ab	73.4 ± 3.6 b	61.0 ± 1.0 a	0.16 ± 0.02 a	9.2 ± 0.3 a
227	225 ± 6 a	23.0 ± 0.3 cd	0.77 ± 0.01 ab	79.4 ± 1.2 ab	73.9 ± 4.7 b	61.4 ± 0.9 a	0.18 ± 0.03 ab	9.7 ± 0.4 ab
C14	226 ± 6 a	24.0 ± 0.3 d	0.76 ± 0.01 ab	79.3 ± 1.0 ab	76.5 ± 3.6 b	61.1 ± 0.8 a	0.18 ± 0.02 ab	9.0 ± 0.3 a
358	204 ± 6 a	22.2 ± 0.3 bc	0.77 ± 0.01 ab	80.4 ± 1.1 ab	72.3 ± 2.8 b	60.5 ± 0.8 a	0.13 ± 0.01 a	9.3 ± 0.3 a
437	235 ± 7 a	23.2 ± 0.3 cd	0.78 ± 0.01 b	80.5 ± 1.2 ab	72.3 ± 3.5 b	60.6 ± 0.9 a	0.18 ± 0.03 ab	9.1 ± 0.4 a
p	<0.001	<0.001	<0.001	<0.001	<0.001	0.536	<0.001	<0.001
Total	286 ± 8	21.9 ± 0.3	0.76 ± 0.01	79.6 ± 0.4	70.9 ± 1.2	61.0 ± 0.3	0.17 ± 0.01	10.4 ± 0.2

Table 3. Average values (±SE) of stomatal density (d) and length (SL) of the occlusive cells, ratio between length and width of the occlusive cells (SW/SL ratio), relative water content at stomatal closure (RWC_c), time elapsed (t_c) before stomatal closure, moisture content at stomatal closure (M_c), cuticular transpiration rate (E_c) and specific leaf area (SLA) at each measurement date. Different letters in each column indicate significant differences between dates.

Measurement Date	d (mm^{-2})	SL (μm)	SW/SL	RWC_c (%)	t_c (min)	M_c (%)	E_c (mmol m^{-2} s^{-1})	SLA (m^2 kg^{-1})
01 Feb. 2015	278 ± 6 bcd	21.6 ± 0.3 bcd	0.74 ± 0.01 a	73.2 ± 1.2 a	86.3 ± 3.3 c	58.7 ± 0.6 ab	0.16 ± 0.01 abc	11.5 ± 0.3 c
01 May 2015	287 ± 7 cd	22.8 ± 0.3 de	0.78 ± 0.01 d	82.9 ± 0.8 c	81.3 ± 4.3 bc	60.4 ± 0.6 bc	0.12 ± 0.01 a	10.1 ± 0.3 bc
25 July 2015	276 ± 10 abcd	20.7 ± 0.4 bc	0.78 ± 0.01 bcd	79.2 ± 1.7 abc	50.0 ± 2.7 a	59.4 ± 1.5 abcd	0.22 ± 0.04 abc	9.9 ± 0.7 abc
01 Nov. 2015	247 ± 8 abc	18.4 ± 0.4 a	0.75 ± 0.01 abc	81.4 ± 2.5 abc	49.0 ± 1.5 a	67.7 ± 1.1 f	0.24 ± 0.03 abc	14.4 ± 0.6 d
01 Feb. 2016	247 ± 7 ab	22.8 ± 0.3 cde	0.76 ± 0.01 abcd	80.8 ± 0.8 c	66.3 ± 1.6 b	62.4 ± 0.6 cd	0.16 ± 0.02 abc	10.6 ± 0.4 bc
01 May 2016	276 ± 10 abcd	24.2 ± 0.3 e	0.77 ± 0.01 bcd	82.6 ± 0.7 c	64.1 ± 2.7 b	56.5 ± 0.4 a	0.14 ± 0.01 abc	8.1 ± 0.2 a
25 July 2016	299 ± 8 de	26.1 ± 0.3 f	0.78 ± 0.01cd	80.5 ± 2.0 abc	74.7 ± 2.5 bc	60.8 ± 0.9 bce	0.18 ± 0.05 abc	9.2 ± 0.3 ab
01 Nov. 2016	229 ± 6 a	20.3 ± 0.3 ab	0.75 ± 0.01 abc	79.4 ± 1.2 bc	70.2 ± 3.4 bc	64.6 ± 0.9 def	0.22 ± 0.02 c	10.7 ± 0.4 bc
01 Feb. 2017	283 ± 7 bcd	20.2 ± 0.3 ab	0.75 ± 0.01 ab	75.5 ± 0.9 ab	73.9 ± 3.0 bc	64.5 ± 0.5 df	0.18 ± 0.02 bc	11.8 ± 0.4 cd
01 May 2017	344 ± 9 e	21.8 ± 0.3 bcd	0.77 ± 0.01 bcd	81.8 ± 0.9 c	77.7 ± 4.8 bc	57.3 ± 0.7 ab	0.12 ± 0.01 ab	8.3 ± 0.3 a
p	<0.001	<0.001	<0.001	<0.001	<0.001	<0.001	<0.001	<0.001
Total	286 ± 8	21.9 ± 0.3	0.76 ± 0.01	79.6 ± 0.4	70.9 ± 1.2	61.0 ± 0.3	0.17 ± 0.01	10.4 ± 0.2

3.2. Cuticular Transpiration

Concerning the RWC_c and the t_c, significant differences were detected among clones (Table 2) and among dates (Table 3). They were also found regarding the interaction between these two factors, with $p < 0.001$ for RWC_c (Figure 3) and $p = 0.001$ for t_c. This indicated that the clones' seasonal evolution patterns differed among themselves.

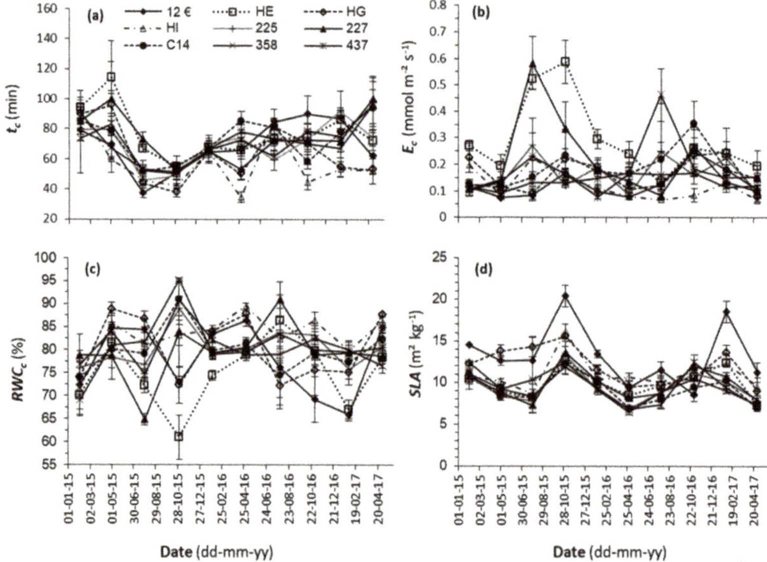

Figure 3. Seasonal evolution (average ±SE) of (**a**) the time elapsed until stomata closure (t_c), (**b**) cuticular transpiration (E_c) based on *LA*, (**c**) relative water content at the time of stomata closure (RWC_c), and (**d**) specific leaf area (*SLA*), of the nine *Eucalyptus* clones studied on the 10 measurement dates.

The E_c differed significantly according to clones and measurement dates; both were calculated based on *DW* and *LA* ($p < 0.001$) (Tables 2 and 3). Furthermore, significant differences were found regarding the Clone × Date interaction ($p < 0.001$, Figure 3). The behavior pattern of E_c, expressed on a leaf weight basis did not differ significantly from that calculated based on *LA*, in terms of differentiation among clones or among dates, with an overall average value of 1.69 ± 0.06 mmol kg^{-1} s^{-1} of H_2O.

Regarding M_c, no significant differences were detected among the different clones (Table 2), but they were detected according to dates (Table 3), as well as to Clone × Date interaction ($p < 0.001$). Regarding a general trend, it is worth noting that the lowest values were obtained from the measurements made in May and July ($M_c = 56.5\%$–60.8%) and the highest values were obtained in November ($M_c = 64.6\%$–67.7%).

Finally, *SLA* differed significantly according to the different clones, dates of measurement, and Clone × Date interaction (Figure 3, Tables 2 and 3). Regarding the measurement dates, the clones of *E. globulus* and HE (9.0–9.9 m^2 kg^{-1}) were different from HI (11.0 m^2 kg^{-1}) and the latter, in turn, from the group formed by HG and 12€ (13.2–13.4 m^2 kg^{-1}).

4. Discussion

4.1. Stomatal Characteristics

The stomata of the *Eucalyptus* clones used in the present study were concentrated on the leaf's abaxial surface. A very small number of stomata were found on the adaxial surface i.e., 2–3 stomata per square millimeter, in all clones. Tuffi Santos et al. [45], when studying d in very young plants of *E. grandis*, *E. urophylla*, *E. saligna*, *Eucalyptus pellita* F. Muell, and *Eucalyptus resinifera* Sm., also found that in the case of the whole species, the adaxial surface (10–80 stomata mm^{-2}) had 10 times fewer stomata than the abaxial surface (600 stomata mm^{-2}), with differences between taxa. In the present study, the average d values of the abaxial face were between 204 and 434 stomata per mm^2. This value is within the value range of sclerophyllous leaves, i.e., 100–500 mm^{-2}, and is typical in species inhabiting rainforests [46] or temperate zones such as *Pinus taeda* L., *Taxodium distichum* (L.) Rich., or *Ilex cassine* L. [17]. The value is, however, below the 750–1050 mm^{-2} value range found in subtropical species such as *Toona ciliata* M. Roem. [47], or the value of 1000 mm^{-2} reached by some oaks and maples proper to humid temperate zones [17].

Although the studied plants were well watered and fertilized at all times, the leaves that grew mainly in spring, specifically from the end of winter to the beginning of summer (May and July measurement dates) tended to present greater d, at least in the case of clones that showed more marked seasonal differences. A higher density would facilitate the water's exit (and the assimilation of CO_2) at times of suitable water availability in the soil and non-excessive atmospheric demand [5,35,42]. However, in the case of leaves developed in the middle-end of summer and early autumn (mainly from late July to late September) which correspond to the measurements taken at the beginning of November since leaves from the 3rd–5th whorl were sampled when atmospheric demand was greater d decreased, allowing plants to save water and better endure droughts. All this indicates that, apart from the availability of soil water and nutrients, the plants responded to other environmental stimuli (photoperiod, solar radiation, air temperature, relative humidity, etc.) [10,13,48]. The latter controlled the d of the new developing leaves at all times, suggesting the existence of an internal mechanism that stimulates and transmits the signal [10]. The clones with the highest d studied here, HI and HG, with an average of around 430 mm^{-2}, almost doubled the d of *E. globulus* clones (358, C14, 437, 227, and 225). The lower d of *E. globulus* among the clones studied could be a drought-adaptation characteristic, although other characteristics should be taken into account as a whole [49]. For the *Azadirachta indica* A. Juss and *Populus* species, for example, d was positively correlated with net photosynthesis and biomass production for *A. indica* [50] or with gs for *Populus* sp. [51]. However, in other studies, d did not significantly affect gs or photosynthetic rates [52]. On the other hand, clones of *E. globulus* and clone HE presented lower plasticity regarding d variation throughout the year ($\Delta d < 150$ mm^{-2}), while the density in the other three clones varied according to the dates, from 160 mm^{-2} (HG) to 300 mm^{-2} (HI). Therefore, interestingly, these latter clones presented greater plasticity regarding this parameter than *E. globulus* clones.

Generally, in this study, stomatal cell size was smaller in the leaves that grew in the middle-end of summer and early autumn (November measurements), and bigger in the leaves that grew in spring. As in the case of d, the reason could be that during the summer and early autumn months, when the temperature, radiation, and photoperiod are higher and relative humidity lower (Table 1), the new developing leaves favored the formation of smaller stomata at warmer and drier times, and vice versa in winter-spring, in order to regulate gas exchange and WUE. The same phenomenon was found in the case of *Sequoia sempervirens* (D. Don) Endl. plants across different plantations in Chile [53]. When comparing the clones, clones containing *Eucalyptus globulus* alleles were significantly larger in size (SL = 21.3–24.0 µm, SW = 15.9–18.2 µm) than 12€ and HI (SL = 18.5–19.1 µm, SW = 14.2–14.3 µm). These values were in a slightly higher range than those found for three other *Eucalyptus* species (*E. delegatensis* R. Baker, *E. pauciflora* Sieb. ex Spreng., and *E. radiata* Sieber ex DC.), presenting a range of 9.8–12.0 µm for OL and 10.0–14.0 µm for SW [49].

On the other hand, considering the size and d combination in our study, the clones with the biggest pore opening surface potential, HG, HE, and HI, could have a greater transpiration potential with full open stomata, while at the same time pose a risk of excessive water loss in situations of water shortage when stomatal control was not optimal. It has been reported that d and occlusive cell length are related to gs and net photosynthesis, as well as other plant physiological characteristics [11,54–56].

Nevertheless, in other recent studies, no significant correlations between stomatal density, size and the rapidity of response have been detected [13,57]. Consequently, we can assume that the seasonal modifications and adjustments of the stomatal size and d found in this study may affect the plants' physiology, to some extent at least. These modifications surely allow each clone to self-adjust to maintain its best level of photosynthetic efficiency, responding to environmental stimuli. However, the results of this study did not find that d and stomatal size, in themselves, were relevant clone selection criteria for water saving, since in the case of well-watered plants (i.e., $\Psi \geq 1.0$ MPa), maximum transpiration rates were similar between clones (Figures S4 and S5). Therefore, when all pores are supposed to be fully open, apart from the maximum potential of open pore surface determined by d and stomatal size, other factors such as mesophilic conductance, boundary layer resistance, etc., should be taken into account. In addition, water loss, which depends on stomatal opening and other environmental factors, can vary greatly even for plants with a good water status, and this multiple control mechanism seems to have a greater effect on the total amount of water transpired every day than d and stomatal size. For instance, the HE, HG, and 12€ clones maintained high transpiration rates when Ψ dropped from −1.0 to −2.0 MPa. Meanwhile, the other studied clones began to reduce their transpiration when Ψ reached −1.0 MPa (Figures S4 and S5), which indicates a more pronounced water saving behavior. Moreover, in this Ψ (−1.0 to −2.0 MPa) range, the photosynthetic rate was reduced to a greater extent than E and gs for HE, HG, and 12€ clones, so WUE and intrinsic water use efficiency (IWUE, Figure S6) decreased in the case of these three clones to a greater extent than for *E. globulus* clones in the Ψ range. The latter also points to the worse behavior of the first three clones under moderate water stress conditions.

Studies of different tree species, such as riverside poplars in a semiarid environment [51], rainforest species [46], hardwood species in a subtropical climate [58], and *Eucalyptus globulus* in sites with varying precipitation [59], reported a reduction in stomatal size as d increased. However, in this study, despite finding a significant negative correlation between d and stomata size in the nine clones as a whole ($p = 0.017$), the result was very weak (r = −0.252) and not significant for each clone separately ($p > 0.10$).

4.2. Cuticular Transpiration and SLA

RWC_c is a notable indicator of the leaves' water state under severe water stress conditions. Water state is closely related to cell turgor and, therefore, it accurately reflects the balance between internal water content, water supply to the leaf and transpiration rate, as well as leaf dehydration tolerance [60,61]. In our study, the seasonal development varied between a minimum of 73.2% (leaves developed in autumn and winter) and a maximum of 82.9% (developed in spring), presenting no significant differences between those developed in spring and in summer. The clones that diverged the most from this general trend and that presented large seasonal oscillations throughout the study were: HE, 227, and 12€. The average values obtained for RWC_c, as well as the seasonal development were within the range obtained by Carevic et al. [43], who studied Holm oak trees (*Quercus ilex* L. spp. *ballota*), a Mediterranean species with sclerophyllous leaves. These results are also compatible with the values obtained by other researchers [62,63] for eucalypts, as a range of 79%–90% was obtained, detecting differences between species and between clones. Andivia et al. [64] studied the drought tolerance of two Holm oak provenances and observed seasonal variations for this parameter: they reflected a water conservation strategy during the summer (with RWC_c close to 90%) and water spending during the rainy season. All the above indicates that although the nine *Eucalyptus* clones under study presented slight differences, they all used physiological adaptations to reduce water loss at times of greatest demand, as they reacted by closing the stomata at higher hydration levels (RWC_c) at the most unfavorable times from the viewpoint of available water. This latter trait is useful for surviving under climates with dry seasons, such as the Mediterranean climate. Following this line of argument, the series of measurement dates revealed that the HE clone showed the least cautious water conservation behavior: it allowed greater dehydration before stomata closure indicating a survival risk as water stress progresses, while the reverse was found for the HI clone, and an intermediate behavior was observed for the rest of the clones.

As the leaf is dehydrated, the t_c may vary according to the following factors: the leaf's age; the size of the stoma, and its location; the species; the individuals within a species; the vegetative state; and the environmental conditions under which the measurements are taken [65]. The HI clone stood out among the clones of the present study due to its shorter stomatal closing time (56 min under the measurement conditions), which may indicate a drought resistance strategy but at the cost of reducing growth, differing significantly from seven other clones (225, 227, 358, 437, C14, 12€, HE), whose closing times varied between 70 min (12€) and 80 min (HE). Furthermore, the HG clone showed an intermediate stomatal closing time (65 min). Regarding seasonal development, t_c was included within a range of 49 min (November 2015 measurement) to 86 min (February 2015). In 2015, leaves that formed during periods of greater atmospheric demand took half the time to close their stomata compared to those that formed during colder and wetter periods. This phenomenon was not observed the following year. Further evidence was thus obtained regarding the plasticity strategies employed to acclimatize to conditions such as the Mediterranean climate, and the necessary acclimatization to the environmental conditions of the moment to prevent excessive water loss in the drier months.

Plants' leaf epidermis, especially the cuticle, acts as an effective protective barrier against uncontrolled water loss. During a water stress period, when the stomata are closed, the plants' survival greatly depends on the amount of water lost through the cuticle in question. The E_c values obtained i.e., 0.17 mmol m^{-2} s^{-1} of H_2O (1.69 mmol kg^{-1} s^{-1}) on average, were higher than those found for full-grown oaks [43] grown in the field but measured also under laboratory conditions (0.06–0.19 mmol kg^{-1} s^{-1}). They were, however, below those obtained by Fernández et al. [66] for seven species (*Dichrostachys cinerea* (L.) Wight & Arn., *Populus* × *euroamericana* (Dode) Guinier "I-214", *Eucalyptus camaldulensis*, *Casuarina cunninghamiana* Miq., *Paulownia fortunei* (Seem) Hemsl., *Salix purpurea* L., and *Leucaena diversifolia* (Schltdl.) Benth.) that had grown under shade cloth in a nursery. In this latter study, the values obtained were of 0.83 mmol m^{-2} s^{-1} for *E. camaldulensis* up to 3.98 mmol m^{-2} s^{-1} for *S. purpurea*. The leaves of *Eucalyptus haemastoma* (Sm.), a species belonging to a semiarid summer climate with cold winter nights were analyzed at different temperatures [67], and the E_c values ranged from 0.04 mmol m^{-2} s^{-1} at low temperature (18 °C), to 0.5 mmol m^{-2} s^{-1} at high temperature (38 °C). All these value ranges are within those found in our study. Considering the seasonal variation obtained, the general tendency was that the leaves developed in spring (May measurements) presented the lowest level of E_c. Regarding differences between clones, despite showing highly similar E_c, one clone (HE), with the highest E_c value for the study period as a whole, differed significantly from five other clones (12€, HG, HI, 225, 358), as it presented an E_c that was 61% above the average of the nine clones. This suggests that when measuring leaf epidermis permeability, the HE clone had a less efficient water saving strategy [64] compared to these five clones in the study and therefore less drought resistance; whereas in the intermediate range, the clones 437, 227, and C14 revealed a moderate water saving strategy. Considering the relationship of E_c with Ψ determined via a water stress test with these clones (Figure S3 for $\Psi < -1.7$ MPa), most stomata are supposed to be closed since the turgor loss point has been overcome (Table S1). Thus, only E_c would remain. Under these water stress conditions, the HE clones along with HG and 12€, the three clones that may have inherited *E. grandis* alleles, showed the highest E_c rates (Figures S4 and S5), indicating a thinner and/or more permeable leaf epidermis, a parameter that is unsuitable for withstanding periods of water stress. Thus, it is worth studying the relationship between E_c and leaf permeability more in depth, addressing not only leaf and cuticle thickness but also other factors such as the presence of waxes and trichomes which could result in stable selection criteria [68,69]. In addition, concerning the water stress test, only for these three clones (HE, HG, 12€) was the relative water content for which they closed stomata (RWC_{cs}) significantly lower than the relative water content in which the loss of cell turgor occurred, RWC_0 (Table S1). This would indicate a lower degree of stomatal control, since the stomata do not close even when cell turgor is lost, indicating an unsuitable behavior when the water stress progresses. However, the latter should be interpreted with caution because RWC_0 and RWC_{cs} are measured using different methods and their physiological interpretation differs a little.

Regarding the *SLA*, the *Eucalyptus globulus* clones and the HE clone had the lowest values, differentiating themselves mainly from hybrids 12€ and HG, both with *Eucalyptus grandis* alleles. All this would indicate that *E. globulus* would have thicker leaves, appropriate for its greater adaptation to dry climates, compared to *E. grandis*, typical of more humid climates and drought tolerant [30,70]. Of the two clones that may have inherited both *E. globulus* and *E. grandis* alleles, the *E. globulus* inheritance seems to have dominated regarding this parameter in the case of HE, while in the case of HG, the *E. grandis* inheritance seems to have dominated. The *SLA* values obtained in this study are within the range reported by other authors for eucalypts e.g., 16.1 m^2 kg^{-1} (*Eucalyptus occidentalis* Endl.) to 25.4 m^2 kg^{-1} (*E. grandis*) [71], and 6.1–8.3 m^2 kg^{-1} for plants of *E. dunnii* and *Corymbia citriodora* subsp. *Variegata* (F. Muell.) A.R. Bean & M.W. McDonald aged 11 years [72]. During the seasonal development of the studied clones, *SLA* decreased progressively over time from autumn-winter leaves to spring-summer leaves. This results from the varying climatic conditions of relative humidity, radiation, and temperature, taking into account that the plants were well watered and fertilized, demonstrating once again their sensitivity and ability to react to climatic variables.

5. Conclusions

The follow-up, over 2.5 years, of the stomata characteristics and the E_c of nine nursery-grown *Eucalyptus* clones led to the following conclusions:

- All the clones under study showed seasonal variations in d and stomatal size, *SLA*, and E_c, as well as in RWC_c and the t_c, despite the substrate being constantly humidified to field capacity and fertilized, in response to stimuli such as light radiation, photoperiod, temperature, and/or the air's evaporative demand to acclimatize to the environmental conditions.

- Each clone adjusted its own d and size values to acclimatize its stomata to the growth conditions. The maximum amount of water transpired with fully open stomata might depend on other internal and external factors. Thus, the criteria of size and d alone are not sufficient to differentiate between clones, at least for this study.

- The HE and HG clones presented poor stomatal control as they showed high transpiration rates when Ψ was between −1.0 and −2.0 MPa, which represent a risk in cases of prolonged drought. The HI clone, on the other hand, conserved internal water very efficiently because it closed stomata sooner and at a lesser degree of dehydration.

- Taking into account only the morpho-physiological parameters measured in this study and the genotypes considered, the clone known to resist to droughts, C14, was characterized by low values of *SLA* and E_c, high RWC_c, low seasonal plasticity regarding d, and good stomatal control. These characteristics were shared by the other *E. globulus* clones, though the clones of hybrids differed as they presented less favorable properties for drought resistance such as lower epidermis impermeability (HE, HG, 12€), higher *SLA* (12€, HG), and lower stomatal control under conditions of moderate water stress (HE, HG).

- These analyzed eucalyptus clones showed they had genetic variability for drought resistance. Gains could thus potentially be obtained through the selection and implantation of improved populations. However, in situations of water stress, other morpho-physiological properties should be studied (e.g., WUE, xylem anatomy and cavitation vulnerability, osmotic adjustment capacity, etc.) together with the characteristics studied here.

Supplementary Materials: The following are available online at http://www.mdpi.com/1999-4907/11/1/9/s1, Figure S1: (left) Stomata observed on leaf prints taken from the abaxial side of a leaf (400×); (right) cross section of a leaf sampled in the summer of 2017, showing the adaxial side to the right and the abaxial side to the left (400×). Regarding the cross-section of the leaf, four leaves per clone were analyzed in summer 2017 and no significant differences were detected between clones in these four parameters measured: cross-sectional thickness ($p = 0.111$; 267.56 ± 27.4 μm); thickness of the adaxial epidermis ($p = 0.160$; 20.6 ± 2.1 μm); thickness of the abaxial epidermis ($p = 0.370$; 17.1 ± 1.6 μm); and palisade parenchyma thickness ($p = 0.500$; 76.7 ± 7.5 μm), Figure S2: The graph generated by all the pairs of values, *FW*-time (continuous line, rhombuses) and the regression line generated with the points marking a leaf's constant weight drop (dashed line, squares). The cut-off point between the curve and the regression line is assumed to be the t_c (arrow). The slope of the regression line reflects the loss of water over time, from which E_c can be deduced, Figure S3: Plants of the nine clones under study, used to measure daily transpiration (by weighing), and instantaneous transpiration using a portable infrared gas analyzer (Model LCi, ADC, London, UK). Daily transpiration was calculated based on the difference in weight between two measurements taken 24 h apart, measured 1 h after dawn on two consecutive days. The containers were wrapped with white plastic to avoid direct evaporation from the substrate. The total *LA* was measured for each plant. The assay started with plants watered to field capacity, but subsequently, they were watered, every day, with half of the water transpired the previous day, to subject the plants to a slow and progressive process of water stress for 30 days. Ψ was measured exactly at dawn (PMS 1000, Corvallis, USA). The instantaneous transpiration rate (E) was measured 2 h after dawn, when plants show maximum daily transpiration rates. This test was carried out during the summer of 2017, using three plants per clone from the additional plants left over from the main assay, Figure S4: Relationship between daily transpiration rate over a 24-h period, and the water potential at dawn of the first day of each measurement date, for the nine clones studied, Figure S5: Relationship between the E measured 2 h after dawn and the water potential at dawn, for the nine clones studied. E was significantly correlated with gs (E = 9.036 gs + 5.547, r = 0.963, $p < 0.001$) and net photosynthetic rate, A (A = $0.038\,E^4 - 0.661\,E^3 + 3.446\,E^2 - 2.525\,E + 0.619$, r = 0.962, $p < 0.001$). E (mmol m^{-2} s^{-1} of H_2O), gs (mol m^{-2} s^{-1} of H_2O), A (μmol m^{-2} s^{-1} of CO_2), Figure S6. Relationship between the intrinsic water use efficiency (IWUE = A/gs) measured 2 h after dawn and the water potential at dawn, for the nine clones studied. A (μmol m^{-2} s^{-1} of CO_2), gs (mol m^{-2} s^{-1} of H_2O), Table S1: Mean value (±SE) of the osmotic potential at full turgor (Ψ_{s100}) and at the point of turgor loss (Ψ_{s0}), the RWC_0 and RWC_{cs} of the nine studied clones. The measurements were made on two dates, first in well-watered plants and then after the plants were subjected to a progressive water stress test for 30 days in the summer of 2017 (see Figure S3), by means of the construction of isothermal pressure-volume curves, using the methodology described by [73]. *p*: level of significance. Different letters in each column indicate significant differences between clones. *: for each clone, asterisk indicates significant differences between RWC_0 and RWC_{cs} ($p < 0.001$, Dunnett's T3 test).

Author Contributions: F.R., M.F. and R.T. designed the experiments; A.C. and M.F. conducted the experiment, and wrote the first draft; A.C., J.V.-P. and M.F. analyzed the data; J.V.-P. and R.T. revised and edited the manuscript. All authors have read and agreed to the published version of the manuscript.

Funding: This study was supported by the Conselho Nacional de Desenvolvimento Científico e Tecnológico (CNPq)—Brazil (grant number 203224/2014-0), the company ENCE, energía y celulosa S.A., (grant number Contrato art. 68/83) and the National Research Programme, reference CTQ2013-46804-C2-1R and CTQ2017-85251-C2-2-R which, in turn, were financed by FEDER.

Acknowledgments: We thank all authors for their contributions to this study. We would like to thank Open Five S.L. services for the English-language revision.

Conflicts of Interest: The authors declare no conflict of interest.

References

1. Schlichting, C.D. The evolution of phenotypic plasticity in plants. *Annu. Rev. Ecol. Syst.* **1986**, *17*, 677–693. [CrossRef]

2. Bussis, D.; Von Groll, U.; Fisahn, J.; Altman, T. Stomatal aperture can compensate altered stomatal density in *Arabidopsis thaliana* at growth light conditions. *Funct. Plant Biol.* **2006**, *33*, 1037–1043. [CrossRef]

3. Poorter, L.; Rozendaal, D.M.A. Leaf size and leaf display of thirty-eight tropical tree species. *Oecologia* **2008**, *158*, 35–46. [CrossRef] [PubMed]

4. Mejía de Tafur, M.S.; Burbano Diaz, R.A.; García Díaz, M.A.; Baena García, D. Respuesta fotosintética de *Eucalyptus grandis* W. Hill a la disponibilidad de agua en el suelo y a la intensidad de luz. *Acta Agron.* **2014**, *63*, 311–317. [CrossRef]

5. Haworth, M.; Scutt, C.P.; Douthe, C.; Marino, G.; Gomes, M.T.G.; Loreto, F.; Flexas, J.; Centritto, M. Allocation of the epidermis to stomata relates to stomatal physiological control: Stomatal factors involved in the diversification of the angiosperms and development of amphistomaty. *Environ. Exp. Bot.* **2018**, *151*, 55–63. [CrossRef]

6. Haworth, M.; Marino, G.; Cosentino, S.L.; Brunetti, C.; De Carlo, A.; Avola, G.; Riggi, E.; Loreto, F.; Centritto, M. Increased free abscisic acid during drought enhances stomatal sensitivity and modifies stomatal behaviour in fast growing giant reed (*Arundo donax* L.). *Environ. Exp. Bot.* **2018**, *147*, 116–124. [CrossRef]

7. García-Mata, C.; Lamattina, L. Nitric oxide induces stomatal closure and enhances the adaptive plant responses against drought stress. *Plant Physiol.* **2001**, *126*, 1196–1204. [CrossRef]

8. Shimazaki, K.; Doi, M.; Assmann, S.M.; Kinoshita, T. Light regulation of stomatal movement. *Annu. Rev. Plant Biol.* **2007**, *58*, 219–247. [CrossRef]

9. Gerardin, T.; Douthe, C.; Flexas, J.; Brendel, O. Shade and drought growth conditions strongly impact dynamic responses of stomata to variations in irradiance in *Nicotiana tabacum*. *Environ. Exp. Bot.* **2018**, *153*, 188–197. [CrossRef]

10. Dai, Z.; Edwards, G.E.; Ku, M.S.B. Control of photosynthesis and stomatal conductance in *Ricinus communis* L. (castor bean) by leaf to air vapor pressure deficit. *Plant Physiol.* **1992**, *99*, 1426–1434.

11. Mott, K.A.; Peak, D. Testing a vapour-phase model of stomatal responses to humidity. *Plant Cell Environ.* **2013**, *36*, 936–944. [CrossRef] [PubMed]

12. Haworth, M.; Elliott-Kingston, C.; McElwain, J.C. Stomatal control as a driver of plant evolution. *J. Exp. Bot.* **2011**, *62*, 2419–2423. [CrossRef] [PubMed]

13. Haworth, M.; Killi, D.; Materassi, A.; Raschi, A. Co-ordination of stomatal physiological behavior and morphology with carbon dioxide determines stomatal control. *Am. J. Bot.* **2015**, *102*, 677–688. [CrossRef] [PubMed]

14. Givnish, T.J. Adaptation to sun and shade: A whole-plant perspective. *Funct. Plant Biol.* **1988**, *15*, 63–92. [CrossRef]

15. Woodward, F.I.; Kelly, C. The influence of CO_2 concentration on stomatal density. *New Phytol.* **1995**, *131*, 311–327. [CrossRef]

16. Hetherington, A.M.; Woodward, F.I. The role of stomata in sensing and driving environmental change. *Nature* **2003**, *424*, 901–908. [CrossRef]

17. Lammertsma, E.I.; de Boer, H.J.; Dekker, S.C.; Dilcher, D.L.; Lotter, A.F.; Wagner-Cremer, F. Global CO_2 rise leads to reduced maximum stomatal conductance in Florida vegetation. *Proc. Natl. Acad. Sci. USA* **2011**, *108*, 4035–4040. [CrossRef]

18. Nouvellon, Y.; Laclau, J.P.; Epron, D.; Kinana, A.; Mabiala, A.; Roupsard, O.; Bonnefond, J.M.; le Marie, G.; Marsden, C.; Bontemps, J.D.; et al. Within-stand and seasonal variations of specific leaf area in a clonal *Eucalyptus* plantation in the Republic of Congo. *For. Ecol. Manag.* **2010**, *259*, 1796–1807. [CrossRef]

19. Canny, M.J.; Huang, C.X. Leaf water content and palisade cell size. *New Phytol.* **2006**, *170*, 75–85. [CrossRef]

20. Ali, I.; Abbas, S.Q.; Hameed, M.; Naz, N.; Zafar, S.; Kanwal, S. Leaf anatomical adaptations in some exotic species of *Eucalyptus* L'Hér. (Myrtaceae). *Pak. J. Bot.* **2009**, *41*, 2717–2727.

21. Oteros, J.; García-Mozo, H.; Vázquez, L.; Mestre, A.; Domínguez-Vilches, E.; Galán, C. Modelling olive phenological response to weather and topography. *Agric. Ecosyst. Environ.* **2013**, *179*, 62–68. [CrossRef]

22. Navarrete-Campos, D.; Bravo, L.A.; Rubilar, R.A.; Emhart, V.; Sanhueza, R. Drought effects on water use efficiency, freezing tolerance and survival of *Eucalyptus globulus* and *Eucalyptus globulus* × *nitens* cuttings. *New For.* **2013**, *44*, 119–134. [CrossRef]

23. FAO. *El Estado de los Bosques del Mundo 2016. Los Bosques y la Agricultura: Desafíos y Oportunidades en Relación con el uso de la Tierra*; FAO: Rome, Italy, 2016; 119p.

24. Nunes, B.H.S.; Rezende, G.D.S.P.; Ramalho, M.A.P.; Santos, J.B. Implicações da interação genótipo × ambientes na seleção de clones de eucalipto. *Cerne* **2002**, *8*, 49–58.

25. Berger, R.; Schneider, P.R.; Finger, C.A.G.; Haseilen, C.R. Efeito do espaçamento e da adubação no crescimento de um clone de *Eucalyptus saligna* Smith. *Cienc. Florest.* **2002**, *12*, 75–87. [CrossRef]

26. Leslie, A.D.; Mencuccini, M.; Perks, M.P.; Wilson, E.R. A review of the suitability of eucalypts for short rotation forestry for energy in the UK. *New For.* **2019**, 1–19. [CrossRef]

27. Almeida, A.C.; Smethurst, P.J.; Siggins, A.; Cavalcante, R.B.L.; Borges, N.J. Quantifying the effects of *Eucalyptus* plantations and management on water resources at plot and catchment scales. *Hydrol. Process.* **2016**, *30*, 4687–4703. [CrossRef]

28. Silva, P.; Campoe, O.; Paula, R.; Lee, D. Seedling growth and physiological responses of sixteen eucalypt taxa under controlled water regime. *Forests* **2016**, *7*, 110. [CrossRef]

29. Costa e Silva, F.; Shvaleva, A.; Broetto, F.; Ortuño, M.F.; Rodrigues, M.L.; Almeida, M.H.; Chaves, M.M.; Pereira, J.S. Acclimation to short-term low temperatures in two *Eucalyptus globulus* clones with contrasting drought resistance. *Tree Physiol.* **2008**, *29*, 77–86. [CrossRef]

30. Silva, C.D.; Nascimento, J.S.; Scarpinati, E.A.; Paula, R.C. Classification of *Eucalyptus urograndis* hybrids under different water availability based on biometric traits. *For. Syst.* **2014**, *23*, 209–215. [CrossRef]

31. Silva, I.M.A.; de Souza, M.W.R.; Rodrigues, A.C.P.; Correia, L.P.S.; Veloso, R.V.S.; dos Santos, J.B.; Titon, M.; Gonçalves, J.F.; de Laia, M.L. Determination of parameters for selection of *Eucalyptus* clones tolerant to drought. *Afr. J. Agric. Res.* **2016**, *11*, 3940–3949.

32. Li, C. Population differences in water-use efficiency of *Eucalyptus microtheca* seedlings under different watering regimes. *Physiol. Plant.* **2000**, *108*, 134–139. [CrossRef]

33. Jiménez, E.; Vega, J.A.; Pérez-Gorostiaga, P.; Fonturbel, T.; Cuiñas, P.; Fernández, C. Evaluación de la transpiración de Eucalyptus globulus mediante la densidad de flujo de savia y su relación con variables meteorológicas y dendrométricas. *Bol. Inf. CIDEU* **2007**, 119–138. Available online: https://www.uhu.es/cideu/PreWeb/Boletin/Boletin3/BolInf3CIDEU119-138.pdf (accessed on 18 March 2019).

34. Almeida, A.C.; Soares, J.V.; Landsberg, J.J.; Rezende, G.D. Growth and water balance of *Eucalyptus grandis* hybrid plantations in Brazil during a rotation for pulp production. *For. Ecol. Manag.* **2007**, *251*, 10–21. [CrossRef]

35. Bhusal, N.; Bhusal, S.J.; Yoon, T.M. Comparisons of physiological and anatomical characteristics between two cultivars in bi-leader apple trees (*Malus × domestica* Borkh.). *Sci. Hortic.* **2018**, *231*, 73–81. [CrossRef]

36. Bhusal, N.; Han, S.G.; Yoon, T.M. Impact of drought stress on photosynthetic response, leaf water potential, and stem sap flow in two cultivars of bi-leader apple trees (*Malus × domestica* Borkh.). *Sci. Hortic.* **2019**, *246*, 535–543. [CrossRef]

37. McDowell, N.G.; Beerling, D.J.; Breshears, D.D.; Fisher, R.A.; Raffa, K.F.; Stitt, M. The interdependence of mechanisms underlying climate-driven vegetation mortality. *Trends Ecol. Evol.* **2011**, *26*, 523–532. [CrossRef]

38. Pita, P.; Cañas, I.; Soria, F.; Ruiz, F.; Toval, G. Use of physiological traits in tree breeding for improved yield in drought-prone environments. The case of *Eucalyptus globulus*. *For. Syst.* **2005**, *14*, 383–393. [CrossRef]

39. Fernández, M.; Tapias, R.; Alesso, P. Adaptación a la sequía y necesidades hídricas de *Eucalyptus globulus* Labill. en Huelva. *Bol. Inf. CIDEU* **2010**, *8*, 31–41. Available online: http://rabida.uhu.es/dspace/handle/10272/4560 (accessed on 20 March 2019).

40. Moraes, C.B.; Freitas, T.C.M.; Pieroni, G.B.; Resende, M.D.V.; Zimback, L.; Mori, E.S. Estimativas de parâmetros genéticos para seleção precoce de clones de *Eucalyptus* para região com ocorrência de geadas. *Sci. For.* **2014**, *42*, 219–227.

41. Flores, T.B.; Alvares, C.A.; Souza, V.C.; Stape, J.L. *Eucalyptus no Brasil—Zoneamento Climático e Guia Para Identificaçao*; IPEF: Piracicaba, Brazil, 2016; 447p.

42. Héroult, A.; Lin, Y.S.; Bourne, A.; Medlyn, B.E.; Ellsworth, D.S. Optimal stomatal conductance in relation to photosynthesis in climatically contrasting *Eucalyptus* species under drought. *Plant Cell Environ.* **2013**, *36*, 262–274. [CrossRef]

43. Carevic, F.; Fernández, M.; Alejano, R.; Vázquez-Piqué, J.; Tapias, R.; Corral, E.; Domingo, J. Plant water relations and edaphoclimatic conditions affecting acorn production in a holm oak (*Quercus ilex* L. ssp. *ballota*) open woodland. *Agrofor. Syst.* **2010**, *78*, 299–308. [CrossRef]

44. Akaike, H. A new look at the statistical model identification. *IEEE Trans Autom. Control* **1974**, *19*, 716–723. [CrossRef]

45. Tuffi Santos, L.D.; Iarema, L.; Thadeo, M.; Ferreira, F.A.; Meira, R.M.S.A. Características da epiderme foliar de eucalipto e seu envolvimento com a tolerância ao glyphosate. *Planta Daninha* **2006**, *23*, 513–520. [CrossRef]

46. Camargo, M.A.B.; Marenco, R.A. Density, size and distribution of stomata in 35 rainforest trees species in Central Amazonia. *Acta Amazon.* **2011**, *41*, 205–212. [CrossRef]

47. Carins Murphy, M.R.; Jordan, G.J.; Brodribb, T.J. Acclimation to humidity modifies the link between leaf size and the density of veins and stomata. *Plant Cell Environ.* **2014**, *37*, 124–131. [CrossRef] [PubMed]

48. Eksteen, A.B.; Grzeskowiak, V.; Jones, N.B.; Pammenter, N.W. Stomatal characteristics of *Eucalyptus grandis* clonal hybrids in response to water stress. *S. For.* **2013**, *75*, 105–111. [CrossRef]

49. Gharun, M.; Turnbull, T.L.; Pfautsch, S.; Adams, M.A. Stomatal structure and physiology do not explain differences in water use among montane eucalypts. *Oecologia* **2015**, *177*, 1171–1181. [CrossRef]

50. Araus, J.L.; Alegre, L.; Tapia, L.; Calafell, R.; Serret, M.D. Relationship between photosynthetic capacity and leaf structure in several shade plants. *Am. J. Bot.* **1986**, *73*, 1760–1770. [CrossRef]

51. Pearce, D.W.; Millard, S.; Bray, D.F.; Rood, S.B. Stomatal characteristics of riparian poplar species in a semi-arid environment. *Tree Physiol.* **2006**, *26*, 211–218. [CrossRef]

52. Bakker, J.C. Effects of humidity on stomatal density and its relation to leaf conductance. *Sci. Hortic.* **1991**, *48*, 205–212. [CrossRef]

53. Toral, M.; Manríquez, A.; Navarro-Cerrillo, R.; Tersi, D.; Naulin, P. Características de los estomas, densidad e índice estomático en secuoya (*Sequoia sempervirens*) y su variación en diferentes plantaciones de Chile. *Bosque* **2010**, *31*, 157–164. [CrossRef]

54. Xu, Z.; Zhou, G. Responses of leaf stomatal density to water status and its relationship with photosynthesis in a grass. *J. Exp. Bot.* **2008**, *59*, 3317–3325. [CrossRef] [PubMed]

55. Drake, P.L.; Froend, R.H.; Franks, P.J. Smaller, faster stomata: Scaling of stomatal size, rate of response, and stomatal conductance. *J. Exp. Bot.* **2013**, *64*, 495–505. [CrossRef] [PubMed]

56. Xiong, D.; Douthe, C.; Flexas, J. Differential coordination of stomatal conductance, mesophyll conductance, and leaf hydraulic conductance in response to changing light across species. *Plant Cell Environ.* **2018**, *41*, 436–450. [CrossRef]

57. Elliott-Kingston, C.; Haworth, M.; Yearsley, J.M.; Batke, S.P.; Lawson, T.; McElwain, J.C. Does size matter? Atmospheric CO_2 may be a stronger driver of stomatal closing rate than stomatal size in taxa that diversified under low CO_2. *Front. Plant Sci.* **2016**, *7*, 1253. [CrossRef]

58. Kröber, W.; Bruelheide, H. Transpiration and stomatal control: A cross-species study of leaf traits in 39 evergreen and deciduous broadleaved subtropical tree species. *Trees* **2014**, *28*, 901–914. [CrossRef]

59. Franks, P.J.; Drake, P.L.; Beerling, D.J. Plasticity in maximum stomatal conductance constrained by negative correlation between stomatal size and density: An analysis using *Eucalyptus globulus*. *Plant Cell Environ.* **2009**, *32*, 1737–1748. [CrossRef]

60. Correia, B.; Pintó-Marijuan, M.; Neves, L.; Brossa, R.; Dias, M.C.; Costa, A.; Castro, B.B.; Araújo, C.; Santos, C.; Chaves, M.M.; et al. Water stress and recovery in the performance of two *Eucalyptus globulus* clones: Physiological and biochemical profiles. *Physiol. Plant.* **2014**, *150*, 580–592. [CrossRef]

61. Hernandez, M.J.; Montes, F.; Ruiz, F.; Lopez, G.; Pita, P. The effect of vapour pressure deficit on stomatal conductance, sap pH and leaf-specific hydraulic conductance in *Eucalyptus globulus* clones grown under two watering regimes. *Ann. Bot.* **2016**, *117*, 1063–1071. [CrossRef]

62. Merchant, A.; Callister, A.; Arndt, S.; Tausz, M.; Adams, M. Contrasting physiological responses of six *Eucalyptus* species to water deficit. *Ann. Bot.* **2007**, *100*, 1507–1515. [CrossRef]

63. Mendes, H.S.J.; Paula, N.F.; Scarpinatti, E.A.; Paula, R.C. Respostas fisiológicas de genótipos de *Eucalyptus grandis* × *E. urophylla* à disponibilidade hídrica e adubação potássica. *Cerne* **2013**, *19*, 603–611. [CrossRef]

64. Andivia, E.; Carevic, F.; Fernández, M.; Alejano, R.; Vázquez-Piqué, J.; Tapias, R. Seasonal evolution of water status after outplanting of two provenances of holm oak nursery seedlings. *New For.* **2012**, *43*, 815–824. [CrossRef]

65. Kozlowski, T.T. *Water Deficits and Plant Growth, Volume III: Plant Responses and Control of Water Balance*, 1st ed.; Academic Press: New York, NY, USA; London, UK, 1972; 382p.

66. Fernández, M.; García-Albalá, J.; Andivia, E.; Alaejos, J.; Tapias, R.; Menéndez, J. Sickle bush (*Dichrostachys cinerea* L.) field performance and physical-chemical property assessment for energy purposes. *Biomass Bioenergy* **2015**, *81*, 483–489. [CrossRef]

67. Eamus, D.; Taylor, D.T.; Macinnis-Ng, C.M.; Shanahan, S.; De Silva, L. Comparing model predictions and experimental data for the response of stomatal conductance and guard cell turgor to manipulations of cuticular conductance, leaf-to-air vapour pressure difference and temperature: Feedback mechanisms are able to account for all observations. *Plant Cell Environ.* **2008**, *31*, 269–277.

68. Li, H.; Madden, J.L.; Potts, B.M. Variation in leaf waxes of the Tasmanian *Eucalyptus* species—I. Subgenus *Symphyomyrtus*. *Biochem. Syst. Ecol.* **1997**, *25*, 631–657. [CrossRef]

69. Migacz, I.P.; Raeski, P.A.; Almeida, V.P.; Raman, V.; Nisgoski, S.; Muniz, G.I.B.; Farago, P.V.; Khan, I.A.; Budel, J.M. Comparative leaf morpho-anatomy of six species of *Eucalyptus* cultivated in Brazil. *Rev. Bras. Farmacogn.* **2018**, *28*, 273–281. [CrossRef]

70. Fernández, M.; Alaejos, J.; Andivia, E.; Vázquez-Piqué, J.; Ruiz, F.; López, F.; Tapias, R. *Eucalyptus* × *urograndis* biomass production for energy purposes exposed to a Mediterranean climate under different irrigation and fertilisation regimes. *Biomass Bioenergy* **2018**, *111*, 22–30. [CrossRef]

71. Sefton, C.A.; Montagu, K.D.; Atwell, B.J.; Conroy, J.P. Anatomical variation in juvenile eucalypt leaves accounts for differences in specific leaf area and CO_2 assimilation rates. *Aust. J. Bot.* **2002**, *50*, 301–310. [CrossRef]

72. Yao, R.L.; Glencross, K.; Nichols, J.D. Difference in shade tolerance affects foliage–sapwood response to thinning by two eucalypts. *S. For.* **2014**, *76*, 93–100. [CrossRef]

73. Tyree, M.T.; Richter, H. Alternative Methods of Analysing Water Potential Isotherms: Some Cautions and Clarifications. *J. Exp. Bot.* **1981**, *32*, 643–653. [CrossRef]

Article

Effects of Temperature Factors on Resistance against Pine Wood Nematodes in *Pinus thunbergii*, Based on Multiple Location Sites Nematode Inoculation Tests

Taiichi Iki [1,*,†], Koji Matsunaga [2,*,†], Tomonori Hirao [3], Mineko Ohira [3], Taro Yamanobe [3], Masakazu G Iwaizumi [4], Masahiro Miura [4], Keiya Isoda [3], Manabu Kurita [2], Makoto Takahashi [3] and Atsushi Watanabe [5]

1   Tohoku Regional Breeding Office, Forest Tree Breeding Center, Forestry and Forest Products Research Institute, 95 Osaki, Takizawa, Iwate 020-0621, Japan
2   Kyushu Regional Breeding Office, Forest Tree Breeding Center, Forestry and Forest Products Research Institute, 2320-5 Suya, Koshi, Kumamoto 861-1102, Japan; mkuri@affrc.go.jp
3   Forest Tree Breeding Center, Forestry and Forest Products Research Institute, 3809-1 Ishii, Juo, Hitachi, Ibaraki 319-1301, Japan; hiratomo@affrc.go.jp (T.H.); sasamine@affrc.go.jp (M.O.); yamanobe@affrc.go.jp (T.Y.); keiso@affrc.go.jp (K.I.); makotot@affrc.go.jp (M.T.)
4   Kansai Regional Breeding Office, Forest Tree Breeding Center, Forestry and Forest Products Research Institute, 1043, Uetsukinaka, Shoo, Katsuta, Okayama 709-4335, Japan; ganchan@affrc.go.jp (M.G.I.); miumasa@affrc.go.jp (M.M.)
5   Department of Forest Environmental Science, Faculty of Agriculture, Kyushu University, 744 Motooka, Nishiku, Fukuoka 819-0395, Japan; nabeatsu@agr.kyushu-u.ac.jp
*   Correspondence: iki@affrc.go.jp (T.I.); makoji@affrc.go.jp (K.M.);
      Tel.: +81-19-688-4517 (T.I.); +81-96-242-3151 (K.M.)
†   These authors contributed equally to this work.

Received: 22 July 2020; Accepted: 20 August 2020; Published: 24 August 2020

Abstract: Pine wilt disease (PWD) caused by the pinewood nematode (PWN) (*Bursaphelenchus xylophilus* (Steiner and Buhrer) Nickle) is a worldwide issue. Infection is considered to be promoted mainly by the increased air temperature, but it is important to investigate whether the effect of high temperature similarly influences the different ranks of resistant clone. In the present study, we conducted PWN inoculation tests using six common open-pollinated families of resistant *Pinus thunbergii* Parl. The tests were conducted at nurseries of five test sites across Japanese archipelago between 2015 and 2017. Our analysis focused specifically on temperature. Firstly, we examined the effects of test sites, inoculation year, and their interaction on unaffected seedling rate and found that the unaffected seedling rate of all tested pine families decreased as the cumulative temperature increased. We found that the unaffected seedling rate decreased as the cumulative temperature increased for all tested pine families. In general, higher cumulative temperatures were required for having an effect on the unaffected seedling rates of higher PWN-resistant families. Typically, early cumulative temperatures, i.e., 19 days after inoculation, had the greatest effect on the unaffected seedling rates of PWN-resistant pines. However, the relationship between cumulative temperature and predicted unaffected seedling rate follow similar rate for all families. Thus, the order of resistance level is maintained in terms of the cumulative temperature required for having an effect.

Keywords: pine wood disease; resistance to pine wood nematode; inoculation test; multisite; cumulative temperature; *Pinus thunbergii*

---

1. Introduction

Pine wilt disease is an epidemic disease caused by the invasive pinewood nematode, *Bursaphelenchus xylophilus* (Steiner and Buhrer) Nickle [1], and vectored by pine sawyer beetles, *Monochamus alternatus* Hope [2]. PWD is currently a worldwide issue [3], and future climate change could lead to further spread of the disease given that its development depends on temperature and drought [4]. *Pinus thunbergii* Parl. and *Pinus densiflora* Sieb. et Zucc., two major planted pine species in Japan, are susceptible to PWN [5]. Thus, Japanese pine forests have been seriously damaged by PWD. The first documented observation of PWD was in Nagasaki Prefecture in 1905 [6]. The disease has subsequently spread to every prefecture in Japan, apart from Hokkaido [7]. As the disease causes widespread damage, Japan began a tree breeding project in 1978 in order to select resistant pine varieties as a countermeasure against PWD. In the first breeding project, conducted from 1978 to 1984 in southwestern Japan, 16 and 92 resistant clones of *P. thunbergii* and *P. densiflora* were selected, respectively [8–10]. Resistance against PWN was conferred by artificial inoculation, wherein higher survival rates represented greater resistance [9]. Up to March 2019, several related projects have led to the selection of 211 and 288 resistant clones of *P. thunbergii* and *P. densiflora*, respectively [11].

The development of PWD symptoms and mortality after inoculation are affected by the climatic factors, including air temperature [12–14], precipitation [9,15] and light conditions [16]. In particular, temperature is strongly related to PWD, as PWD only occurs in areas where the average temperature exceeds 20 °C for several weeks [17]. In non-resistant *P. thunbergii* seedlings grown in phytotrons and temperature-controlled greenhouses, PWD symptoms develop faster and mortality rates increase as temperatures rise [13,14]. It appears that the effect of temperature on the PWN propagation rate affects the development of PWD symptoms, leading to tree death [13,14]. In nursery-based inoculation tests using open-pollinated families of resistant *P. thunbergii*, survival rates varied substantially among inoculation years and families, suggesting that differences in climate variation at the period of inoculation can influence tree mortality [18–20]. Variation in resistance exist among resistant *P. thunbergii* clones [9]; however, no studies have yet been conducted to investigate whether the effect of high temperature similarly influences the different ranks of resistant clones. It is important to understand how temperature affects the resistance, especially given the potential effects of future climate change.

To clarify how temperature factors affect the different ranks of resistance against PWN, inoculation tests were conducted over multiple years, from 2015 to 2017, at the nurseries of five test sites in regions with different climates. In addition, we used common open-pollinated families of six resistant *P. thunbergii* clones with different levels of resistance to PWN.

2. Materials and Methods

2.1. Test Sites

Inoculation tests were conducted at the nurseries of five test sites: Forest Tree Breeding Center (FTBC) Head Quarters in Hitachi, Ibaraki (36.69° N, 140.69° E); Tohoku Regional Breeding Office, FTBC (TBO) in Takizawa, Iwate (39.83° N, 141.14° E); Kansai Regional Breeding Office, FTBC (KABO) in Shou-cho, Okayama (35.06° N, 134.11° E); Shikoku Breeding Stock Garden, KABO, FTBC (SSG) in Kami, Kochi (33.61° N, 133.70° E); and Kyushu Regional Breeding Office, FTBC (KYBO) in Koshi, Kumamoto (32.88° N, 130.74° E). Their locations are shown in Figure 1. The mean temperatures for the current inoculation test periods at each test site are shown in Figure S1.

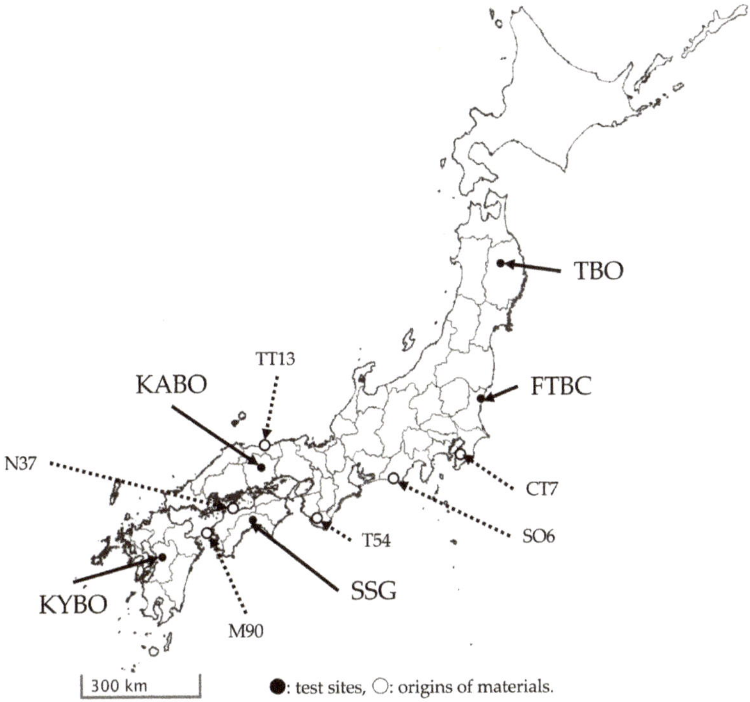

Figure 1. Location of each test site and origin of materials used in the present study. This map was created from the blank map published by Geospatial Information Authority of Japan (https://www.gsi.go.jp/tizu-kutyu.html). Test site abbreviations: Tohoku Regional Breeding Office (TBO), Forest Tree Breeding Center (FTBC), Kansai Regional Breeding Office (KABO), Shikoku Breeding Stock Garden (SSG), and Kyushu Regional Breeding Office (KYBO). Family name abbreviations: Misaki 90 (M90), Namikata 37 (N37), Tanabe 54 (T54), Shizuoka (Oosuka) 6 (SO6), Chiba (Tomiura) 7 (CT7), and Tottori (Tottori) 13 (TT13).

2.2. Plant Material and PWN Inoculation

Open-pollinated families of six resistant *P. thunbergii* clones were used in the present study (Table 1), and the same seed-lots were used for all test sites across the three-year inoculations. The six families used in this study were Misaki 90 (M90), Namikata 37 (N37), Tanabe 54 (T54), Shizuoka (Oosuka) 6 (SO6), Chiba (Tomiura) 7 (CT7), and Tottori (Tottori) 13 (TT13). The origins of these families are shown in Figure 1. Of these, three clones (M90, N37, and T54) were selected in the first breeding project and had already been evaluated for their resistance rank [9]. The other clones (SO6, CT7, and TT13) were recently selected from other regions, and their resistance against PWN had not yet been evaluated.

Seedlings were grown using the same procedure until the inoculation test, except for at TBO. The standard method was as follows: seeds were sown on the seedbed in spring one year before the inoculation test; the seedlings were then transplanted to a nursery in the following spring; then the inoculation test was conducted with two replications. As the TBO location is colder than the other test sites, the growth of seedlings was slower; therefore, to promote the growth of seedlings and allow them to be inoculated at the 2-year-old stage, the seeds were sown individually in a nursery with two replications, and the seedlings were not transplanted for about 16 months, until the inoculation test was conducted.

Table 1. Resistant *P. thunbergii* families used in this study.

Family Name	Resistance Ranking *	Number of Seedlings Used in the Inoculation Tests at Each Study Site in Each Year													
		2015					2016					2017			
		TBO	FTBC	KABO	SSG	KYBO	TBO	FTBC	KABO	SSG	KYBO	TBO	FTBC	KABO	KYBO
M90	4	27	37	22	31	32	35	40	20	40	22	33	34	25	24
N37	4	28	32	23	29	32	27	32	28	35	16	30	29	23	19
T54	2	44	40	33	40	39	57	39	41	37	37	55	40	21	36
SO6	-	37	40	28	40	32	50	40	37	40	34	44	37	28	17
CT7	-	37	37	32	40	40	35	40	7	40	27	35	39	32	29
TT13	-	40	40	27	19	39	50	39	11	40	32	28	39	31	26

* Resistance ranking was evaluated by using a least squares means method based on the survival rates following inoculation tests on open-pollinated families. -: Resistance not evaluated. Test site abbreviations: Tohoku Regional Breeding Office (TBO), Forest Tree Breeding Center (FTBC), Kansai Regional Breeding Office (KABO), Shikoku Breeding Stock Garden (SSG), Kyushu Regional Breeding Office (KYBO). Family name abbreviations: Misaki 90 (M90), Namikata 37 (N37), Tanabe 54 (T54), Shizuoka (Oosuka) 6 (SO6), Chiba (Tomiura) 7 (CT7), and Tottori (Tottori) 13 (TT13).

The number of seedlings at each test site is shown in Table 1. At SSG, the growth of seedlings was poor due to insect damage (unrelated to PWN) in 2017; hence, we could not conduct inoculation tests at SSG in 2017. In general, the number of seedlings in each of the families ranged from 7 to 57. It has been reported that the size of *P. thunbergii* seedlings affects their post-inoculation survival [20,21]; therefore, we also measured seedling height immediately before inoculation testing.

The virulent isolate of the PWN, Ka4 [22], which has been widely used in the resistance breeding program in Japan, was used in the present study. A 50-µL suspension containing 5000 PWNs (a density of 100,000 PWN/mL) was inoculated onto wounds in the basal axes of subjects, which were made by peeling the cortex with a knife and scratching the xylem with a fine saw. At each test site, inoculation was conducted in early July in each of the three years from 2015 to 2017 (Table S1). The seedling age at the time of inoculation was approximately 15–16 months (based on time after sowing).

2.3. Symptom Observation

Symptoms were observed approximately ten weeks after inoculation (Table S1) and classified by visual judgement for each inoculated seedling class as follows: 0: no symptom, 1: browning of needles on one or more branches, 2: browning of all needles. From the results, the unaffected seedling rate was calculated using the following equation:

$$\text{Unaffected seedling rate (\%)} = \text{number of seedlings in symptom class 0/number inoculated seedlings} \times 100$$

2.4. Temperature Data

The temperature data used in the present study were collected from the Automated Meteorological Data Acquisition System (AMeDAS) provided by the Japanese Meteorological Agency [23]. We collected climate data from the AMeDAS station nearest to each of test site; Morioka (39.70° N, 141.16° E) for TBO, Hitachi (36.58° N, 140.65° E) for FTBC, Tsuyama (35.07° N, 134.02° E) for KABO, Kochi (33.57° N, 133.55° E) for SSG and Kumamoto (32.82° N, 130.70° E) for KYBO.

Hourly temperature data were collected from each station. Temperature data for analysis were labelled *CT*, *CT20*, *CT25*, and *CT30*, which represent cumulative temperature with different thresholds (0 °C, 20 °C, 25 °C, and 30 °C) from the day of inoculation to the *n* day after inoculation (DAI). They were calculated using the following the formulae:

$$CT \text{ at } n \text{ DAI} = \sum_{j=1}^{n} \sum_{i=1}^{24} \left(TH_{ij} \times \delta_{ij} \right) / n \qquad \delta_{ij} = \begin{cases} 1, \left(TH_{ij} > 0 \right) \\ 0, \left(TH_{ij} \leq 0 \right) \end{cases}$$

$$CT20 \text{ at } n \text{ DAI} = \sum_{j=1}^{n}\sum_{i=1}^{24}\left(\left(TH_{ij} - 20\right) \times \delta_{ij}\right)/n \qquad \delta_{ij} = \begin{cases} 1, \left(\left(TH_{ij} - 20\right) > 0\right) \\ 0, \left(\left(TH_{ij} - 20\right) \leq 0\right) \end{cases}$$

$$CT25 \text{ at } n \text{ DAI} = \sum_{j=1}^{n}\sum_{i=1}^{24}\left(\left(TH_{ij} - 25\right) \times \delta_{ij}\right)/n \qquad \delta_{ij} = \begin{cases} 1, \left(\left(TH_{ij} - 25\right) > 0\right) \\ 0, \left(\left(TH_{ij} - 25\right) \leq 0\right) \end{cases}$$

$$CT30 \text{ at } n \text{ DAI} = \sum_{j=1}^{n}\sum_{i=1}^{24}\left(\left(TH_{ij} - 30\right) \times \delta_{ij}\right)/n \qquad \delta_{ij} = \begin{cases} 1, \left(\left(TH_{ij} - 30\right) > 0\right) \\ 0, \left(\left(TH_{ij} - 30\right) \leq 0\right) \end{cases}$$

The inoculation day was defined as Day 0, and the climate data from this day were not used in our analysis. Temperature data were therefore analyzed from the 1 DAI to the 35 DAI.

2.5. Statistical Analysis

All statistical analyses were conducted in R version 3.6.1 [24]. To examine the effects of test site and inoculation year on the unaffected seedling rate of *P. thunbergii* families, logistic regression analysis was conducted using a generalized mixed linear model via the "glmer" function of the lme4 package [25]. Appropriate models were selected using Akaike information criterion (AIC) values obtained with the "dredge" function of the MuMIn package [26]. In this analysis, test site, inoculation year, seedling height (mean value in replication), and the interaction between test site and inoculation year were fixed effects; family, replication, the interaction between family and test site, and the interaction between family and inoculation year were the random effects. Since the unaffected seedling rate showed a binomial distribution, the family was set as "binomial" using the logit link function in R. To identify significant differences in the fixed effects, the "ANOVA" function in the car package [27] was used to conduct deviance analysis (Type II test, level of significance was $p < 0.01$) on the test site, inoculation year, and interaction between the test site and inoculation year. In addition, for test site and inoculation year, multiple comparison analysis was performed using the Tukey method (level of significance was $p < 0.05$) via the "glht" and "cld" functions in the multcomp package [28]. Furthermore, the variance component of each random effect was calculated using the "VarCorr" function in the lme4 package. The best linear unbiased prediction (BLUP) value for each family was calculated using the "ranef" function in the lme4 package for unaffected seedling rate.

The correlation coefficients between each temperature factor and the unaffected seedling rate of each family were calculated using product–moment correlation via the "cor.test" function. The mean of correlation coefficients for all families was calculated in each temperature factor.

To analyze the post-inoculation temperature factors that affected the unaffected seedling rates, logistic regression analysis was conducted using a generalized mixed linear model via the "glmer" function of the lme4 package. The model was again selected using AIC values obtained with the "dredge" function in the MuMIn package according to a model that excluded the temperature factor (i.e., the "Null" model). In this analysis, the unaffected seedling rate was the response variable; seedling height (mean value in replication) and individual post-inoculation temperature factors were the fixed effects among the explanatory variables; family, test site, inoculation year, repetition, the interaction between family and test site, the interaction between inoculation family and inoculation year, and the interaction between test site and inoculation year were the random effects among the explanatory variables. Family was again set as "binomial" using the logit link function. Following logistic regression analyses, the AIC values of all models were calculated using the "AIC" function. We also used the function "predict", to predict the unaffected seedling rate under the optimal climatic conditions selected by the analysis. At that time, the prediction was performed using the same seedling height (overall mean across all test sites, years, and families), considering the effect of seedling height on unaffected seedling rate.

3. Results

3.1. Seedling Heights

Seedling heights (with standard deviations) prior to the inoculation tests are shown in Figure 2. The overall mean of seedling heights was 26.6 cm across five test sites, three years, and six families. Mean seedling heights for each family were 28.3 cm for M90, 27.2 cm for N37, 23.4 cm for T54, 25.6 cm for SO6, 27.7 cm for CT7, and 28.0 cm for TT13 across five test sites and three years. Seedlings tended to be taller in TBO compared to in the other test sites.

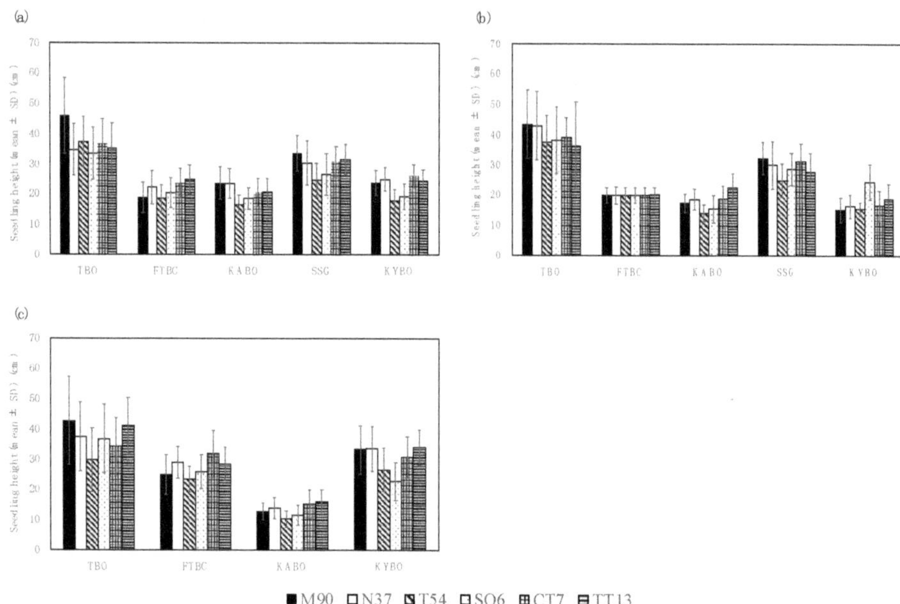

Figure 2. Mean seedling heights of *P. thunbergii* families before inoculation with PWNs in: (**a**) 2015, (**b**) 2016, (**c**) 2017. Error bars represent standard deviations (SD). Test site abbreviations: Tohoku Regional Breeding Office (TBO), Forest Tree Breeding Center (FTBC), Kansai Regional Breeding Office (KABO), Shikoku Breeding Stock Garden (SSG), and Kyushu Regional Breeding Office (KYBO). Family name abbreviations: Misaki 90 (M90), Namikata 37 (N37), Tanabe 54 (T54), Shizuoka (Oosuka) 6 (SO6), Chiba (Tomiura) 7 (CT7), and Tottori (Tottori) 13 (TT13).

3.2. Unaffected Seedling Rates

The unaffected seedling rates are shown in Figure 3. The mean of unaffected seedling rates for the six families in 2015 were 72.3% for TBO, 56.2% for FTBC, 21.5% for KABO, 29.0% for SSG, and 20.5% for KYBO. In 2016, the means of unaffected seedling rates for all families were 74.3% for TBO, 65.8% for FTBC, 7.4% for KABO, 23.7% for SSG, and 6.7% for KYBO. The mean of unaffected seedling rates in 2017 for all families were 39.1% for TBO, 51.6% for FTBC, 2.9% for KABO, and 13.8% for KYBO. The mean unaffected seedling rates for each family were 52.7% for M90, 46.9% for N37, 27.1% for T54, 24.6% for SO6, 14.5% for CT7, and 42.2% for TT13 across five test sites and three inoculation years. Across the three years, the unaffected seedling rate was higher at TBO and FTBC than at KABO, SSG, and KYBO.

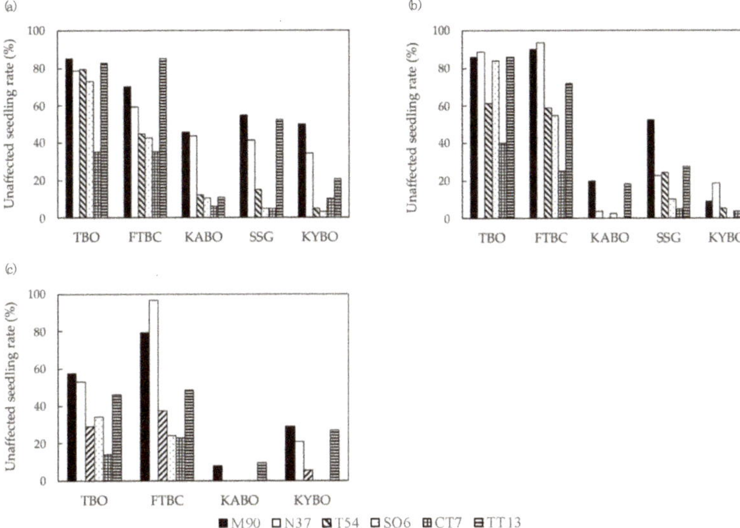

Figure 3. Unaffected seedling rate of *P. thunbergii* families before inoculation with PWNs in: (**a**) 2015, (**b**) 2016, (**c**) 2017. Test site abbreviations: Tohoku Regional Breeding Office (TBO), Forest Tree Breeding Center (FTBC), Kansai Regional Breeding Office (KABO), Shikoku Breeding Stock Garden (SSG), and Kyushu Regional Breeding Office (KYBO). Family name abbreviations: Misaki 90 (M90), Namikata 37 (N37), Tanabe 54 (T54), Shizuoka (Oosuka) 6 (SO6), Chiba (Tomiura) 7 (CT7), and Tottori (Tottori) 13 (TT13).

The deviance and significance of the test site, inoculation year, and interaction between test site and inoculation year are shown in Table 2. Significant differences ($p < 0.01$) in unaffected seedling rates were observed according to test site, inoculation year, and the interaction between test site and inoculation year. Multiple comparison analysis (using the Tukey method) among the test sites showed that the unaffected seedling rates at TBO and FTBC were significantly higher than the rates at KABO, SSG, and KYBO ($p < 0.05$). A similar analysis of inoculation years showed that the unaffected seedling rates in 2015 and 2016 were significantly higher than the rate in 2017 ($p < 0.05$). From the estimated proportions of the variance components of each random effect, family accounted for >90% of all variance components; in contrast, replication and the interaction between family and inoculation year were almost zero. The BLUP values of unaffected seedling rate in each family were as follows: 1.06 for M90, 0.73 for N37, −0.25 for T54, −0.52 for SO6, −1.40 for CT7, and 0.42 for TT13.

Table 2. Effects of site and inoculation year on unaffected seedling rate analyzed using a generalized linear mixed model.

	Deviance	Degree of Freedom	Significance Pr ($>\chi^2$)	
Seedling height	10.369	1	0.0013	$p < 0.01$
Test site	242.816	4	$<2.2 \times 10^{-16}$	$p < 0.01$
Year	101.763	2	$<2.2 \times 10^{-16}$	$p < 0.01$
Test site × Year	29.450	7	0.0001	$p < 0.01$

3.3. Analysis of Temperature Factors Affecting Unaffected Seedling Rate

For all temperature factors, all correlation coefficients values were negative values (Figure 4). The correlation coefficients of *CT*, *CT20*, and *CT25* were higher than those of *CT30*. The highest correlation coefficient in all temperature factors was *CT* at the 6 DAI ($r = -0.804$). For *CT20*, the highest correlation coefficient was observed on the 13 DAI ($r = -0.786$). For *CT25*, the highest correlation coefficient was observed on the 14 DAI ($r = -0.746$). In *CT*, *CT20*, and *CT25*, the correlation coefficients tended to gradually decrease after showing the highest value.

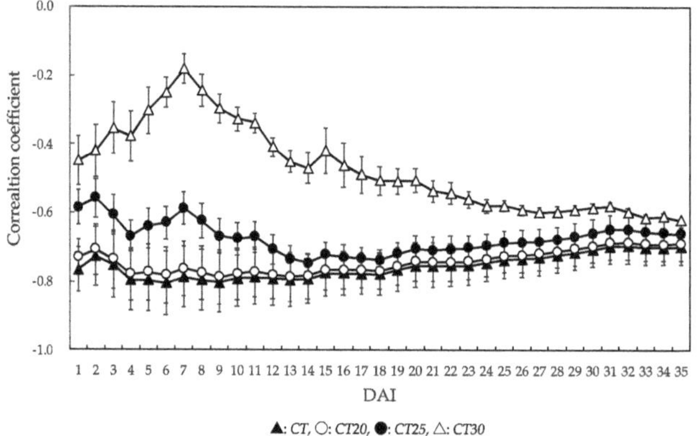

Figure 4. Changes in the correlation coefficients for each temperature factor according to the unaffected seedling rate of each *P. thunbergii* family. The value of the correlation coefficient in this figure is the mean value of six families. Error bars represent SD. DAI: days after inoculation. Temperature factor abbreviations: *CT*, *CT20*, *CT25*, and *CT30* are represented cumulative temperature with different thresholds (0 °C, 20 °C, 25 °C, and 30 °C) from the day of inoculation to the *n* DAI.

In the analysis of the effects of post-inoculation temperature factors on the unaffected seedling rate of *P. thunbergii* ten weeks after inoculation (Figure 5), the AICs of *CT*, *CT20*, and *CT25* were lower than that of *CT30*. The lowest AIC for all temperature factors for *CT25* was on the 19 DAI (AIC = 665). For *CT20*, the lowest AIC was observed on the 19 DAI (AIC = 666). The correlation coefficients 19 DAI were −0.766 for *CT*, −0.756 for *CT20*, and −0.719 for *CT25*. In all temperature factors, the AICs tended to gradually increase after reaching the lowest value.

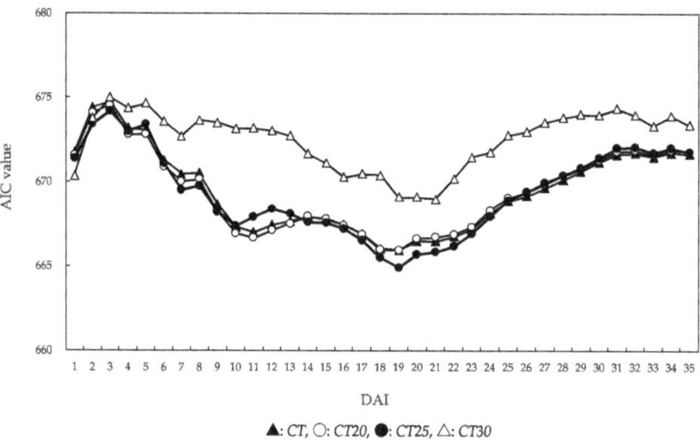

Figure 5. Changes in Akaike information criterion (AIC) values calculated from a generalized linear mixed model including temperature. DAI: days after inoculation. Temperature factor abbreviations: *CT*, *CT20*, *CT25*, and *CT30* represented cumulative temperature with different thresholds (0 °C, 20 °C, 25 °C, and 30 °C) from the day of inoculation to the *n* DAI.

When the family-level unaffected seedling rate was predicted by the model including *CT25* on the 19 DAI, negative relationships were shown in all families (Figure 6). The observed correlation coefficients were −0.760 for M90, −0.774 for N37, −0.792 for T54, −0.800 for SO6, −0.767 for CT7, and −0.789 for TT13. Unaffected seedling rates for each family apparently decrease as the cumulative temperature rises. At all ranges of cumulative temperature, the unaffected seedling rate was predicted to be higher in the order of M90, N37, TT13, T54, SO6, and CT7. This order of families matched that of the BLUP results.

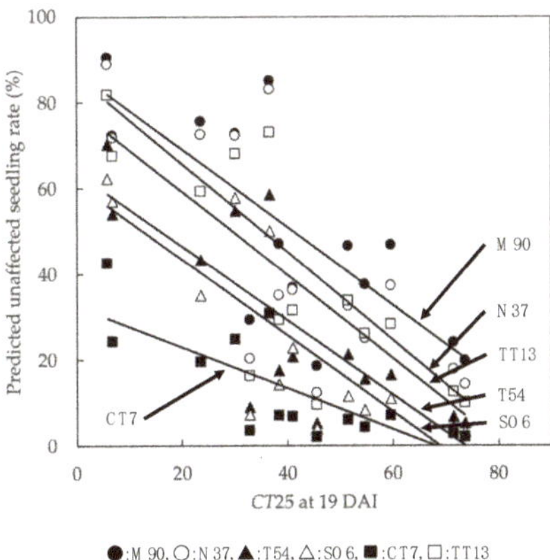

Figure 6. Relationship between optimum temperature factor and predicted unaffected seedling rate. The predicted unaffected seedling calculated from a generalized linear mixed model including optimum temperature factor (*CT25* at 19 DAI). DAI: days after inoculation. Family name abbreviations: Misaki 90 (M90), Namikata 37 (N37), Tanabe 54 (T54), Shizuoka (Oosuka) 6 (SO6), Chiba (Tomiura) 7 (CT7), and Tottori (Tottori) 13 (TT13).

4. Discussion

4.1. Differences in PWN-Resistance among Test Sites and Inoculation Years

In inoculation tests conducted across multiple nurseries, it has been suggested that the unaffected seedling rates of *P. thunbergii* resistant families vary greatly depending on climatic factors related to the test site and the inoculation year [18–20]. Our results support this suggestion: the unaffected seedling rates differed among test sites and inoculation years, thus the PWN resistance of *P. thunbergii* was apparently affected by climatic factors at the test site and during the inoculation year. However, the variance component of resistant families was substantially larger for the pine family than for the interaction between the family and the test site or the interaction between the family and the inoculation year, thus the rank of the unaffected seedling rate of six families was stable even though the climatic conditions were different in the three years studied. The ranking of unaffected seedling rate was in the following order: M90, N37, TT13, T54, SO6, and CT7. The highest levels of resistance to PWN also apparently followed this order. The variation in unaffected seedling rate in the present study was large, ranging from 2.9% to 74.3%. This wide range may have been created by the differing temperature factors at the five sites over three years.

4.2. Temperature Factors Affecting PWN-Resistance

Previous studies have shown that PWN migration and propagation in *P.thunbergii* trees are important for symptom development of PWD and for mortality [12–14]. It has also been shown that cumulative temperature, especially at 25–30 °C, is important for the propagation of inoculated PWNs in pine trees [13,14]. In the present study, the AIC value was lowest in *CT25* on the 19 DAI, and the correlation coefficient also showed a high value. Our results suggest that a cumulative temperature of 25 °C or higher affects the unaffected seedling rate of resistant *P. thunbergii* after inoculation, which is the same temperature range with propagation of PWN after inoculation [13,14].

Not only *CT25* at 19 DAI but also *CT* and *CT20* at 19 DAI showed low AICs. From this result, the temperature factor, the cumulative temperature until 19 DAI, seems to be considerably related to the unaffected seedling rate. Hirao et al. [29] investigated gene expression after inoculation using grafted clones of PWN-resistant and PWN-susceptible *P. thunbergii* and reported that genes related to cell wall strength were significantly higher in resistant *P. thunbergii* clones at 14 DAI than in susceptible pine clones at 7 DAI. Kusumoto et al. [30] also investigated histological response, tissue damage expansion, and PWN distribution after inoculation using grafted clones of PWN-resistant and PWN-susceptible *P. thunbergii,* and reported that PWN propagation was suppressed immediately after inoculation, and that proteins related to cell wall strength were highly expressed in resistant pine clones relative to susceptible pine clones. According to our findings, in general, cumulative temperature has a greater effect on seedling rate soon after inoculation, e.g., 19 DAI. Therefore, in resistant *P. thunbergii,* we suggest that if the cumulative temperature is low soon after inoculation, the propagation of PWNs in the *P. thunbergii* tree is likely to be insufficient to cause death in the infested trees due to the protective reaction.

As the correlation coefficients between the predicted unaffected seedling rate and the *CT25* at 19 DAI were negative for all families, unaffected seedling rate apparently decreases as the cumulative temperature (25 °C or higher) rises. However, the cumulative temperature of 25 °C or higher that affected the unaffected seedling rate differed among resistant pine families. The more resistant the family was, the higher the cumulative temperature required to effect the unaffected seedling rate. We suggest that these family differences in the cumulative temperature that affect the unaffected seedling rate are related to the propagation of PWNs in the trees after inoculation. Although the resistance mechanism to PWNs in *P. thunbergii* has yet to be confirmed, it has been suggested that post-inoculation migration and propagation of PWNs is restricted in resistant pine trees [21,31,32]. Among the families studied here, for example, M90 is highly resistant to PWN, so they perhaps restricted the propagation of PWN after inoculation more than other families. Therefore, higher cumulative temperatures were required for sufficient propagation for the disease develop.

4.3. Application to the Resistance Breeding Program

The results of the present study suggest that the unaffected seedling rates of PWN-resistant *P. thunbergii* families decrease as the cumulative temperature (25 °C or higher) increases. In addition, the effects of cumulative temperature appear to occur soon after PWN inoculation, i.e., at 19 DAI. However, these effects differed in each of the families tested here. That is, the higher the resistance to PWN, the higher the cumulative temperature needed.

In Japan, resistant clones selected from the field (first-generation) are crossed, and second-generation resistant *P. thunbergii* are then selected. Candidate trees from the second generation of resistant clones are estimated to have higher levels of resistance than those of the first generation [33,34]. Given our results, it will be necessary to increase the selection criteria in future, when selecting the second generation, in order to effectively evaluate the resistance level. For example, we suggest that inoculation tests should be conducted when high temperatures are expected.

5. Conclusions

We conducted PWN inoculation tests on six common open-pollinated families of resistant *P. thunbergii* in the nurseries of five test sites from 2015 to 2017, in order to consider the effects of temperature factors on PWN resistance. The results suggested that the unaffected seedling rates of PWN-resistant *P. thunbergii* families decrease as the cumulative temperature increases. However, the cumulative temperatures that affected unaffected seedling rate differed among families. Namely, higher cumulative temperatures were required for the effect in more highly PWN-resistant pines.

In addition, early cumulative temperatures had greater effects on the unaffected seedling rate of PWN-resistant *P. thunbergii*.

Supplementary Materials: The following are available online at http://www.mdpi.com/1999-4907/11/9/922/s1, Figure S1: Mean temperatures for the current inoculation test period at each test site in 2015, 2016, and 2017. This Figure was created from data obtained from AMeDAS (Automated Meteorological Data Acquisition System) of Japan Meteorological Agency (http://www.data.jma.go.jp/obd/stats/etrn/index.php). Test site abbreviations: Tohoku Regional Breeding Office (TBO), Forest Tree Breeding Center (FTBC), Kansai Regional Breeding Office (KABO), Shikoku Breeding Stock Garden (SSG), and Kyushu Regional Breeding Office (KYBO). Table S1: Dates of Inoculation and investigations.

Author Contributions: Conceptualization, all authors; material management and investigation, T.I., K.M., T.Y., M.O., M.G.I., M.M., K.I., M.K.; data curation and writing—original draft preparation, T.I. and K.M.; writing—review and editing, K.M., T.H., T.Y., M.O., M.G.I., M.M., M.T. and A.W.; project administration, T.H., M.T. and A.W. All authors have read and agreed to the published version of the manuscript.

Funding: The present study is part of the project on 'Project to advance the development of technology for varieties of Japanese black pine and red pine resistant to the pine wood nematode' supported by Forestry Agency, Ministry of Agriculture, Forestry and Fisheries, Japan.

Acknowledgments: We thank Hiroshi Hoshi (FTBC, FFPRI) for his well coordination of the research project, Jin'ya Nasu (TBO, FTBC, FFPRI) and Michinari Matsushita (FTBC, FFPRI) for advice of statistical analysis. We thank our colleagues in the field management section of TBO, FTBC, KABO, SSG and KYBO, in FTBC, FFPRI for management of materials in each nursery.

Conflicts of Interest: The authors declare no conflict of interest.

References

1. Kiyohara, T.; Tokushige, Y. Inoculation experiments of a nematode, *Bursaphelenchus* sp., onto pine trees. *J. Jpn. For. Soc.* **1971**, *5*, 210–218, (In Japanese with English summary).

2. Mamiya, Y.; Endo, N. Transmission of *Bursaphelenchus lignicolus* (Nematoda: Aphelenchoididae) by *Monochamus alternatus* (Coleoptera: Cerambycidae). *Nematologica* **1972**, *18*, 159–162. [CrossRef]

3. Zhao, B.G.; Futai, K.; Sutherland, J.R.; Takeuchi, Y. *Pine Wilt Disease*; Springer: New York, NY, USA, 2008.

4. Hirata, A.; Nakamura, K.; Nakao, K.; Kominami, Y.; Tanaka, N.; Ohashi, H.; Takano, K.T.; Takeuchi, W.; Matsui, T. Potential distribution of pine wilt disease under future climate change scenarios. *PLoS ONE* **2017**, *12*, e0182837. [CrossRef] [PubMed]

5. Futai, K.; Furuno, T. The variety of resistance among pine-species to pine wood nematode, *Bursaphelenchus lignicolus*. *Bull. Kyoto Univ. For.* **1979**, *51*, 23–36, (In Japanese with English summary).

6. Yano, M. Investigation on the causes of pine mortality in Nagasaki Prefecture. *Sanrinkoho* **1913**, *4*, 1–14. (In Japanese)

7. Nakamura, K.; Otsuka, I. *Forest Pest Management Leading to Establishment of Forestry Business: Pine Wilt Disease Unites the Locality*; Japan Forest Investigation Company: Tokyo, Japan, 2019. (In Japanese)

8. Fujimoto, Y.; Toda, T.; Nishimura, K.; Yamate, H.; Fuyuno, S. Breeding project on resistance to the pine Wood nematode—An outline of the research and the achievement of the project for ten years. *Bull. For. Tree. Breed. Center.* **1989**, *7*, 1–84, (In Japanese with English summary).

9. Toda, T. Studies on the breeding for resistance to the pine wilt disease in *Pinus densiflora* and *P. thunbergii*. *Bull. For. Tree. Breed. Inst.* **2004**, *20*, 83–217, (In Japanese with English summary).

10. Kurinobu, S. Current status of resistance breeding of Japanese pine species to pine wilt disease. *For. Sci. Tec.* **2008**, *4*, 51–57. [CrossRef]

11. Forest Tree Breeding Center. *Annual Report*; Forest Tree Breeding Center: Hitachi, Japan, 2020. (In Japanese)

12. Kiyohara, T. Effect of temperature on the disease incidence of pine seedlings inoculated with *Bursaphelenchus lignicolus*. *Trans. Ann. Meet. Jpn. For. Soc.* **1973**, *84*, 334–335, (In Japanese with English summary).

13. Ichihara, Y.; Fukuda, K.; Suzuki, K. Early symptom development and histological changes associated with migration of *Bursaphelenchus xylophilus* in seedling tissues of *Pinus thunbergii*. *Plant Dis.* **2000**, *84*, 675–680. [CrossRef]

14. Yamaguchi, R.; Matsunaga, K.; Watanabe, A. Influence of temperature on pine wilt disease progression in *Pinus thunbergii* seedlings. *Eur. J. Plant. Pathol.* **2020**, *156*, 581–590. [CrossRef]

15. Suzuki, K.; Kiyohara, T. Influence of water stress on development of pine wilting disease caused by Bursaphelenchus lignicolus. *Eur. J. For. Pathol.* **1978**, *8*, 97–107. [CrossRef]

16. Kaneko, S. Effect of light intensity on the development of pine wilt disease. *Can. J. Bot.* **1989**, *67*, 1861–1864. [CrossRef]

17. Rutherford, T.; Webster, J. Distribution of pine wilt disease with respect to temperature in North America, Japan, and Europe. *Can. J. For. Res.* **1987**, *17*, 1050–1059. [CrossRef]

18. Toda, T.; Kurinobu, S. Genetic improvement in pine wilt disease resistance in *Pinus thunbergii*: The effectiveness of pre-screening with an artificial inoculation at the nursery. *J. For. Res.* **2001**, *6*, 197–201. [CrossRef]

19. Toda, T.; Kurinobu, S. Realized genetic gains observed in progeny tolerance of selected red pine (*Pinus densiflora*) and black pine (*P. thunbergii*) to pine wilt disease. *Silvae. Gent.* **2002**, *51*, 42–44.

20. Hakamata, T. Studies on efficient production technique of *Pinus thunbergii* seedlings to *Bursaphelenchus xylophilus*. *Tech. Bull. Shizuoka. Res. Ins. Agric. For.* **2013**, *10*, 1–46, (In Japanese with English summary).

21. Kuroda, K.; Ohira, M.; Okamura, M.; Fujisawa, Y. Migration and population growth of the pine wood nematode (*Bursaphelenchus xylophilus*) related to the symptom development in the seedlings of Japanese black pine (*Pinus thunbergii*) families selected as resistant to pine wilt. *J. Jpn. For. Soc.* **2007**, *89*, 241–248. (In Japanese with English summary) [CrossRef]

22. Aikawa, T.; Kikuchi, T.; Kosaka, H. Demonstration of interbreeding between virulent and avirulent populations of *Bursaphelenchus xylophilus* (*Nematoda: Aphelenchoididae*) by PCR-RFLP method. *Appl. Entomol. Zool.* **2003**, *38*, 565–569. [CrossRef]

23. Japan Meteorological Agency. 2020 Climate Statistics. Available online: http://www.data.jma.go.jp/obd/stats/etrn/index.php (accessed on 10 December 2017).

24. R Core Team. *R: A Language and Environment for Statistical Computing*; R Foundation for Statistical Computing: Vienna, Austria, 2019.

25. Bates, D.; Maechler, M.; Bolker, B.; Walker, S. Fitting linear mixed-effects models using LME4. *J. Stat. Softw.* **2015**, *67*, 1–48. [CrossRef]

26. Fox, J.; Weisberg, S. *An R Companion to Applied Regression*, 3rd ed.; Sage: Thousand Oaks, CA, USA, 2019.

27. Hothorn, T.; Bretz, F.; Westfall, P. Simultaneous inference in general parametric models. *Biom. J.* **2018**, *50*, 346–363. [CrossRef] [PubMed]

28. Bartoń, K. *MuMIn: Multi-Model Inference*; R Package Version 1.43.17; 2020. Available online: https://CRAN.R-project.org/package=MuMIn (accessed on 1 June 2020).

29. Hirao, T.; Fukatsu, E.; Watanabe, A. Characterization of resistance to pine wood nematode infection in *Pinus thunbergii* using suppression subtractive hybridization. *BMC Plant. Biol.* **2012**, *12*, 13. [CrossRef] [PubMed]

30. Kusumoto, D.; Yonemichi, T.; Inoue, H.; Hirao, T.; Watanabe, A.; Yamada, T. Comparison of histological responses and tissue damage expansion between resistant and susceptible *Pine thunbergii* infected with pine wood nematode *Bursaphelenchus xylophilus*. *J. For. Res.* **2014**, *19*, 285–294. [CrossRef]

31. Kuroda, K. Inhibiting factors of symptom development in several Japanese red pine (*Pinus densiflora*) families selected as resistant to pine wilt. *J. For. Res.* **2004**, *9*, 217–224. [CrossRef]

32. Nakajima, G.; Iki, T.; Yamanobe, T.; Nakamura, K.; Aikawa, T. Spatial and temporal distribution of Bursaphelenchus xylophilus inoculated in grafts of a resistant clone of Pinus thunbergii. *J. For. Res.* **2019**, *24*, 93–99. [CrossRef]

33. Matsunaga, K.; Watanabe, A. Outline of the results of the "Project to advance the development of technology for varieties of Japanese black pine and red pine resistant to the pine wood nematode" commissioned by the Forestry Agency, Ministry of Agriculture, Forestry and Fisheries of Japan—Characterization of pine wood nematode and development of 2nd generation resistant varieties. *For. Genet. Breed.* **2018**, *7*, 115–119. (In Japanese)

34. Iwaizumi, M.G. The development of 2nd generation resistant Japanese red pine varieties in collaboration with Prefecture. *For. Genet. Breed.* **2018**, *7*, 159–161. (In Japanese)

Article

Do Seedlings Derived from Pinewood Nematode-Resistant *Pinus thunbergii* Parl. Clones Selected in Southwestern Region Perform Well in Northern Regions in Japan? Inferences from Nursery Inoculation Tests

Koji Matsunaga [1,*,†], Taiichi Iki [2,†], Tomonori Hirao [3], Mineko Ohira [3], Taro Yamanobe [3], Masakazu G. Iwaizumi [4], Masahiro Miura [4], Keiya Isoda [3], Manabu Kurita [1], Makoto Takahashi [3] and Atsushi Watanabe [5]

[1] Kyushu Regional Breeding Office, Forest Tree Breeding Center, Forestry and Forest Products Research Institute, 2320-5 Suya, Koshi, Kumamoto 861-1102, Japan; mkuri@affrc.go.jp

[2] Tohoku Regional Breeding Office, Forest Tree Breeding Center, Forestry and Forest Products Research Institute, 95 Osaki, Takizawa, Iwate 020-0621, Japan; iki@affrc.go.jp

[3] Forest Tree Breeding Center, Forestry and Forest Products Research Institute, 3809-1 Ishii, Juo, Hitachi, Ibaraki 319-1301, Japan; hiratomo@affrc.go.jp (T.H.); sasamine@affrc.go.jp (M.O.); yamanobe@affrc.go.jp (T.Y.); keiso@affrc.go.jp (K.I.); makotot@affrc.go.jp (M.T.)

[4] Kansai Regional Breeding Office, Forest Tree Breeding Center, Forestry and Forest Products Research Institute, 1043, Uetsukinaka, Shoo, Katsuta, Okayama 709-4335, Japan; ganchan@affrc.go.jp (M.G.I.); miumasa@affrc.go.jp (M.M.)

[5] Department of Forest Environmental Science, Faculty of Agriculture, Kyushu University, 744 Motooka, Nishiku, Fukuoka 819-0395, Japan; nabeatsu@agr.kyushu-u.ac.jp

* Correspondence: makoji@affrc.go.jp; Tel.: +81-96-242-3151

† These authors contribute equally.

Received: 24 July 2020; Accepted: 28 August 2020; Published: 1 September 2020

Abstract: Background and Objectives: To determine whether the progeny of pinewood nematode-resistant *Pinus thunbergii* Parl. clones selected in the southwestern region of Japan could be successful in reforestation in the northern region, we investigated the magnitude of the genotype–environment interaction effect on the resistance against *Bursaphelenchus xylophilus* (Steiner and Buhrer) Nickle in *P. thunbergii*. Materials and Methods: We inoculated *P. thunbergii* seedlings of six full-sib families, with various resistance levels, with *B. xylophilus* in nurseries at three experimental sites in the northern and southern regions of Japan. All parental clones of the tested families originated from southwestern Japan, and selection of parental clones for resistance was performed in the same region. Sound rates after nematode inoculation were calculated, and survival analysis, correlation analysis and variance component analysis were performed. Results and Conclusions: Families with high sound rate in the southern region also showed a high sound rate in the northern region. In almost all cases, Spearman's correlation coefficients for sound rates were more than 0.698 among sites. The variance component of the interaction between site and family was small compared to that of site and family separately. Thus, we conclude that the resistant clones selected in the southern region would retain their genetic resistance in the northern regions.

Keywords: pine wilt disease; *Bursaphelenchus xylophilus*; genotype by environment interaction; Japanese black pine; variance component

---

1. Introduction

Japanese black pine *Pinus thunbergii* Parl. is one of the major forestry species in Japan. Pine seedlings have been planted across a wide coastal area of Japan, from the northern part of Honshu island to the southern part of Kyushu island, to protect land and houses against strong winds and sand movement inland [1]. After the invasion of the pinewood nematode, *Bursaphelenchus xylophilus* (Steiner and Buhrer) Nickle, from North America to Kyushu island in the early 20th century causing pine wilt disease (PWD) in *P. thunbergii* forests, the disease has spread to the northern part of Japan [2–4]. Currently, the disease has been reported in all the prefectures of Japan except Hokkaido, the northern most prefecture [5]. From a global perspective, PWD in East Asia (Japan, South Korea, China and Taiwan) has now spread to southwestern Europe (Portugal and Spain) [4–9], and there is a risk that the disease will spread to neighboring countries [10,11].

To combat PWD, a national resistance breeding program of *Pinus densiflora* Sieb. et Zucc. (Japanese red pine) and *P. thunbergii* was started in southwestern Japan in 1978 as a part of an integrated pest management. In the program, 92 *P. densiflora* and 16 *P. thunbergii* resistant clones were selected [12]. The selected clones were propagated by grafting and used in PWD-resistant seed orchards. As PWD spread into the eastern and northern parts of Japan, supplemental resistance breeding programs were started in the Tohoku and Kanto regions [13,14]. Although many resistant clones were selected and resistant seed orchards were established in the eastern and northern regions in the programs, these eastern and northern orchards included resistant clones selected in the southern region of Japan to supplement shortages of resistant clones in the surrounding regions. Japan has a large geographic extension from north to south, with a highly variable climate. Until now, the genetic capability of resistant clones selected in southern Japan, or their progeny, in the northern regions of Japan has not been examined.

Species, provenance and family variation in resistance or susceptibility to pinewood nematode has been reported in artificial inoculation experiments using graftings, half- or full-sib families of pine species, and studies have shown the relatively high heritability of resistance or susceptibility in *P. thunbergii*, *P. densiflora* and *Pinus pinaster* Ait. (maritime pine) [15–20]. On the other hand, environmental factors also affect PWD development in infected trees. High air temperature, dry soil conditions and low-light intensity promote disease development, shorten the time until death and increase mortality [21–23]. However, there is limited knowledge of the effect of the interaction between genotype by environmental factors (G × E) on resistance. A previous study, based on a six-year *B. xylophilus* inoculation experiment using open-pollinated families of *P. thunbergii*, showed that the family-by-year effect for resistance level is smaller than the family effect [16]. In *P. pinaster*, a G × E interaction was reported based on a greenhouse inoculation test using seedlings from six provenances [18].

Resistance breeding against invasive pests generally begins at the site of the pest introduction. When there is a risk of pest expansion to neighboring regions with different climates, and if the clones or gene pool selected by resistance breeding display resistance in other regions with different climates, those genetic resources and breeding materials can be used in pest control strategies in other regions. In recent years, PWD has invaded southwestern Europe, and resistance breeding of *P. pinaster* has begun in Portugal and Spain [19,20]. In Japan, the first PWD outbreak occurred in Nagasaki, Kyushu in the southwestern region, and resistance breeding began in the southwestern region.

Here, to clarify if the progeny of southern resistant clones retain their genetic resistance to PWD in the northern region of Japan, the seedlings of six *P. thunbergii* families with various resistance levels were inoculated with an isolate of *B. xylophilus* at three sites with different climates. Then, the external symptom was assessed and analyzed.

2. Materials and Methods

2.1. Experimental Sites

Nurseries in the Tohoku Regional Breeding Office (TBO), Forest Tree Breeding Center (FTBC), and Kyushu Regional Breeding Office (KYBO) were used as the three sites for the experiment (Figure 1).

TBO (39°49′4.8″ N, 141°8′13.2″ E) is located in Iwate Prefecture in the Tohoku region, in the northern part of Honshu island. FTBC (36°41′31.2″° N, 140°41′24″ E) is located in Ibaraki Prefecture in the Kanto region, central Honshu island. KYBO (32°52′51.6″ N, 130°44′9.6″ E) is located in Kumamoto Prefecture in the Kyushu region, Kyushu island. The distance between TBO and KYBO is about 1200 km. The climate around TBO is cool and the monthly average temperature in winter is below 0 °C (Figure 2) [24]. On the other hand, the climate around KYBO is warm and the monthly average temperature in summer exceeds 25 °C. The climate around FTBC is intermediate between the two; in summer, the temperature is close to that of TBO, and in winter it is close to that of KYBO. Precipitation in June and July is high as it is the rainy season around KYBO.

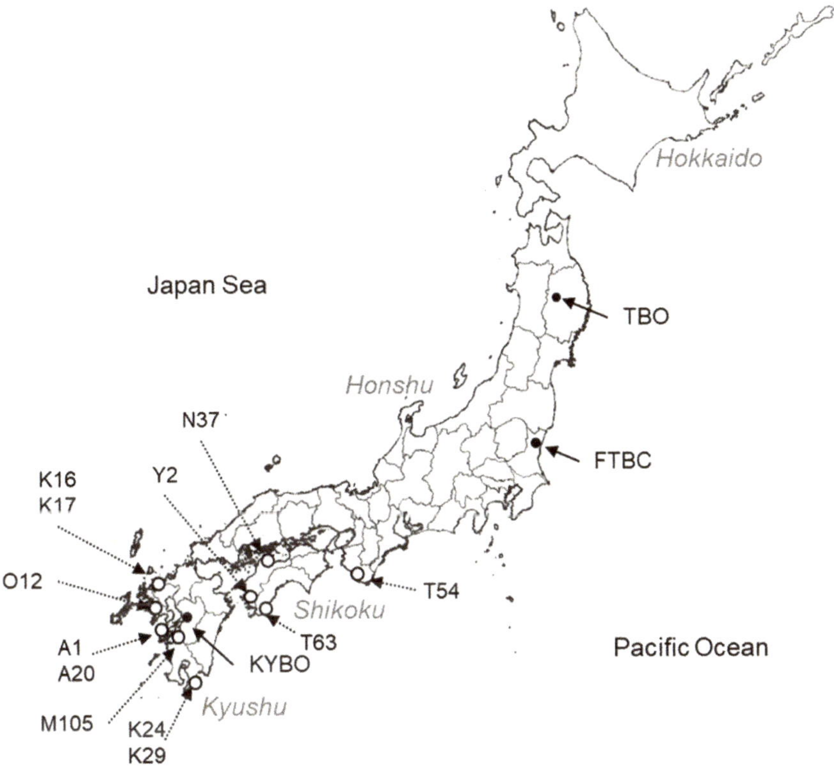

Figure 1. The experimental sites and the origin of materials used in this study. Filled circles indicate the three experimental sites, Tohoku Regional Breeding Office (TBO), Forest Tree Breeding Center (FTBC) and Kyushu Regional Breeding Office (KYBO). Open circles indicate the origins of 12 parental clones crossed to produce full-sib families used in this study. Italicize letters indicate the names of the four main islands of Japan. This map was created from the blank map published Geospatial Information Authority of Japan. Clone name abbreviations: Amakusa20 (A20), Namikata37 (N37), Yoshida2 (Y2), Tanabe54 (T54), Kimotuki24 (K24), Minamatasho105 (M105), Karatsu17 (K17), Karatsu16 (K16), Tosashimizu63 (T63), Oseto12 (O12), Kimotsuki29 (K29), and Amakusa1 (A1).

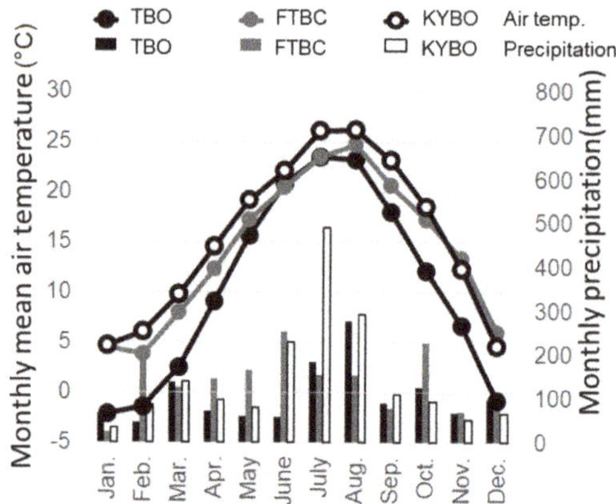

Figure 2. Monthly mean air temperature and monthly precipitation of three experimental sites in 2014. Data was obtained from the AMeDAS (Automated Meteorological Data Acquisition System) of Japan Meteorological Agency. The values for TBO, FTBC, and KYBO sites were measured at Morioka, Hitachi, and Kikuchi observatories, which are close to the experimental sites.

2.2. Pine Seedlings

Six *P. thunbergii* full-sib families produced by artificial crossing were used in the experiment (Table 1). In order to include pine trees with a large variation in their resistance to PWD, we used six PWD-resistant clones with high or intermediate resistance (Amakusa20 (A20), Namikata37 (N37), Yoshida2 (Y2), Karatsu17 (K17), Karatsu16 (K16), and Tosashimizu63 (T63)), two PWN-resistant clones with relatively low resistance (Tanabe54 (T54) and Oseto12 (O12)), and four plus-tree clones (Kimotuki24 (K24), Kimotsuki29 (K29), Minamatasho105 (M105), and Amakusa1 (A1)). The resistant clones were selected using *B. xylophilus* artificial inoculation tests and plus-tree clones were selected by phenotypically superior evaluation for growth and stem form. All parental clones originated in southwestern Japan (Figure 1) and were propagated by grafting and stored in KYBO. The inoculation test for resistant clone selection was also performed in southwestern Japan. The resistance levels of eight parental resistant clones, based on the nematode inoculation test using their open-pollinated families, have already been reported [25,26]. The resistance rank of parental clones is described in Table 1. The other four plus-tree clones were not selected for their resistance and the resistance levels of their open-pollinated or crossed families were low, based on preliminary inoculation tests.

Table 1. Details of the *Pinus thunbergii* full-sib families used in this study.

Mother		Father		No. of Seedlings, Height (Mean ± SD)					
Clone (Abbreviation)	Rank*	Clone (Abbreviation)	Rank*		TBO		FTBC		KYBO
Amakusa20 (A20)	3/13	Karatsu17 (K17)	1/13	39	31.9 ± 7.2 a	70	30.1 ± 4.0 b	92	26.6 ± 5.4 b
Namikata37 (N37)	3/16	Karatsu16 (K16)	5/13	42	30.2 ± 7.6 ab	69	32.7 ± 5.2 a	77	30.9 ± 5.1 a
Yoshida2 (Y2)	4/16	Tosashimizu63 (T63)	6/16	39	29.7 ± 6.5 ab	71	26.6 ± 3.9 d	52	21.1 ± 3.0 c
Tanabe54 (T54)	10/16	Oseto12 (O12)	14/16	25	22.5 ± 6.0 c	68	25.4 ± 4.6 e	76	15.2 ± 4.3 d
Kimotsuki24 (K24)	-	Kimotsuki29 (K29)	-	22	25.8 ± 7.0 bc	70	28.5 ± 5.3 c	58	19.7 ± 4.0 c
Minamatasho105 (M105)	-	Amakusa1 (A1)	-	33	32.7 ± 7.5 a	71	26.7 ± 5.2 d	72	20.6 ± 4.5 c
Total				200	29.4 ± 7.7	419	28.3 ± 5.3	427	22.7 ± 7.0

Rank* shows the rank of clone resistance based on the progeny test by Matsunaga et al, [26]. The denominator and the numerator indicate the number of resistant clones evaluated at the same time and the clone rank among them, respectively. -: no evaluation. Means followed by a common letter are not significantly different at 5% level of significance.

One hundred seeds belonging to each of the six families were sown in the TBO, FTBC and KYBO nurseries in the spring of 2013. The following spring, the seedlings of each family were transplanted to another location in the nursery, with a random block design of family with two replicates at 20 cm × 20 cm spacing in FTBC and KYBO. In the TBO nursery, our preliminary test results showed that 1.5-year-old seedlings were not large enough for use in the inoculation experiment. In the present study, seedlings were not transplanted in the TBO nursery; instead, 2–3 seeds were sown at 20 cm × 20 cm spacing, and extra seedlings were removed to ensure only one remained in each 20 cm × 20 cm grid. Prior to inoculation, there were 200, 419, and 427 seedlings in the TBO, FTBC, and KYBO sites, respectively (Table 1). The height of each seedling was measured during the week prior to inoculation.

2.3. Nematode Inoculation and Symptom Observation

For inoculation, an isolate of *B. xylophilus* (Ka4) obtained from dead *P. densiflora* in Ibaraki Prefecture in 1999 [27] and sub-cultured in the laboratory at FTBC was used. After an incubation of approximately 10 d on *Botrytis cinerea* Pers., a fungus, on barley grains, the nematodes were separated from the media using the Baermann funnel method. The nematode suspension was adjusted to 200,000 nematodes/mL of water. Nematode incubation and suspension adjustment were conducted at each site.

Nematode inoculation was conducted on 1 July 2014 in all sites. A 5 cm length of seedling stem was peeled with a sharp knife at approximately 5–10 cm above the ground, and the wound was scratched with small sow before inoculation with 50 µL of suspension containing 10,000 nematodes using a micropipette.

The inoculated seedlings were observed weekly and external symptoms were classified into three categories (0: no symptoms, 1: browning of needles on one or more branches, 2: browning of all needles). Seedlings with an external symptom level of 1 were considered as diseased, and the seedlings with a level of 2 as dead. Subsequently, sound seedling rate and survival rate were calculated as follows:

Sound seedling rate = No. of seedlings in symptom class 0/No. of inoculated seedlings
Survival rate = No. of seedlings in symptom class 0 and 1/No. of inoculated seedlings

The sound seedling rate was the rate of seedlings without external symptoms, and focused on the seedlings with higher resistant level. On the other hand, the survival rate was the rate of surviving seedlings that included not only sound seedlings but also diseased and partially dead ones. From the viewpoint of preventive counteracts, we used the sound seedling rate as a major indicator and the survival rate as the supplemental result, as we considered that no symptoms were more important than surviving. Observations were carried out for 10 weeks after inoculation (WAI); however, in TBO the 4-week and 8-week survey was not conducted.

2.4. Statistical Analysis

R version 4.0.0 [28] was used for all statistical analyses. Seedling height was analyzed with a linear mixed model using the lmer function of the lme4 package [29] to determine the size variation of seedlings. In the model, mean height of each replicate of each of the six families from the three sites was calculated and used as the response variable; while family, site and their interaction were used as explanatory variables with fixed effects and replication within site was used as an explanatory variable with random effects. As model selection based on the AIC value with the function dredge in the MuMIn package [30] selected the model with an interaction between family and site (Table S1), we separated the data of each site and conducted multiple comparisons among families using the glht function in the multcomp package [31].

To compare the disease development process among sites and families, we conducted a two-step survival analysis using Kaplan–Meier estimators. For comparison among sites, Kaplan–Meier estimators were calculated for sound rate of seedlings and log-rank test with Bonferroni-adjusted *p* values was applied for multiple comparisons among sites. Since the composition ratio of the six families did not differ significantly by site (Chi-square test, X^2-value: 15.56, *d. f.*: 10, *p*-value: 0.1130), no weighting was applied to the number of seedlings for each family in each site. Then, to compare the disease development process among families within sites, Kaplan–Meier estimators were calculated and the log-rank test with Bonferroni-adjusted *p* values was also applied. For the survival analysis, the functions survfit and survdiff in the survival package [32] were used.

Pairwise Spearman's correlation coefficients among sites were calculated to compare the order of resistance level among the six families.

To compare the relative effects of the site, family and their interaction on the variance of sound seedling rate, variance components of the factors were estimated using generalized linear mixed models with the glmer function of the lme4 package, with family assumed as a binomial error structure and logit link function [29]. Number of diseased seedlings (symptom class 1 and 2) and number of sound seedlings (symptom class 0) in a family in each replicate in a site was used as the response variable, and site, family, their interaction and replicate within each site were used as explanatory variables with random effects. Mean seedling height was added to the model as an explanatory variable with fixed effects. This analysis was applied to the data for 5, 6, 7, 9 and 10 WAI only, because no data was collected weeks 4 and 8 in TBO and there were few diseased seedlings before 3 WAI in FTBC.

Survival analysis, calculation of Spearman's correlation coefficient and variance component analysis were also conducted on survival rate.

3. Results

3.1. Seedling Height

Overall, mean seedling height was 26.4 cm across the three sites. Mean height across the six families at sites TBO, FTBC and KYBO were 29.4, 28.3, and 22.7 cm, respectively (Table 1). A model including the interaction between site and family was selected as the best model for seedling height (Table S1). After separating the data according to site, the model selected for each site included the family component. Multiple comparisons showed that seedling height significantly varied among families in all sites (Table 1). The height of T54 × O12 was always significantly lower than that of the other families. M105 × A1 was the tallest family in TBO, but was the fourth tallest family in FTBC and KYBO.

3.2. Sound Seedling Rate

The Kaplan–Meier estimators for sound seedling rates showed that the disease developmental process varied among the three sites (Figure 3). Diseased seedlings were first observed at 2 WAI at TBO and KYBO, and one week later at FTBC. The sound seedling rate sharply decreased until 6, 5 and 4 WAI in TBO, FTBC and KYBO, respectively, and then decreased more gradually. Total sound seedling rate at 10 WAI across the three sites was 0.27 ± 0.45 (mean $\pm$ SD) and 0.24 ± 0.43, 0.50 ± 0.50, and 0.07 ± 0.25 in TBO, FTBC, and KYBO, respectively. Pairwise log-rank tests showed that the survival curves of the three sites significantly differed from each other (X^2: 97.7, *d.f.*: 1, *p*: <0.001 for TBO vs. FTBC; X^2: 60.9, *d.f.*: 1, *p*: <0.001 for TBO vs. KYBO; X^2: 399, *d.f.*: 1, *p*: <0.001 for FTBC vs. KYBO).

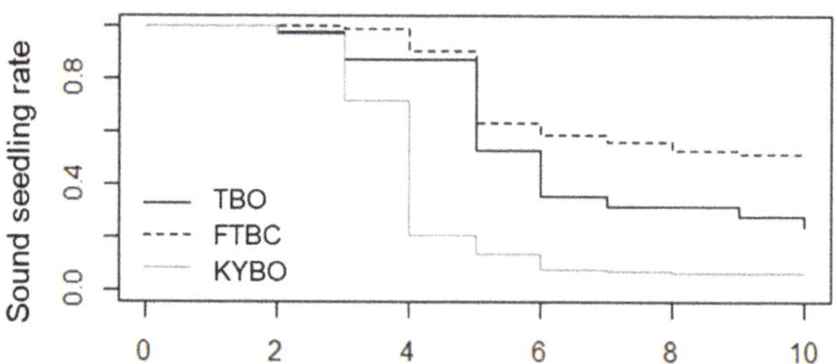

Figure 3. Kaplan–Meier estimator for sound seedling rate of six *Pinus thunbergii* full-sib families inoculated with *Bursaphelenchus xylophilus* in the three experimental sites. Black, dashed and gray lines indicate TBO, FTBC, and KYBO respectively.

The Kaplan–Meier estimators showed that the disease developmental process varied among families in all sites (Figure 4). Diseased seedlings were observed at 2 WAI in three (T54 × O12, K24 × K29, and M105 × A1) of the six families at TBO and in four (Y2 × T63, T54 × O12, K24 × K29, and M105 × A1) of the six families at KYBO (Table S2). At FTBC, disease development in inoculated seedlings appeared at 3 WAI in two families (T54 × O12 and K24 × K29). Pairwise log-rank tests showed that families derived from high- and intermediate-resistance parental clones had a significantly lower risk of disease development than the families derived from low-resistance and plus-tree parental clones (Figure 4). The curves of the disease development process were more clearly divergent among families in FTBC and TBO than in KYBO.

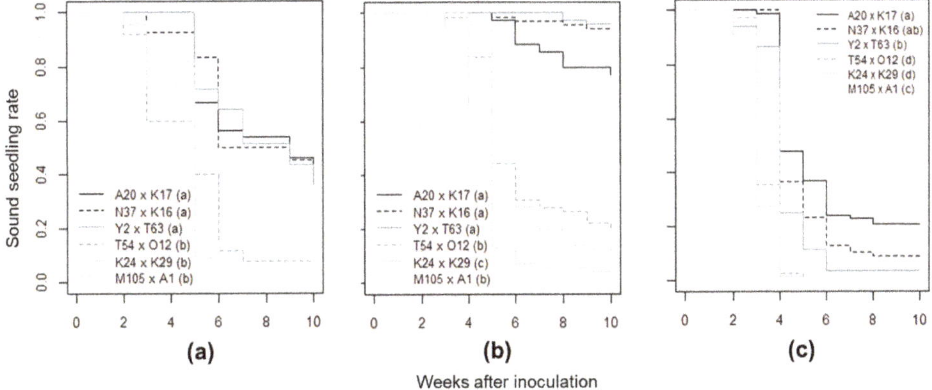

Figure 4. Kaplan–Meier estimator for sound seedling rate of six *Pinus thunbergii* full-sib families inoculated with *Bursaphelenchus xylophilus* in each of the three experimental sites. (**a**) TBO; (**b**) FTBC; and (**c**) KYBO. See Table 1 for abbreviated names of pine families. Family names followed by a common letter are not significantly different at the 5% level of significance

Spearman correlation coefficients for sound rate were higher than 0.698 at three or more WAI in each pair of sites (Table 2). At 2 WAI, when disease development was in the initial phase, the coefficient was relatively low: 0.400 between TBO and KYBO.

Table 2. Spearman's correlation coefficient for sound seedling rate of *Pinus thunbergii* inoculated with *Bursaphelenchus xylophilus*.

Week	Spearman's Correlation Coefficient (p-Value)		
	TBO vs. FTBC	**TBO vs. KYBO**	**FTBC vs. KYBO**
2		0.400 (0.419)	
3	0.789 (0.058)	0.754 (0.103)	0.845 (0.033)
4			0.759 (0.058)
5	0.943 (0.003)	0.759 (0.058)	0.698 (0.136)
6	0.928 (0.017)	0.770 (0.058)	0.698 (0.136)
7	0.812 (0.058)	0.893 (0.017)	0.698 (0.136)
8			0.698 (0.136)
9	0.754 (0.103)	0.955 (0.003)	0.698 (0.136)
10	0.812 (0.058)	0.893 (0.017)	0.698 (0.136)

Correlation coefficients were calculated for data after the occurrence of disease development in seedlings.

For the variance components from 5–10 WAI, the family component consistently occupied the largest proportion (Figure 5, Table S3). The proportion of the variance component of the interaction between site and family was consistently small compared to that of both site and family separately.

Variance component proportions of replication within site were consistently very small (less than 1% in all examined weeks).

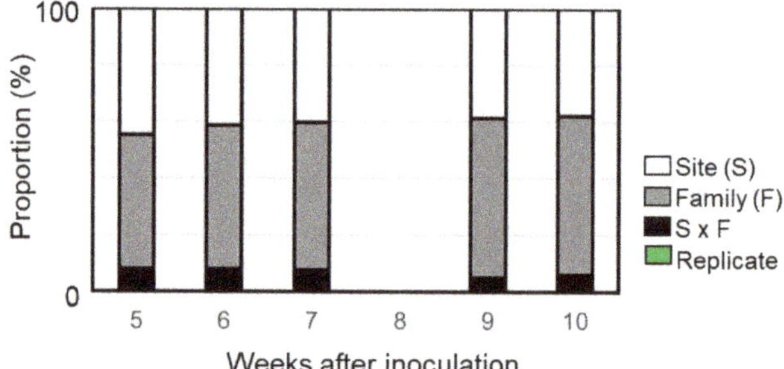

Figure 5. Proportion of variance components of site, family, their interaction, and replication within site for sound seedling rate of *Pinus thunbergii* after inoculation with *Bursaphelenchus xylophilus*. Variance component analysis was not conducted for 8 weeks after inoculation due to missing data in TBO.

Survival rate data was similar to that of the sound seedlings rate. All survival rate results are shown in supplemental figures and tables (survival analysis among sites: Figure S1, survival analysis among families: Figure S2, Correlation: Table S4, and variance component analysis: Figure S3 and Table S5).

4. Discussion

In this study, we inoculated six *P. thunbergii* families with variable resistance to PWD with a *B. xylophilus* isolate in nurseries at three different sites in the northern and southern regions of the Japanese archipelago. Consequently, families with higher sound seedling rate in the KYBO site exhibited higher sound seedling rate in the TBO and FTBC sites. Spearman correlation coefficients for family sound seedling rate among sites were relatively high and positive. Moreover, variance component analyses revealed only a small contribution of the interaction between site and family to total variance in sound seedling rate. These results show that the *P. thunbergii* seedlings obtained from the selected resistant clones with high resistance level in southern Japan may retain their high resistance in northern Japan.

Previous studies have described the G × E interaction of resistance or susceptibility of pine seedlings to *B. xylophilus*. A six-year inoculation experiment using the 16 half-sib families of resistant *P. thunbergii* clones showed that the variance component of the interaction between year and family was less than one half of the family variance component [16]. G × E interactions in the susceptibility to *B. xylophilus* was reported in *P. pinaster*, based on greenhouse inoculation tests using seedlings derived from six provenances [18]. Although the magnitude of the effect of the interaction was not clearly described in the paper, seedlings from a particular provenance may exhibit some degree of interaction. The results of the previous studies and the present study suggest that the effect of the G × E interaction of resistance to *B. xylophilus* could be small in the half-sib or full-sib families of pine seedlings, although certain genetic groups may be more sensitive to the ambient environment.

In this study, the sound seedling rate was highest in FTBC, followed by TBO and KYBO. TBO is located northward of FTBC, with a cooler climate (specifically, climatological standard normal of the average monthly temperature in July is 21.8 °C for TBO and 22.8 °C for FTBC). Since low temperature suppresses the progress of PWD development [33,34], we expected that the sound seedling rate of TBO would be the highest, but based on the inoculation test results of TBO this was not the case. Close examination of the meteorological data in the experimental year, 2014, revealed that the average temperature in July was 23.5 °C in both TBO and FTBC, and the average temperature during the week after inoculation was 22.8 °C in TBO and 21.6 °C in FTBC. The low sound seedling rate in TBO may have been affected by the slightly higher temperature just after inoculation.

By March 2020, 54 first-generation and 40 second-generation PWD-resistant *P. thunbergii* clones had been selected in southwestern Japan. If progeny of the resistant clones selected in the southern region were to be planted in the northern region, the following factors should be considered: growth, snow-resistance, reproductive traits of clones, administrative seed transfer zones [35] and genetic structure of *P. thunbergii* throughout Japan [36]. However, the present study focused on the most important factor, which is the resistance to pinewood nematode. Using the most- and least-resistant seedlings available from southern *P. thunbergii* clones, we showed that the possibility of southern high-resistance clones could be used in the eastern and northern regions of Japan. Introduction of the southern resistant clones into the production population in eastern and northern regions could enable the promotion of resistance breeding programs in those regions. Conversely, the possibility of utilizing northern resistant clones in the southern region should also be considered.

5. Conclusions

Inoculation tests for six *P. thunbergii* families with different resistance levels were carried out using an isolate of *B. xylophilus* at three sites with different climates. We indicated that the resistant rank of the families was relatively stable regardless of different climates among three sites. The results obtained in this study suggest that the resistant *P. thunbergii* clones selected in the southern region of Japan may relatively perform their high genetic resistance well in the northern region of Japan.

Supplementary Materials: The following are available online at http://www.mdpi.com/1999-4907/11/9/955/s1, Figure S1: Kaplan–Meier estimator for survival rate of seedlings of six *P. thunbergii* full-sib families inoculated with *B. xylophilus* in three experimental sites, Figure S2: Kaplan–Meier estimator for survival rate of six *P. thunbergii* full-sib families inoculated with *B. xylophilus* in each of three experimental site, Figure S3:Propertion of variance components of site, family, their interaction and replication within site for survival rate of *P. thunbergii* after inoculation of *B. xylophilus*, Table S1: Model selection table for mean height of *P. thunbergii* seedlings, Table S2: Time trend in mean sound seedling rate and survival rate of seedlings of six *P. thunbergii* families inoculated with *B. xylophilus* in three experimental sites, Table S3: Estimated variance components for sound seedling rate of *P. thunbergii* after inoculation of *B. xylophilus*, Table S4: Spearman's correlation coefficients for survival rate of *P. thunbergii* seedlings inoculated with *B. xylophilus*, Table S5: Estimated variance components for survival rate of *P. thunbergii* seedlings inoculated with *B. xylophilus*.

Author Contributions: Conceptualization, K.M.; materials management and investigation, T.I., M.O., T.Y., M.G.I., M.M., K.I., M.K.; materials production, K.M., M.K.; methodology, T.H.; data compilation and analysis, T.I, K.M.; writing—original draft preparation, K.M., T.I.; writing—review and editing, T.Y., T.H., M.O., M.G.I., M.M., M.K., M.T. and A.W.; project administration, T.H., M.T. and A.W. All authors have read and agreed to the published version of the manuscript.

Funding: This research was funded by "Project to advance the development of technology for varieties of Japanese black pine and red pine resistant to the pinewood nematode" by Forestry Agency, Ministry of Agriculture, Forestry and Fisheries of Japan.

Acknowledgments: We thank H. Hoshi of FTBC, FFPRI for their coordination of the research project. We also thank our colleagues in the field management section of FTBC, TBO and KYBO for the production and cultivation of plant materials. We also thank TMB for nematode management.

Conflicts of Interest: The authors declare no conflict of interest.

References

1. Forest Tree Breeding Center (Japan). *Statistics and States of Implementation in Forest Tree Breeding Projects in Japan*; Forest Tree Breeding Center: Hitachi, Japan, 2020; pp. 68–69.
2. Kiyohara, T.; Tokushige, Y. Inoculation experiments of a nematode, *Bursaphelenchus* sp., onto pine trees. *J. Jpn. For. Soc.* **1971**, *53*, 105–114. (In Japanese with English Summary)
3. Mamiya, Y. History of pine wilt disease in Japan. *J. Nematol.* **1988**, *20*, 219–226. [PubMed]
4. Kazuyoshi, F. Pine wilt in Japan from first incidence to the present. In *Pine Wilt Disease*; Zhao, B.G., Futai, K., Sutherland, J.R., Takeuchi, Y., Eds.; Springer: New York, NY, USA, 2008; pp. 5–12.
5. Nakajima, G.; Iki, T.; Yamanobe, T.; Nakamura, K.; Aikawa, T. Spatial and temporal distribution of *Bursaphelenchus xylophilus* inoculated in grafts of a resistant clone of *Pinus thunbergii*. *J. For. Res.* **2019**, *24*, 93–99. [CrossRef]

6. Mota, M.M.; Braasch, H.; Bravo, M.A.; Penas, A.C.; Burgermeister, W.; Metge, K.; Sousa, E. First report of *Bursaphelenchus xylophilus* in Portugal and in Europe. *Nematology* **1999**, *1*, 727–734. [CrossRef]

7. Abelleira, A.; Picoaga, A.; Mansilla, J.P.; Aguin, O. Detection of *Bursaphelenchus xylophilus*, causal agent of pine wilt disease on *Pinus pinaster* in northwestern Spain. *Plant Dis.* **2011**, *95*, 776. [CrossRef] [PubMed]

8. Robertson, L.; Cobacho Arcos, S.; Escuer, M.; Santiago Merino, R.; Esparrago, G.; Abelleira, A.; Navas, A. Incidence of the pinewood nematode *Bursaphelenchus xylophlius* Steiner & Buhrer, 1934 (Nickle, 1970) in Spain. *Nematology* **2011**, *13*, 755–757. [CrossRef]

9. Inácio, M.L.; Nóbrega, F.; Vieira, P.; Bonifácio, L.; Naves, P.; Sousa, E.; Mota, M. First detection of *Bursaphelenchus xylophilus* associated with *Pinus nigra* in Portugal and in Europe. *For. Path.* **2015**, *45*, 235–238. [CrossRef]

10. Vicente, C.; Espada, M.; Vieira, P.; Mota, M. Pine Wilt Disease: A threat to European forestry. *Eur. J. Plant Pathol.* **2012**, *133*, 89–99. [CrossRef]

11. Hirata, A.; Nakamura, K.; Nakao, K.; Kominami, Y.; Tanaka, N.; Ohashi, H.; Takano, K.T.; Takeuchi, W.; Matsui, T. Potential distribution of pine wilt disease under future climate change scenarios. *PLoS ONE* **2017**, *12*, e0182837. [CrossRef]

12. Fujimoto, Y.; Toda, T.; Nishimura, K.; Yamane, H.; Fuyuno, S. Breeding project on resistance to the pine-wood nematode an outline of the research and the achievement of the project for ten year. *Bull. For. Tree Breed. Inst.* **1989**, *7*, 1–84.

13. Kurinobu, S. Current status of resistance breeding of Japanese pine species to pine wilt disease. *Forest Sci. Technol.* **2008**, *4*, 51–57. [CrossRef]

14. Nose, M.; Shiraishi, S. Breeding for resistance to pine wilt disease. In *Pine Wilt Disease*; Zhao, B.G., Futai, K., Sutherland, J.R., Takeuchi, Y., Eds.; Springer: New York, NY, USA, 2008; pp. 334–350. [CrossRef]

15. Futai, K. Furuno Tooshu The variety of resistances among pine-species to pinewood nematode, Bursaphelenchus lignicolus. *Bull. Kyoto Univ. For.* **1979**, *51*, 23–36, (In Japanese with English Summary).

16. Toda, T.; Kurinobu, S. Genetic improvement in pine wilt disease resistance in *Pinus thunbergii*: The effectiveness of pre-screening with an artificial inoculation at the nursery. *J. For. Res.* **2001**, *6*, 197–201. [CrossRef]

17. Iwaizumi, M.G.; Tamaki, S.; Isoda, K.; Kubota, M. Evaluation of combining ability for pine wood nematode-resistance based on half-diallele lines of *Pinus densiflora* resistant clones. *For. Genet. For. Breed* **2019**, *8*, 121–130. [CrossRef]

18. Menéndez-Gutiérrez, M.; Alonso, M.; Toval, G.; Díaz, R. Variation in pinewood nematode susceptibility among *Pinus pinaster* Ait. provenances from the Iberian Peninsula and France. *Ann. For. Sci.* **2017**, *74*, 76. [CrossRef]

19. Menéndez-Gutiérrez, M.; Alonso, M.; Toval, G.; Díaz, R. Testing of selected *Pinus pinaster* half-sib families for tolerance to pinewood nematode (*Bursaphelenchus xylophilus*). *Forestry* **2018**, *91*, 38–48. [CrossRef]

20. Carrasquinho, I.; Lisboa, A.; Inácio, M.L.; Gonçalves, E. Genetic variation in susceptibility to pine wilt disease of maritime pine (*Pinus pinaster* Aiton) half-sib families. *Ann. For. Sci.* **2018**, *73*, 45–67. [CrossRef]

21. Kiyohara, T. Effect of temperature on the disease incidence of pine seedlings inoculated with *Bursaphelenchus lignicolus*. *Forest Soc. Ann. Trans. Meet. Jpn.* **1973**, *84*, 334–335. (In Japanese)

22. Kazuo, S.; Tomoya, K. Influence of water stress on development of pine wilting disease caused by *Bursaphelenchus lignicolus*. *Eur. J. For. Pathol.* **1978**, *8*, 97–107. [CrossRef]

23. Shigeru, K. Effect of light intensity on the development of pine wilt disease. *Can. J. Bot.* **1989**, *67*, 1861–1864. [CrossRef]

24. Japan Meteorological Agency. 2020 Climate Statistics. Available online: http://www.data.jma.go.jp/obd/stats/etrn/index.php (accessed on 26 February 2020).

25. Toda, T. Studies on the breeding for resistance to the pine wilt disease in *Pinus densiflora* and *P. thunbergii*. *Boll. Tree Breed. Cent.* **2004**, *20*, 83–217.

26. Matsunaga, K.; Ohira, M.; Kuramoto, N.; Kurahara, Y.; Hoshi, H.; Yamada, H.; Koyama, T.; Miyazato, M. Evaluation of pinewood-nematode resistance in open pollinated seedlings from additionally selected *Pinus thunbergii* clones Kyushu. *J. For. Res.* **2011**, *64*, 84–86.

27. Aikawa, T.; Kikuchi, T.; Kosaka, H. Demonstration of interbreeding between virulent and avirulent populations of *Bursaphelenchus xylophilus* (Nematoda: Aphelenchoididae) by PCR-RFLP method. *Appl. Entomol. Zool.* **2003**, *38*, 565–569. [CrossRef]

28. R Core Team. *R: A Language and Environment for Statistical Computing*; R Foundation for Statistical Computing: Vienna, Austria, 2020.

29. Bates, D.; Mächler, M.; Bolker, B.M.; Walker, S.C. Fitting linear mixed-effects models using lme4. *J. Stat. Softw.* **2015**, *67*, 1–48. [CrossRef]

30. Barton, K. MuMIn: Multi-Model Inference. R Package Version 1.43.17 2020. Available online: https://cran.r-project.org/web/packages/MuMIn/index.html (accessed on 22 May 2020).

31. Hothorn, T.; Bretz, F.; Westfall, P. Simultaneous inference in general parametric models. *Biometr. J.* **2008**, *50*, 346–363. [CrossRef]

32. Therneau, M. A Package for Survival Analysis in R. R Package Version 3.2-3. 2020. Available online: https://CRAN.R-project.org/package=survival (accessed on 22 May 2020).

33. Ichihara, Y.; Fukuda, K.; Suzuki, K. Early symptom development and histological changes associated with migration of *Bursaphelenchus xylophilus* in seedling tissues of *Pinus thunbergii*. *Plant Dis.* **2000**, *84*, 675–680. [CrossRef]

34. Yamaguchi, R.; Matsunaga, K.; Watanabe, A. Influence of temperature on pine wilt disease progression in *Pinus thunbergii* seedlings. *Eur. J. Plant Pathol.* **2020**, *156*, 581–590. [CrossRef]

35. Masahiro, K. Breeding district and seed transfer zone. *For. Genet. Tree Breed.* **2015**, *4*, 12–15. (In Japanese)

36. Iwaizumi, M.G.; Miyata, S.; Hirao, T.; Tamura, M.; Watanabe, A. Historical seed use and transfer affects geographic specificity in genetic diversity and structure of old planted *Pinus thunbergii* populations. *For. Ecol. Manag.* **2018**, *408*, 211–219. [CrossRef]

forests

MDPI

Article

Ten Years of Provenance Trials and Application of Multivariate Random Forests Predicted the Most Preferable Seed Source for Silviculture of *Abies sachalinensis* in Hokkaido, Japan

Ikutaro Tsuyama [1,*], Wataru Ishizuka [2], Keiko Kitamura [1], Haruhiko Taneda [3] and Susumu Goto [4]

[1] Hokkaido Research Center, Forestry and Forest Products Research Institute, 7 Hitsujigaoka, Toyohira, Sapporo, Hokkaido 062-8516, Japan; kitamq@ffpri.affrc.go.jp

[2] Forestry Research Institute, Hokkaido Research Organization, Koushunai, Bibai, Hokkaido 079-0198, Japan; wataru.ishi@gmail.com

[3] Department of Biological Sciences, Graduate School of Science, The University of Tokyo, 7-3-1, Hongo, Bunkyo, Tokyo 113-0033, Japan; taneda@bs.s.u-tokyo.ac.jp

[4] Education and Research Center, The University of Tokyo Forests, Graduate School of Agricultural and Life Sciences, The University of Tokyo, 1-1-1 Yayoi, Bunkyo-ku, Tokyo 113-8657, Japan; gotos@uf.a.u-tokyo.ac.jp

* Correspondence: itsuyama@affrc.go.jp

Received: 10 August 2020; Accepted: 27 September 2020; Published: 30 September 2020

Abstract: Research highlights: Using 10-year tree height data obtained after planting from the range-wide provenance trials of *Abies sachalinensis*, we constructed multivariate random forests (MRF), a machine learning algorithm, with climatic variables. The constructed MRF enabled prediction of the optimum seed source to achieve good performance in terms of height growth at every planting site on a fine scale. Background and objectives: Because forest tree species are adapted to the local environment, local seeds are empirically considered as the best sources for planting. However, in some cases, local seed sources show lower performance in height growth than that showed by non-local seed sources. Tree improvement programs aim to identify seed sources for obtaining high-quality timber products by performing provenance trials. Materials and methods: Range-wide provenance trials for one of the most important silvicultural species, *Abies sachalinensis*, were established in 1980 at nine transplanting experimental sites. We constructed an MRF to estimate the responses of tree height at 10 years after planting at eight climatic variables at 1 km × 1 km resolution. The model was applied for prediction of tree height throughout Hokkaido Island. Results: Our model showed that four environmental variables were major factors affecting height growth—winter solar radiation, warmth index, maximum snow depth, and spring solar radiation. A tree height prediction map revealed that local seeds showed the best performance except in the southernmost region and several parts of northern regions. Moreover, the map of optimum seed provenance suggested that deployment of distant seed sources can outperform local sources in the southernmost and northern regions. Conclusions: We predicted that local seeds showed optimum growth, whereas non-local seeds had the potential to outperform local seeds in some regions. Several deployment options were proposed to improve tree growth.

Keywords: local adaptation; Sakhalin fir; silviculture; seed zone; tree improvement program

---

1. Introduction

Plant species often show local adaptation, which is a process by which populations genetically diverge in response to natural selection specific to their habitat [1]. Therefore, maladaptation is often observed when plants are transplanted to different growth environments [2]. This is also observed for long-lived forest tree species with a wide distribution range which are often genetically adapted to local climatic environments, despite their extensive gene flow [3]. Traditionally, provenance trials of forest trees have aimed to identify optimal seed provenances to ensure successful tree planting [4–6]. These attempts often result in accepting local seeds to avoid maladaptation caused by environmental mismatch between the afforestation site and the seed origin [7]. In contrast, in some cases, the best performance was achieved by introducing seeds of several species at several planting sites. For example, range-wide provenance trials of *Pinus sylvestris* revealed that progeny derived from warmer climates outgrew local seed sources in central and northern sites, whereas local seeds grew best in southern sites [8].

Range-wide provenance trials are necessary for evaluating the validity of seed zones, choosing appropriate seed sources, and providing transfer guidelines in forest improvement programs [5,7,9–11]. Seed zones are generally established in geo-topographically and climatically distinct regions [12]. Conifers have a long history of provenance trials, such as descriptions of the seed zones for pines in the southern USA based on a series of long-term trials of *Pinus echinata*, *P. elliottii*, *P. palustris*, and *P. taeda* [13]. In British Columbia to Minnesota, seed zones for forestry species have been evaluated and modified based on provenance trials, genotypes, and phenotypes [14]. Seed zones for *P. densifolia* were validated by long-term provenance trials in Japan [15].

While range-wide provenance trials are fundamental for identifying appropriate seed sources for reforestation programs, these trials are costly, time-consuming, and can only handle a limited number of provenances [7]. Recently, application of statistical models has been considered as relevant for establishing seed zones and seed transfer guidelines [11,14,16,17]. One of the pioneering examples is that fine-scale seed transfer guidelines were developed based on multivariate models for white spruce in Alberta, Canada [11]. Recent studies used statistical models (e.g. species distribution models) to predict the potential distribution of forestry species under various climatic conditions [18–21]. These studies suggested solutions for forest conservation management such as future vulnerability of core and buffer conservation areas [22–27]. However, statistical models have not been applied to improve forestry production such as in predicting future growth of plantations. Together with data from provenance trials and environmental factors, statistical models are practical for evaluating seed zones and seed transfer guidelines based on the prediction of traits and local adaptation under given climatic conditions.

Abies sachalinensis is a major component of natural forests in Hokkaido, northern Japan. The geographical distribution of *A. sachalinensis* includes Sakhalin, the southern Kuril Islands, and Hokkaido, the northernmost island of the Japanese Archipelago [28]. As one of the most important commercial timber species in Hokkaido, the proportion of timber volume of artificial plantations is approximately 50% and the seedling stock is 25–30%. The breeding program for *A. sachalinensis* began in the 1950s in Hokkaido. During the program, a total of 782 "plus trees", showing good performance in growth and stem straightness, were selected from natural and artificial forests throughout Hokkaido. The initial seed zones were determined in 1985 based on the results of common garden and provenance trials (e.g., [29]) and the local climate. According to the seed zones, breeding programs have been started to establish seed orchards in different regional zones. To improve timber production, these seed zones must be validated. Initial evaluation of seed zones was based on the results obtained from three provenance test sites [30]. For further improvement of future timber production, comprehensive assessment using a range-wide provenance test is necessary to validate the seed zones.

In this study, we applied a statistical model with climatic variables to predict tree height at 10 years after planting of different provenances of *A. sachalinensis* throughout Hokkaido. Furthermore, we proposed appropriate seed sources for achieving the best performance in terms of height growth. Finally, we described modifications to seed zones and seed transfer guidelines.

2. Materials and Methods

2.1. Study Area

Hokkaido is the northernmost island of the Japanese Archipelago and ranges from N 41°21′–45°3′ to E 139°20′–145°49′ (Figure 1). It has upper temperate forest in the southern peninsula and sub-boreal forest in the northern and eastern parts. Climatic conditions in the western region are affected by the coastal climate of the Sea of Japan, characterized by heavy snowfall in winter (Figure 2). In contrast, the eastern part of the island has cold and dry winters. There are volcanic mountain ranges of approximately 2000 m altitude at the centre of the island, which comprises alpine forests. The natural distribution of *A. sachalinensis* covers most of the montane forests in Hokkaido, where upper temperate to sub-boreal and alpine forests are found.

2.2. Regional Groups and Seed Provenances

The different commercial seed zones for *A. sachalinensis* were recently updated by Nakada et al. [31], and five zones were recognized—the West, North, East, Eastern edge, and South. In this study, we evaluated seven regional groups—W, N, EN, EE, ES, S, and SS groups (Table S1, Figure 1). In the East zone, two sub-zones were suggested based on the climatic difference between the Sea of Okhotsk and the Pacific Ocean sides [32]. We adopted this suggestion and subdivided the East zone into the EN and ES groups. Moreover, the Oshima Peninsula in the West and South zones have a warmer climate and different vegetation from other parts of Hokkaido in which wild *A. sachalinensis* is scarcely distributed [33]. Additionally, our previous study revealed that the southern populations showed low genetic diversity and were genetically differentiated from other populations [34]. Thus, we discriminated SS as a regional group referring to the southern part of the South seed zone.

In the 1970s, open pollinated seeds were collected from 88 trees throughout Hokkaido. Most of the selected trees were plus trees. Seeds were distinguished as a "family" based on each mother tree. Meanwhile, selected breeding materials were absent in SS, where open pollinated seeds produced from local trees were bulked and used for planting. Thus, these local seeds were used as a single family. The resulting 89 families and their regional groups are shown in Table S1. The geographical coordinates of the origins of these families were curated from the register books deposited at the Forestry Research Institute, Hokkaido Research Organization (HRO) as provenance locations (Figure 1). Notably, a provenance location of the local family in the SS group was set to the location of a natural stand, assuming the historical origin of the local seeds used. The collected seeds were sown in 1975 in several nurseries in Hokkaido.

2.3. Range-Wide Provenance Tests

In 1980, the HRO established range-wide provenance tests at nine localities [35–38]. Test sites were established in seven regional groups including three boundary areas and were managed by code numbers A30 to A38 (Table 1; Figure 1). Six-year-old seedlings were transplanted to these sites in the autumn of 1980. Because of the limited number of seedlings, the number of planted families differed among sites, ranging from 41 to 82 (56 families on average) (Table 1). The average number of planted sites for a single family was 5.7, indicating an effective number of repetitions for the provenance tests.

Three replicates were established at each planting site. Thirty trees per family were planted within each replicate, whereas 40 trees were planted in A35. The tree density ranged from 2200 to 5200 trees per ha according to the conditions of the sites. A complete dataset of tree height and mortality was available until the measurements done in 1989—10 years after transplantation. The measurements at several sites were abandoned after 1990 because of severe meteorological and/or biological damage, which made the range-wide model prediction difficult. We then used the tree height at 10 years after transplantation as a single time point for subsequent analysis.

Table 1. Summary of provenance tests for *A. sachalinensis*. The values for tree height and survival rate at 10 years after planting.

Site Code	Regional Group	Latitude (°N)	Longitude (°E)	No. of Planted Families	Tree Height (m)	Survival Rate (%)
A30	SS	41.8488	141.1192	53	3.47 ± 0.70	90.5
A31	W	42.7481	140.6308	41	2.79 ± 1.41	65.6
A32	S	42.4586	142.4664	45	3.32 ± 0.96	63.7
A33	W	43.2803	141.8785	82	3.50 ± 0.90	85.7
A34	N	44.5550	142.2576	55	1.56 ± 0.75	76.8
A35	EN	44.2273	142.9343	51	3.17 ± 0.91	77.1
A36	EN	43.6542	143.7846	49	2.97 ± 0.98	59.4
A37	ES	42.7309	143.4939	64	3.71 ± 0.84	93.7
A38	EE	43.0415	144.9744	64	2.37 ±0.91	70.7
Average				56	3.14 ± 1.08	75.9

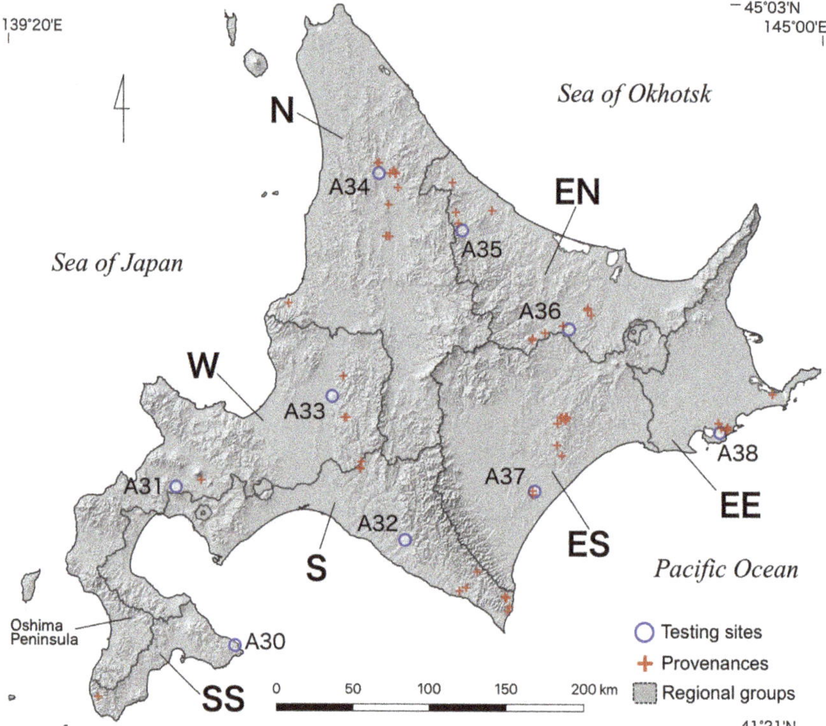

Figure 1. A map of study area showing locations of testing sites, provenances, and regional groups.

2.4. Climatic Data

We used climatic data for grid cells from the Japan Meteorological Agency [39], which had a spatial resolution of 30″ N × 45″ E (approximately 1 km × 1 km). The following eight climatic variables, which were considered as important for the growth of *A. sachalinensis* according to previous studies [29,32], were calculated (Figure 2). The warmth index (WI) (°C month) was defined as the annual sum of positive differences between the monthly mean temperature and +5 °C [40], indicating an effective heat quantity for the growth of plants. The monthly mean daily minimum temperature of the coldest month (TMC) (°C) provides a measure of extreme coldness. Precipitation during summer

(PRS) (mm) is a sum of precipitation from May to September and represents the water supply during the growing season. The maximum snow depth (MSD) (m) is a measure of snow accumulation. Solar radiation in winter (October to April, WinSR), spring (May, SprSR), summer (June to August, SumSR), and autumn (September, AutSR) (0.1 MJ/m^2/day) are measures of energy distributions in each season. Spring, summer, autumn, and winter months for each variable were determined by variables of each month showing mutually positive relationships.

To evaluate the effects of distances in climatic environments between 89 seed provenances (Table S1) and nine testing sites, we calculated the differences in temperatures and solar radiation as well as the relative ratios of precipitation variables including PRS and MSD between the seed provenances and the testing sites. Previous studies also revealed that environmental distances gave better results than the geographic distances for predicting the fitness of other species [2]. Our preliminary analyses used actual climate values as explanatory variables, by which the models did not fit better than the present models (data not shown). Thus, climatic distances between seed sources and planting sites were valid and used as explanatory variables for model construction. Relative ratios were calculated as follows:

$$Rp_i = ((Tp_i - Sp_j)/Sp_j) \times 100 \tag{1}$$

Rp_i: relative ratio of precipitation variables (PRS and MSD) between testing site i and seed provenance j, Tp_i: precipitation variables at testing site i, Sp_j: precipitation variables at seed provenance j.

These climatic distance data were used for model construction. To project the result of prediction throughout the study area, the differences and relative ratios of the climatic variables between each grid cell and mean values among seed provenances in each regional group were calculated.

2.5. Statistical Model for the Height Growth of A. sachalinensis

To identify the effects of climatic distances between seed provenances and testing sites on the height growth of *A. sachalinensis*, we constructed multivariate random forests (MRF) [41]. Random forest (RF) is a machine learning method that assembles the results of base learners such as tree-based models with a randomization process that enables high learning performance [42]. MRF extended the RF method to treat unified cases including multivariate response regression. Tree heights at 10 years after planting among seven regional groups were used as response variables (Code S1). The distances of eight climatic variables between seed provenances and testing sites were used as explanatory variables. We applied the MRF to every grid cell throughout Hokkaido and predicted the tree heights in the seven regional groups. Optimum provenances for every grid cell were projected based on the prediction throughout Hokkaido. We excluded the area showing a WI of less than 35 from the projection because these areas are out of the *A. sachalinensis* plantation range (data not shown). "randomForestSRC" package [43,44] on R version 3.6.2 [45] was used to construct the MRF, and QGIS version 3.4 [46] was used for projection.

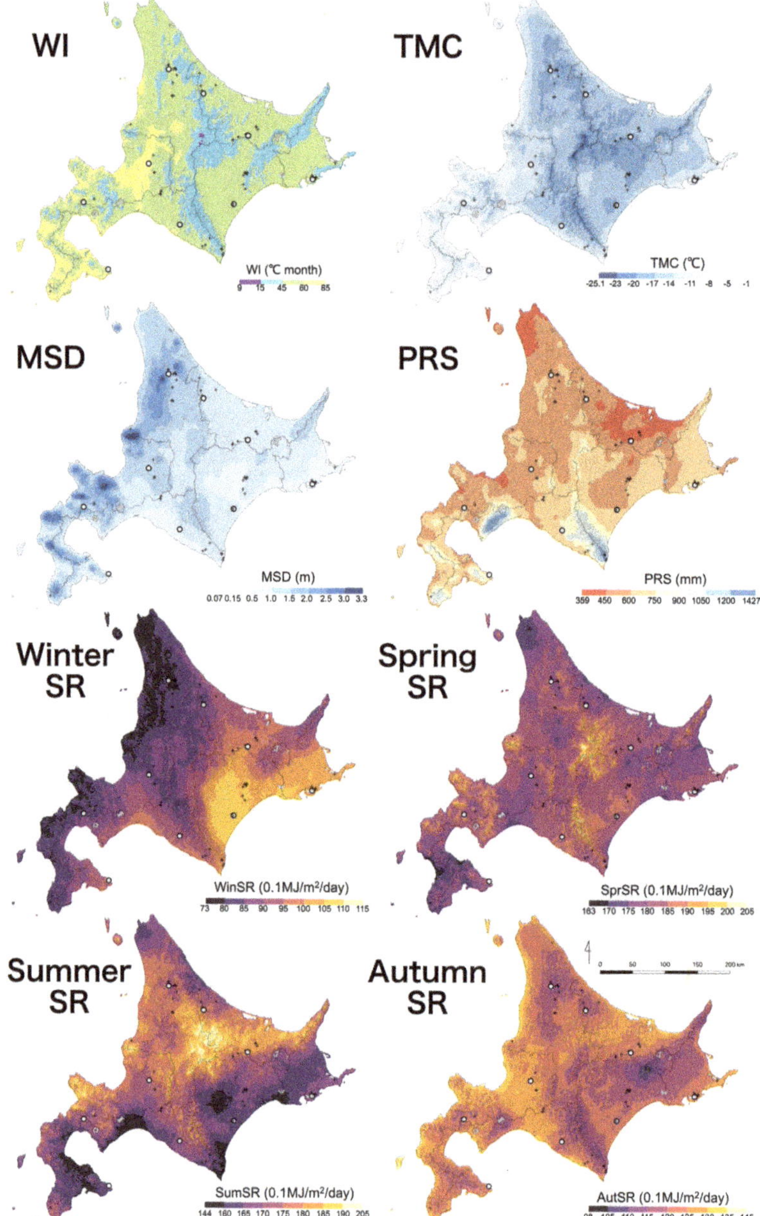

Figure 2. Maps for climatic variables used as explanatory variables. Open circles indicate testing sites, and plus signs indicate provenances.

3. Results

3.1. Measured Height Growth Performance among Testing Sites

The average tree height was 3.16 m among all planting sites (Table 1). The average of tree heights in A37 was 2.4-fold of that in A34. Tree height in A34 was severely compromised by *Scleroderris* canker disease. For many transplants at this site, a reduced height was frequently observed because of the death of branches by the disease, whereas death of the tree was caused by the disease in a few cases. Tree height was also negatively affected by several meteorological factors at other sites; for example, snow pressure broke branches in A31, late frost damaged young shoots in A36, and winter cold injury or desiccation occurred in A38 where transplants were not covered by snow because of the shallowest snow depth among all testing sites.

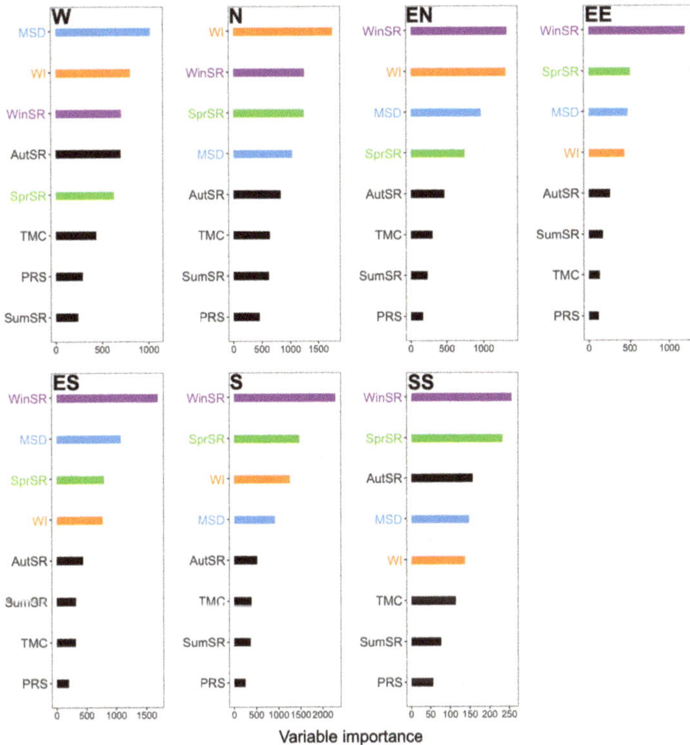

Figure 3. Importance of explanatory climatic variables in each regional group in multivariate random forests (MRF). The importance of each variable was identified based on increased mean square errors in the MRF.

3.2. Model Accuracy and Climatic Conditions Controlling Height Growth of A. sachalinensis

The MRF for the height growth of *A. sachalinensis* with climatic distances between seed provenances and testing sites showed the following coefficient of determinations (R^2)—0.32 for W, 0.36 for N, 0.31 for EN, 0.33 for EE, 0.38 for ES, 0.38 for S, and 0.45 for SS (Figure S1). These R^2 values and Figure S1 show that the prediction accuracy of the MRF was not high; however, the overall trends in height growth were reproduced successfully. Variable importance analysis of the regional groups in the MRF showed that WinSR, WI, MSD, and SprSR had relatively important effects on the height growth of *A. sachalinensis*, whereas the order of variable importance differed among the regional groups (Figure 3). WinSR showed the highest importance in most regional groups, except in W and N. MSD showed the highest importance in the W region, which has the heaviest snowfall. In contrast, PRS had the lowest importance, except in the W region.

The response of predicted tree height to the important four climatic variables in the MRF showed that trends in the response to the climatic distances generally differed by region (Figure 4). However, there was a common trend in the responses to WinSR in which the tree height growth exhibited peaks close to the local point across all regional groups. In addition, two patterns were mainly identified along a geographic gradient; W, N, and SS in the western part of Hokkaido had unimodal peaks in height growth on the positive side of the WinSR distances, whereas EE and ES in the eastern part of Hokkaido retained peaks after the distances when WinSR increased to positive. As for MSD, height growth peaked around zero in the distances (i.e., local environment) among all the regional groups in common. Two patterns in height growth were identified in response to the distances in the MSD; height growth peaked when the distances in MSD decreased to negative values (W, N, and SS) or increased to positive values (EN, EE, ES, and S). The response patterns to distances in the WI and SprSR were consistent among regional groups except SS for WI and W and SS for SprSR, respectively.

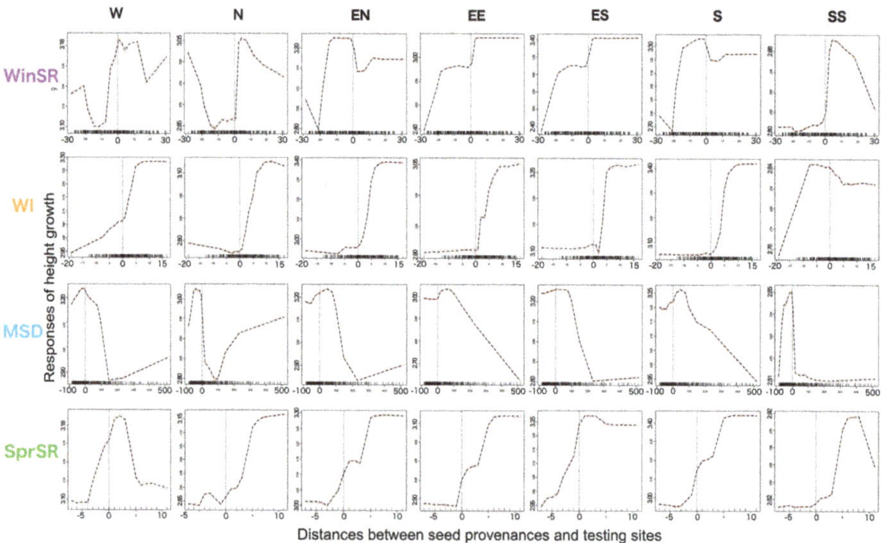

Figure 4. Responses of height growth to important climatic variable distances in each regional group according to multivariate random forests. Dashed black lines indicate expected values, dashed red lines indicate mean confidence intervals (expected values $\pm 2 \times$ standard errors), and rugs on x-axes indicate data existence.

3.3. Prediction Maps for Tree Height

Assuming that local seeds were planted in each regional group, we predicted the tree height by region based on the prediction of the MRF (Figure 5a). We also projected maps for tree height which assumed that seeds derived from each regional group were planted throughout Hokkaido individually (Figure S2). These maps showed that predicted heights differed among the planted seeds in regional groups because of the difference in the importance and responses to climatic distances (Figures 3 and 4). Trees derived from the ES and S regions grew taller (larger than 3.7 m) in the local region. In contrast, trees were relatively shorter (less than 3.4 m) when they were derived from marginal regional groups (i.e., EE and SS) than those from other groups.

The maximum tree heights for every grid cell were projected to show a potential tree height map which assumed that the best seeds were used for planting, regardless of whether they were local or non-local (Figure 5b). Seeds derived from the ES, S, and EE regions showed similar heights and distribution patterns between local and optimum seed provenance cases (Figures 5a,b and S3). However, in the N and W regions, tree height growth was clearly improved when seeds derived from optimum seed provenances were used.

A map of the optimum regional groups as seed sources for each grid cell showed that seeds derived from local regional groups generally showed the best height growth (Figure 6). The proportions of local seeds were the highest among the W, EN, EE, and ES regional groups. Particularly, local seeds exhibited the best performance in 95% of the area in S regional group. S was the only regional group selected as an optimum seed source among all regional groups. In the N region, however, the proportions of non-local seeds derived from neighboring regional groups (S, EN, and W) were higher than those of local seeds. In addition, local seeds were typically not selected as the best seed provenance in the SS regional group.

4. Discussion

In this study, we successfully estimated the responses of height growth of *A. sachalinensis* to climatic conditions corresponding to seed sources (i.e., regional groups) using MRF, data from range-wide provenance tests, and fine scale (approximately 1 km) climatic data. Prediction maps for the tree height of *A. sachalinensis* were successfully projected by assuming that seeds derived from local or optimum seed sources were used. Moreover, we evaluated the validity of the current framework of seed zones, which used local seed sources and proposed appropriate options to improve the height growth of *A. sachalinensis*.

4.1. Important Climatic Factors Affecting the Height Growth of A. sachalinensis

The MRF showed that WinSR, WI, MSD, and SprSR were important factors affecting the height growth of *A. sachalinensis*, although the order of importance differed among regional groups (Figure 3). The model also showed that responses of height growth to climatic variables differed among regional groups, which may reflect different selection regimes to local environments (Figure 4). Genetic differentiation along environmental gradients was circumstantial evidence of local adaptation to the respective environment [34].

For five of the seven regional groups, WinSR was the most important climatic factor. WI and MSD were the most important factors at two sites (N and W, respectively) (Figure 3). The responses to WinSR and MSD showed that the height growth had peaks close to the local environments (Figure 4). These results suggest that WinSR and MSD, climatic factors of the winter, are the main drivers of local adaptation in the height growth of *A. sachalinensis*. Furthermore, these two variables showed another common trend; the response pattern of the height growth to the variables differed between western (W, N, and SS) and eastern regional groups (EE and ES). Hatakeyama [29] also revealed that WinSR and MSD affected height variation among seed provenances. Okada et al. [47] indicated that the number of layers of winter buds, which may be correlated with climatic factors such as WinSR, was significantly different between western and eastern seed provenances in Hokkaido. In addition, our previous study revealed that WinSR was positively related to the genetic differentiation of natural populations along longitudinal gradients, whereas WI, MSD, and SprSR were negatively related to these differences [34]. Therefore, the regional ecotypes of *A. sachalinensis* would be genetically adapted to the local climate along a longitudinal gradient.

In general, the photosynthetic activity of evergreen conifers including *A. sachalinensis* is ceased or extremely low during winter [48]. However, our results suggested WinSR as an important factor affecting the height growth of *A. sachalinensis*. The months of October, November, and April were included in WinSR, in which the photosynthetic activity of *A. sachalinensis* was reported to be active [48]. In addition, some studies suggested that solar radiation in early spring was important for the shoot growth of *A. sachalinensis* and coniferous trees [49,50]. Our results suggest that solar radiation during early winter and spring are important in the local adaptation for height growth of *A. sachalinensis* through photosynthesis, although unmeasured variables, which are highly correlated with winter solar radiation, can be important.

The results of examination of the MSD response revealed sharp phenotypic adaptation in all provenances. Too much snowfall shortens growing season by causing a long snow cover period and increases mechanical damages due to snow pressure [29]; these effects may decrease the height growth of *A. sachalinensis* derived from eastern regions with low snowfall, as these plants were not adapted to heavy snow. Therefore, MSD is an important selective driver of the local adaptation of *A. sachalinensis* [29]. Ecological divergence at the ecotone between the Sea of Japan side (heavy snowfall) and the Pacific Ocean side (poor snowfall) was also demonstrated in other species such as *Fagus crenata* [51] and *Cryptomeria japonica* [52].

In contrast, PRS was not selected as an important factor in this study (Figure 3). However, PRS is considered as a critical factor determining the distribution of species in the genus *Abies* because *Abies* is generally less drought-resistant than other coniferous species are [53,54]. Our results suggest that the range of PRS in Hokkaido is sufficient for the height growth of *A. sachalinensis* in all regions.

4.2. Optimum Regions and Seed Sources for Height Growth of A. sachalinensis

Tree height prediction maps revealed that local seeds generally showed the optimum performance, although the predicted tree heights differed among planting regions (Figures 5 and 6). The height growth of local seeds was fairly good in the western ES and eastern S regional groups (Figure 5a). This suggests that the climatic conditions in these areas are suitable for the height growth of *A. sachalinensis*. In contrast, local seeds did not perform well in the EE region, and even deployment of other seed sources did not improve the performance. These results indicate that climatic conditions in the EE region are unfavorable for *A. sachalinensis* height growth at young ages, although productivity at the mature stage in this region is known to be high. Seed sources were selected as optimum only in the local region except for that in the eastern part of ES, which is adjacent to the EE region. Clear local adaptation to the Pacific Ocean side may be responsible for this trend, which has also been observed for other woody species [51,52].

In some regions including W, N, and SS, seeds from distant sources outperformed those from local sources (Figures 5b and 6). In these regions, tree height growth was clearly improved when non-local but optimum seed sources were used (Figure S3). Particularly, seeds from the S region were the only materials found to be optimum across all regional groups, suggesting that they have high plasticity to grow in a wide range of climatic conditions. The geographic location of the region may be a reason for this; the S region lies on an ecotone between the Sea of Japan side with heavy snow and low solar radiation and the Pacific Ocean side with poor snow and high solar radiation in winter (Figure 2), which is a common characteristic across the Japanese Archipelago. Trees derived from the S region were also known to show intermediate values in some key traits with regional variations, for example freezing tolerance [29,32]. Therefore, seed sources in S may be universal materials useful for wide-range planting (Figure 6, Figure S2). For planting in the N region, seed sources derived from adjacent regional groups were selected as optimum materials as well as local seeds. Unsuitability of distant non-local regional groups seems to be a common pattern of this species, since the consistent result was also recognized in EE and ES regions, even though they had contrasting climatic characters to the N region. Therefore, overall responses to climatic factors were considered to be robust including the N region. However, when considering the optimum materials and their performances in this region, our model seems to have a limit of prediction accuracy. The test site in the N region (A34) severely suffered from *Scleroderris* canker disease. The infection rate was reported to as high as 87.4% until the seventh year after transplantation [38]. The height growth that was affected by this disease was observed to be the smallest in average among all sites, whereas the survival rate (76.5%) was not greatly affected (Table 1). Biological factors, including such diseases, were not incorporated in this study, which may cause decrease in prediction accuracy especially in this region. Further studies are needed to assess the growth potential of local and adjacent non-local materials.

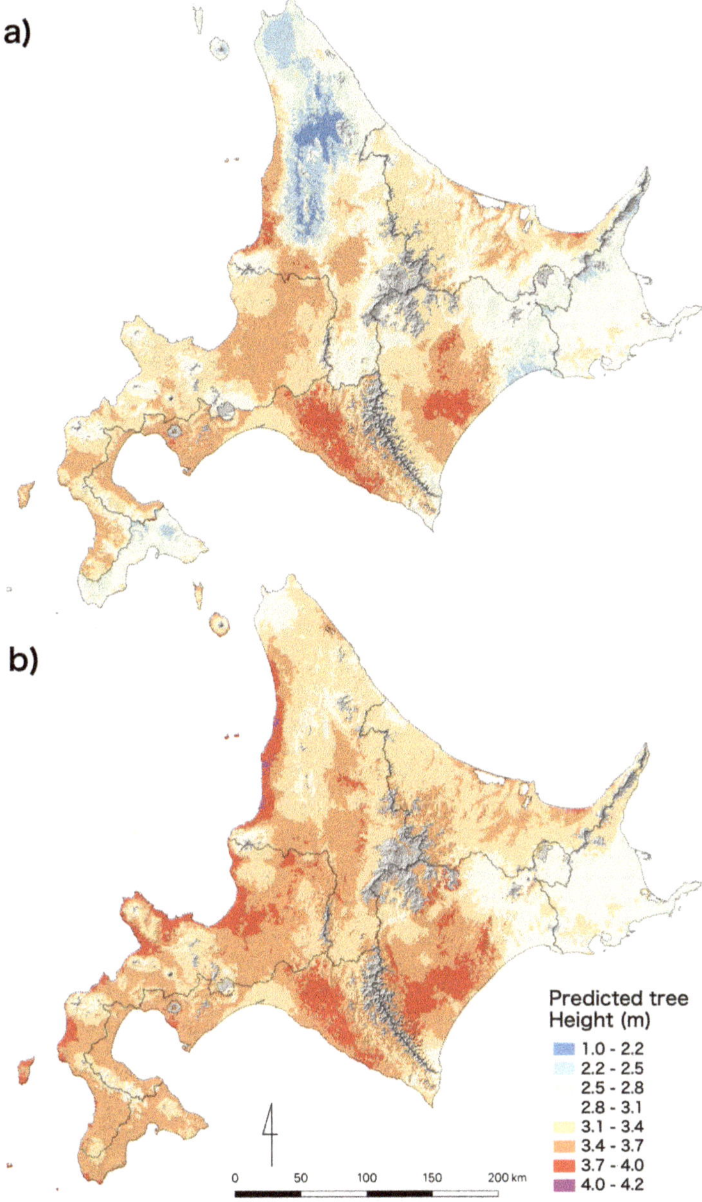

Figure 5. Predicted tree height for (**a**) local provenance and (**b**) optimum seed provenance. Gray area indicates alpine zone with less than 35 in the WI, which was excluded from the projection.

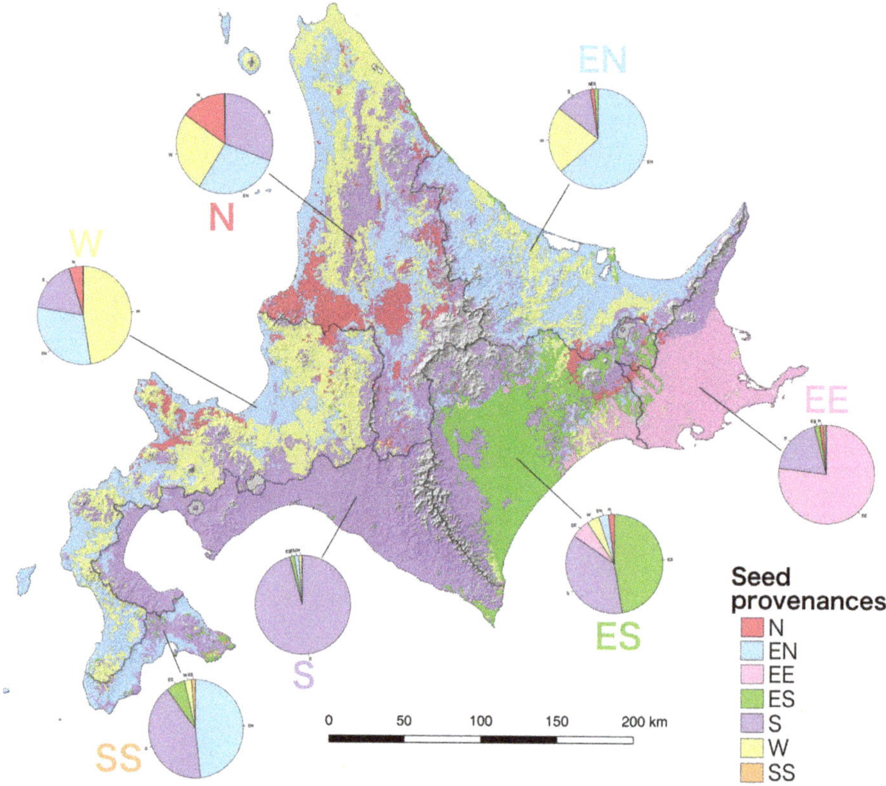

Figure 6. Optimum provenance for planting based on predicted tree height by multivariate random forests. Pie charts show the fraction of areas for which each seed provenance was predicted as optimum in each regional group. Gray area indicates alpine zone with less than 35 in the WI which was excluded from the projection.

Similar to the N regional group, some difficulties remain in comprehensively evaluating seed sources. A seed source in the SS regional group was indicated to be an inferior material for local planting. However, further validation of SS is required because only one provenance was evaluated in the present study, and unlike in other regions, the seeds were not derived from selected breeding materials but originated from a local natural stand. Furthermore, the mechanisms of local adaptation of *A. sachalinensis* remain unclear. To achieve successful silviculture, we should consider not only tree growth but also other factors such as mortality [55].

Average 10-year survival rates for each of the regional groups at each testing site showed that the local group or the adjacent groups showed relatively high survival rate in most sites (Figure S4). This pattern was similar to the overall trends shown by MRF in this study (Figures 5 and 6), that is also observed in the averaged tree height in Figure S4. Moreover, our preliminary analysis on mortality showed consistency with the results on tree height in this study, such as common important variables. Therefore, we concluded that the validity of our current results was confirmed, which suggested appropriate deployment of seed sources. It has been widely acceptable for major conifers to use tree height for their evaluation as optimum seed sources, such as *Pinus sylvestris* [56], *P. glauca* [11], and *Picea abies* [57]. Further analysis is also expected to quantify both the growth and survival potentials of each seed sources, since non-local seeds represented complex responses (Figure S4).

4.3. Implications for Improving the Height Growth of A. sachalinensis

Range-wide provenance tests have been used to establish and modify seed zones and seed transfer guidelines to improve the tree growth of many forestry species [55,58–60]. There are three conceptual options for improving the current frameworks: (i) re-set the current seed zones, (ii) facilitate appropriate seed transfer, (iii) facilitate assisted gene flow beyond the seed zones to avoid a reduction in tree growth caused by a mismatch to future climatic conditions [7,61,62]. The present study incorporated MRF and can thoroughly propose the former two options for *A. sachalinensis*.

For the first option, our result demonstrated the relevance of the current seed zones of this species, as local seed sources represented optimum growth at most planting sites. Particularly, the validity of local seed sources was clear for the S, ES, and EE regions on the Pacific Ocean side of Hokkaido. In the present study, we set seven regional groups against the current five seed zones based on genetic differentiation in several traits and climatic differences [32]. The East zone was then subdivided into the EN and ES groups between the Sea of Okhotsk and the Pacific Ocean sides. The geographical distribution of the optimum ranges for these two groups clearly indicated that the current inclusive management by one East zone was insufficient (Figure 6) and supported the subdivision of the East zone to use independent seed sources according to climate differences. Because *A. sachalinensis* exhibited local adaptation across the environmental gradient even in a small geographic range [63], fine-scale seed zones should be useful for this species.

For the second option, the novel seed transfer guidelines of *A. sachalinensis* can be predicted to effectively improve tree growth. Adaptation to WinSR and MSD are critical factors, as indicated in Figure 4 and previous studies [29,32]. Therefore, distant transfer over the climatic gradient, such as the transfer between the heavy snowfall region in the western part of Hokkaido and region with lower snowfall and more abundant sun in winter in the eastern part of Hokkaido should be avoided. Alternatively, transfer to a warmer climate is a relevant option because the increase in WI at planting sites from that in seed sources contributed to improving tree growth. Indeed, the model prediction demonstrated that some non-local seed sources have the potential to show increased tree height when at different planting sites (Figure 6). Transferring seed sources derived from S and EN into the SS region and from W into EN are candidates for increasing height growth. Among these candidates, seed sources in the S regional group for using SS is recommended because this follows the adjacent and warmer transfer.

Recently, prediction of survival and growth of trees under future climate becomes important [62], although we focused on the evaluation of current seed zones and seed transfer guidelines in this study. In fact, MRF enables us to predict future height growth across Hokkaido by applying the model to future climate scenarios. These data must be useful for mitigation and adaptation to the climate change.

5. Conclusions

The application of MRF to obtain results from provenance trials was found to be relevant for improving the seed transfer guidelines of *A. sachalinensis*. MRF revealed that each provenance had a different selection regime against climatic factors. Particularly, solar radiation in winter was the most important explanatory variable affecting tree height. The optimum map enables the fine tuning of seed sources for a given planting site to improve timber production. As a result, height growth was predicted to be improved using optimum seed sources rather than local seeds at some planting sites. Thus, seed source deployment can be recommended in some localities to improve timber production. However, silviculture practice requires a long time to yield harvests, and thus approaches for mitigating the risk of maladaptation of deployment materials during cultivation should be considered.

Supplementary Materials: The following are available online at http://www.mdpi.com/1999-4907/11/10/1058/s1, Table S1: List of seed zones and regional groups of *A. sachalinensis* in Hokkaido, Japan, Table S2: Climatic conditions of testing sites, Figure S1: Relationships between observed and predicted tree heights, Figure S2: Predicted tree heights at 10 years after planting in each provenance region, Figure S3: Difference of predicted tree height between using optimum and local seed sources, Figure S4: Averaged survival rate (left panels) and tree height with standard deviation (right panels) of transplants for each of the regional groups at each test site, Code S1: The code for developing a multivariate random forests (MRF) in R.

Author Contributions: Conceptualization, I.T. and W.I.; methodology, I.T. and W.I.; formal analysis, I.T. and W.I.; Visualization, I.T.; Investigation, W.I.; writing—original draft preparation, I.T. and W.I.; writing—review and editing, I.T., W.I., K.K., H.T., S.G. All authors have read and agreed to the published version of the manuscript.

Funding: This research was funded by Grants-in-Aid for Scientific Research from the Japan Society for Promotion of Science (JSPS KAKENHI), grant numbers 16H02554 and 20H03021.

Acknowledgments: We thank the staff at HRO and Hokkaido for the establishment of provenance tests and field measurements, M. Kuromaru for helpful suggestions for the analysis, and Y. Hasegawa for database construction.

Conflicts of Interest: The authors declare no conflict of interest.

Abbreviations

The following abbreviations are used in this manuscript:

HRO	Hokkaido Research Organization
WI	Warmth Index
TMC	Monthly mean daily minimum temperature
PRS	Precipitation during summer
MSD	Maximum snow depth
WinSR	Solar radiation in winter
SprSR	Solar radiation in spring
SumSR	Solar radiation in summer
AutSR	Solar radiation in autumn
MRF	Multivariate random forests
RF	Random forests
QGIS	Quantum Geographic Information System

References

1. Hufford, K.M.; Mazer, S.J. Plant ecotypes: Genetic differentiation in the age of ecological restoration. *Trends Ecol. Evol.* **2003**, *18*, 147–155. [CrossRef]
2. Montalvo, A.M.; Ellstrand, N.C. Transplantation of the subshrub *Lotus scoparius*: Testing the home-site advantage hypothesis. *Conserv. Biol.* **2000**, *14*, 1034–1045. [CrossRef]
3. Savolainen, O.; Pyhäjärvi, T.; Knürr, T. Gene flow and local adaptation in trees. *Annu. Rev. Ecol. Evol. Syst.* **2007**, *38*, 595–619. [CrossRef]
4. Turesson, G. The genotypical response of the plant species to the habitat. *Hereditas* **1922**, *3*, 211–350. [CrossRef]
5. Morgenstern, E.K. *Geographic Variation in Forest Trees: Genetic Basis and Application of Knowledge in Silviculture*; UBC Press: Vancouver, BC, Canada, 1996.
6. Isik, F.; Keskin, S.; McKeand, S.E. Provenance variation and provenance-site interaction in *Pinus brutia* Ten.: Consequences of defining breeding zones. *Silvae Genet.* **2000**, *49*, 213–223.
7. White, T.; Adams, W.; Neale, D. *Forest Genetics*; CAB International: Wallingford, UK, 2007.
8. Reich, P.B.; Oleksyn, J. Climate warming will reduce growth and survival of Scots pine except in the far north. *Ecol. Lett.* **2008**, *11*, 588–597. [CrossRef]
9. Kitzmiller, J.H. Provenance trials of ponderosa pine in northern California. *For. Sci.* **2005**, *51*, 595–607.
10. Ying, C.C.; Yanchuk, A.D. The development of British Columbia's tree seed transfer guidelines: Purpose, concept, methodology, and implementation. *For. Ecol. Manag.* **2006**, *227*, 1–13. [CrossRef]
11. Gray, L.K.; Hamann, A.; John, S.; Rweyongeza, D.; Barnhardt, L.; Thomas, B.R. Climate change risk management in tree improvement programs: Selection and movement of genotypes. *Tree Genet. Genomes* **2016**, *12*, 23. [CrossRef]

12. Lindgren, D.; Ying, C.C. A model integrating seed source adaptation and seed use. *New For.* **2000**, *20*, 87–104. [CrossRef]
13. Schmidtling, R.C.; Carroll, E.; LaFarge, T. Allozyme diversity of selected and natural loblolly pine populations. *Silvae Genet.* **1999**, *48*, 35–45.
14. Hamann, A.; Gylander, T.; Chen, P.Y. Developing seed zones and transfer guidelines with multivariate regression trees. *Tree Genet. Genomes* **2011**, *7*, 399–408. [CrossRef]
15. Nagamitsu, T.; Shimada, K.i.; Kanazashi, A. A reciprocal transplant trial suggests a disadvantage of northward seed transfer in survival and growth of Japanese red pine (*Pinus densiflora*) trees. *Tree Genet. Genomes* **2015**, *11*, 813. [CrossRef]
16. Potter, K.M.; Hargrove, W.W. Determining suitable locations for seed transfer under climate change: A global quantitative method. *New For.* **2012**, *43*, 581–599. [CrossRef]
17. Isaac-Renton, M.G.; Roberts, D.R.; Hamann, A.; Spiecker, H. Douglas-fir plantations in Europe: A retrospective test of assisted migration to address climate change. *Glob. Chang. Biol.* **2014**, *20*, 2607–2617. [CrossRef]
18. Matsui, T.; Yagihashi, T.; Nakaya, T.; Tanaka, N.; Taoda, H. Climatic controls on distribution of *Fagus crenata* forests in Japan. *J. Veg. Sci.* **2004**, *15*, 57–66. [CrossRef]
19. Tsuyama, I.; Horikawa, M.; Nakao, K.; Matsui, T.; Kominami, Y.; Tanaka, N. Factors controlling the distribution of a keystone understory taxon, dwarf bamboo of the section *Crassinodi*, at a national scale: Application to impact assessment of climate change in Japan. *J. For. Res.* **2012**, *17*, 137–148. [CrossRef]
20. Higa, M.; Tsuyama, I.; Nakao, K.; Nakazono, E.; Matsui, T.; Tanaka, N. Influence of nonclimatic factors on the habitat prediction of tree species and an assessment of the impact of climate change. *Landsc. Ecol. Eng.* **2013**, *9*, 111–120. [CrossRef]
21. Nakao, K.; Higa, M.; Tsuyama, I.; Lin, C.T.; Sun, S.T.; Lin, J.R.; Chiou, C.R.; Chen, T.Y.; Matsui, T.; Tanaka, N. Changes in the potential habitats of 10 dominant evergreen broad-leaved tree species in the Taiwan-Japan archipelago. *Plant Ecol.* **2014**, *215*, 639–650. [CrossRef]
22. Nakao, K.; Higa, M.; Tsuyama, I.; Matsui, T.; Horikawa, M.; Tanaka, N. Spatial conservation planning under climate change: Using species distribution modeling to assess priority for adaptive management of *Fagus crenata* in Japan. *J. Nat. Conserv.* **2013**, *21*, 406–413. [CrossRef]
23. Tsuyama, I.; Higa, M.; Nakao, K.; Matsui, T.; Horikawa, M.; Tanaka, N. How will subalpine conifer distributions be affected by climate change? Impact assessment for spatial conservation planning. *Reg. Environ. Chang.* **2015**, *15*, 393–404. [CrossRef]
24. Serra-Diaz, J.M.; Franklin, J.; Ninyerola, M.; Davis, F.W.; Syphard, A.D.; Regan, H.M.; Ikegami, M. Bioclimatic velocity: The pace of species exposure to climate change. *Divers. Distrib.* **2014**, *20*, 169–180. [CrossRef]
25. Carroll, C.; Lawler, J.J.; Roberts, D.R.; Hamann, A. Biotic and climatic velocity identify contrasting areas of vulnerability to climate change. *PLoS ONE* **2015**, *10*, e0140486. [CrossRef] [PubMed]
26. Hamann, A.; Roberts, D.R.; Barber, Q.E.; Carroll, C.; Nielsen, S.E. Velocity of climate change algorithms for guiding conservation and management. *Glob. Chang. Biol.* **2015**, *21*, 997–1004. [CrossRef]
27. Barber, Q.E.; Nielsen, S.E.; Hamann, A. Assessing the vulnerability of rare plants using climate change velocity, habitat connectivity, and dispersal ability: A case study in Alberta, Canada. *Reg. Environ. Chang.* **2016**, *16*, 1433–1441. [CrossRef]
28. Yamazaki, T. Pinaceae. In *Flora of Japan: Pteridophyta and Gymnospermae*; Iwatsuki, K., Yamazaki, T., Boufford, D.E., Ohba, H., Eds.; Kodansha: Tokyo, Japan, 1995; pp. 266–277.
29. Hatakeyama, S. Genetical and breeding studies on the regional differences of interprovenance variation in *Abies sachalinensis* Mast. *Bull. Hokkaido For. Res. Inst.* **1981**, *19*, 1–19. (In Japanese with English Summary)
30. Kuromaru, M. Effect of seed zoning for breeding seed-lot of *Abies sachalinensis*. *Kosyunai Kihou* **1989**, *76*, 1–3. (In Japanese)
31. Nakada, R.; Sakamoto, S.; Nishioka, N.; Hanaoka, S.; Kita, K.; Kon, H.; Ishizuka, W.; Kuromaru, M. Selection of superior genotypes from the population of plus trees in *Abies sachalinensis* by the results from progeny trials in Hokkaido. *Bull. FFPRI* **2018**, *17*, 155–174. (In Japaneses with English Summary)
32. Eiga, S.; Sakai, A. *Regional Variation in Cold Hardiness of Sakhalin fir* (Abies sachalinensis *Mast.*) *in Hokkaido, Japan*; Li, P.H., Ed.; Plant Cold Hardiness; Alan R. Liss, Inc.: New York, NY, USA, 1987; pp. 169–182.
33. Tatewaki, M. Forest ecology of the islands of the North Pacific Ocean. *J. Fac. Agric. Hokkaido Univ.* **1958**, *50*, 371–486.

34. Kitamura, K.; Uchiyama, K.; Ueno, S.; Ishizuka, W.; Tsuyama, I.; Goto, S. Geographical gradients of genetic diversity and differentiation among the southernmost marginal populations of *Abies sachalinensis* revealed by EST-SSR polymorphism. *Forests* **2020**, *11*, 233. [CrossRef]

35. Hirosawa, T.; Sasaki, K.; Watanabe, K.; Imoto, M.; Kuromaru, M. Inter-provenance differences in growth conditions of *Abies sachalinensis* plantations in heavy snow zone. *For. Tree Breed. Hokkaido* **2000**, *43*, 9–11. (In Japanese)

36. Ishizuka, W.; Kon, H.; Kita, K. Selection for 2nd generation plus tree in *Abies sachalinensis* in eastern Hokkaido, Japan. *Kosyunai Kihou* **2015**, *176*, 9–14. (In Japanese)

37. Kuromaru, M. The difference of adaptability to heavy snow zone among provenances and families of *Abies sachalinensis*. *For. Tree Breed.* **1988**, *146*, 6–9. (In Japanese)

38. Kuromaru, M. Inter-provenance variation in resistance to *Scleroderris* canker of using progeny of plus trees in *Abies sachalinensis*. *For. Tree Breed. Hokkaido* **1994**, *37*, 20–23. (In Japanese)

39. Japan Meteorological Agency. *Mesh Climate Data of Japan*; Japan Meteorological Business Support Center: Tokyo, Japan, 2002. (In Japanese)

40. Kira, T. Forest ecosystems of east and southeast Asia in a global perspective. *Ecol. Res.* **1991**, *6*, 185–200. [CrossRef]

41. Segal, M.; Xiao, Y. Multivariate random forests. *Wiley Interdiscip. Rev. Data Min. Knowl. Discov.* **2011**, *1*, 80–87. [CrossRef]

42. Breiman, L. Random Forests. *Mach. Learn.* **2001**, *45*, 5–32. [CrossRef]

43. Ishwaran, H.; Kogalur, U.; Blackstone, E.; Lauer, M. Random survival forests. *Ann. Appl. Stat.* **2008**, *2*, 841–860. [CrossRef]

44. Ishwaran, H.; Kogalur, U. Fast Unified Random Forests for Survival, Regression, and Classification (RF-SRC). R Package Version 2.9.3. 2020. Available online: https://cran.r-project.org/web/packages/randomForestSRC/randomForestSRC.pdf (accessed on 30 July 2020).

45. R Core Team. *R: A Language and Environment for Statistical Computing*; R Foundation for Statistical Computing: Vienna, Austria, 2020.

46. QGIS Geographic Information System. QGIS.org. Open Source Geospatial Foundation Project. 2020. Available online: https://qgis.org/en/site/forusers/download.html (accessed on 30 July 2020).

47. Okada, S.; Mukaide, H.; Sakai, A. Genetic variation in Saghalien fir from different areas of Hokkaido. *Silvae Genet.* **1973**, *22*, 1–2.

48. Sakagami, Y.; Fujimura, Y. Seasonal changes in the net photosynthetic and respiratory rate of *Abies sachalinensis* and *Picea glehnii* seedlings. *J. Jpn. For. Soc.* **1981**, *63*, 194–200. (In Japanese with English Summary)

49. Kitao, M.; Kitaoka, S.; Harayama, H.; Agathokleous, E.; Han, Q.; Uemura, A.; Furuya, N.; Ishibashi, S. Sustained growth suppression in forest-floor seedlings of Sakhalin fir associated with previous-year springtime photoinhibition after a winter cutting of canopy trees. *Eur. J. For. Res.* **2019**, *138*, 143–150. [CrossRef]

50. Hirano, Y.; Saitoh, T.M.; Fukatsu, E.; Kobayashi, H.; Muraoka, H.; Shen, Y.; Yasue, K. Effects of climate factors on the tree-ring structure of *Cryptomeria japonica* in central Japan. *J. Jpn. Wood Res. Soc.* **2020**, *66*, 117–127. (In Japanese with English Summary) [CrossRef]

51. Koyama, Y.; Takahashi, M.; Murauchi, Y.; Fukatsu, E.; Watanabe, A.; Tomaru, N. Japanese beech (*Fagus crenata*) plantations established from seedlings of non-native genetic lineages. *J. For. Res.* **2012**, *17*, 116–120. [CrossRef]

52. Tsumura, Y.; Uchiyama, K.; Moriguchi, Y.; Ueno, S.; Ihara-Ujino, T. Genome scanning for detecting adaptive genes along environmental gradients in the Japanese conifer, *Cryptomeria japonica. Heredity* **2012**, *109*, 349–360. [CrossRef] [PubMed]

53. Farjon, A. Pinaceae. In *Drawings and Descriptions of the Genera Abies, Cedrus, Pseudolarix, Keteleeria, Nothotsuga, Tsuga, Cathaya, Pseudotsuga, Larix and Picea*; Koeltz Scientific Books: Königstein, Germany, 1990.

54. Semerikova, S.; Semerikov, V. Molecular phylogenetic analysis of the genus *Abies* (Pinaceae) based on the nucleotide sequence of chloroplast DNA. *Rus. J. Genet.* **2014**, *50*, 7–19. [CrossRef]

55. Etterson, J.R.; Cornett, M.W.; White, M.A.; Kavajecz, L.C. Assisted migration across fixed seed zones detects adaptation lags in two major North American tree species. *Ecol. Appl.* **2020**, *30*, e02092. [CrossRef]

56. Rehfeldt, G.E.; Tchebakova, N.M.; Parfenova, Y.I.; Wykoff, W.R.; Kuzmina, N.A.; Milyutin, L.I. Intraspecific responses to climate in *Pinus sylvestris. Glob. Chang. Biol.* **2002**, *8*, 912–929. [CrossRef]

57. Kapeller, S.; Lexer, M.J.; Geburek, T.; Hiebl, J.; Schueler, S. Intraspecific variation in climate response of Norway spruce in the eastern Alpine range: Selecting appropriate provenances for future climate. *For. Ecol. Manag.* **2012**, *271*, 46–57. [CrossRef]

58. Rehfeldt, G.E.; Crookston, N.L.; Warwell, M.V.; Evans, J.S. Empirical analyses of plant-climate relationships for the western United States. *Int. J. Plant Sci.* **2006**, *167*, 1123–1150. [CrossRef]

59. Bower, A.D.; Aitken, S.N. Ecological genetics and seed transfer guidelines for *Pinus albicaulis* (Pinaceae). *Am. J. Bot.* **2008**, *95*, 66–76. [CrossRef]

60. Chakraborty, D.; Schueler, S.; Lexer, M.J.; Wang, T. Genetic trials improve the transfer of Douglas-fir distribution models across continents. *Ecography* **2019**, *42*, 88–101. [CrossRef]

61. O'Neill, G.; Wang, T.; Ukrainetz, N.; Charleson, L.; McAuley, L.; Yanchuk, A.; Zedel, S. A Proposed Climate-Based Seed Transfer System for British Columbia. Technical Report, BC Tech. Rep. 099. 2017. Availaible online: www.for.gov.bc.ca/hfd/pubs/Docs/Tr/Tr099.htm (accessed on 30 July 2020).

62. Aitken, S.N.; Bemmels, J.B. Time to get moving: Assisted gene flow of forest trees. *Evol. Appl.* **2016**, *9*, 271–290. [CrossRef] [PubMed]

63. Ishizuka, W.; Goto, S. Modeling intraspecific adaptation of *Abies sachalinensis* to local altitude and responses to global warming, based on a 36-year reciprocal transplant experiment. *Evol. Appl.* **2012**, *5*, 229–244. [CrossRef] [PubMed]

 *forests*

Article

Effects of Light Intensity and Girdling Treatments on the Production of Female Cones in Japanese Larch (*Larix kaempferi* (Lamb.) Carr.): Implications for the Management of Seed Orchards

Michinari Matsushita [1,*,†], Hiroki Nishikawa [2,†], Akira Tamura [1] and Makoto Takahashi [1]

[1] Forest Tree Breeding Center, Forestry and Forest Products Research Institute, 3809-1 Ishi, Juo, Hitachi, Ibaraki 319-1301, Japan; akirat@affrc.go.jp (A.T.); makotot@affrc.go.jp (M.T.)
[2] Fujiyoshida Examination Garden, Yamanashi Forest Research Institute, 1-18-2 Shinnishihara, Fuji-Yoshida, Yamanashi 403-0017, Japan; nishikawa-vvu@pref.yamanashi.lg.jp
* Correspondence: matsushita@affrc.go.jp
† Both authors contributed equally to this work.

Received: 7 September 2020; Accepted: 15 October 2020; Published: 19 October 2020

Abstract: To ensure sustainable forestry, it is important to establish an efficient management procedure for improving the seed production capacity of seed orchards. In this study, we evaluated the effects of girdling and increasing light intensity on female cone production in an old *L. kaempferi* (Lamb.) Carr. seed orchard. We also evaluated whether there is a genotype-specific reproductive response to these factors among clones. The results showed that female cone production was augmented by girdling and increasing light intensity. There was a difference in the effectiveness of girdling treatment levels, and the probability of producing female cones increased markedly at higher girdling levels. At light intensities where the relative photosynthetic photon flux density was higher than 50%, more than half of the trees tended to produce female cones, even in intact (ungirdled) trees, and the genotype-specific response to light intensity was more apparent in less-reproductive clones. These findings suggested that girdling less-reproductive trees combined with increasing light intensity was an effective management strategy for improving cone production in old seed orchards.

Keywords: breeding; genotype × environment interaction; mast seeding; seed production; thinning

1. Introduction

Japanese larch (*Larix kaempferi*) is a major plantation species in central and northern Japan [1]. The species has been introduced widely in China, Europe and North America, because of its rapid growth and sparse branching characteristics. *Larix kaempferi* often shows superior growth compared to other larch species (e.g., *L. decidua* Mill. and *L. laricina* (Du Roi) K. Koch) and has therefore been widely used in breeding programs involving hybrid breeding [2,3]. As a result of these efforts, hybrids between *L. kaempferi* and other larches have been used commercially in North America [4], Europe [3] and Japan [5,6].

In Japan, a breeding program for *L. kaempferi* was initiated in the 1950s. As part of the program, more than 500 first-generation plus trees were selected and used to establish clonal seed orchards [1]. The selection of second-generation plus trees started in the 2010s, and this is ongoing [1,7]. Compared to traditional seed sources, the superior growth of seedlings derived from the clonal seed orchards of the plus trees has been demonstrated [1], and significant variations in growth and wood property traits among families were reported [7,8]. Data analyses based on several provenance tests clarified that genotype × environment (G × E) interactions in several of these growth traits were not small [9].

However, genotype-specific responses in the reproductive traits (e.g., cone production) to environmental conditions have not yet been clarified in detail for *L. kaempferi*.

Reforestation using *L. kaempferi* has increased over the last decade in Japan, and *L. kaempferi* is now the second most important forestry species for plantation in Japan and occupies about 25% of newly planted forest areas [10]. Owing to its high juvenile growth performance, the demand for improved seeds and seedlings of *L. kaempferi* has been increasing [11]. However, the clonal seed orchards of the second-generation plus trees are currently too young to produce enough seeds. On the other hand, the clonal seed orchards of the first-generation plus trees, which were established mainly in the 1960s and 1970s, are now more than 50 years old. Due to the high density of old stems and branches, these old seed orchards are too tall to be managed efficiently, and too dark to produce sufficient quantities of good-quality seeds. In many conifers, mast seeding is a limiting factor for sustainable seed production [12,13]. To overcome this limitation, numerous efforts have been made to enhance seed production, e.g., improving light intensity [14–17], girdling [18–23], fertilizer [19] and drought [24,25]. To establish an efficient management system for the old seed orchards of Japanese larch, it is necessary to quantitatively evaluate such treatment effects on seed production.

Clear reproductive responses to changes in light intensity have been reported in forest plants [26–28]. Previous studies that examined the effect of increasing light intensity found that thinning operations often offer increased flowering [14,29,30]. Trees located at well-lit sites tended to have more flowers, and the quality and quantity of their seeds were positively correlated with magnitudes of flowering [31]. Previous studies on larch orchards have also reported that increasing light intensity improved cone production [14–17]. Uchiyama et al. [14] quantified a relationship between light levels and female cone production in an old orchard of *L. gmelinii* var. *japonica*. However, quantitative estimation of the effect of light intensity is still limited for old seed orchards of *L. kaempferi*.

In horticulture, girdling is commonly used to promote flower production [32] and improve fruit quality [33] and yield [32,34]. Girdling interrupts phloem transport by removing a part of the bark and cambium without affecting xylem transport [35]. When the main stem (trunk) of a tree is girdled, the translocation of assimilates to the below-ground parts of the tree is interrupted, resulting in a super-abundance of assimilates in the above-ground parts above the girdle [36,37]. Girdling has therefore been used to control the resource allocation, and to promote cone production [18–23]. It is, however, rarely quantified how different girdling levels (severity) affect the cone production of *L. kaempferi*.

To ensure the sustainability of forestries, establishing an efficient management procedure for improving the seed production in seed orchards is necessary. In this context, studies on the effects of light intensity and girdling manipulation on the reproductive performance of mast seeding conifer species, such as *Larix* and *Picea*, have attracted the interest of silviculturists and forest managers. In this study, we quantitatively evaluated the effects of girdling manipulation and increasing light intensity on female cone production in an old *L. kaempferi* seed orchard. We also evaluated genotype-specific responses in reproduction among *L. kaempferi* clones after making changes in light intensity. Finally, we examined the relationship between female cone production and tree sizes.

2. Materials and Methods

2.1. Study Site

The study site was the Fujisan Seed Orchard (35.42° N, 138.74° E; 1320–1350 m a.s.l.; total area 10 ha), managed by Yamanashi prefecture. The orchard, located on the northeastern slope of Mount Fuji, is divided into several plots and the largest plot (#9; 2.4 ha) was used for this study. The orchard was established between 1961 and 1962 and contains 526 stems (stem density: 223.8 stems/ha in 2017) comprising 44 first-generation *L. kaempferi* clones that were selected mainly from forests in Yamanashi and Nagano prefectures. The mean annual temperature and precipitation were 10.6 °C and 1568 mm, respectively. The soil of the area is comprised primarily of weathered volcanic sediments derived from Mount Fuji.

2.2. Girdling Treatments

We examined whether girdling could be used to enhance the reproductive performance of *L. kaempferi*. Girdling is a procedure that involves removing a 2 cm-wide semilunar ring of bark and cambium at a height of approximately 80–100 cm on the main stem (trunk) using a knife (Figure A1). We clarified the effects of the following four types of girdling treatments: one semilunar ring of bark and cambium was removed (referred as to level 1); two semilunar rings of bark and cambium were removed, with each ring oriented in opposite directions (level 2); three semilunar rings of bark and cambium with the same orientation were removed (level 3); and no girdling was performed (level 0; i.e., intact). We randomly assigned 8, 14 and 12 trees to levels 1, 2 and 3, respectively, and all of the remaining trees were assigned to level 0. Girdling was conducted in May 2016. As two of 12 trees from the level 3 treatment group were dead in autumn, these trees were removed from the analysis.

2.3. Light Intensity

The photosynthetic photon flux density (PPFD) was used as an indicator of light intensity in this study. Measurements were conducted on a cloudy day in July using LI-250 light meters (LI-COR, Lincoln, Dearborn, MI, USA). The measurements were performed four times (in four directions) around the crown of each tree at the approximate midpoint of the height of the crown (5–6 m). The mean relative PPFD (rPPFD) was calculated for each planting position, as follows: rPPFD = PPFD above the crown of each tree/PPFD above forest canopy (i.e., open sky).

2.4. Tree size and Reproductive Status

We scored the reproductive status of each tree based on the extent of female cone production, as follows: trees that did not produce female cones (hereafter referred to as index 1), trees with female cones that were very sparsely distributed within the crown (index 2), trees with female cones that were sparsely distributed within the crown, or trees produced numerous female cones on a few branches only (index 3) and trees producing cones abundantly on several branches (index 4).

The number of female cones produced per stem was counted manually, and all counts were performed in triplicate. To investigate the relationship between reproductive status and tree size, all living tree stems (trunks) within the orchard were mapped and their diameter at breast height (DBH), tree height and crown radius was measured.

2.5. Data Analysis

To analyze the reproductive performance of *L. kaempferi*, we used generalized linear mixed-effect models [38], by using R 3.2.5 [39]. Based on the reproductive score data, the effects of light intensity and girdling on the probability of female cone production were analyzed using ordered logit and binomial logit functions. The fixed-effect explanatory variables were rPPFD and girdling treatment; the rPPFD was used as a covariate, while the girdling was treated as a main factor. We treated "block within site" as a random-effect factor, to account for spatial pseudo-replication. To test whether the genotype-specific reproductive response to light intensity varied among clones, we included "clone" and the interaction with rPPFD (i.e., "clone:rPPFD") as random effects.

The relationship between female cone production and tree size traits was analyzed by ANCOVA-like linear mixed models. The DBH, height, crown radius and height/crown radius ratio for tree stems were treated as fixed-effect covariates. We also included "clone" and the interaction terms with traits (e.g., "clone:DBH") as random effects. In this study, each trait was analyzed separately because of the multicollinearity among size traits.

3. Results

3.1. Status of the L. kaempferi Orchard

There was a marked variation in the light environment in the study orchard, and the mean rPPFD was 50.1 ± 15.2% (Figure 1A). A total of 526 stems belonging to 44 clones were investigated. Of these stems, 31.3% (164/526) produced female cones (Figure 1B). Of the 164 trees that reproduced, reproductive scores of 76.8% (126) and 15.2% (25) were obtained for indices 2 and 3, respectively, while only 7.9% (13) was obtained for index 4. In total, 18,114 cones were produced in the orchard, with the mean and maximum numbers of cones per stem being 34.4 and 5960, respectively (Figure 1C). The mean (±SD) DBH, height and crown radius were 38.6 ± 6.2 cm, 10.3 ± 2.2 m and 4.3 ± 0.9 m, respectively.

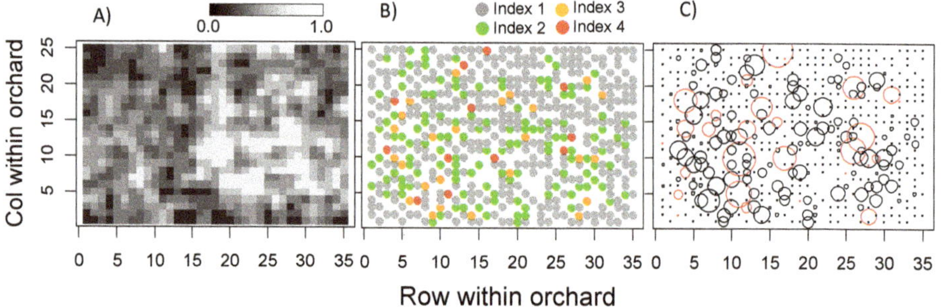

Figure 1. Within the orchard, spatial variation in (**A**) relative photosynthetic photon flux density. (**B**) Fruiting index of each tree. Index 1: trees that did not produce female cones, index 2: trees with female cones that were very sparsely distributed within the crown, index 3: trees with female cones that were sparsely distributed within the crown, or trees produced numerous female cones on a few branches only and index 4: trees that produced female cones abundantly on several branches. (**C**) Number of female cones produced per tree. Red circles indicate girdled trees and black circles indicate intact trees.

3.2. Relationship between Female Cone Production and Light Intensity or Girdling Treatments

The probability of female cone production varied markedly in response to light intensity (rPPFD) across different girdling treatments (Figure 2, Table 1). In girdling level 0 (i.e., intact stems), the probability of producing female cones (green: index 2 and orange: indices 3 and 4) increased markedly with increasing light intensity. In the situation where rPPFD was greater than approximately 0.5 (i.e., 50% sunlight), the probability of producing female cones (green and orange) was larger than that of not producing cones (blue: index 1). Two out of 12 trees in girdling level 3 died, but none of the trees in girdling levels 0–2 died.

Table 1. Summary of generalized linear mixed-effect model for estimating the probability of reproduction in Japanese larch. Relative photosynthetic photon flux density (rPPFD) was used as a fixed-effect covariate, while the effect of girdling treatment was used as a fixed-effect main factor.

	Coef.	S.E.	Z-Value	*p*-Value
Intercept	−1.84	0.45	−4.05	<0.001
Girdling level 1	1.25	0.78	1.61	0.109
Girdling level 2	2.39	0.71	3.39	<0.001
Girdling level 3	3.09	1.12	2.77	0.006
rPPFD	1.73	0.79	2.20	0.028

In darker environments, such as where rPPFD = 0.2, fewer than 20% of trees in the control group (i.e., intact stems) produced female cones (Figure 2). On the other hand, in the girdling level 3 group, more than 70% of trees produced female cones, even at similar light intensities.

Figure 3 shows the genotype-specific reproductive response to light conditions. Gray lines indicate response curves estimated for each of the 44 clones. The among-clone variation in the probability of female cone production was more apparent in darker areas, and less apparent under more brightly lit conditions (rPPFD > 0.8).

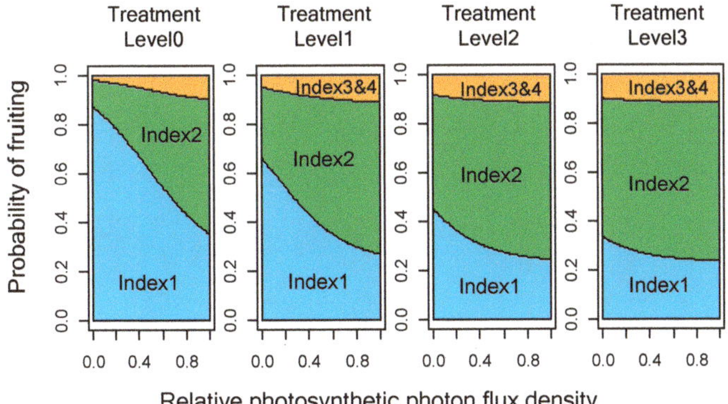

Figure 2. Relationship between fruiting probability of trees and light intensity (relative photosynthetic photon flux density) for different girdling treatment levels. Blue: fruiting index 1, trees not producing female cones. Green: fruiting index 2, trees producing female cones sparsely within their crown. Orange: fruiting index 3, trees producing abundant female cones within their crown.

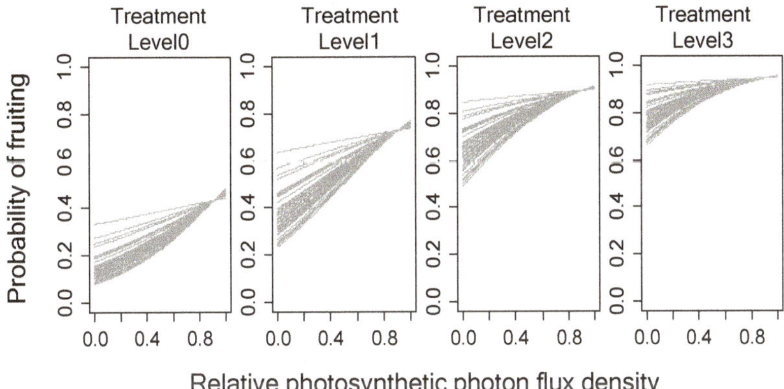

Figure 3. Genotype-specific relationship between fruiting probability and light intensity (relative photosynthetic photon flux density) across different girdling treatment levels. Each gray line indicates each *L. kaempferi* genotype (clone).

Based on the coefficient of variances (Table 2), among-clone variances in the probability of producing cones were much evident under conditions of lower light intensities and girdling levels, while the variation among clones decreased with increasing light intensities.

3.3. Relationship between Female Cone Production and Tree Size

When examining the relationships between female cone production and traits of the trees (size and shape), a significant negative relationship was observed between female cone production and the height/crown radius ratio ($p < 0.05$; bold black line in Figure 4D), and smaller trees with a relatively wider crown radius tended to produce more female cones. No significant relationships were observed between female cone production and the other traits (Figure 4A–C).

Table 2. The coefficient of variance (CV) for among-clone differences in the probability of producing female cones.

	rPPFD: 0.2	rPPFD: 0.5	rPPFD: 0.8
Girdling level 0	26.4%	13.1%	2.9%
Girdling level 1	16.8%	7.4%	1.4%
Girdling level 2	8.3%	3.3%	0.6%
Girdling level 3	4.8%	1.8%	0.3%

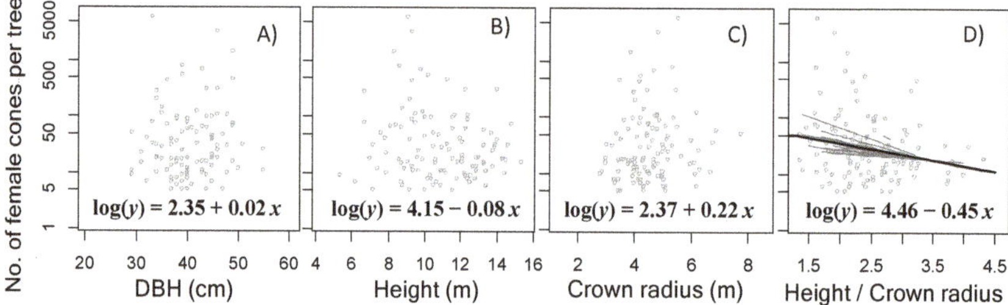

Figure 4. Relationship between number of female cones per tree and (**A**) tree diameter, (**B**) tree height, (**C**) crown radius and (**D**) height/crown radius ratio. In panel D, the thick black line indicates a significant regression relationship across all genotypes ($p < 0.05$), while gray lines indicate relationships for each genotype. There were no significant relationships between female cone production and the other traits (**A–C**).

4. Discussion

In northern conifer species including Japanese larch, mast seeding, i.e., low frequent abundant fruiting events, is a limiting factor for sustainable seed production, and numerous efforts have been made to overcome this limitation [12,13]. Several trials for increasing cone production have been conducted on larch species, such as *L. kaempferi*, *L decidua* and *L. laricina*, and some success has been reported for treatments involving girdling [18–23], gibberellins [40], nitrogen fertilizer [19], drought [24] and branch bending [41]. However, the results obtained from yet other studies using similar treatments have often been inconsistent, such as ineffective stimulation by gibberellins [42–44], nitrogen fertilizer [15] and root pruning [24]. These disparities could be attributed to differences in the age and growing conditions (natural stands, seed orchards or pots in greenhouses, etc.). In old seed orchards containing trees that are not well managed, restoring the reproductive capacity could be considered to be relatively difficult. However, the findings of our study showed that female cone production of old *L. kaempferi* trees can be enhanced by girdling and increasing the light intensity.

As in previous studies [18–23], our study confirmed that girdling was effective for improving female cone production, even in the old *L. kaempferi* orchard. In an orchard of 42-year-old *L. kaempferi*, more than 90% (19/20) of the girdled trees totally produced about 8000 cones, while only 25% (5/20) of the intact trees yielded 42 cones [22]. Similarly, female cone production in girdled trees increased more than ten times in natural stands of 70- [45] and 90-year-old [19] western larch, compared to intact trees. However, when girdling was conducted on young trees (17 years old), the effects of girdling on the proportion of trees which produced cones and the cone production per tree were less apparent [40], suggesting that age-dependent sexual maturity may affect the efficiency of girdling. Although the efficiency of girdling at different ages has not yet been clarified, our study found that girdling is a low-cost method that can be used to efficiently restore the cone production capacity of old *L. kaempferi* seed orchards.

Girdling has been regarded as a cost-efficient and useful technique for disrupting phloem transport while limiting detrimental effects [46]. However, it has also been reported that the positive effects of girdling are relatively limited in duration, and repeated girdling might adversely affect tree vigor and decrease total reproductive output [47]. In our study, there was considerable variation in the magnitude of the girdling effect among the different girdling treatment levels (see the coefficients in Table 1: larger coefficient values indicate a larger positive effect); the probability of producing cones was significantly increased as the treatment level was higher (Figure 2). However, 2 out of 12 girdled trees in the level 3 group died, while none of the girdled trees in the level 1 and 2 groups died. These results suggest that level 2 girdling might be better for balancing tree vigor and reproduction, while level 3 might be too severe. Although the sample size of severe girdling levels in the present study may be slightly small and the long-term effects of girdling on cone production are still unclear, the short-term effect on increasing female cone production of *L. kaempferi* was confirmed in the old seed orchard.

In the present study, improvements in light intensity also had a positive influence on the probability of producing cones. Previous studies similarly reported that trees on southern slopes that received higher levels of insolation tended to produce abundant cones, and the branches in a crown that received full sunlight achieved the highest cone production [48–50]. In our estimates, even in intact (ungirdled) trees, more than half of the trees tended to produce cones when the light conditions reached rPPFD > 50%. In a study on *L. gmelinii* var. *japonica*, Uchiyama et al. [14] recommended that light intensity at over 50% rPPFD in an orchard is optimal. Since there was a large spatial variation in light intensity in our studied orchard, thinning the darker areas to reduce the stem density appears to be an efficient means of improving the reproductive status of *L. kaempferi* trees and for improving the seed production capacity of the seed orchards.

Moreover, our quantitative estimates demonstrated the existence of a genotype-specific response to increases in light intensity. When light intensity was sufficient, all of the clones had similar reproductive potentials. However, among-clone differences in reproductive potential were much evident under conditions of insufficient light intensity, and less-reproductive *L. kaempferi* clones were susceptible to light conditions. Uchiyama et al. [14] also reported a significant G × E interaction among eight *L. gmelinii* var. *japonica* clones. Large among-clone variation in fecundity often makes it difficult to achieve panmixia, as seed orchards are designed on the premise of an equal reproductive contribution of each constitutive clone [51]. In some cases, less than half of the mother trees were responsible for more than half of the parentage in the seed orchard, resulting in an uneven genetic contribution among seedlings [52,53]. It has been reported that treatments often have a greater effect on less-reproductive trees, improving their genetic contribution across an orchard [52]. Improving light conditions by thinning can therefore be an effective method for ensuring that mating in a seed orchard more closely approximates panmixia through increasing the participation of clones in reproduction. In this context, improving the light conditions and collecting information on genotype-specific responses to light intensity may be useful for optimizing management regimes for improving seed production capacity of old seed orchards, not only in terms of seed quantity but also in quality.

In an experiment on young *L. kaempferi* trees [41], bending the upright branches so they were oriented downward had the effect of increasing cone bud initiation, especially on the lower surfaces of the horizontal shoots. In the early stages of orchard management, top pruning (cutting the main trunk and upright leader shoots) is typically conducted at a height of 3–4 m. However, in the less-intensively managed old seed orchards of *L. kaempferi*, it has been found that upright branches often re-sprouted dense adventitious shoots from the cut-off position, and the tree height was recovered. Based on our findings, ensuring that trees have a wide crown radius and relatively low height could be a better management strategy for improving female cone production in old orchards.

5. Conclusions

This study quantitatively evaluated the effects of girdling and increasing light intensity on female cone production in an *L. kaempferi* seed orchard. The findings showed that female cone production was augmented by both girdling and increasing light intensity. The probability of producing cones was markedly increased when higher girdling levels were applied. When the light intensity reached rPPFD > 50%, more than half of the mother trees tended to produce cones, even intact (ungirdled) trees, and the genotype-specific response to light intensity was more apparent in less-reproductive clones. A significant negative relationship was observed between female cone production and the height/crown radius ratio. Taken together, these findings suggested that a management procedure that combines girdling less-reproductive trees and improving light intensity could be used to optimize and improve the cone production capacity of old *L. kaempferi* seed orchards.

Author Contributions: M.M., H.N., A.T. and M.T. conceived and designed the research. M.M., H.N. and A.T. performed the field survey. H.N. managed the study site. M.M. conducted data analyses and wrote the original draft. A.T. and M.T. conducted supervision and project administration. M.M. and M.T.; funding acquisition. All authors have read and agreed to the published version of the manuscript.

Funding: This research was supported by grants from the Project of the NARO Bio-oriented Technology Research Advancement Institution (the special scheme project on regional developing strategy; Forestry C105) and JSPS KAKENHI Grant Number 17K15291.

Acknowledgments: We thank the staff of the Fujisan Orchard for their assistance with field investigations.

Conflicts of Interest: The authors declare no conflict of interest.

Appendix A

Girdling level 1 **Girdling level 2** **Girdling level 3**

Figure A1. Example photos for girdling levels 1–3.

References

1. Kurinobu, S. Forest Tree Breeding for Japanese larch. *Eurasian J. For. Res.* **2005**, *8*, 127–134.
2. Park, Y.S.; Fowler, D.P. A provenance test of Japanese larch in eastern Canada, including comparative data on European larch and tamarack. *Silvae Genet.* **1983**, *32*, 96–101.
3. Pâques, L.E. Roles of European and Japanese larch in the genetic control of growth, architecture and wood quality traits in interspecific hybrids (*Larix* × *eurolepis* Henry). *Ann. For. Sci.* **2004**, *61*, 25–33. [CrossRef]
4. Baltunis, B.S.; Greenwood, M.S.; Eysteinsson, T. Hybrid vigor in *Larix*: Growth of intra-and interspecific hybrids of *Larix decidua*, *L. laricina*, and *L. kaempferi* after 5-years. *Silvae Genet.* **1998**, *47*, 288–293.
5. Kita, K.; Fujimoto, T.; Uchiyama, K.; Kuromaru, M.; Akutsu, H. Estimated amount of carbon accumulation of hybrid larch in three 31-year-old progeny test plantations. *J. Wood Sci.* **2009**, *55*, 425–434. [CrossRef]
6. Kita, K.; Sugai, T.; Fujita, S.; Koike, T. Breeding effort on hybrid larch F1 and its responses to environmental stresses. *For. Gen. Tree Breed.* **2018**, *7*, 107–114, (In Japanese with English Summary).
7. Fukatsu, E.; Hiraoka, Y.; Matsunaga, K.; Tsubomura, M.; Nakada, R. Genetic relationship between wood properties and growth traits in *Larix kaempferi* obtained from a diallel mating test. *J. Wood Sci.* **2015**, *61*, 10–18. [CrossRef]
8. Fukatsu, E.; Tsubomura, M.; Fujisawa, Y.; Nakada, R. Genetic improvement of wood density and radial growth in *Larix kaempferi*: Results from a diallel mating test. *Ann. For. Sci.* **2013**, *70*, 451–459. [CrossRef]
9. Nagamitsu, T.; Nagasaka, K.; Yoshimaru, H.; Tsumura, Y. Provenance tests for survival and growth of 50-year-old Japanese larch (*Larix kaempferi*) trees related to climatic conditions in central Japan. *Tree Genet. Genomes* **2014**, *10*, 87–99. [CrossRef]

10. Forestry Agency, Ministry of Agriculture, Forestry and Fisheries of Japan. *Annual Report on Forest and Forestry in Japan for FY 2018*; Forestry Agency, Ministry of Agriculture, Forestry and Fisheries of Japan: Tokyo, Japan, 2019.

11. Forest Tree Breeding Center. *The Current States and Statistics in Forest Tree Breeding in Japan*; Forest Tree Breeding Center, Forestry and Forestry Product Research Institute: Hitachi, Japan, 2019.

12. Bonnet-Masimbert, M. Flower induction in conifers: A review of available techniques. *For. Ecol. Manag.* **1987**, *19*, 135–146. [CrossRef]

13. Crain, B.A.; Cregg, B.M. Regulation and management of cone induction in temperate conifers. *For. Sci.* **2017**, *64*, 82–101. [CrossRef]

14. Uchiyama, K.; Kuromaru, M.; Kita, K. Effect of light intensity and girdling on seed production of *Larix gmelinii* var. japonica clones. *Bull. Hokkaido For. Res. Inst.* **2007**, *44*, 119–127, (In Japanese with English Summary).

15. Asakawa, S.; Fujita, K.; Nagao, A.; Yokoyama, T. The effect of girdling on the coning of larch seed trees as affected by stand density. *J. Jpn. For. Soc.* **1966**, *48*, 245–249, (In Japanese with English Summary).

16. Tamura, A.; Ubukata, M.; Yamada, H.; Fukuda, Y.; Yano, K.; Orita, H. Effect of line thinning on stimulation of flowering in a Japanese larch orchard. *Jpn. For. Soc. Cong.* **2015**, *126*, 334.

17. Shearer, R.C.; Schmidt, W.C. Cone production and stand density in young *Larix occidentalis*. *For. Ecol. Manag.* **1987**, *19*, 219–226. [CrossRef]

18. Bonnet-Masimbert, M. Effect of growth regulators, girdling, and mulching on flowering of young European and Japanese larches under field conditions. *Can. J. For. Res.* **1981**, *12*, 276–279. [CrossRef]

19. Prill, R. Cone induction on western larch seed trees. *B.C. Min. For. Silv. Br. Prog. Rep.* **1990**, *SX87601-10*, 29.

20. Mikami, S.; Asakawa, S.; Iizuka, M.; Yokoyama, T.; Nagao, A.; Takehana, S.; Kaneko, T. Flower induction in Japanese larch, *Larix leptolepis* Gord. *Bull. FFPRI* **1979**, *307*, 9–24, (In Japanese with English summary).

21. Miller, L.K.; Debell, J. Current seed orchard techniques and innovations. In *National Proceedings: Forest and Conservation Nursery Associations 2012*; USDA Forest Service: Fort Collins, CO, USA, 2013; pp. 80–86.

22. Lee, W.Y.; Lee, J.S.; Lee, J.H.; Noh, E.W.; Park, E.-J. Enhanced seed production and metabolic alterations in *Larix leptolepis* by girdling. *For. Ecol. Manag.* **2011**, *261*, 1957–1961. [CrossRef]

23. Markiewicz, P. Problems with seed production of European larch in seed orchards in Poland. In *Proceedings of a Seed Orchard Conference*; Lindgren, D., Ed.; Swedish University of Agricultural Sciences: Umeå, Sweden, 2007; pp. 161–164.

24. Philipson, J.J. Effects of cultural treatments and gibberellin A4/17 on flowering of container-grown European and Japanese larch. *Can. J. For. Res.* **1995**, *25*, 184–192. [CrossRef]

25. Colas, F.; Perron, M.; Tousignant, D.; Parent, C.; Pelletier, M.; Lemay, P. A novel approach for the operational production of hybrid larch seeds under northern climatic conditions. *For. Chron.* **2008**, *84*, 95–104. [CrossRef]

26. Verkaik, I.; Espelta, J.M. Post-fire regeneration thinning, cone production, serotiny and regeneration age in *Pinus halepensis*. *For. Ecol. Manag.* **2006**, *231*, 155–163. [CrossRef]

27. Peters, G.; Sala, A. Reproductive output of ponderosa pine in response to thinning and prescribed burning in western Montana. *Can. J. For. Res.* **2008**, *38*, 844–850. [CrossRef]

28. Matsushita, M.; Nakagawa, M.; Tomaru, N. Sexual differences in year-to-year flowering trends in the dioecious multi-stemmed shrub *Lindera triloba*: Effects of light and clonal integration. *J. Ecol.* **2011**, *99*, 1520–1530. [CrossRef]

29. Lindh, B.C. Flowering of understory herbs following thinning in the western Cascades, Oregon. *For. Ecol. Manag.* **2008**, *256*, 929–936. [CrossRef]

30. Matsushita, M.; Setsuko, S.; Tamaki, I.; Nakagawa, M.; Nishimura, N.; Tomaru, N. Thinning operations increase the demographic performance of the rare subtree species *Magnolia stellata* in a suburban forest landscape. *Landsc. Ecol. Eng.* **2016**, *12*, 179–186. [CrossRef]

31. Setsuko, S.; Tamaki, I.; Ishida, K.; Tomaru, N. Relationships between flowering phenology and female reproductive success in the Japanese tree species *Magnolia stellata*. *Botany* **2008**, *86*, 248–258. [CrossRef]

32. Levin, A.G.; Lavee, S. The influence of girdling on flower type, number, inflorescence density, fruit set, and yields in three different olive cultivars (Barnea, Picual, and Souri). *Austral J. Agric. Res.* **2005**, *56*, 827–831. [CrossRef]

33. Brar, H.S.; Singh, Z.; Swinny, E.; Cameron, I. Girdling and grapevine leafroll associated viruses affect berry weight, colour development and accumulation of anthocyanins in 'Crimson Seedless' grapes during maturation and ripening. *Plant Sci.* **2008**, *175*, 885–897. [CrossRef]

34. Rivas, F.; Gravina, A.; Agustí, M. Girdling effects on fruit set and quantum yield efficiency of PSII in two *Citrus* cultivars. *Tree Physiol.* **2007**, *27*, 527–535. [CrossRef]

35. Van Kleunen, M.; Stuefer, J.F. Quantifying the effects of reciprocal assimilate and water translocation in a clonal plant by the use of steam girdling. *Oikos* **1999**, *85*, 135–145. [CrossRef]

36. Isogimi, T.; Matsushita, M.; Watanabe, Y.; Nakagawa, M. Sexual differences in physiological integration in the dioecious shrub *Lindera triloba*: A field experiment using girdling manipulation. *Ann. Bot.* **2011**, *107*, 1029–1037. [CrossRef] [PubMed]

37. Isogimi, T.; Matsushita, M.; Nakagawa, M. Species-specific sprouting pattern in two dioecious *Lindera* shrubs: The role of physiological integration. *Flora* **2014**, *209*, 718–724. [CrossRef]

38. Bolker, B.M.; Brooks, M.E.; Clark, C.J.; Geange, S.W.; Poulsen, J.R.; Stevens, M.H.H.; White, J.S.S. Generalized linear mixed models: A practical guide for ecology and evolution. *Trends Ecol. Evol.* **2009**, *24*, 127–135. [CrossRef] [PubMed]

39. R Core Team. *R: A Language and Environment for Statistical Computing*; R Foundation for Statistical Computing; R Core Team: Vienna, Austria, 2016.

40. Ross, S.D. *Promotion of Flowering in Western Larch by Girdling and Gibberellin A4/7, and Recommendations for Selection and Treatment of Seed Trees*; BC Ministry of Forests, Research Laboratory: Victoria, BS, Canada, 1991; p. 13.

41. Longman, K.A.; Nasr, T.A.; Wareing, P.F. Gravimorphism in trees. 4. The effect of gravity on flowering. *Ann. Bot.* **1965**, *29*, 105–111. [CrossRef]

42. Hashizume, H. Studies on flower bud formation, flower sex differentiation and their control in conifers. *Bull. Tottori Univ. For.* **1973**, *7*, 1–139, (In Japanese with English Summary).

43. Katsuta, M.; Saito, M.; Yamamoto, C.; Kaneko, T.; Itoo, M. Effect of gibberellins on the promotion of strobilus production in *Larix leptolepis* Gord. and *Abies hornolepis* Sieb. and Zucc. *Bull. FFPRI* **1981**, *313*, 37–45, (In Japanese with English Summary).

44. Philipson, J.J. Effects of girdling and gibberellin A4/17 on flowering of European and Japanese larch grafts in an outdoor clone bank. *Can. J. For. Res.* **1996**, *26*, 355–359. [CrossRef]

45. Wheeler, N.C.; Cade, S.C.; Masters, C.J.; Ross, S.D.; Keeley, J.W.; Hsin, L.Y. Girdling: A safe, effective and practical treatment for enhancing seed yields in Douglas-fir seed orchards. *Can. J. For. Res.* **1985**, *15*, 505–510. [CrossRef]

46. Graham, R.T. Effect of nitrogen fertilizer and girdling on cone and seed production of western larch. In *Proc. Conifer Tree Seed in the Inland Mountain West Symposium General Technical Report*; Shearer, R.C., Ed.; USDA Forest Service Intermountain Research Station: Ogden, UT, USA, 1986; pp. 166–170.

47. Owens, J.N.; Blake, M.D. Forest seed tree production. In *Information Report PI-X-53*; Petawawa National Forestry Institute, Canada Forest Service: Chalk River, ON, Canada, 1985; p. 161.

48. Chałupka, W.; Giertych, M.; Kopcewicz, J. Effect of polyethylene covers on the flowering of Norway spruce (*Picea abies* (L.) Karst.) grafts. *Physiol. Plant.* **1982**, *54*, 79–81. [CrossRef]

49. Matthews, J.D. Factors affecting the production of seed by forest trees. *For. Abstr.* **1963**, *24*, 1–13.

50. Despland, E.; Houle, G. Aspect influences cone abundance within the crown of *Pinus banksiana* Lamb. trees at the limit of the species distribution in northern Quebec (Canada). *Écoscience* **1997**, *4*, 521–525. [CrossRef]

51. Funda, T.; El-Kassaby, Y.A. Seed orchard genetics. *Cab. Rev.* **2012**, *7*, 1–23. [CrossRef]

52. Chałupka, W. Do we need flower stimulation in seed orchards? In *Proceedings of a Seed Orchard Conference*; Lindgren, D., Ed.; Swedish University of Agricultural Sciences: Umeå, Sweden, 2007; pp. 37–42.

53. Funda, T. Population Genetics of Conifer Seed Orchards. Ph.D. Thesis, University of British Columbia, Vancouver, BS, Canada, 2012.

Publisher's Note: MDPI stays neutral with regard to jurisdictional claims in published maps and institutional affiliations.

Article

Evaluation of Responsivity to Drought Stress Using Infrared Thermography and Chlorophyll Fluorescence in Potted Clones of *Cryptomeria japonica*

Yuya Takashima [1,*], Yuichiro Hiraoka [2], Michinari Matsushita [1] and Makoto Takahashi [1]

[1] Forest Tree Breeding Center, Forestry and Forest Products Research Institute, Forest Research and Management Organization, 3809-1 Ishi, Juo, Hitachi 319-1301, Ibaraki, Japan; matsushita@affrc.go.jp (M.M.); makotot@affrc.go.jp (M.T.)

[2] Faculty of Production and Environmental Management, Shizuoka Professional University of Agriculture, 678-1 Tomigaoka, Iwata 438-8577, Shizuoka, Japan; hiraoka.yuichiro@spua.ac.jp

[*] Correspondence: ytakashima@ffpri.affrc.go.jp

Abstract: As climate change progresses, the breeding of drought-tolerant forest trees is necessary. Breeding drought-tolerant trees requires screening for drought stress using a large number of individuals and a high-throughput phenotyping method. The aim of this study was therefore to establish high-throughput methods for evaluating the clonal stress responses to drought stress using infrared thermography and chlorophyll fluorescence methods in *Cryptomeria japonica*. The stomatal conductance index (Ig), maximum photochemical quantum yield of photosystem II (Fv/Fm), and axial growth of four plus-tree clones of *C. japonica* planted in pots were measured weekly for 85 days after irrigation was stopped. The phenotypic trait responsivity to drought stress was estimated by a nonlinear mixed model and by introducing the cumulative water index, which considers the past history of the soil water environment. These methods and procedures enabled us to evaluate the clonal stress responses in *C. japonica* and could be applied to large-scale clone materials to promote the breeding program for drought tolerance.

Keywords: infrared thermography; chlorophyll fluorescence; cumulative drought stress; high-throughput phenotyping; *Cryptomeria japonica*

Citation: Takashima, Y.; Hiraoka, Y.; Matsushita, M.; Takahashi, M. Evaluation of Responsivity to Drought Stress Using Infrared Thermography and Chlorophyll Fluorescence in Potted Clones of *Cryptomeria japonica*. *Forests* **2021**, *12*, 55. https://doi.org/10.3390/f12010055

Received: 13 November 2020
Accepted: 30 December 2020
Published: 2 January 2021

Publisher's Note: MDPI stays neutral with regard to jurisdictional claims in published maps and institutional affiliations.

1. Introduction

The frequency of extreme climatic events, such as droughts and heatwaves, is expected to increase as the global climate change progresses [1]. Higher temperatures and more frequent, longer droughts are expected in Japan [2]. Stress due to drought and high temperatures would have a negative impact on the growth and survival of forest trees [3,4]. In particular, drought stress would impair the physiological mechanisms of forest trees, which are dependent on water [5]. Thus, to adapt to these emerging circumstances, drought tolerance has become an important target trait in tree breeding and genetic improvement [6,7].

Breeding for drought resistance has been assessed based on sustained growth or yield or suppressed mortality under water-deficient conditions in crop plants, such as rice [8], wheat [9], and soybeans [10], as in forest trees such as *Pinus pinaster* [11] and *Eucalyptus globulus* [12]. When breeding for drought resistance, morphological traits, such as growth, yield, and mortality, and physiological traits, such as water use efficiency (WUE), stomatal conductance, cavitation of conductive tissue, photosynthetic ability, leaf wilting, leaf water potential, and osmotic regulation, are used as target traits. However, most of these traits require considerable effort and time to measure and are thus not suitable for large-scale screening in breeding programs.

Among the physiological traits involved in the drought stress response of plants, leaf transpiration depending on stomatal conductance and photosynthetic ability can be evaluated with high-throughput by using infrared thermography and chlorophyll fluorescence methods, respectively [13]. In trees under drought stress, stomatal conductance, which is affected by the transpiration rate, is changed so as to maintain optimal water conditions within trees [14–16]. In addition, the leaf temperature also changes due to evaporative heat loss, which changes with transpiration. A method for estimating stomatal conductance by measuring the change in leaf temperature using infrared thermography has been developed [13,17]. Infrared thermography has been applied to some tree species to evaluate drought stress responses, because it enables to obtain data from single leaves, as well as tree crowns, expeditiously (*Jatropha curcas* [18], *Pinus sylvestris* [19], *Firmiana platanifolia* [20], and *Vitis vinifera* [21]). Chlorophyll fluorescence is another method that has been widely used for to measure photosynthesis [22]. Chlorophyll fluorescence occurs when light energy is absorbed by the chlorophyll antenna but is not used for photosynthesis or converted into heat and which is then re-emitted as red fluorescence. Chlorophyll fluorescence is closely related to photosynthetic carbon metabolism and leaf gas exchange [23,24]. In plants, drought stress increases the water deficit and progressively decreases the carbon assimilation by photosynthesis as a result of both stomatal and metabolic limitations [25–29]. Consequently, the measurement of chlorophyll fluorescence, which is noninvasive, can be used to infer plant viability and performance in response to drought stress [30]. The measurement of stomatal conductance using infrared thermography and the measurement of photosynthesis using chlorophyll fluorescence are thus potentially useful tools for preforming high-throughput phenotyping in tree breeding experiments focused on improving drought resistance.

For plants, soil drought is typically not a transient stress but a cumulative one that is also influenced by past conditions. It is therefore considered that the drought stress responses in plants can be better understood by introducing a cumulative water index. In forest ecology, forest remote sensing, and dendrochronology, the Standardized Precipitation Evapotranspiration Index (SPEI) is used to estimate how the "cumulative stress" associated with drought stress affects the decline and recovery of tree growth in the field [31,32]. However, few reports have examined the relationship between tree responses and cumulative stress under artificially controlled soil water conditions in potted plants. When examining the drought stress responses in potted plants, the SPEI not easily applicable to potted plants, because the SPEI uses the monthly (or weekly) difference between precipitation and potential evapotranspiration [33]. To solve this difficulty, we introduced the Cumulative Water Index (CWI), which is a novel index defined as the soil moisture content cumulated for any number of past days, which can consider the past history of the soil moisture status. Then, we examined the responsivity and clonal variations of phenotypic traits to the CWI.

Cryptomeria japonica is a major forestry species in Japan, accounting for about 44% of the area used for plantations [34]. It grows relatively quickly, the wood is light and soft with excellent workability, and it is used for structural materials, such as pillars and interior building materials. This species is highly sensitive to drought [35] due to its high water demand and high transpiration [36]. Consequently, water stress has a marked negative effect on its growth.

The purpose of this study was to clarify the phenotypic variation in stomatal responses and photosynthesis activity associated with drought stress in *C. japonica*. To achieve this goal, we investigated the clonal variation in growth, stomatal response, and photosynthesis activity in response to drought stress among plus-tree clones of *C. japonica* using a high-throughput phenotyping method based on infrared thermography and chlorophyll fluorescence. Moreover, a statistical model that considered the cumulative soil moisture environment was used to clarify the drought resistance characteristics among the clones. The relationships between the growth, stomatal response, and chlorophyll fluorescence, as a proxy of photosynthesis activity, of *C. japonica* in response to drought stress and the soil water environment, considering the past history of soil conditions, are also discussed.

2. Materials and Methods

2.1. Tree Materials and Drought Treatments

Cuttings of four *C. japonica* plus-tree clones (GO1, KA7, TE11, and TS1) were used as the experimental materials in this study. For clonal propagation of cuttings, scions were collected and placed in rooting medium (Kanuma soil) in March 2014. After rooting, cuttings were planted in 3-L pots containing a mixed-culture soil consisting of Kanuma soil, Akadama soil, and gardening soil amended with fertilizer (2:3:4.4) in March 2015 and grown for one year in a greenhouse. In February 2016, the plants were transplanted to 13.5-L plastic pots containing the same mixed-culture soil and were reared in a greenhouse with sufficient water until the drought stress experiments were started. After three months of acclimatization by keeping the same watering scheme, cuttings were randomly assigned to one of two treatments: no watering (drought) and normal watering (control). Table 1 shows the seedling height of each clone at the time the experiment was started. The experiment lasted for 85 days, from 9 May to 1 August 2016. In the drought treatment, watering was withheld from 9 May until the end of the experiment, while the control cuttings were watered three times a week for the duration of the experiment. Figure S1 shows the temporal changes in the soil water contents in the two treatments.

Table 1. Mean height of the four clones at the start of the experiment.

Clone Name	Treatment	n	Height (cm)	SD
GO1	Control	5	38.5	5.0
	Drought	5	37.7	5.9
KA7	Control	5	42.9	9.9
	Drought	5	44.5	3.7
TE11	Control	5	39.2	3.6
	Drought	5	34.8	4.4
TS1	Control	5	50.3	5.1
	Drought	5	48.9	3.6

2.2. Measurement of Leaf Temperature and Calculation of Stomatal Conductance Index

Leaves of *C. japonica* are composed of blanchlets with many small needles; the "leaf" in this manuscript means an about 10–20-cm-long blanchlet composed of the current year shoot and many needles. Leaf temperature was estimated using an infrared thermal camera (InfReC R300SR, Nippon Avionics, Yokohama, Japan), which measured the infrared emissions at wavelengths 8–14 μm at a thermal resolution of 0.03 °C and images with a spatial resolution of 320 × 240 pixels. When measuring individual leaves, two reference leaves were prepared from another individual that was not used in the experiment, because it was confirmed in previous test that there was no difference in the temperature of reference leaves depending on the clone or individual used for the reference: leaves with fully opened stomata (wet reference leaves) and those with fully closed stomata (dry reference leaves). Wet reference leaves were leaves that were sprayed regularly with water to maintain their moisture level. Dry reference leaves were coated with a mixtures of petroleum jelly and liquid paraffin (1:2 by weight). The reference leaves were then removed and placed in the same image capture frame as the leaves targeted for measurements (Figure S2). One pair of the reference leaves was used for all individuals in each measurement day. Thermal images were obtained once a week at 10:00–12:00 in a greenhouse at an adjusted photosynthetically active radiation (PAR) (range of about 200 to 380 μmol m^{-2} s^{-1}). When it was fine or slightly cloudy weather, PAR was controlled by covering the ceiling of the green house using a nonwoven curtain (about 80% shading rate) to avoid an excess of increasing leaf temperature because of direct solar radiation. Thermal images were obtained 11 times in experimental period, and one leaf of each individual was measured each time.

Software (InfReC Analyzer NS9500 Lite, Nippon Avionics) was used for image analysis and data extraction. Pixels of current and the previous year leaves were manually selected in the images, and the average value of the temperature of those pixels was used as the leaf temperature of an individual plant (T_p). The temperatures of the dry and wet reference leaves (T_d and T_w, respectively) were obtained using the same procedure as that used to calculate the T_p. The stomatal conductance index (Ig) was calculated using the following formula [17]:

$$Ig = (T_d - T_p)/(T_p - T_w) \tag{1}$$

To clarify the effect of drought stress on stomatal conductance, the Ig ratio (RIg) for each clone was calculated as follows:

$$RIg_i = Ig_{di}/Ig_{ci} \tag{2}$$

where RIg_i is the Ig ratio of the ith clone, Ig_{di} is the mean value of Ig for the ith clone in the drought treatment, and Ig_{ci} is the mean value of Ig for the ith clone in the control treatment. The RIg values of less than unity mean the decrease of stomatal conductance due to drought treatment.

2.3. Measurement of Actual Stomatal Conductance

To confirm the accuracy of the Ig, during the drought experiment, we checked the relationship between the Ig obtained by the infrared thermography and actual stomatal conductance obtained by a gas exchange method using the other individuals of same four clones that were not used in the drought experiment. The individual samples were grown under varied soil water conditions and were measured under several environmental conditions (soil water content (SWC): 0.05–0.48 cm^3 cm^{-3}, ambient temperature: 24.9–34.5 °C, and relative humidity: 52.5–70%, PAR: 300–1500 μmol m^{-2} s^{-1}) to a cover wider range of stomatal conductance. The actual stomatal conductance was measured by using a portable gas exchange system (LI-6400, LI-COR, Lincoln, NE, USA) and the conifer chamber (6400-22L, LI-COR) with the LED light source intensity, temperature, relative humidity, and CO_2 concentration in the chamber set to the same values as the ambient conditions. After allowing the stomatal conductance to reach a steady state, stomatal conductance was recorded five times every minute. After the measurements, the leaf surface temperature of the same individual was measured immediately by infrared thermography under the same conditions of measuring the actual stomatal conductance. Then, the leaves whose actual stomatal conductance was measured were cut off and dried in an oven at 105 °C for 2 days; the oven-dried weight was measured, and the stomatal conductance per biomass (mol g^{-1} s^{-1}) was calculated. The average of these values was used as the actual stomatal conductance of an individual cutting under ambient conditions. After that, the actual stomatal conductance and the Ig were compared.

2.4. Measurement of Maximum Photochemical Quantum Yield (Fv/Fm)

The maximum photochemical quantum yield of photosystem II (PS II) (Fv/Fm), which is an index of the photosynthetic ability under water deficit conditions, was measured by the chlorophyll fluorescence method. Chlorophyll fluorescence parameters were obtained using a pulse-amplitude modulation fluorometer (MINI-PAM, Walz, Bayern, Germany). Thirty minutes after sunset, the minimum fluorescence level (Fo) was determined with a low-intensity measuring light. The maximum fluorescence level (Fm) was measured after a 0.5 s saturating pulse at 4000 μmol m^{-2} s^{-1}. Fv/Fm was calculated as follows:

$$Fv/Fm = (Fm - Fo)/Fm \tag{3}$$

Fv/Fm was measured once a week at the same time as the measurements of the infrared thermograph were conducted, and an average of three measurements per individual was used as the Fv/Fm of the focal individual.

$$RFv/Fm_i = (Fv/Fm_{di})/(Fv/Fm_{ci}) \tag{4}$$

where RFv/Fm_i is the Fv/Fm ratio of the ith clone, and Fv/Fm_{di} and Fv/Fm_{ci} are the mean values of the Fv/Fm of the ith clone in the drought and control treatments, respectively.

2.5. Measurement of Growth

Axial growth was measured as a growth trait. Before the beginning of the experiment, a line was drawn at a point 5 cm below the apex of the main shoot with a black marker. The distance from the mark to the shoot apex was then measured weekly, and the length differentials were regarded as the amount of shoot growth or growth rate. To clarify the effect of drought stress on growth, the growth rate ratio (GRR) was calculated for each clone as follows:

$$GRR_i = GR_{di}/GR_{ci} \tag{5}$$

where GRR_i is the growth rate ratio of the ith clone, and GR_{di} and GR_{ci} are the mean values of the growth rate of the ith clone in the drought and control treatments, respectively.

2.6. Measurement of Soil Condition

The soil water content (SWC; $cm^3\ cm^{-3}$) during the experiment was measured using a soil moisture sensor (SM150 Soil Moisture Kit, Delta-T Devices, Cambridge, UK). The SWC was measured by inserting the sensor into the soil at three points in each pot, and the average value was used as the SWC of the pot. Measurement of the SWC was conducted every one or two days. Relationship between the soil water content and soil water potential of the mixed culture soil used in the experiment is shown in Figure S3.

2.7. Statistical Analysis

To investigate the tree responses to cumulative drought stress, the novel CWI was defined as follows:

$$CWI_{dp} = \sum_{k\ =\ d\ +\ 1\ -\ p}^{d} SWC_k \tag{6}$$

where SWC_k is the soil water content ($cm^3\ cm^{-3}$) of kth day; d is the number of days since the start of the experiment ($0 \leq d \leq 84$), p is an arbitrary number of days from d ($p = 1, 2, 3, 4, 5, 6, 7, 10, 14, 21, 28, 35, 42, 49, 56, 63, 70, 77,$ and 84); and CPDs is the cumulative number of past days p days from $d + 1 - p$ to d day, in which 1 CPD corresponds to a simple SWC. Since the soil water conditions of the two treatment classes were similar before starting the drought stress experiments, when calculating the CWI before the starting day of the experiment (i.e., $d + 1 - p < 0$), the SWCs of the drought treatment class before the experiment ($d = 0$) were assumed to be similar to the mean value of SWC in the control class during the overall experimental period ($0.461\ cm^3\ cm^{-3}$). Since SWC was measured every 1 or 2 days, these data were not available for every day of the experimental period. Therefore, the SWC for nonmeasurement days was estimated by linear interpolation.

To simulate *C. japonica* responses to drought stress, the responsiveness of RIg, GRR, and RFv/Fm to the CWI and the clonal effects were modeled using a nonlinear mixed-effect model (NLMM). The Gompertz function was used to fit the responses of RIg and GRR to the CWI (Equation (7)). Additionally, the von Bertalanffy function was used to fit the response of RFv/Fm to the CWI (Equation (8)).

$$y_{ij} = \exp(-\alpha \times \exp(-\beta \times CWI_{ij})) + e_{ij} \tag{7}$$

$$y_{ij} = 1 - \exp(-\beta \times (CWI_{ij} - \alpha)) + e_{ij} \tag{8}$$

where y_{ij} is the response variable (RIg, GRR, or RFv/Fm) of jth time point of the ith individual; α and β are the parameters of the Gompertz and von Bertalanffy functions; CWI_{ij} is the explanatory variable of the jth time point for the ith individual; and e_{ij} is the random residual. Each of the parameters α and β can be described by a linear mixed-effects model. The full models for observation j of individual i are

$$y_{ij} = \exp(-(b_{\alpha ij} + c_{\alpha i}) \times \exp(-(b_{\beta ij} + c_{\beta i}) \times \text{CWI}_{ij})) + e_{ij} \tag{9}$$

$$y_{ij} = 1 - \exp(-(b_{\alpha ij} + c_{\alpha i}) \times (\text{CWI}_{ij} - (b_{\beta ij} + c_{\beta i}))) + e_{ij} \tag{10}$$

where b is a vector of the fixed effects, and c is a vector of the random clonal effects.

The application of NLMMs was used to clarify the clonal response in three traits: RIg, GRR, and RFv/Fm against drought stress. The CWI values were used as explanatory variables in NLMMs to consider the cumulative effect of past soil water conditions. The CPD, which was used to calculate the CWI, was serially changed from 1 to 84 days, and the optimum CPD was determined based on the Akaike's Information Criterion (AIC) for each of the three traits. NLMM analysis was performed using the LME4 package in R [37,38], and the statistical significance of the random effect parameters (i.e., among-clone difference) were tested using the analysis of deviance.

3. Results

3.1. Ig and Actual Stomatal Conductance

Figure 1 shows the relationship between the actual stomatal conductance and the Ig. Both traits were measured on 27 July, 2 August, and 1 October, and the relationships between the two traits were determined on each measurement day. The relationships between the actual stomatal conductance and the Ig on all measured days were significantly and positively correlated (r = 0.83, 0.66, and 0.78 and p < 0.01, 0.001, and 0.05, respectively).

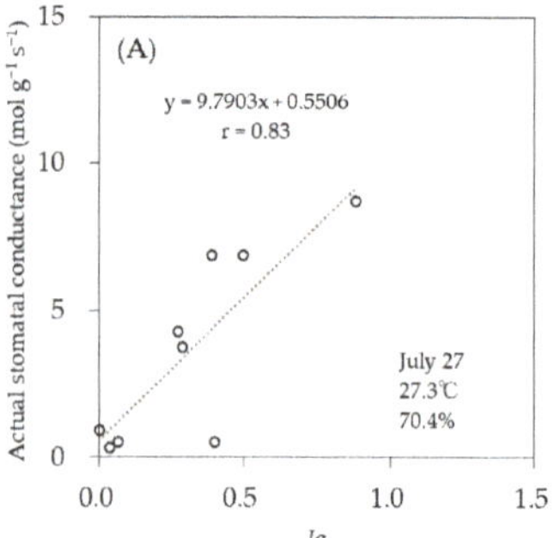

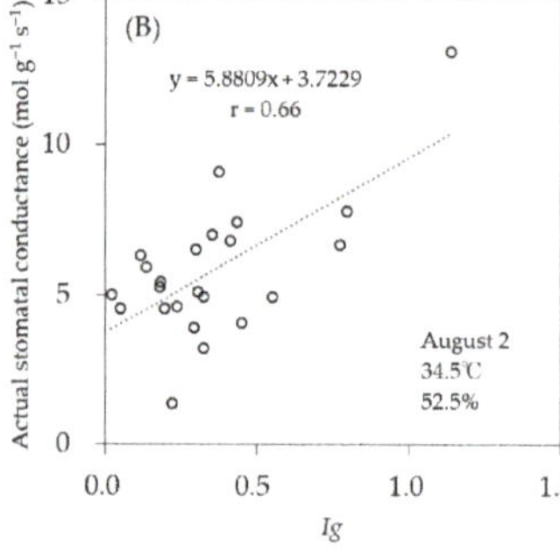

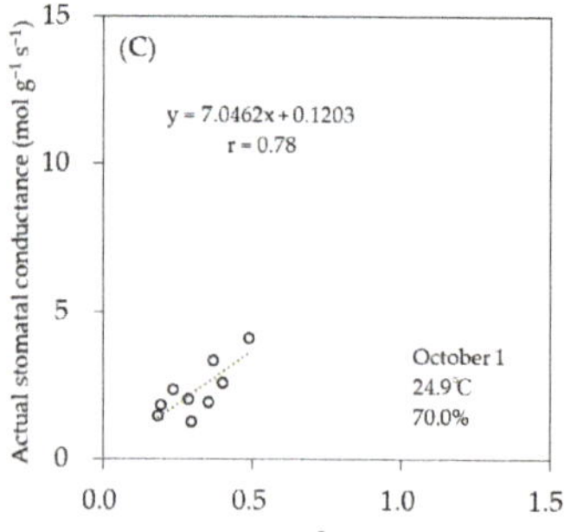

Figure 1. Relationship between the stomatal conductance index (Ig) estimated by the leaf temperature and actual stomatal conductance measured by gas exchange. The measurements in (**A–C**) were measured on different days: 27 July, 2 August, and 1 October, respectively.

3.2. Stomatal Conductance Response

Figure 2 shows the weekly changes in Ig from 9 May to 1 August. The Ig values for GO1 and KA7 in the drought treatment were significantly lower than those of the controls at 21 days post-treatment (dpt), while, in both TE11 and TS1, the first significant differences between the control and drought treatments were observed at 14 dpt. As the experiment progressed, Ig approached zero at 28 dpt in KA7 and TS1 and at 56 dpt in GO1 and TE11. As shown in Figure 3, the RIg tended to decrease with the SWC under the conditions of drought treatment. In *C. japonica*, the RIg tended to decrease towards zero when the SWC decreased from about 0.25 cm^3 cm^{-3} to 0.15 cm^3 cm^{-3}.

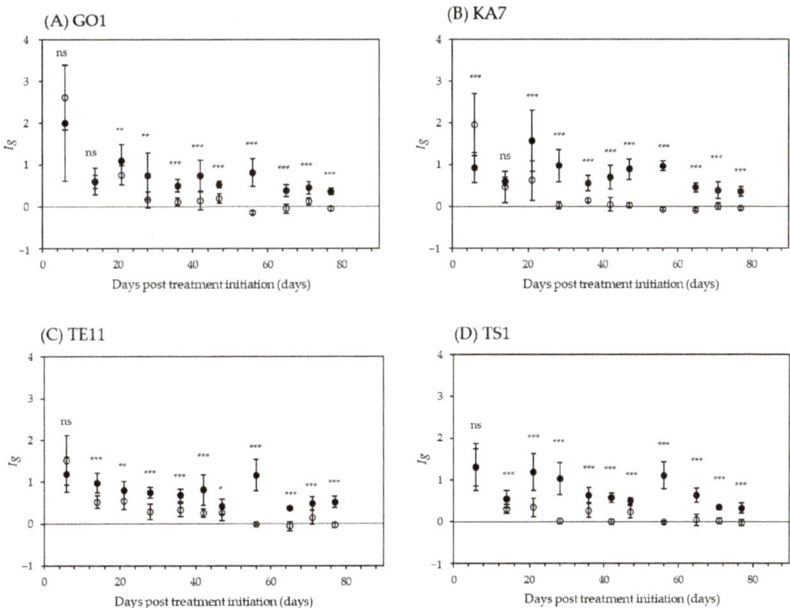

Figure 2. Changes in the stomatal conductance index (Ig) in the four clones. (**A–D**) show the results obtained for GO1, KA7, TE11, and TS1, respectively. Open and closed circles denote the drought and control treatments. Statistically significant differences in the Ig between treatments at each day post-treatment (dpt) for each clone were evaluated by Student's *t*-test (ns: not significant, *: $p < 0.1$, **: $p < 0.05$, and ***: $p < 0.01$).

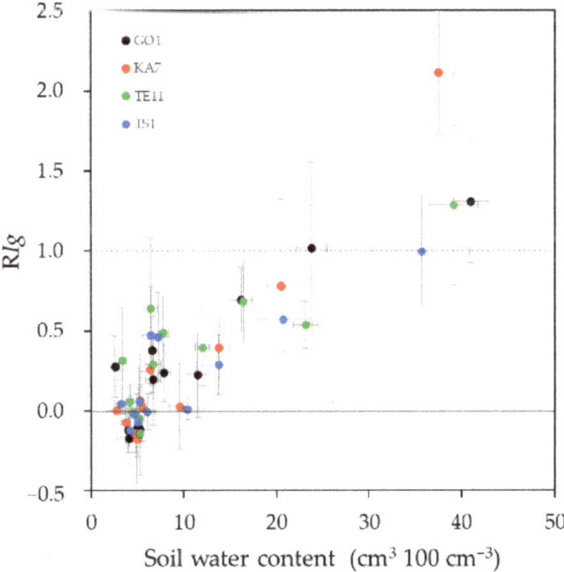

Figure 3. Relationship between the stomatal conductance index ratio (RIg) and the soil water content under the conditions of drought treatment. Black, red, green, and blue correspond to the clones GO1, KA7, TE11, and TS1, respectively.

3.3. Growth Rate

Figure 4 shows the change in the temporal growth rates of four clones during the experiment. In the control treatment, the growth in all four clones peaked at around 21 dpt (30 May) before decreasing. The growth in KA7 and TE11 ceased at 77 dpt (25 July), whereas the growth in GO1 and TS1 was maintained at 0.52 and 2.14 cm/week, respectively. As in the control treatment, the growth peaked at around 21 dpt in the drought treatment. The growth rate of TS1 in the drought treatment decreased significantly after 21 dpt compared to the control, and the reduction became evident earlier than in the other clones. GO1, KA7, and TE11 exhibited significantly reduced growth after 28 dpt, 28 dpt, and 35 dpt, respectively. In the drought treatment, KA7 and TS1 ceased growing at 49 dpt, GO1 at 56 dpt, and TE11 at 63 dpt. As shown in Figure 5, the GRRs tended to decrease with the SWC under the conditions of drought treatment. In *C. japonica*, the GRR tended to decrease towards zero when the SWC decreased from about $0.20 \text{ cm}^3 \text{ cm}^{-3}$ to $0.10 \text{ cm}^3 \text{ cm}^{-3}$.

3.4. Fv/Fm

The *Fv/Fm* exhibited similar values until the middle of the experiment, whereafter an abrupt decrease in *Fv/Fm* became evident in TS1 and KA7 at 77 dpt and in GO1 and TE11 after 84 dpt (Figure 6). As the drought conditions progressed, at 84 dpt, the *Fv/Fm* value of some Ts1 and KA7 cuttings decreased to zero. Figure 7 shows the relationship between the soil water content and R*Fv/Fm*. The R*Fv/Fm* started to decrease sharply at around $0.05 \text{ cm}^3 \text{ cm}^{-3}$ of the SWC.

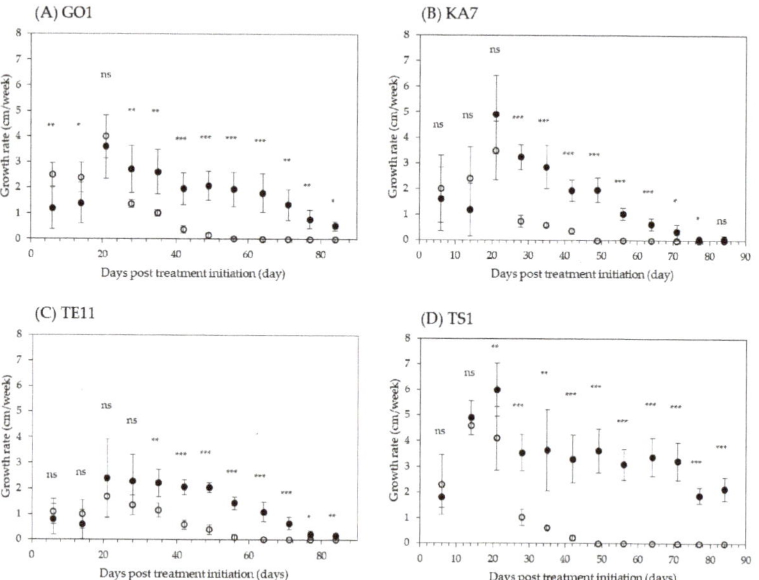

Figure 4. Temporal changes in the growth rates in the four clones. (**A–D**) show the results of GO1, KA7, TE11, and TS1, respectively. Open and closed circles denote the drought and control treatments. Statistical significance of the growth rates between treatments at each dpt for each clone was evaluated by Student's *t*-test (ns: not significant, *: $p < 0.1$, **: $p < 0.05$, and ***: $p < 0.01$).

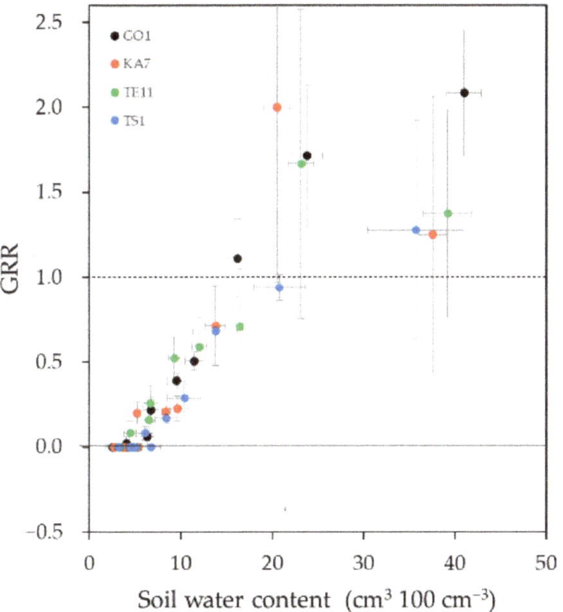

Figure 5. Relationship between the soil water content and growth rate ratio (GRR). Black, red, green, and blue symbols correspond to the clones GO1, KA7, TE11, and TS1, respectively.

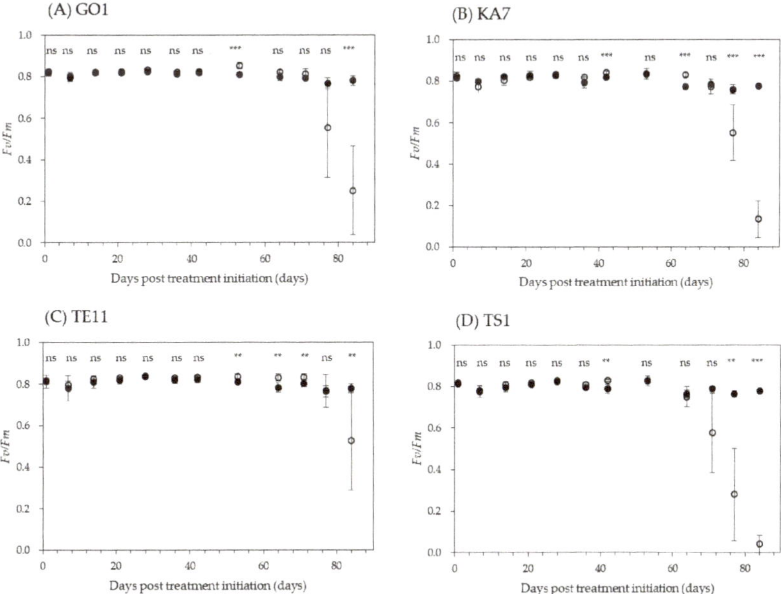

Figure 6. Temporal changes in the maximum photochemical quantum yield of photosystem II (Fv/Fm) in the four clones. (**A–D**) show the results obtained for GO1, KA7, TE11, and TS1, respectively. Open and closed circles denote the drought and control treatments. Statistical significance of the Fv/Fm between treatments at each dpt for each clone was evaluated by Student's t-test (ns: not significant, **: $p < 0.05$, and ***: $p < 0.01$).

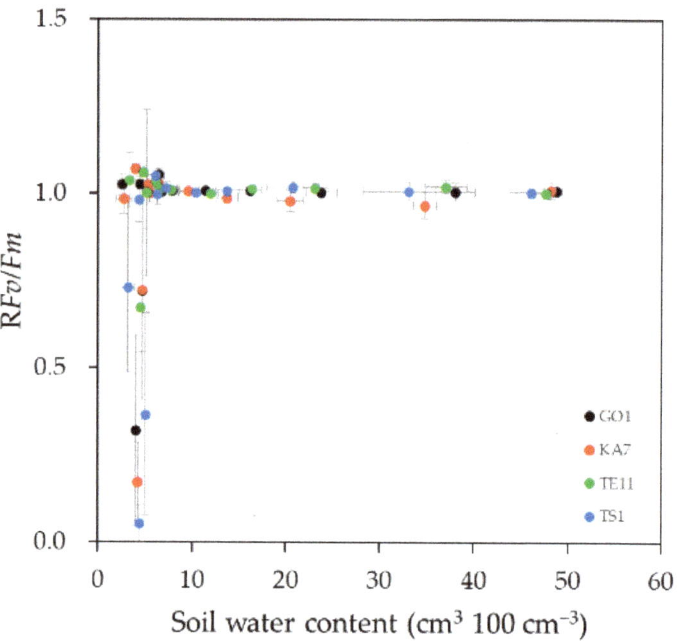

Figure 7. Relationship between the soil water content and RFv/Fm. The black, red, green, and blue symbols correspond to the clones GO1, KA7, TE11, and TS1, respectively.

3.5. Optimization of CPD for CWI Using an NLMM

According to our NLMM analyses incorporating random effects for the responses of different clones in the model, the optimum CPDs were estimated on the basis of the lowest AIC values for each trait (Table 2). The models were applied to the pooled data of the four clones. The patterns of the changes in the AICs differed among the three traits. The AIC of the RIg reached a minimum value at 2 CPD, before increasing again thereafter (Table 2). The AIC of the GRR decreased until 21 CPD before increasing thereafter (Table 2). The AIC of the RFv/Fm values decreased gradually and then reached a minimum value at 70 CPD (Table 2). These findings imply that the response in the RFv/Fm was best explained by longer cumulative soil water conditions (i.e., the previous 70 days), whereas the RIg was best explained by shorter cumulative soil water conditions (i.e., the previous two days). We estimated the responses of the three traits in the four clones using an NLMM and the CWI at the optimized CPD. Regarding the RIg, KA7 responded the fastest, with an increase in the CWI, followed by TS1, GO1, and TE11 in order (Figure 8A). Thus, KA7 exhibited the most sensitive reduction in stomatal conductance under conditions of drought stress. Conversely, TE11 responded the slowest to drought stress. In terms of the GRR, the clonal response was thus in the order of TS1, KA7, GO1, and TE11 (Figure 8B), and in terms of the RFv/Fm, it was TS1, GO1, KA7, and TE11 (Figure 8C). The random effect parameters α and β representing the differences among the clones in responsivity to drought stress were significant, except for the β of RIg (Table 3).

Table 2. Parameter estimates and Akaike's information criteria (AIC) used in the models. For the maximum photochemical quantum yield of photosystem II (RFv/Fm), model algorism did not converge for $p = 1, 2, 3, 4$, and 5. RIg: stomatal conductance index ratio and GRR: growth rate ratio. CWI: Cumulative Water Index.

Traits	p	Model Expression	Fixed Effects		SD of Random Effects			AIC
			α	β	α	β	Residual	
RIg	1	$\mathrm{CWI}_{dp} = \sum_{k=d+1-p}^{d} \mathrm{SWC}_k$ $y_{ij} = \exp(-\alpha \times \exp(-\beta \times \mathrm{CWI}_{ij})) + e_{ij}$	3.839	4.771	0.804	0.004	0.276	74.346
	2		3.758	4.719	0.778	0.021	0.273	70.747
	3		3.546	4.439	0.761	0.000	0.273	70.997
	4		3.396	4.326	0.745	0.000	0.274	71.748
	5		3.306	4.198	0.721	0.010	0.274	72.307
	6		3.263	4.082	0.716	0.006	0.275	72.954
	7		3.424	3.896	0.728	0.001	0.275	73.606
	10		3.285	3.558	0.736	0.016	0.276	75.027
	14		3.300	3.180	0.785	0.002	0.276	75.658
	21		3.464	2.758	0.914	0.002	0.277	77.202
	28		3.725	2.536	1.072	0.029	0.280	82.932
	35		4.084	2.462	1.163	0.000	0.280	84.219
	42		4.536	2.476	1.305	0.063	0.280	83.060
	49		5.533	2.619	1.537	0.003	0.279	81.112
	56		7.396	2.898	2.031	0.141	0.278	80.770
	63		8.739	3.067	0.290	0.393	0.279	81.607
	70		11.031	3.302	0.286	0.388	0.279	81.484
	77		11.996	3.386	0.284	0.386	0.279	81.442
	84		12.319	3.413	0.284	0.386	0.279	81.429
GRR	1	$\mathrm{CWI}_{dp} = \sum_{k=d+1-p}^{d} \mathrm{SWC}_k$ $y_{ij} = \exp(-\alpha \times \exp(-\beta \times \mathrm{CWI}_{ij})) + e_{ij}$	7.578	9.995	1.177	0.000	0.155	−181.317
	2		7.276	9.731	1.242	0.001	0.153	−186.424
	3		6.974	9.453	1.257	0.004	0.151	−189.880
	4		6.759	9.324	1.252	0.016	0.151	−190.601
	5		6.538	9.290	1.156	0.000	0.151	−191.492
	6		6.051	9.031	1.029	0.000	0.151	−191.246
	7		5.566	8.404	0.976	0.006	0.151	−192.244
	10		5.316	7.655	0.911	0.119	0.149	−197.066
	14		4.667	6.284	0.750	0.379	0.148	−200.229
	21		4.412	4.738	0.625	0.549	0.146	−203.677
	28		4.626	3.954	0.003	0.556	0.149	−193.514
	35		6.224	3.908	0.003	0.471	0.151	−187.846
	42		11.424	4.452	0.118	0.427	0.153	−184.530
	49		23.552	5.181	0.140	0.408	0.153	−184.231
	56		46.862	5.873	0.144	0.392	0.152	−184.386
	63		92.042	6.550	0.145	0.381	0.152	−184.564
	70		167.363	7.150	0.157	0.373	0.152	−184.683
	77		255.696	7.575	0.156	0.368	0.152	−184.752
	84		323.788	7.811	0.156	0.366	0.152	−184.786
RFv/Fm	1	$\mathrm{CWI}_{dp} = \sum_{k=d+1-p}^{d} \mathrm{SWC}_k$ $y_{ij} = 1 - \exp(-\beta \times (\mathrm{CWI}_{ij} - \alpha)) + e_{ij}$						-
	2							-
	3							-
	4							-
	5							-
	6		−0.040	22.590	0.035	0.215	0.213	−42.095
	7		−0.028	27.666	0.025	0.209	0.209	−50.825
	10		−0.024	32.737	0.026	0.201	0.203	−63.955
	14		0.000	57.847	0.012	0.172	0.172	−141.472
	21		0.003	69.171	0.009	0.138	0.138	−245.302
	28		0.002	63.303	0.009	0.135	0.135	−255.212
	35		0.002	53.494	0.010	0.129	0.129	−278.650
	42		0.001	43.527	0.011	0.124	0.124	−296.940
	49		−0.001	32.963	0.014	0.119	0.119	−314.384

Table 2. *Cont.*

Traits	p	Model Expression	Fixed Effects		SD of Random Effects			AIC
			α	β	α	β	Residual	
	56		−0.002	25.681	0.016	0.117	0.117	−326.460
	63		−0.002	18.286	0.021	0.111	0.114	−339.369
RFv/Fm	70		−0.004	12.711	0.032	0.109	0.111	−348.159
	77		−0.002	9.774	0.039	0.116	0.115	−332.181
	84		0.002	8.631	0.055	0.127	0.120	−310.388

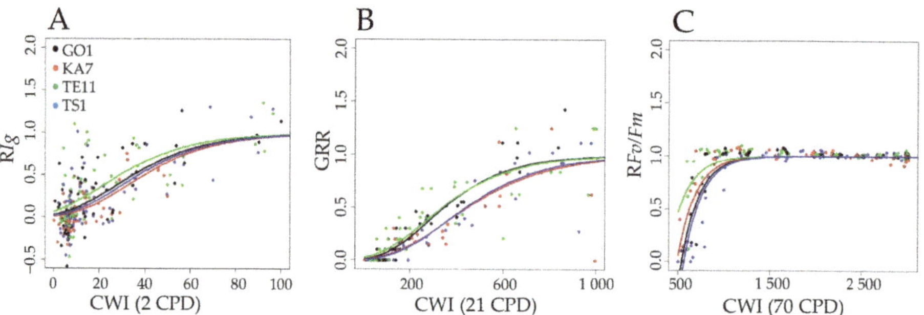

Figure 8. Responses of traits to the Cumulative Water Index (CWI) at the optimum cumulative number of past days (CPD). (**A**) Response to the CWI at 2 CPD in the stomatal conductance index ratio (R*Ig*) model. (**B**) Response to the CWI at 21 CPD in the growth rate ratio (GRR) model. (**C**) Response to the CWI at 70 CPD in the maximum photochemical quantum yield of photosystem (PS) II (*RFv/Fm*) model. Data points are the measured values, and the curves are the predicted clonal response estimated by the nonlinear mixed-effect model (NLMM) fitted to the Gompertz function (**A**,**B**) or the von Bertalanffy function (**C**). The black, red, green, and blue symbols correspond to the clones GO1, KA7, TE11, and TS1, respectively.

Table 3. Parameter estimates and variance components (VC) of the clones. The percentages of VCs are shown in parentheses, and the statistical significances of the random effects are shown as asterisks. CPD: cumulative number of past days.

Traits	CPD	Estimates of Random Effects in α				Estimates of Random Effects in β				Variance of Random Effects		
		GO1	KA7	TE11	TS1	GO1	KA7	TE11	TS1	α	β	Residual
RIg	2	−0.1624	0.7303	−1.0108	0.2428	0.0006	−0.0004	0.0001	−0.0004	0.6047 (88.9%) *	0.0005 (0.1%)	0.0748
GRR	21	−0.0311	0.0976	−0.7285	0.5117	0.5522	−0.5550	0.1700	−0.2895	0.3906 (2.1%) **	0.3012 (3.0%) **	0.0212
RFv/Fm	70	0.0196	−0.0008	−0.0455	0.0292	0.0198	0.0197	0.0136	−0.0532	0.0010 (4.1%) **	0.0119 (46.9%) **	0.0124

4. Discussion

4.1. Application of Ig Measured by Infrared Thermography in C. japonica

In this study, the *Ig* was obtained by preparing dry and wet reference leaves according to the method of Leinonen and Jones [39]. For crops and broad-leaved trees, which often have flat leaves, dry reference leaves were prepared by coating them with petroleum jelly [17,20,39]. In the case of *C. japonica*, which has needles with a complex steric structure, coating the leaves with a thin layer of petroleum jelly is difficult because of the high consistency of the jelly. To overcome this problem, we added paraffin, an involatile oil, to the petroleum jelly to adjust the consistency, and this enabled us to coat thoroughly the surface of the *C. japonica* needles. Yu et al. [20] measured the *Ig* of leaves from *Firmiana platanifolia* trees under drought conditions using dry reference leaves prepared using only petroleum jelly and found a positive correlation $r = 0.85$ between the *Ig* values and actual stomatal conductance. Similarly, Leinonen and Jones [39] observed a positive correlation $r = 0.44$ between the *Ig* values and actual stomatal conductance in *Vicia faba*. The findings of this study revealed comparable positive correlations $r = 0.66$ to 0.83 between the *Ig* values and stomatal conductance, which varied depending on the environment on the measured day. In this study, R*Ig*, which is the ratio of the *Ig* value in the drought treatment to the value in the control, was adopted as the trait to evaluate the stomatal drought responses. Therefore, the stomatal responsivity to drought is successfully evaluated, even if the intercept and slope in the relationship between the *Ig* value and actual stomatal conductance varied among the measured days. Thus, provided that dry reference leaves are prepared with care, high-throughput measurements using infrared thermography are well-suited for estimating stomatal conductance in *C. japonica* with its complex steric needles.

4.2. Estimation of Phenotypic Trait Responses to CWI by NLMM

To promote the breeding programs aiming at improving drought resistance, it is essential to evaluate appropriately the clonal values of traits tightly related to the physiological responses against drought stress. Drought stress has been shown to have a cumulative effect on trees [31,32], and the drought intensity in the soil, which is affected by the rate of water consumption, varies among samples depending on the size and physiological condition of the seedlings. Therefore, because seedlings' water consumption rates vary from pot to pot, it is not appropriate to use the trait values measured at a particular time point as if they would be under the same intensity of stress. Previous studies (e.g., Nanayakkara et al. [40] and Bigras [41]) used a controlled SWC or within-tissue water potential as an indicator of the intensity of the drought stress and then evaluated the performance of the stress resistance at the individual or race levels. In this study, however, as we could not perfectly control the water consumption rates of all growing seedlings, we overcame this difficulty by flexibly modeling the responses of clones to the CWI, which was introduced as an indicator of cumulative drought stress. The clonal responses to the CWI were fitted by using the Gompertz or von Bertalanffy functions; then, the optimal CPDs were searched for each trait by the AIC score basal model selection. When achieving the lowest AIC score, the good estimates of the model parameters α and β were obtained, and the CWI at the optimal CPD adequately described the trait response to drought stress. Our modeling approach successfully provided an indicator reflecting both the stress duration and intensity by estimating the optimum CPD.

Functionally, parameters α and β reflect the x-intercept and slope (attenuation rate) of the curves, respectively. When the β values are smaller, biologically, the trait value (such as GRR) more gradually approaches zero with drought stress. On the other hand, when the α values are larger, the trait value rapidly drops to zero, even under weak drought stress conditions. We estimated the α and β values as random effect variables of the clones in the NLMM, and this modeling approach allows us to capture the trait-response curves of each clones to drought stress.

4.3. Relationship between CPDs and Responsivity of Phenotypic Traits

In this study, as the soil drying progressed, the phenotypic traits (R*Ig*, GRR, and R*Fv/Fm*) also declined. However, the responses to drought stress varied among the traits in the temporal order of R*Ig*, GRR, and R*Fv/Fm*, and the corresponding optimum CPDs were estimated as two, 21, and 70, respectively. Drought stress typically reduced the leaf water potential by inducing a decrease in stomatal opening via an abscisic acid (ABA)-mediated signal transduction [42,43]. Stomatal closure decreases transpiration and CO_2 influx and leads to a decrease in photosynthesis, limits CO_2 fixation, and inhibits plant growth [44]. Stomatal closure also increases the number of electrons that are not used for photosynthesis, and a surplus of electrons generates reactive oxygen species (ROS), which damage the reaction center of photosystem (PS) II [43,45,46]. The *Fv/Fm* (i.e., the maximum efficiency at which the light absorbed by PS II is used to reduce the primary quinone electron acceptor) decreases when PS II is damaged. However, a recent study of *Acer* species [47] discussed that only using the *Fv/Fm* as an indicator of drought stress detection is not suitable. Our results show that significant differences among the clones in R*Fv/Fm* reduction only appeared at the end of the drought treatment, suggesting that this situation (when the CWI at 70 CPD was the optimum) might be too severe under natural conditions for *C. japonica*. The response order observed in this study was concordant with the general response of tree species under drought conditions, i.e., the reduction of *Ig* was very sensitive to drought and, the reduction of the growth rate was moderate, while the *Fv/Fm* was one of relatively insensitive traits [43,45]. In the present study, by introducing the CWI, the parameter of cumulative stress intensity defined by the CPD, it is suggested that the sensitivity of the measured traits to drought were adequately reflected by our modeling approach using the NLMM.

4.4. Differences in Clonal Responsivity to Drought Stress

In this study, our modeling approach allowed us to capture the trait response curves of each clones to drought stress. TE11 seemed to be most insensitive clone among the four tested clones in the responses of the three traits measured in this study (see green lines in Figure 8). Strategies of the drought resistance include drought avoidance and drought tolerance [43,46,48,49]. Drought avoidance is based on the ability to maintain the tissue water potential through stomatal closure, root elongation, and high water use efficiency [43,46,48,49]. On the other hand, drought tolerance is an ability to endure low tissue water potential by maintaining enzyme activities and osmotic adjustments [43,46,48,49]. Among the four tested clones, TE11 was the most insensitive clone in *RIg* and, also, the most sustainable clone in GRR against drought stress, suggesting that TE11 might be a superior clone in terms of drought tolerance. There is a possibility of clonal variation in responsivity to drought stress, and therefore, a drought-resistant strategy may differ from clone to clone in *C. japonica*. In order to get deeper insights, it is necessary to conduct a larger scale drought stress experiment using a larger number of clones in the future.

5. Conclusions

As an adaptation to global-scale climate change, tree breeding for drought tolerance is necessary. However, because a traditional approach such as using a gas exchange analyzer is difficult to evaluate the stomatal response of abundant clones to drought stress, mass sample evaluation and the procedures used to accomplish such evaluations are therefore needed in order to increase the efficiency of breeding programs. In this study, we (1) established an evaluation method for estimating the stomatal response against drought stress by measuring leaf temperatures using infrared thermography, (2) evaluated the clonal growth responses and (3) *Fv/Fm* under conditions of drought stress, and (4) modeled the clonal responses to cumulative drought stress by introducing the CWI. Compared to the traditional approach to evaluate the stomatal response to drought stress using a gas exchange analyzer, the method using an infrared thermography is faster. These methods and findings enabled us to evaluate the clonal stress responses in *C. japonica* to drought. As a next step, these methods should be applied to large-scale clone materials to promote breeding programs. To accelerate breeding, it is also important to examine the feasibility of genome-wide association studies and genomic selection for assessing drought tolerance in *C. japonica*.

Supplementary Materials: The following are available online at https://www.mdpi.com/1999-490 7/12/1/55/s1: Figure S1: Changes of the soil water content over the course of the drought experiment. Open and solid circles denote the drought and control treatments, respectively. Black, red, green, and blue correspond to GO1, KA7, TE11, and TS1, respectively. Figure S2: The thermograph taken to obtain the stomatal conductance index (*Ig*). Target: Target individual to measure the *Ig* used in the experiment. Wet reference: The leaves that were sprayed regularly with water to maintain their moisture levels. Dry reference: The leaves that were coated with a mixture of petroleum jelly and liquid paraffin (1:2 by weight). Figure S3: Relationship between the soil water content and soil water potential ($\log \Psi_w$) of the mixed-culture soil used in the experiment.

Author Contributions: Y.T. and Y.H. conceived and designed the experiments; Y.T. and Y.H. performed the experiments; Y.T., Y.H., and M.M. analyzed the data; Y.T. wrote the manuscript; and Y.T., Y.H., M.M. and M.T. revised the manuscript. All authors have read and agreed to the published version of the manuscript.

Funding: The present study is part of the project "Development of adaptation techniques to the climate change in the sectors of agriculture, forestry, and fisheries" supported by the Ministry of Agriculture, Forestry and Fisheries, Japan.

Institutional Review Board Statement: Not applicable.

Informed Consent Statement: Not applicable.

Data Availability Statement: The data presented in this study are available on request from the corresponding author.

Acknowledgments: We are grateful to Hiroshi Hoshi (FTBC, FFPRI) for his well-done coordination of the research project.

Conflicts of Interest: The authors declare that the research was conducted in the absence of any commercial or financial relationships that could be construed as a potential conflict of interests.

References

1. IPCC. *Climate Change 2013: The Physical Science Basis: Contribution of Working Group I to the Fifth Assessment Report of the Intergovernmental Panel on Climate Change*; Stocker, T.F., Qin, D., Plattner, G.-K., Tignor, M., Allen, S.K., Boschung, J., Nauels, A., Xia, Y., Bex, V., Midgley, P.M., Eds.; Cambridge University Press: Cambridge, UK; New York, NY, USA, 2013.
2. Japan Meteorological Agency. *Climate Change Monitoring Report 2017*; Japan Meteorological Agency: Tokyo, Japan, 2018.
3. Van Mantgem, P.J.; Stephenson, N.L.; Byrne, J.C.; Daniels, L.D.; Franklin, J.F.; Fulé, P.Z.; Harmon, M.E.; Larson, A.J.; Smith, J.M.; Taylor, A.H.; et al. Widespread increase of tree mortality rates in the western united states. *Science* **2009**, *323*, 521–524. [CrossRef] [PubMed]
4. Allen, C.D.; Macalady, A.K.; Chenchouni, H.; Bachelet, D.; McDowell, N.; Vennetier, M.; Kitzberger, T.; Rigling, A.; Breshears, D.D.; Hogg, E.H.T.; et al. A global overview of drought and heat-induced tree mortality reveals emerging climate change risks for forests. *For. Ecol. Manag.* **2010**, *259*, 660–684. [CrossRef]
5. Hamanishi, E.T.; Campbell, M.M. Genome-wide responses to drought in forest trees. *Forestry* **2011**, *84*, 273–283. [CrossRef]
6. Cortés, A.J.; Restrepo-Montoya, M.; Bedoya-Canas, L.E. Modern strategies to assess and breed forest tree adaptation to changing climate. *Front. Plant Sci.* **2020**, *11*, 583323. [CrossRef]
7. Alcaide, F.; Solla, A.; Mattioni, C.; Castellana, S.; Martín, M.Á. Adaptive diversity and drought tolerance in *Castanea sativa* assessed through EST-SSR genic markers. *Forestry* **2019**, *92*, 287–296. [CrossRef]
8. Kumar, A.; Dixit, S.; Ram, T.; Yadaw, R.B.; Mishra, K.K.; Mandal, N.P. Breeding high-yielding drought-tolerant rice: Genetic variations and conventional and molecular approaches. *J. Exp. Bot.* **2014**, *65*, 6265–6278. [CrossRef]
9. Mwadzingeni, L.; Shimelis, H.; Tesfay, S.; Tsilo, T.J. Screening of bread wheat genotypes for drought tolerance using phenotypic and prolin analysis. *Front. Plant. Sci.* **2016**, *7*, 1276. [CrossRef]
10. Yan, C.; Song, S.; Wang, W.; Wang, C.; Li, H.; Wang, F.; Li, S.; Sun, X. Screening diverse soybean genotypes for drought tolerance by membership function value based on multiple traits and drought-tolerant coefficient of yield. *BMC Plant Biol.* **2020**, *20*, 321. [CrossRef]
11. Gasparm, M.J.; Velasco, T.; Feito, I.; Alía, R.; Majada, J. Genetic variation of drought tolerance in *Pinus pinaster* at three hierarchical levels: A comparison of induced osmotic stress and field testing. *PLoS ONE* **2013**, *8*, e79094.
12. Pita, P.; Cañas, I.; Soria, F.; Ruiz, F.; Toval, G. Use of physiological traits in tree breeding for improved yield in drought-prone environments. The case of *Eucalyptus globulus*. *Investig. Agrar. Sist. Recur. For.* **2005**, *14*, 383–393. [CrossRef]
13. Ludovisi, R.; Tauro, F.; Salvati, R.; Khoury, S.; Mugnozza, G.S.; Harfouche, A. UAV-based thermal imaging for high-throughput field phenotyping of black poplar response to drought. *Front. Plant Sci.* **2017**, *10*, 3389. [CrossRef] [PubMed]
14. Berry, J.A.; Beerling, D.J.; Franks, P.J. Stomata: Key players in the earth system, past and present. *Curr. Opin.Plant Biol.* **2010**, *13*, 233–240. [CrossRef] [PubMed]
15. Nemeskéri, E.; Molnár, K.; Vígh, R.; Nagy, J.; Dobos, A. Relationships between stomatal behavior, spectral traits and water use and productivity of green peas (*Pisum sativum* L.) in dry seasons. *Acta Physiol. Plant.* **2015**, *37*, 1–16. [CrossRef]
16. Pirasteh-Anosheh, H.; Saed-Moucheshi, A.; Pakniyat, H.; Pessarakli, M. Stomatal responses to drought stress. In *Water Stress and Crop Plants: A Sustainable Approach*, 1st ed.; Ahmad, P., Ed.; John Wiley & Sons, Ltd.: New York, NY, USA, 2016; Volume 1, pp. 24–40.
17. Jones, H.G. Use of infrared thermometry for estimation of stomatal conductance as a possible aid to irrigation scheduling. *Agri. For. Meteorol.* **1999**, *95*, 139–149. [CrossRef]
18. Maes, W.H.; Achten, W.M.J.; Reubens, B.; Muys, B. Monitoring stomatal conductance of *Jatropha curcas* seedlings under different levels of water shortage with infrared thermography. *Agri. For. Meteorol.* **2011**, *151*, 554–564. [CrossRef]
19. Seidel, H.; Schunk, C.; Matiu, M.; Menzel, A. Diverging drought resistance of Scots pine provenances revealed by infrared thermography. *Front. Plant Sci.* **2016**, *7*, 1247. [CrossRef] [PubMed]
20. Yu, M.H.; Ding, G.D.; Gao, G.L.; Zhao, Y.Y.; Yan, L.; Sai, K. Using plant temperature to evaluate the response of stomatal conductance to soil moisture deficit. *Forests* **2015**, *6*, 3748–3762. [CrossRef]

21. Grant, O.M.; Tronina, Ł.; Jones, H.G.; Chaves, M.M. Exploring thermal imaging variables for the detection of stress responses in grapevine under different irrigation regimes. *J. Exp. Bot.* **2016**, *58*, 815–825. [CrossRef]

22. Bolhar-Nordenkampf, H.R.; Long, S.P.; Baker, N.R.; Oquist, G.; Schreiber, U.; Lechner, E.G. Chlorophyll fluorescence as a probe of the photosynthetic competence of leaves in the field: A review of current instrumentation. *Funct. Ecol.* **1989**, *3*, 497–514. [CrossRef]

23. Krause, G.H. Chlorophyll fluorescence and photosynthesis: The basics. *Ann. Rev. Plant Physiol. Plant Mol. Biol.* **1991**, *42*, 313–349. [CrossRef]

24. Baker, N.R. Chlorophyll fluorescence: A probe of photosynthesis in vivo. *Ann. Rev. Plant Biol.* **2008**, *59*, 89–113. [CrossRef] [PubMed]

25. Tezara, W.; Mitchell, V.J.; Driscoll, S.D.; Lawlor, D.W. Water stress inhibits plant photosynthesis by decreasing coupling factor and ATP. *Nature* **1999**, *401*, 914–917. [CrossRef]

26. Cornic, G. Drought stress inhibits photosynthesis by decreasing stomatal aperture—Not by affecting ATP synthesis. *Trends Plant Sci.* **2000**, *5*, 187–188. [CrossRef]

27. Flexas, J.; Medrano, H. Drought-inhibition of photosynthesis in C_3 plants: Stomatal and non-stomatal limitations revisited. *Ann. Bot.* **2002**, *89*, 183–189. [CrossRef]

28. Lawlor, D.W.; Cornic, G. Photosynthetic carbon assimilation and associated metabolism in relation to water deficits in higher plants. *Plant Cell Environ.* **2002**, *25*, 275–294. [CrossRef]

29. Camisón, Á.; Martín, M.Á.; Dorado, F.J.; Moreno, G.; Solla, A. Changes in carbohydrates induced by drought and waterlogging in *Castanea sativa. Trees* **2020**, *34*, 579–591. [CrossRef]

30. Woo, N.S.; Dadger, M.R.; Pogson, B.J. A rapid, non-invasive procedure for quantitative assessment of drought survival using chlorophyll fluorescence. *Plant Methods* **2008**, *4*, 27. [CrossRef]

31. Navarro-Cerrillo, R.M.; Rodríguez-Vallejo, C.; Silveiro, E.; Hortal, A.; Palacios-Rodriguez, G.; Duque-Lazo, J.; Camarero, J.J. Cumulative drought stress leads to a loss of growth resilience and explains higher mortality in planted than in naturally regenerated *Pinus pinaster* stands. *Forests* **2018**, *9*, 358. [CrossRef]

32. Zhao, A.; Yu, Q.; Feng, L.; Zhang, A.; Pei, T. Evaluating the cumulative and time-lag effects of drought on grassland vegetation: A case study in the Chinese loess plateau. *J. Environ. Manag.* **2020**, *261*, 110214. [CrossRef]

33. Vicente-Serrano, S.M.; Beguería, S.; López-Moreno, J.I. A Multiscalar drought index sensitive to global warming: The standardized precipitation evapotranspiration index. *J. Clim.* **2010**, *23*, 1696–1718. [CrossRef]

34. Forestry Agency, Ministry of Agriculture, Forestry and Fisheries of Japan. *Statistical Handbook of Forest and Forestry 2019*; Forestry Agency, Ministry of Agriculture, Forestry and Fisheries of Japan: Tokyo, Japan, 2019. (In Japanese)

35. Hayashi, Y. *An Illustrated Book of Useful Trees*; Seibundosinkosha: Tokyo, Japan, 1969; 472p. (In Japanese)

36. Nagakura, J.; Shigenaga, H.; Akama, A.; Takahashi, M. Growth and transpiration of Japanese cedar (*Cryptomeria japonica*) and Hinoki cypress (*Chamaecyparis obtusa*) seedlings in response to soil water content. *Tree Physiol.* **2004**, *24*, 1203–1208. [CrossRef] [PubMed]

37. Bates, D.; Mächler, M.; Bolker, B.M.; Walker, S.C. Fitting Linear Mixed-Effects Models Using lme4. *J. Stat. Softw.* **2015**, *67*, 1–51. [CrossRef]

38. R Core Team. *R: A Language and Environment for Statistical Computing*; R Foundation for Statistical Computing; R Core Team: Vienna, Austria, 2019.

39. Leinonen, I.; Jones, H.G. Combining thermal and visible imagery for estimating canopy temperature and identifying plant stress. *J. Exp. Bot.* **2004**, *55*, 1423–1431. [CrossRef] [PubMed]

40. Nanayakkara, B.; Lagane, F.; Hodgkiss, P.; Dibley, M.; Smaill, S.; Riddell, M.; Harrington, J.; Cown, D. Effects of induced drought and tilting on biomass allocation, wood properties, compression wood formation and chemical composition of young *Pinus radiata* genotypes (clones). *Holzforschung* **2014**, *68*, 455–465. [CrossRef]

41. Bigras, F.J. Photosynthetic response of white spruce families to drought stress. *New For.* **2005**, *29*, 135–148. [CrossRef]

42. Pei, Z.-M.; Murata, Y.; Benning, G.; Thomine, S.; Klüsener, B.; Allen, G.J.; Grill, E.; Schroeder, J.I. Calcium channels activated by hydrogen peroxide mediate abscisic acid signalling in guard cells. *Nature* **2000**, *406*, 731–734. [CrossRef]

43. Aroca, R. *Plant Responses to Drought Stress: From Morphological to Molecular Features*; Springer: Berlin, Germany, 2012.

44. Peñuelas, J.; Filella, I.; Llusià, J.; Siscart, D.; Piñol, J. Comparative field study of spring and summer leaf gas exchange and photobiology of the mediterranean trees *Quercus ilex* and *Phillyrea latifolia. J. Exp. Bot.* **1998**, *49*, 229–238. [CrossRef]

45. Hsiao, T.C. Plant Responses to Water Stress. *Ann Rev. Plant Physiol.* **1973**, *24*, 519–570. [CrossRef]

46. Osakabe, Y.; Osakabe, K.; Shinozaki, K.; Tran, L.-S.P. Response of plants to water stress. *Front Plant Sci.* **2014**, *5*, 1–8. [CrossRef]

47. Banks, J.M. Chlorophyll fluorescence as a tool to identify drought stress in *Acer* genotypes. *Eviron. Exp. Bot.* **2018**, *155*, 118–127. [CrossRef]

48. Touchette, B.W.; Iannacone, L.R.; Turner, G.E.; Frank, A.R. Drought tolerance versus drought avoidance: A comparison of plant-water relations in herbaceous wetland plants subjected to water withdrawal and repletion. *Wetlands* **2007**, *3*, 656–667. [CrossRef]

49. Moran, E.; Lauder, J.; Musser, C.; Stathos, A.; Shu, M. The genetics of drought tolerance in conifer. *New Phytol.* **2017**, *216*, 1038–1048. [CrossRef] [PubMed]

MDPI

Article

Transcriptome Analysis in Male Strobilus Induction by Gibberellin Treatment in *Cryptomeria japonica* D. Don

Manabu Kurita [1,†], **Kentaro Mishima** [2,*,†], **Miyoko Tsubomura** [2], **Yuya Takashima** [2], **Mine Nose** [2], **Tomonori Hirao** [2] and **Makoto Takahashi** [2]

[1] Kyushu Regional Breeding Office, Forest Tree Breeding Center, Forestry and Forest Products Research Institute, Forest Research and Management Organization, 2320-5 Suya, Koshi, Kumamoto 861-1102, Japan; mkuri@affrc.go.jp

[2] Forest Tree Breeding Center, Forestry and Forest Products Research Institute, Forest Research and Management Organization, 3809-1 Ishi, Juo, Hitachi, Ibaraki 319-1301, Japan; mtsubo@affrc.go.jp (M.T.); ytakashima@ffpri.affrc.go.jp (Y.T.); minenose@affrc.go.jp (M.N.); hiratomo@affrc.go.jp (T.H.); makotot@affrc.go.jp (M.T.)

*   Correspondence: mishimak@affrc.go.jp

†   Contributed equally.

Received: 13 May 2020; Accepted: 1 June 2020; Published: 3 June 2020

Abstract: The plant hormone gibberellin (GA) is known to regulate elongating growth, seed germination, and the initiation of flower bud formation, and it has been postulated that GAs originally had functions in reproductive processes. Studies on the mechanism of induction of flowering by GA have been performed in *Arabidopsis* and other model plants. In coniferous trees, reproductive organ induction by GAs is known to occur, but there are few reports on the molecular mechanism in this system. To clarify the gene expression dynamics of the GA induction of the male strobilus in *Cryptomeria japonica*, we performed comprehensive gene expression analysis using a microarray. A GA-treated group and a nontreated group were allowed to set, and individual trees were sampled over a 6-week time course. A total of 881 genes exhibiting changed expression was identified. In the GA-treated group, genes related to 'stress response' and to 'cell wall' were initially enriched, and genes related to 'transcription' and 'transcription factor activity' were enriched at later stages. This analysis also clarified the dynamics of the expression of genes related to GA signaling transduction following GA treatment, permitting us to compare and contrast with the expression dynamics of genes implicated in signal transduction responses to other plant hormones. These results suggested that various plant hormones have complex influences on the male strobilus induction. Additionally, principal component analysis (PCA) using expression patterns of the genes that exhibited sequence similarity with flower bud or floral organ formation-related genes of *Arabidopsis* was performed. PCA suggested that gene expression leading to male strobilus formation in *C. japonica* became conspicuous within one week of GA treatment. Together, these findings help to clarify the evolution of the mechanism of induction of reproductive organs by GA.

Keywords: gibberellin; male strobilus induction; transcriptome; conifer; *Cryptomeria japonica*

---

1. Introduction

Gibberellins (GAs), a class of terpenoid plant hormones, regulate various important plant physiological processes, such as plant elongation, seed germination, and floral initiation [1,2]. The endogenous active GA_4 has been detected in the Lycopsida (*Selaginella moellendorffii*) but not in mosses (*Physcomitrella patens*) [3–5]. It is thought that GA signaling was acquired in the vascular plant lineage after divergence of the bryophytes [5,6]. In ferns, GAs are involved in microspore formation and sex determination, and GAs are inferred to have originally functioned in reproductive processes [7,8].

Studies of the model plant *Arabidopsis thaliana* identified six flowering pathways, known as age, autonomous, vernalization, photoperiod, temperature, and GA [9–16]. GA-mediated floral transitions in angiosperms have been the subject of multiple studies performed in model plants, including *A. thaliana* [9,15,17,18].

Previous studies using *A. thaliana* reported several key observations. First, GAs are necessary for flowering under short day conditions [15,19,20]. Second, GAs promote the expression of *LEAFY* (*LFY*), a well-known floral meristem identity gene, via *cis*-elements (located within the *LFY* promoter) that can be bound by the GAMYB protein [18,21,22]. Third, *LFY* regulates GA levels through the activation of the GA catabolism gene and functions coordinately with *DELLA* (negative gibberellin-response regulator) and *SQUAMOSA PROMOTER BINDING PROTEIN-LIKE 9* (*SPL9*), thereby activating the *APETALA 1* (*AP1*) gene and inducing flowering [23].

In coniferous species, the dynamics of GAs and GA-related genes associated with the development of reproductive organs has been described [24–26]. Although physiological analyses have been performed to examine the effects of GA treatment [27], the mechanisms underlying the regulation of reproductive organ induction or differentiation by GAs remain unknown [28].

In many conifers, reproduction begins 5 to 10 years after planting [29]. However, in Japanese cedar (*Cryptomeria japonica* D. Don), the GA_3 treatment onto seedlings, even 1-year-old seedlings, facilitates male strobilus induction [30], indicating that the species possesses high reactivity to GAs. Therefore, *C. japonica* could be a useful model coniferous tree species for understanding the mechanism underlying the flowering induction by GAs.

The effects of GA_3 treatment on male strobilus induction in *C. japonica* have been investigated, particularly via phenotypic and physiological analyses [31]. The influence of the concentration and seasonal timing of GA_3 treatment on the induction of reproductive organs and the associated changes in carbohydrate and nitrogen content of the shoot have been studied [31]. It was shown that the male strobili were strongly induced by GA treatment in July, and the C–N ratio was significantly increased by GA treatment [31].

To clarify the molecular mechanism of male strobilus induction by GA_3 treatment in *C. japonica*, we conducted comprehensive gene expression analysis using the microarray method. GA_3-treated and non-treated individuals (as controls) were prepared and their current year shoots were sampled along a time course. The RNA from the shoot samples were subjected to the microarray analysis to obtain gene expression data. The extensive expression data obtained from the treated and non-treated samples at each time point were compared to determine when and which classes of gene transcripts were enriched following GA_3 treatment. Furthermore, we examined changes in the expression patterns of genes associated with the GA signaling pathway, other plant hormone signaling pathways, or flowering in other species, aiming at providing insights into GA functional mechanisms in male strobilus induction in *C. japonica*.

2. Materials and Methods

2.1. Plant Material and GA Treatment

Six individuals of three different plus-trees (plus-tree codes 1725, 840, and 1503) that had been planted in 1995 in Hitachi, Ibaraki, Japan (36°69 N, 140°69 E; elevation 52 m), were used for the gene expression analysis (Figure S1). One individual from each clone was designated as a GA-treated individual (1725_GA, 840_GA and 1503_GA) and subjected to the GA_3 treatment. The others were designated as non-treated individuals (1725_CT, 840_ CT and 1503_ CT). The GA_3 spraying was conducted at approximately 10:00 h on July 14, 2015. The branches of GA-treated individuals were sprayed with 100 ppm GA_3 (Kyowa-Hakko, Japan) solution. On July 13, 2015, approximately 3-cm-long shoots tips were sampled from the six individuals. This sampling period was designated as

−1 d (pre-dose). In the post-dose sampling periods, samples were collected after 3 h (14 July 2015), 1 day (1 d; 15 July 2015), 3 days (3 d; 17 July 2015), 1 week (1 w; 21 July 2015), 2 weeks (2 w; 27 July 2015), 4 weeks (4 w; August 11, 2015), and 6 weeks (6 w; 24 August 2015). Moreover, at each time point, samples were collected from the non-treated individuals (designated as CT; for example 1725_CT_3 h is the sample collected from the non-treated 1725 clone 3 h after the treatment period; Figure S1). All the samples, including the pre-dose ones, were collected at approximately 13:00 h. All 48 samples were stored at −80 °C until RNA extraction and analysis.

2.2. Extraction of Total RNA

Total RNA was extracted from each of the 48 samples using Plant RNeasy Mini Kits (QIAGEN, Hilden, Germany) according to the manufacturer's instructions, including on-column (i.e., prior to elution) DNase treatment with an RNase-Free DNase set (QIAGEN) according to the manufacturer's instructions. The quality of the total RNA in the samples was assessed using an Agilent 2100 Bioanalyzer and RNA 6000 Nano kit (Agilent Technologies, Mississauga, ON, Canada). The all samples exhibited an RNA integrity number (RIN) exceeding 7.7.

2.3. Microarray

The custom microarray was designed based on isotigs from next-generation sequencing (NGS) data as described in previous reports [32–34]. A set of 19,360 probes was selected and accommodated in the Agilent 8 × 60 K format (Agilent). In this format, 19,360 probes were accommodated in at least triplicate in our custom array, as described in a previous report (GEO accession: GPL21366, [35]). Gene annotations represent the top-scoring BLASTX hits using each sequence's predicted protein product as a query against The Arabidopsis Information Resource (TAIR; http://www.arabidopsis.org) Arabidopsis protein database TAIR10-pep-20101214 using the CLC Genomic Workbench version 4.1.1 software package (CLC bio, Aarhus, Denmark) as described in a previous report (Fukuda et al., 2018). Gene expression data were acquired using microarray analysis (Agilent Technologies). Total RNA (200 ng) from all 48 samples were amplified and labeled using a Low Input Quick-Amp Labeling Kit (Agilent Technologies). Hybridization and washing were performed according to the manufacturer's recommendations. Labeled and hybridized slides were scanned using a SureScan Microarray Scanner G4900DA (Agilent Technologies), and the dataset was trimmed using Agilent Feature Extraction Software 11.5.1.1 (Agilent Technologies). The data presented in this study have been deposited in NCBI's Gene Expression Omnibus and are accessible through GEO Series as accession number GSE120227.

2.4. Analyses of Gene Expression Patterns and Identification of Differentially Expressed Genes

Expression analysis of the genes was carried out using the Subio platform (Subio, Inc., Kagoshima, Japan, http://www.subio.jp) [36]. The raw signal data were converted to processed signal data using the following steps: (1) Global normalization was performed at the 75th percentile. (2) Log transformation was performed by converting the data to the $\log_2$ data. The genes used for analysis then were extracted using the following steps: (1) Among 19,360 isotigs, the isotigs that did not yield reliable data (i.e., those for which reliable measured values were not obtained with more than 91.7% of samples (in 44 out of 48 samples)) were excluded with filter glsWellAboveBG = 0 (QC1: 17,162 isotigs). (2) For all 48 samples, we excluded probes whose processed signal was in the range of −1 to 1 (QC2: 11,738 isotigs). Following these filtering steps, an average (mean) value was calculated for each of the three biological replicates (plus-tree codes 1725, 840, and 1503). Differentially expressed genes (DEGs) were detected via a two-step process, as follows: In Step 1, seven comparisons (DEG1 to 7) were performed, and gene groups whose expression levels differed by more than 2-fold and yielded *p*-values <0.05 (by two-tailed non-paired Student's *t*-test) were extracted using the Subio platform basic plug-in (Subio, Inc.). The comparisons (DEG1 to 7) were as follows (respectively): GA-treated (GA)_3 h vs. Control (CT)_3 h, GA_1 d vs. CT _1 d, GA_3 d vs. CT _3 d, GA_1 w vs. CT _1 w, GA_2 w vs. CT _2 w, GA_4 w vs. CT _4 w, and GA_6 w vs. CT _6 w. In Step 2, another seven comparisons were performed against the Step-1 results. These Step-2 comparisons (DEG a to g) were as follows (respectively): GA_−1 d (Sample for subtracting genes specifically expressed in GA-treated individual) vs. DEG1, GA_−1 d vs. DEG2, GA_−1 d vs. DEG3, GA_−1 d vs. DEG4, GA_−1 d vs. DEG5, GA_−1 d vs. DEG6, and GA_−1 d vs. DEG7. A total of 881 DEGs were isolated from these comparisons. Tree clustering analyses of DEGs by Pearson correlation as a similarity measurement were performed using the Subio platform basic plug-in (Subio, Inc.) according to the corresponding instructional videos. Extraction of patterns of genes that had sequence similarity to plant hormone signal transduction genes of the KEGG pathway database (http://www.kegg.jp/kegg/pathway.html) were performed using the pathway edit tool of the Subio platform advanced plug-in (Subio Inc.) according to the corresponding instructional videos. Principal component analysis (PCA) of MADS-box genes in DEGs also was performed using the Subio platform basic plug-in according to the corresponding instructional videos.

2.5. Gene Ontology (GO) Analysis

GO analyses using the TAIR ID of the top hit for each DEG and for selected genes of our microarray (genes with E-values < 1E−5) were performed using the GO annotation search tool of TAIR in 8 April 2018. Enrichment analysis was carried out by comparing the GO analysis result of each cluster with the GO analysis result of all genes on our microarray. Enrichment analysis using the TAIR ID of the top hit for all DEGs with E-values < 1E−5 also was performed using DAVID Bioinformatics Resources 6.8 (https://david.ncifcrf.gov/) [37,38].

2.6. Real-Time PCR

To validate our microarray data, 16 samples were tested using the real-time PCR (RT-PCR) method. For RT-PCR analysis, RNA samples from each time point of GA-treated and nontreated plus-tree code 840 specimens were analyzed. A High-Capacity RNA-to-cDNA Kit (Thermo Fisher Scientific, Waltham, MA, USA) was used, and cDNA synthesis (20 μL) was performed using 450 ng of total RNA according to the kit's instruction manual. Primers, which were designed using Primer3Plus [39], were intended to have melting temperatures (Tm) between 60 °C and 65 °C, and to produce amplicons of 80 to 150 bp. Specific primer pairs are listed below. *AGAMOUS-like 9* (*AGL9, reCj28306:M—:isotig28158*): forward 5′-ATCTTCGTAAAAGGGAGACTTTGCT-3′, reverse 5′-GGGTCTGGAGTCTTGTTGAGTTG-3′); *UNUSUAL FLORAL ORGANS* (*UFO, reCj27181:—:isotig27033*): forward 5′-TGTGCTGTCTGTCGGAGAAC-3′, reverse 5′-CGATGACCTTGTATGTCTTGGTG-3′); *LEAFY 3* (*LFY3, reCj22786:—:isotig22638*): forward 5′-TGGCAAGTTTCTGCTGGATG-3′, reverse 5′-CATTTTCCCCTCGTTCTTTGTAG-3′); *PISTILLATA* (*PI, reCj29951:M—:isotig29803*): forward 5′-AAGAATGCCTCTGGAGGACG-3′, reverse 5′-TTCTTTGCTGCAAGCACAAGAGC-3′); and the endogenous control *Ubiquitin 10* (*UBQ10*): forward 5′-CGTTAAAGCCAAGATCCAGGACAA-3′, reverse 5′-TCCATCCTCAAGCTGTTTCCCA-3′) [34]. For each sample, triplicate RT-PCR assays were performed by using 1 μL of cDNA and Power SYBR Green PCR master mix (Thermo Fisher Scientific) according to the manufacturer's protocol. Amplification was carried out with a StepOnePlus system (Thermo Fisher Scientific). After an initial 10 min activation step at 95 °C, reactions were performed as 40 cycles at 95 °C for 15 s and 60 °C for 1 min, and a single fluorescence reading was obtained after each cycle (immediately following the annealing/elongation step at 60 °C). Preliminary RT-PCR assays were performed to evaluate primer pair efficiency. A melting curve analysis was performed at the end of cycling to ensure that a single product had been amplified. For relative quantification and comparisons, we used the delta-delta-Ct method [40] with the *UBQ10* transcript as the internal normalization control.

3. Results

3.1. DEG Enrichment in Response to GA₃ Treatment

Male strobili of *C. japonica* were induced by GA₃ spraying onto the shoots. We collected 48 samples using three different plus-trees as biological repeats; one individual in each plus-tree was treated with GA or non-treated individual (CT), and samples were collected at each of the eight time points (1 pre-dose, 7 post-dose up to 6 w). Male strobili are formed at the axil of shoots. Male strobili were phenotypically confirmed on the shoots of treated individuals on August 24, 2015 (6 w; Figure 1). The differentially expressed genes (DEGs) were extracted to permit assessment of the transcriptional response to GA₃ treatment. Using the Subio platform tree clustering tool, the resulting 881 DEGs were sorted into several clusters according to their expression profiles (Pearson correlation, Figure 2), and were organized into three primary clusters (Clusters D, U1, and U2) (Figure 2, Table S1). Cluster D comprised of genes that were downregulated in GA-treated samples (379 DEGs) compared with those in CT samples. Conversely, the remaining two clusters U1 and U2 comprised of genes that were upregulated in the GA-treated samples (104 DEGs and 398 DEGs, respectively) compared with those in CT samples. Cluster U1 was characterized by genes that were upregulated at earlier time points following the GA₃ treatment, whereas cluster U2 was characterized by genes that were gradually upregulated as the time course progressed.

Figure 1. Morphology of the shoot top of *Cryptomeria japonica*: (**a**) nontreated control sample at 6 weeks (CT 6 w); (**b**) GA3-treated sample at 6 weeks (GA 6 w). Male strobili (MS: indicated by arrow) are formed at the axil of the shoot.

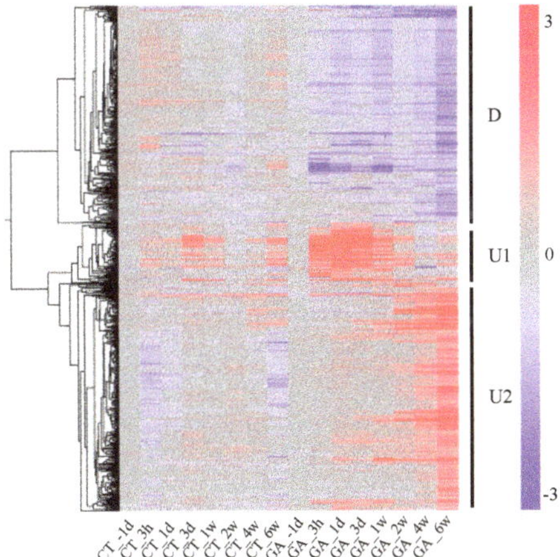

Figure 2. Tree clustering analysis of 881 DEGs. D, U1, and U2 on the right indicate three different clusters of differentially expressed genes (DEGs) used for further analysis. CT_−1 d, CT_3 h, CT_1 d, CT_3 d, CT_1 w, CT_2 w, CT_4 w, CT_6 w, GA_−1 d, GA_3 h, GA_1 d, GA_3 d, GA_1 w, GA_2 w, GA_4 w, and GA_6 w are sample names.

3.2. Functional Analysis of DEGs

To categorize the genes included in each cluster, we selected the annotated DEGs with low E-values (<1E−5) and performed GO enrichment analysis (Figure 3). Our data suggested that 'response to stress', 'response to abiotic or biotic stimulus' (Biological Process), 'cell wall', 'extracellular' (Cellular Component), and 'kinase activity' (Molecular Function) -related genes were enriched in Cluster U1 (90 DEGs); 'transcription, DNA-dependent' (Biological Process), 'cell wall', 'extracellular' (Cellular Component), 'transcription factor activity', and 'nucleic acid binding' (Molecular Function) -related genes were enriched in Cluster U2 (336 DEGs); and 'electron transport or energy pathway' (Biological Process), 'plastid' (Cellular Component), and 'receptor binding or activity' (Molecular Function) -related genes were enriched in Cluster D (324 DEGs). Enrichment analysis, performed using DAVID, yielded similar patterns (Table S2).

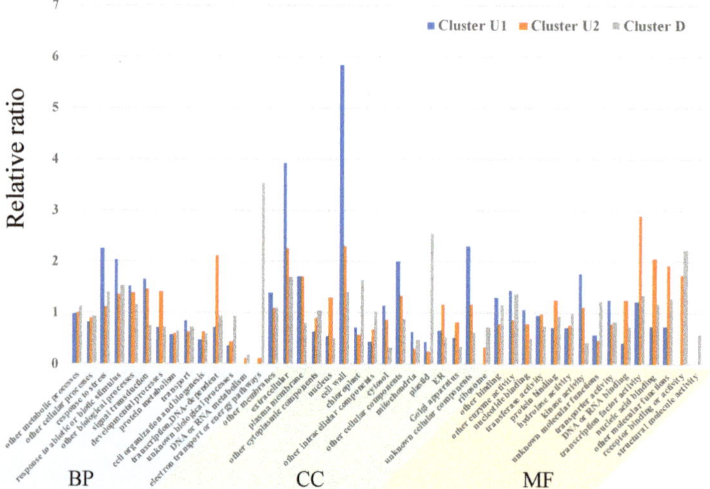

Figure 3. Gene ontology (GO) categories of DEGs that encoded proteins with sequence similarity (E-value <1E−5) to proteins in the TAIR database. The longitudinal axis shows the relative ratio of the genes of Cluster U1 (blue), U2 (orange), and D (gray) against GO analysis results for all genes on the custom microarray (GEO accession: GPL21366). BP: biological process, CC: cellular component, MF: molecular function.

3.3. Expression Patterns of GA Signaling Pathway Genes

To clarify the expression patterns of GA signaling pathway-related genes following the GA$_3$ treatment of *C. japonica*, the expression patterns of genes exhibiting high sequence similarity (E-value <1E−5) with plant hormone signal transduction genes (KEGG Pathway Database) were extracted using the Subio Platform pathway edit tool (Figure 4, Table 1). Among the remaining genes, *reCj11694* exhibiting high sequence similarity to *RGA-LIKE 2* (*RGL2*), which encodes the DELLA protein, was observed in Cluster U2, whereas *reCj34040* and *reCj28549* exhibiting high sequence similarity to *SLEEPY1* (*SLY1*), the rice ortholog of *GID2* [41], were observed in Cluster D. Specifically, the *reCj11694* transcript (encoding a DELLA-like protein) accumulated gradually in the GA-treated samples (Figure 4, Table 1), whereas the levels of the *reCj34040* and *reCj28549* transcripts (encoding SLY1-like proteins) decreased in the GA-treated samples (Figure 4, Table 1). Other potentially relevant genes included *reCj21012* [encoding a protein with sequence similarity to GA INSENSITIVE DWARF 1A (GID1A), a GA receptor], *reCj28386* and *reCj11695* (encoding proteins with sequence similarity to RGL2), *reCj09118* [encoding a protein with sequence similarity to GIBBERELLIC ACID INSENSITIVE (GAI), a DELLA protein], *reCj31917* and *reCj15916* (encoding proteins with sequence similarity to SLY1), and *reCj23186* [encoding a protein with sequence similarity to PHYTOCHROME INTERACTING FACTOR 3 (PIF3)]. These members of the QC1 gene pool showed relatively constant expression with little fluctuation in their expression levels following GA$_3$ treatment (Figure S2).

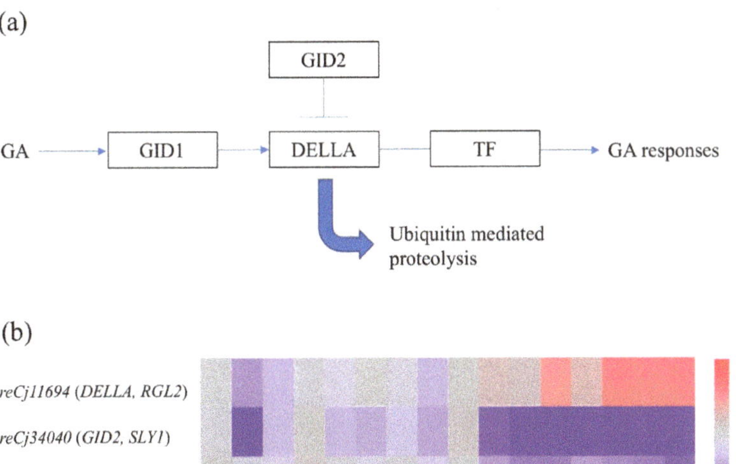

Figure 4. The DEG analysis associated with gibberellin (GA) signal transduction. (**a**) GA signal transduction pathway modified KEGG plant hormone signal transduction reference pathway (https://www.genome.jp/kegg-bin/show_pathway?map04075). GID1: GA INSENSITIVE DWARF 1, DELLA: DELLA protein, GID2: GIBBERELLIN INSENSITIVE DWARF 2, TF: transcription factor; (**b**) The heat-map of DEGs associated with GA signal transduction of each sample. The relative expression values were log$_2$ transformed. Red indicates high expression; blue indicates low expression. CT_−1 d, CT_3 h, CT_1 d, CT_3 d, CT_1 w, CT_2 w, CT_4 w, CT_6 w, GA_−1 d, GA_3 h, GA_1 d, GA_3 d, GA_1 w, GA_2 w, GA_4 w, and GA_6 w are sample names. Details of these genes are provided in Table 1.

Table 1. The DEGs associated with plant hormone signaling pathways after GA$_3$ treatment.

SEQ_ID	Short SEQ_ID	Arabi_ID	Symbol	Description	E-Value
reCj11694:-S–:isotig11549	reCj11694	AT3G03450	RGL2	RGA-like_2	7.86E−56
reCj34040:—-:isotig33892	reCj34040	AT4G24210	SLY1	F-box_family_protein	6.88E−24
reCj28549:-SWR:isotig28401	reCj28549	AT4G24210	SLY1	F-box_family_protein	1.02E−24
reCj31635:—-:isotig31487	reCj31635	AT2G21220	SAUR12	SAUR-like_auxin-responsive_protein_family	7.47E−30
reCj30256:-S–:isotig30108	reCj30256	AT5G20810	SAUR70	SAUR-like_auxin-responsive_protein_family	6.82E−26
reCj27704:-SWR:isotig27556	reCj27704	AT4G14550	IAA14	SLR_indole − 3-acetic_acid_inducible_14	9.46E−76
reCj19503:M–R:isotig19355	reCj19503	AT1G28130	GH3.17	Auxin-responsive_GH3_family_protein	0
reCj19606:M–R:isotig19458	reCj19606	AT5G54510	GH3.6	Auxin-responsive_GH3_family_protein	0
reCj27651:–WR:isotig27503	reCj27651	AT3G57040	ARR9	ATRR4_response_regulator_9	9.29E−39
reCj26644:–W-:isotig26496	reCj26644	AT2G29380	HAI3	highly_ABA-induced_PP2C_gene_3	1.19E−74
reCj25311:–WR:isotig25163	reCj25311	AT2G29380	HAI3	highly_ABA-induced_PP2C_gene_3	2.06E−94
reCj12520:MS-R:isotig12375	reCj12520	AT5G20900	JAZ12	jasmonate-zim-domain_protein_12	2.83E−16
reCj27869:-SWR:isotig27721	reCj27869	AT1G19180	JAZ1	jasmonate-zim-domain_protein_1	1.48E−15
reCj27238:-SWR:isotig27090	reCj27238	AT1G70700	JAZ9	TIFY_domain/Divergent_CCT_motif_family_protein	3.15E−17
reCj23299:MSWR:isotig23151	reCj23299	AT3G50070	CYCD3	CYCLIN_D3;3	2.35E−62
reCj28941:—-:isotig28793	reCj28941	AT5G67260	CYCD3	CYCLIN_D3;2	9.03E−40
reCj32790:—-:isotig32642	reCj32790	AT4G33720	PR1	CAP_(Cysteine-rich_secretory_proteins,_Antigen_5,_and_Pathogenesis-related_1_protein)_superfamily_protein	8.24E−62
reCj20357:–W-:isotig20209	reCj20357	AT2G41370	BOP2	Ankyrin_repeat_family_protein_/_BTB/POZ_domain-containing_protein	1.11E−158
reCj31882:-S–:isotig31734	reCj31882	AT4G33720	PR1	CAP_(Cysteine-rich_secretory_proteins,_Antigen_5,_and_Pathogenesis-related_1_protein)_superfamily_protein	9.89E−63
reCj31173:—R:isotig31025	reCj31173	AT4G33720	PR1	CAP_(Cysteine-rich_secretory_proteins,_Antigen_5,_and_Pathogenesis-related_1_protein)_superfamily_protein	5.87E−62

3.4. Expression Patterns of Genes Encoding Components of Other Plant Hormone Signaling Pathways

The expression patterns of genes encoding components of other plant hormone signaling pathways were also examined. The 17 genes exhibiting high sequence similarity to those listed in the plant hormone signal transduction pathway in the KEGG Pathway Database were extracted from the DEGs identified in the present study (Figure 5, Table 1). Five of the extracted DEGs corresponded to components of the auxin signal transduction pathway. Specifically, the expression of *reCj31635* and *reCj30256*, which encode proteins with sequence similarity to members of the auxin-responsive SAUR protein family, was repressed compared to expression of the respective genes in the non-treated samples. On the other hand, the expression of *reCj27704*, which encodes a protein with sequence similarity to IAA14 (a negative regulator of *Auxin response factor 7*), was upregulated at 4 weeks after GA$_3$ treatment (GA_4 w). The expression of *reCj19503* and *reCj19606*, which encode proteins with sequence similarity to GH3 (indole–3-acetic acid (IAA) -amido synthase), was upregulated at GA_3 d and GA_1 w, respectively. The *reCj26644* and *reCj25311* were extracted from the abscisic acid signal transduction pathway based on sequence similarity to *Highly-ABA induced PPC2 gene 3 (HAI3)*. The expression of these genes was upregulated at GA_1 w and GA_4 w, respectively. The *reCj12520*, *reCj27869*, and *reCj27238* were extracted from the jasmonic acid signal transduction pathway based on sequence similarity to *JAZ*, which encodes a negative regulator. The expression of these genes was upregulated as follows after GA$_3$ treatment: *reCj12520* gradually increased, *reCj27869* from GA_3 h, and *reCj27238* at GA_1 w (intensively) and at GA_6 w. *reCj20357*, *reCj32790*, *reCj31882*, and *reCj31173* were extracted from the salicylic acid signal transduction pathway based on the sequence similarity of *reCj20357* to *BLADE ON PETIOLE2 (BOP2)* (*NONEXPRESSER OF PR GENES 1 (NPR1)*-like; encoding a member of the ankyrin repeat family) and of *reCj32790*, *reCj31882*, and *reCj31173* to *pathogenesis-related protein 1 (PRI)*. The *reCj20357* exhibited a gradual increase of expression starting from GA_3 h, whereas *reCj32790* was gradually downregulated; the expression of both *reCj31882* and *reCj31173* was strongly upregulated after GA_1 d.

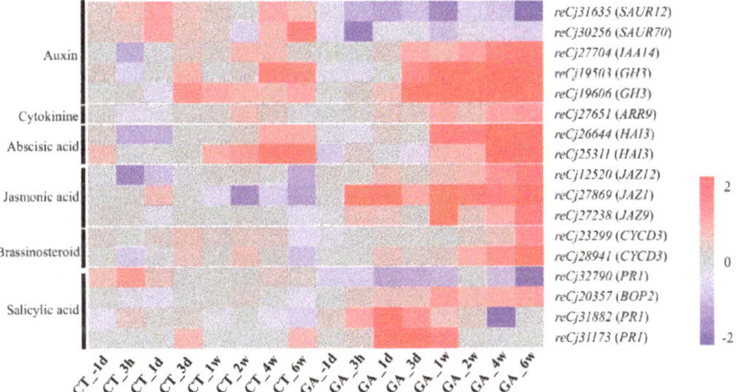

Figure 5. The heat map analysis associated with plant hormone signal transduction pathways without GA. The relative expression values were log$_2$ transformed. Red indicates high expression; blue indicates low expression. CT_–1 d, CT_3 h, CT_1 d, CT_3 d, CT_1 w, CT_2 w, CT_4 w, CT_6 w, GA_–1 d, GA_3 h, GA_1 d, GA_3 d, GA_1 w, GA_2 w, GA_4 w, and GA_6 w are sample names. Details of these genes are provided in Table 1.

3.5. Expression Patterns of MADS-Box Genes

MADS-box genes are well known as floral homeotic genes [42,43]. To clarify changes in the expression patterns of MADS-box genes after GA$_3$ treatment in *C. japonica*, 18 DEGs encoding MADS-box proteins (E-value <1E−5) were extracted (Figure 6, Table 2). These genes were divided into two classes based on the associated expression patterns. The first class consisted of genes whose expression increased gradually following the GA$_3$ treatment, and included *reCj29105*, *reCj27161*, *reCj32389*, *reCj31907*, *reCj28306*, *reCj31827*, *reCj29951*, *reCj33073*, *reCj30226*, *reCj30596*, *reCj17268*, and *reCj25811*. The second class consisted of genes whose expression gradually decreased following the GA$_3$ treatment such as *reCj271510*, *reCj15424*, *reCj29820*, *reCj30835*, *reCj15467*, and *reCj29690*. Based on sequence similarity, the upregulated genes included the following: *reCj31097*, *reCj33073*, and *reCj30226* resembled *AGL6* (which encodes a protein that activates the florigen-encoding locus *FLOWERING LOCUS T* (*FT*) and downregulates the floral repressor-encoding *FLC/MAF*-clade genes, including *FLOWERING LOCUS C* (*FLC*)); *reCj29105* and *reCj28306* resembled *SEPALLATA 1* (*SEP1*) and *SEPALLATA 3* (*SEP3*), respectively, known floral organ identity genes; *reCj31827*, *reCj30596*, *reCj27161*, and *reCj29951* resembled *PISTILLATA* (*PI*), a known floral organ identity gene; and *reCj32389*, *reCj17268*, and *reCj25811* resembled *AGL16*, *AGL15* and *SHP1*, respectively. On the other hand, the downregulated genes included the following: *reCj15424*, *reCj29690*, and *reCj15467* had sequence similarity to *FRUITFULL* (*FUL*), which is downregulated by *APETALA 1* (*AP1*); *reCj27510* had sequence similarity to *AGL22*, a known floral repressor; *reCj29820* had sequence similarity to *AGL20*, which is known to act with *AGL24* to promote flowering and floral meristem identity; and *reCj30835* had sequence similarity to *AGL19*, a known floral activator. Then PCA analysis was carried out using the expression patterns of these 18 genes to estimate the transition from vegetative growth phase to reproductive growth phase (Figure 7). The variance contribution of first and second components of PCA was 94.3% and 1.9%, respectively. Large temporal change of expression patterns was observed in the direction along with the first principal component. After 1 week, divergence in the expression of these genes was observed between the GA-treated and CT samples, and it became evident with the time course.

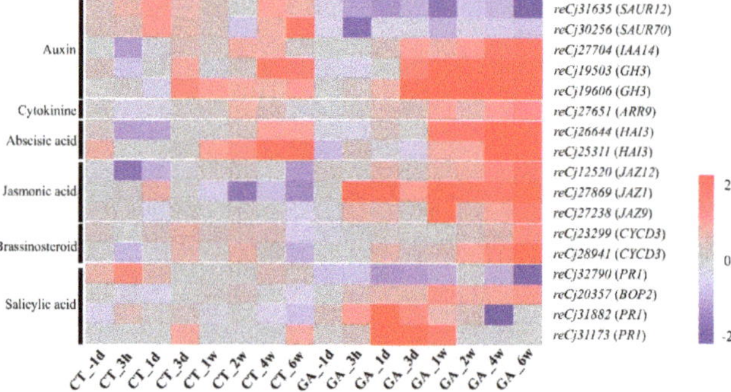

Figure 6. The heat map analysis associated with MADS-box genes. The relative expression values were log$_2$ transformed. Red indicates high expression; blue indicates low expression. CT_−1 d, CT_3 h, CT_1 d, CT_3 d, CT_1 w, CT_2 w, CT_4 w, CT_6 w, GA_−1 d, GA_3 h, GA_1 d, GA_3 d, GA_1 w, GA_2 w, GA_4 w, and GA_6 w are sample names. Details of these genes are provided in Table 2.

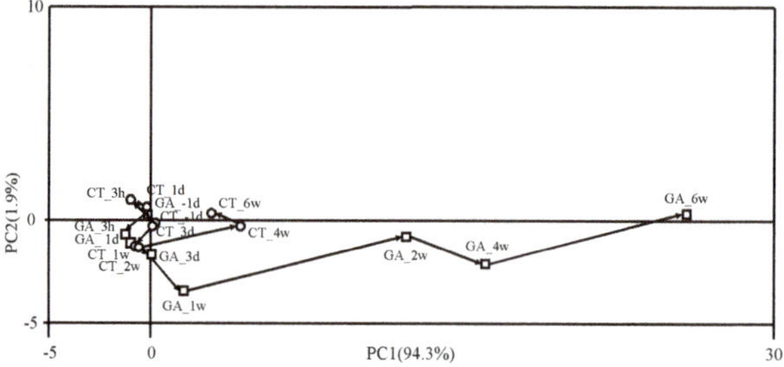

Figure 7. Principal component analysis of all samples for 18 DEGs, including the MADS-box genes listed in Table 2; expression patterns were described in Figure 7. Open circles indicate GA-nontreated control samples. Open squares indicate GA-treated samples.

Table 2. The DEGs associated with MADS-box genes after GA$_3$ treatment.

SEQ_ID	Short SEQ_ID	Arabi_ID	Symbol	Description	E-Value
reCj29105:-S-:isotig28957	reCj29105	AT5G15800	SEP1, AGL2	K-box_region_and_MADS-box_transcription_factor_family_protein	5.59E−30
reCj27161:MS-:isotig27013	reCj27161	AT5G20240	PI	K-box_region_and_MADS-box_transcription_factor_family_protein	1.63E−48
reCj32389:M—:isotig32241	reCj32389	AT3G57230	AGL16	AGAMOUS-like_16	4.66E−24
reCj31097:M—:isotig30949	reCj31097	AT2G45650	AGL6	AGAMOUS-like_6	8.94E−33
reCj28306:M—:isotig28158	reCj28306	AT1G24260	SEP3, AGL9	K-box_region_and_MADS-box_transcription_factor_family_protein	4.47E−54
reCj31827:M—:isotig31679	reCj31827	AT5G20240	PI	K-box_region_and_MADS-box_transcription_factor_family_protein	2.06E−33
reCj29951:M—:isotig29803	reCj29951	AT5G20240	PI	K-box_region_and_MADS-box_transcription_factor_family_protein	1.35E−31
reCj33073:M—:isotig32925	reCj33073	AT2G45650	AGL6	AGAMOUS-like_6	1.51E−21
reCj30226:M—:isotig30078	reCj30226	AT2G45650	AGL6	AGAMOUS-like_6	5.87E−33
reCj30596:M—:isotig30448	reCj30596	AT5G20240	PI	K-box_region_and_MADS-box_transcription_factor_family_protein	2.00E−36
reCj17268:—:isotig17123	reCj17268	AT5G13790	AGL15	AGAMOUS-like_15	3.50E−18
reCj25811:MS-:isotig25663	reCj25811	AT3G58780	SHP1	K-box_region_and_MADS-box_transcription_factor_family_protein	4.57E−47
reCj27510:-SW-:isotig27362	reCj27510	AT2G22540	AGL22	K-box_region_and_MADS-box_transcription_factor_family_protein	2.05E−40
reCj15424:-SW-:isotig15279	reCj15424	AT5G60910	AGL8	AGAMOUS-like_8	7.64E−34
reCj29820:-W-:isotig29672	reCj29820	AT2G45660	AGL20	AGAMOUS-like_20	1.88E−39
reCj30835:—:isotig30687	reCj30835	AT4G22950	AGL19	AGAMOUS-like_19	9.09E−25
reCj15467:—:isotig15322	reCj15467	AT5G60910	AGL8	AGAMOUS-like_8	4.67E−34
reCj29690:-SW-:isotig29542	reCj29690	AT5G60910	AGL8	AGAMOUS-like_8	4.52E−37

3.6. Validation Using Real-Time PCR

To validate our microarray data, 64 test reactions (4 DEGs, in both GA-treated and -nontreated samples, from each of the eight time points) were tested by RT-PCR (Figure 8). Its result demonstrated that the microarray data obtained were highly reproducible and reliable.

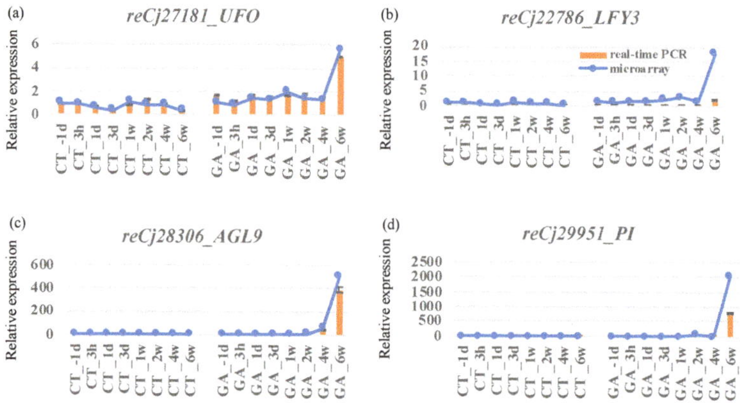

Figure 8. Real-time PCR validation of DEGs data. Bar graphs show relative expression from real-time PCR and line graphs show raw data signals from microarray analysis. Data are presented as mean + standard deviation (n = 3). (**a**) *reCj27181_UFO*, (**b**) *reCj22786_LFY3*, (**c**) *reCj28306_AGL9*, (**d**) *reCj29951_PI*. CT_−1 d, CT_3 h, CT_1 d, CT_3 d, CT_1 w, CT_2 w, CT_4 w, CT_6 w, GA_−1 d, GA_3 h, GA_1 d, GA_3 d, GA_1 w, GA_2 w, GA_4 w, and GA_6 w are sample names.

4. Discussion

4.1. Comprehensive Gene Expression Dynamics Following GA₃ Treatment

Our research clarified changes in gene expression patterns during male strobilus induction following GA_3 treatment. We used the microarray method for comparative analyses of gene expression patterns between GA-treated and non-treated samples.

Overall, the analyses identified 881 DEGs that showed >2-fold changes in expression for a given time point (when comparing GA-treated samples to nontreated samples) or when comparing expression before and after GA_3 treatment. Cluster analyses revealed that these 881 DEGs were grouped into three clusters (U1, U2, and D) depending on up- or down-regulation along the time course. In the following paragraphs, based on the expression patterns of the DEGs, we discussed their potential role in the mechanism of GA-induced male strobilus formation or other functions in *C. japonica*.

4.2. The Expression of GA Signal Transduction-Related Genes

Genes that had sequence similarity with GA signal transduction-related genes were detected among the DEGs. A *DELLA*-like gene (*reCj11694*) was upregulated by GA_3 treatment, whereas *SLY1*-like genes (*reCj34040* and *reCj28549*) were downregulated by GA_3 treatment (Figure 4). Detailed research on the molecular mechanism of GA signal transduction has been carried out in model plants like *Oryza sativa* and *A. thaliana*. In those systems, GAs bind to the GA receptor (GID1), enabling GID1 to interact with DELLA repressor proteins, which are negative regulators of GA signaling [44–46]. DELLA interacts with transcription factors, either impairing transcription factor function (by inhibiting factor ability to bind DNA) or enhancing DNA binding (by acting as a transcriptional coactivator) [47–50]. These GA-induced GID1-DELLA interactions lead to the degradation of DELLA repressors through the Skp, Cullin, F-box complex [3,46,51]. Concordantly with the results found in the present study, the upregulation of *DELLA*-like genes in response to GA treatment has been reported in grape and jatropha, suggesting the possibility that the upregulation of *DELLA* is the result of feedback regulation via GA signaling [52,53]. Regulation by a GA feedback mechanism may be also able to apply in *C. japonica*. Validation of this hypothesis will require further analysis, including determination of quantitative changes of DELLA protein levels in response to GA treatment.

4.3. Expression of Male Strobilus Formation-Related Genes and the Growth Phase Transition from Vegetative to Reproductive Phase by GA Treatment

In *A. thaliana*, GAs promote expression of both *SUPPRESSOR OF OVEREXPRESSION OF CO1* (*SOC1*) and *LFY* directly, and regulate the expression of *SOC1* and *LFY* indirectly (via *GAMYB*) [15]. In the present study, DEGs also included various flower bud formation-related genes, including genes with higher sequence similarity to *LFY*. These DEGs also consisted of genes with high homology to *SHORT VEGETATIVE PHASE* (*SVP*) and *FUL*, genes that are known to be floral meristem identity genes. *SVP* encodes a floral repressor [54] and, like *FUL*, is negatively regulated by *AP1*, which is also a floral meristem identity gene [55,56]. Expression of these genes in *C. japonica* was suppressed during male strobilus formation (Figure 6). In *C. japonica*, these genes may function as suppressors of male strobilus formation. DEGs identified in the present study also included genes with sequence similarity to *SEP1*, *SEP3*, and *PI*, all of which are known as floral organ identity genes [42]. These DEGs were activated during male strobilus formation in *C. japonica* (Figure 6). In addition, activation of genes whose expression in flower buds and floral organs has been reported in another species was observed during male strobilus formation in *C. japonica*. Notably, Tsubomura et al. [57] comprehensively analyzed the expression of genes in the male strobilus and pollen development processes in *C. japonica*, revealing the expression patterns of genes associated with these developmental stages. Those authors observed that these *C. japonica* genes showed similarities in both sequence and expression pattern compared to the corresponding *A. thaliana* genes employed in tapetum development, indicating that these genes play an important role in processes that are fundamental to the maintenance of reproduction in the respective plant species. In the present study, we clarified the comprehensive gene expression dynamics of male strobilus induction, at an earlier stage than the morphology of male strobilus characterized by Tsubomura et al. [57]. Despite structural differences in the development of reproductive organs in *C. japonica* and *A. thaliana*, the expression patterns of relevant genes in these two species were similar, suggesting that the function of these genes have been preserved during evolution. In the present study, based on the sequence similarity to *A. thaliana* MADS-box genes, the corresponding *C. japonica* genes were extracted from the DEGs and annotated. Based on the expression pattern of these *C. japonica* genes, we inferred the transitional timing from the vegetative phase to the reproductive phase in male strobilus formation process as to be 1 week after GA_3 treatment because the divergence in expression dynamics along the first principal component between the GA-treated and -nontreated samples was exhibited at 1 week after the GA_3 treatment and increased with the time course (Figure 7).

4.4. Crosstalk with Auxin Signal Transduction During Male Strobilus Formation

Various studies have reported on possible crosstalk between the pathways regulated by GA and by other plant hormones [58–60]. In *A. thaliana*, it has been reported that auxin signal transduction is involved in the activation of the *LFY* gene; expression of the *LFY* gene is activated by the addition of auxin, resulting in flower bud formation [61,62]. Among the DEGs isolated in the present study, five kinds of genes showing sequence similarity to genes of the auxin signal transduction pathway were identified. Notably, *reCj31635* and *reCj30256*, which encode proteins with sequence similarity to SAURs (members of an auxin-responsive family of proteins) were downregulated in GA-treated samples compared to nontreated samples. On the other hand, *reCj27704*, a gene with sequence similarity to *IAA14* [a negative regulator of *Auxin response factor 7* [63]] was upregulated in GA-treated samples compared to nontreated samples. Moreover, *reCj19503* and *reCj19606*, which encode proteins with sequence similarity to GH3 (IAA-amido synthetase) also were upregulated in GA-treated samples compared to non-treated samples. These results suggested that the auxin signaling system may be suppressed upon GA_3 treatment, an observation that would be consistent with results reported in jatropha [53]. In *C. japonica*, it was reported that auxin alone does not yield male strobilus induction, although the combination of auxin and GA was reported to increase the ratio of female cones [31].

4.5. Growth Control by GA Treatment

Enrichment analysis suggested that Cluster U1 contained many genes related to the response to stress, response to abiotic or biotic stimulus, extracellular, cell wall, other cellular components and unknown cellular components (Figure 3). Hou et al. [49] analyzed the genes targeted by DELLA during flower development in *A. thaliana* and reported that cell wall proteins were among the genes downregulated by RGA. GA has been hypothesized to have roles in both growth regulation and reproductive control. Notably, it has been reported that the GA_3 treatment promoted principal axis elongation and suppressed lateral branch elongation in *C. japonica* [64]. The accumulation of the transcripts of cell wall-related genes and cellular component genes, including WAKs-like genes (*reCj12226*, *reCj12227*, and *reCj32277*; Table S1; [65]), as revealed in the present study, might indicate the role of GA in growth control.

4.6. Senescence-Like Gene Expression Patterns Following GA Treatment

Enrichment analysis showed that Cluster D contained genes related to electron transport or energy pathway (related to photosystem I or photosystem II, etc.), plastids, chloroplasts, and receptor binding or activity (Figure 3, Table S2). This result suggested a decrease in the expression of photosynthesis-related genes. The decreased expression of photosynthesis-related genes is known as a leaf senescence-related phenomenon [66]. The *A. thaliana* NAC and WRKY53 proteins are known as transcription factors that play important roles in the process of senescence [67,68]. Among the DEGs identified in the present work, a gene (*reCj22492*) with sequence similarity to the *A. thaliana* WIP5 gene (a putative target of WRKY53 [68]) was detected within Cluster U2, indicating that this *C. japonica* gene is upregulated during the male strobilus formation process (Figure S3). In addition, plant hormones have been reported to participate in leaf senescence [69–71]. The abscisic acid signal transduction gene *SAG113* (*PP2C*) has been shown to be induced in the process of senescence [71,72]. The DEGs of *C. japonica* included two genes (*reCj26644* and *reCj25311*) with sequence similarity to *PP2C*; expression patterns placed these genes in Cluster U2, such that expression was increased during male strobilus formation (Figure 5). The U1 cluster of DEGs was enriched for genes annotated to be involved in response to stress and response to abiotic or biotic stimulus (Figure 3). These observations together suggested that the phenomena of senescence and stress response may overlap with each other and/or with the response to GA treatment in *C. japonica*; further study will be needed to clarify these inferences.

5. Conclusions

In conclusion, gene expression analysis facilitated better understanding of the molecular dynamics of the induction by GA_3 treatment of male strobilus formation in *C. japonica*. Our study identified various *C. japonica* genes with sequence similarity to genes implicated in GA signaling in other plant species. Our results revealed that the dynamics of gene expression for male strobilus formation became conspicuous from seven days after GA_3 treatment. In addition, we were able to capture the behavior of genes that may explain other phenomena resulting from GA_3 treatment. These data are expected to permit clarification of the molecular mechanism of the induction by GA_3 treatment of male strobilus formation in *C. japonica*, providing detailed information at the protein and metabolite levels. This information for *C. japonica*, a coniferous species, might provide new knowledge of the basic mechanism whereby evolution acquired a GA-regulated pathway for use in the induction of plant reproductive organs.

Supplementary Materials: The following are available online at http://www.mdpi.com/1999-4907/11/6/633/s1, Figure S1: Sampling scheme of this study, Figure S2: The heat map analysis associated with GA signal transduction pathways, Figure S3: The heat map of *reCj22492*, Table S1: The list of DEGs, Table S2: Results of enrichment analysis using DAVID.

Author Contributions: K.M. and M.K. conceived and designed the experiments; K.M. and M.K. performed the experiments; M.K. and K.M. analyzed the data; M.T. (Miyoko Tsubomura), Y.T., M.N. and T.H. contributed EST data/materials/analysis tools; M.K. wrote the manuscript; K.M., M.T. (Miyoko Tsubomura) and M.T. (Makoto Takahashi) revised the manuscript. All authors have read and agreed to the published version of the manuscript.

Funding: The present study is part of the project on 'Development of adaptation techniques to the climate change in the sectors of agriculture, forestry, and fisheries' supported by the Ministry of Agriculture, Forestry and Fisheries, Japan.

Acknowledgments: We are grateful to Hiroshi Hoshi (FTBC, FFPRI) for his well coordination of the research project.

Conflicts of Interest: The authors declare that the research was conducted in the absence of any commercial or financial relationships that could be construed as a potential conflict of interest.

References

1. Richards, D.E.; King, K.E.; Ait-Ali, T.; Harberd, N.P. How gibberellin regulates plant growth and development: A molecular genetic analysis of gibberellin signaling. *Annu. Rev. Plant Physiol. Plant Mol. Biol.* **2001**, *52*, 67–88. [CrossRef] [PubMed]

2. Sun, T.-P.; Gubler, F. Molecular Mechanism of Gibberellin Signaling in Plants. *Annu. Rev. Plant Boil.* **2004**, *55*, 197–223. [CrossRef]

3. Hirano, K.; Nakajima, M.; Asano, K.; Nishiyama, T.; Sakakibara, H.; Kojima, M.; Katoh, E.; Xiang, H.; Tanahashi, T.; Hasebe, M.; et al. The GID1-Mediated Gibberellin Perception Mechanism Is Conserved in the Lycophyte *Selaginella moellendorffii* but Not in the Bryophyte *Physcomitrella patens*. *Plant Cell* **2007**, *19*, 3058–3079. [CrossRef] [PubMed]

4. Vandenbussche, F.; Fierro, A.C.; Wiedemann, G.; Reski, R.; Van Der Straeten, D. Evolutionary conservation of plant gibberellin signalling pathway components. *BMC Plant Boil.* **2007**, *7*, 65. [CrossRef] [PubMed]

5. Hayashi, K.-I.; Horie, K.; Hiwatashi, Y.; Kawaide, H.; Yamaguchi, S.; Hanada, A.; Nakashima, T.; Nakajima, M.; Mander, L.N.; Yamane, H.; et al. Endogenous diterpenes derived from *ent*-kaurene, a common gibberellin precursor, regulate protonema differentiation of the moss *Physcomitrella patens*. *Plant Physiol.* **2010**, *153*, 1085–1097. [CrossRef] [PubMed]

6. Hirano, K.; Ueguchi-Tanaka, M.; Matsuoka, M. GID1-mediated gibberellin signaling in plants. *Trends Plant Sci.* **2008**, *13*, 192–199. [CrossRef] [PubMed]

7. Aya, K.; Hiwatashi, Y.; Kojima, M.; Sakakibara, H.; Ueguchi-Tanaka, M.; Hasebe, M.; Matsuoka, M. The Gibberellin perception system evolved to regulate a pre-existing GAMYB-mediated system during land plant evolution. *Nat. Commun.* **2011**, *2*, 544. [CrossRef]

8. Tanaka, J.; Yano, K.; Aya, K.; Hirano, K.; Takehara, S.; Koketsu, E.; Ordonio, R.L.; Park, S.-H.; Nakajima, M.; Ueguchi-Tanaka, M.; et al. Antheridiogen determines sex in ferns via a spatiotemporally split gibberellin synthesis pathway. *Science* **2014**, *346*, 469–473. [CrossRef]

9. Conti, L. Hormonal control of the floral transition: Can one catch them all? *Dev. Boil.* **2017**, *430*, 288–301. [CrossRef]

10. Mouradov, A.; Cremer, F.; Coupland, G. Control of flowering time: Interacting pathways as a basis for diversity. *Plant Cell* **2002**, *14*. [CrossRef]

11. Simpson, G.G.; Dean, C. *Arabidopsis*, the Rosetta Stone of Flowering Time? *Science* **2002**, *296*, 285–289. [CrossRef] [PubMed]

12. Andrés, F.; Coupland, G.; Andr, F. The genetic basis of flowering responses to seasonal cues. *Nat. Rev. Genet.* **2012**, *13*, 627–639. [CrossRef] [PubMed]

13. Bäurle, I.; Dean, C. The Timing of Developmental Transitions in Plants. *Cell* **2006**, *125*, 655–664. [CrossRef] [PubMed]

14. Srikanth, A.; Schmid, M. Regulation of flowering time: All roads lead to Rome. *Cell. Mol. Life Sci.* **2011**, *68*, 2013–2037. [CrossRef] [PubMed]

15. Mutasa-Gottgens, E.; Hedden, P. Gibberellin as a factor in floral regulatory networks. *J. Exp. Bot.* **2009**, *60*, 1979–1989. [CrossRef]

16. Huijser, P.; Schmid, M. The control of developmental phase transitions in plants. *Development* **2011**, *138*, 4117–4129. [CrossRef]

17. Okamuro, J.K.; Boer, B.G.W.D.; Lotys-Prass, C.; Szeto, W.; Jofuku, K.D. Flowers into shoots: Photo and hormonal control of a meristem identity switch in *Arabidopsis*. *Proc. Natl. Acad. Sci. USA* **1996**, *93*, 13831–13836. [CrossRef]

18. Gocal, G.F.W.; Sheldon, C.C.; Gubler, F.; Moritz, T.; Bagnall, D.J.; MacMillan, C.P.; Li, S.F.; Parish, R.W.; Dennis, E.S.; Weigel, D.; et al. GAMYB-like Genes, Flowering, and Gibberellin Signaling in *Arabidopsis*. *Plant Physiol.* **2001**, *127*, 1682–1693. [CrossRef]

19. Wilson, R.N.; Heckman, J.W.; Somerville, C.R. Gibberellin Is Required for Flowering in *Arabidopsis thaliana* under Short Days. *Plant Physiol.* **1992**, *100*, 403–408. [CrossRef]

20. Moon, J.; Suh, S.-S.; Lee, H.; Choi, K.-R.; Hong, C.B.; Paek, N.-C.; Kim, S.-G.; Lee, I. TheSOC1MADS-box gene integrates vernalization and gibberellin signals for flowering in *Arabidopsis*. *Plant J.* **2003**, *35*, 613–623. [CrossRef]

21. Eriksson, S.; Böhlenius, H.; Moritz, T.; Nilsson, O. GA4 Is the Active Gibberellin in the Regulation of *LEAFY* Transcription and *Arabidopsis* Floral Initiation. *Plant Cell* **2006**, *18*, 2172–2181. [CrossRef] [PubMed]

22. Blazquez, M.; Green, R.; Nilsson, O.; Sussman, M.; Weigel, D. Gibberellins promote flowering of *Arabidopsis* by activating the *LEAFY* promoter. *Plant Cell* **1998**, *10*, 791–800. [CrossRef] [PubMed]

23. Yamaguchi, N.; Winter, C.M.; Wu, M.-F.; Kanno, Y.; Seo, M.; Wagner, D.; Yamaguchi, A. Gibberellin Acts Positively Then Negatively to Control Onset of Flower Formation in *Arabidopsis*. *Science* **2014**, *344*, 638–641. [CrossRef] [PubMed]

24. Fernández, H.; Fraga, M.; Bernard, P.; Revilla-Bahillo, M.A.M. Quantification of GA1, GA3, GA4, GA7, GA9, and GA20 in vegetative and male cone buds from juvenile and mature trees of *Pinus radiata*. *Plant Growth Regul.* **2003**, *40*, 185–188. [CrossRef]

25. Niu, S.; Yuan, L.; Zhang, Y.; Chen, X.; Li, W. Isolation and expression profiles of gibberellin metabolism genes in developing male and female cones of *Pinus tabuliformis*. *Funct. Integr. Genom.* **2014**, *14*, 697–705. [CrossRef]

26. Duan, D.; Jia, Y.; Yang, J.; Li, Z.-H. Comparative Transcriptome Analysis of Male and Female Conelets and Development of Microsatellite Markers in *Pinus bungeana*, an Endemic Conifer in China. *Genes* **2017**, *8*, 393. [CrossRef]

27. Kong, L.; Von Aderkas, P.; Zaharia, L.I. Effects of Exogenously Applied Gibberellins and Thidiazuron on Phytohormone Profiles of Long-Shoot Buds and Cone Gender Determination in Lodgepole Pine. *J. Plant Growth Regul.* **2015**, *35*, 172–182. [CrossRef]

28. Crain, B.A.; Cregg, B.M. Regulation and Management of Cone Induction in Temperate Conifers. *For. Sci.* **2017**, *64*, 82–101. [CrossRef]

29. Williams, C.G. *Conifer Reproductive Biology*; Springer: Berlin, Germany, 2009; p. 169.

30. Hashizume, H. The effect of gibberellin upon flower formation in *Cryptomeria japonica*. *J. Jpn. For. Soc.* **1959**, *41*, 375–381, (In Japanese with English summary).

31. Hashizume, H. Studies on flower bud formation, flower sex differentiation and their control in conifers. *Bull. Tottori Univ. For.* **1973**, *7*, 1–139, (In Japanese with English summary).

32. Mishima, K.; Hirao, T.; Tsubomura, M.; Tamura, M.; Kurita, M.; Nose, M.; Hanaoka, S.; Takahashi, M.; Watanabe, A. Identification of novel putative causative genes and genetic marker for male sterility in Japanese cedar (*Cryptomeria japonica* D.Don). *BMC Genom.* **2018**, *19*, 277. [CrossRef] [PubMed]

33. Mishima, K.; Fujiwara, T.; Iki, T.; Kuroda, K.; Yamashita, K.; Tamura, M.; Fujisawa, Y.; Watanabe, A. Transcriptome sequencing and profiling of expressed genes in cambial zone and differentiating xylem of Japanese cedar (*Cryptomeria japonica*). *BMC Genom.* **2014**, *15*, 219. [CrossRef] [PubMed]

34. Nose, M.; Watanabe, A. Clock genes and diurnal transcriptome dynamics in summer and winter in the gymnosperm Japanese cedar (*Cryptomeria japonica* (L.f.) D.Don). *BMC Plant Boil.* **2014**, *14*, 308. [CrossRef]

35. Fukuda, Y.; Hirao, T.; Mishima, K.; Ohira, M.; Hiraoka, Y.; Takahashi, M.; Watanabe, A. Transcriptome dynamics of rooting zone and aboveground parts of cuttings during adventitious root formation in *Cryptomeria japonica* D. Don. *BMC Plant Boil.* **2018**, *18*, 201. [CrossRef] [PubMed]

36. Fukui, K.; Wakamatsu, T.; Agari, Y.; Masui, R.; Kuramitsu, S. Inactivation of the DNA Repair Genes *mut*S, *mut*L or the Anti-Recombination Gene *mut*S2 Leads to Activation of Vitamin B1 Biosynthesis Genes. *PLoS ONE* **2011**, *6*, e19053. [CrossRef]

37. Huang, D.W.; Sherman, B.T.; Tan, Q.; Kir, J.; Liu, D.; Bryant, D.; Guo, Y.; Stephens, R.; Baseler, M.W.; Lane, H.C.; et al. DAVID Bioinformatics Resources: Expanded annotation database and novel algorithms to better extract biology from large gene lists. *Nucleic Acids Res.* **2007**, *35*, W169–W175. [CrossRef]

38. Huang, D.W.; Sherman, B.; Tan, Q.; Collins, J.R.; Alvord, W.G.; Roayaei, J.; Stephens, R.M.; Baseler, M.; Lane, H.C.; Lempicki, R. The DAVID Gene Functional Classification Tool: A Novel Biological Module-Centric Algorithm to Functionally Analyze Large Gene Lists. *Genome Boil.* **2007**, *8*, R183. [CrossRef]

39. Untergasser, A.; Cutcutache, I.; Koressaar, T.; Ye, J.; Faircloth, B.C.; Remm, M.; Rozen, S.G. Primer3—new capabilities and interfaces. *Nucleic Acids Res.* **2012**, *40*, e115. [CrossRef]

40. Pfaffl, M.W. A new mathematical model for relative quantification in real-time RT-PCR. *Nucleic Acids Res.* **2001**, *29*, 45. [CrossRef]

41. Gomi, K.; Sasaki, A.; Itoh, H.; Ashikari, M.; Kitano, H.; Matsuoka, M.; Ueguchi-Tanaka, M. GID2, an F-box subunit of the SCF E3 complex, specifically interacts with phosphorylated SLR1 protein and regulates the gibberellin-dependent degradation of SLR1 in rice. *Plant J.* **2004**, *37*, 626–634. [CrossRef]

42. Krizek, B.A.; Fletcher, J.C. Molecular mechanisms of flower development: An armchair guide. *Nat. Rev. Genet.* **2005**, *6*, 688–698. [CrossRef] [PubMed]

43. Becker, A.; Theißen, G. The major clades of MADS-box genes and their role in the development and evolution of flowering plants. *Mol. Phylogenet. Evol.* **2003**, *29*, 464–489. [CrossRef]

44. Ueguchi-Tanaka, M.; Ashikari, M.; Nakajima, M.; Itoh, H.; Katoh, E.; Kobayashi, M.; Chow, T.-Y.; Hsing, Y.-I.C.; Kitano, H.; Yamaguchi, I.; et al. GIBBERELLIN INSENSITIVE DWARF1 encodes a soluble receptor for gibberellin. *Nature* **2005**, *437*, 693–698. [CrossRef]

45. Nakajima, M.; Shimada, A.; Takashi, Y.; Kim, Y.-C.; Park, S.-H.; Ueguchi-Tanaka, M.; Suzuki, H.; Katoh, E.; Iuchi, S.; Kobayashi, M.; et al. Identification and characterization of *Arabidopsis* gibberellin receptors. *Plant J.* **2006**, *46*, 880–889. [CrossRef] [PubMed]

46. Ueguchi-Tanaka, M.; Nakajima, M.; Katoh, E.; Ohmiya, H.; Asano, K.; Saji, S.; Hongyu, X.; Ashikari, M.; Kitano, H.; Yamaguchi, I.; et al. Molecular Interactions of a Soluble Gibberellin Receptor, GID1, with a Rice DELLA Protein, SLR1, and Gibberellin. *Plant Cell* **2007**, *19*, 2140–2155. [CrossRef] [PubMed]

47. Fukazawa, J.; Teramura, H.; Murakoshi, S.; Nasuno, K.; Nishida, N.; Ito, T.; Yoshida, M.; Kamiya, Y.; Yamaguchi, S.; Takahashi, Y. DELLAs function as coactivators of GAI-ASSOCIATED FACTOR1 in regulation of gibberellin homeostasis and signaling in *Arabidopsis*. *Plant Cell* **2014**, *26*, 2920–2938. [CrossRef]

48. Cao, N.; Cheng, H.; Wu, W.; Soo, H.M.; Peng, J. Gibberellin Mobilizes Distinct DELLA-Dependent Transcriptomes to Regulate Seed Germination and Floral Development in *Arabidopsis*. *Plant Physiol.* **2006**, *142*, 509–525. [CrossRef]

49. Hou, X.; Hu, W.-W.; Shen, L.; Lee, L.Y.C.; Tao, Z.; Han, J.-H.; Yu, H. Global Identification of DELLA Target Genes during *Arabidopsis* Flower Development. *Plant Physiol.* **2008**, *147*, 1126–1142. [CrossRef]

50. Zentella, R.; Zhang, Z.-L.; Park, M.; Thomas, S.G.; Endo, A.; Murase, K.; Fleet, C.M.; Jikumaru, Y.; Nambara, E.; Kamiya, Y.; et al. Global Analysis of DELLA Direct Targets in Early Gibberellin Signaling in *Arabidopsis*. *Plant Cell* **2007**, *19*, 3037–3057. [CrossRef]

51. Sun, T.-P. Gibberellin-GID1-DELLA: A Pivotal Regulatory Module for Plant Growth and Development. *Plant Physiol.* **2010**, *154*, 567–570. [CrossRef]

52. Jung, C.J.; Hur, Y.Y.; Yu, H.-J.; Noh, J.-H.; Park, K.-S.; Lee, H.J. Gibberellin Application at Pre-Bloom in Grapevines Down-Regulates the Expressions of *VvIAA9* and *VvARF7*, Negative Regulators of Fruit Set Initiation, during Parthenocarpic Fruit Development. *PLoS ONE* **2014**, *9*, e95634. [CrossRef] [PubMed]

53. Hui, W.-K.; Wang, Y.; Chen, X.-Y.; Zayed, M.; Wu, G. Analysis of Transcriptional Responses of the Inflorescence Meristems in *Jatropha curcas* Following Gibberellin Treatment. *Int. J. Mol. Sci.* **2018**, *19*, 432. [CrossRef] [PubMed]

54. Hartmann, U.; Hohmann, S.; Nettesheim, K.; Wisman, E.; Saedler, H.; Huijser, P. Molecular cloning of *SVP*: A negative regulator of the floral transition in *Arabidopsis*. *Plant J.* **2000**, *21*, 351–360. [CrossRef] [PubMed]

55. Wellmer, F.; Alves-Ferreira, M.; Dubois, A.; Riechmann, J.L.; Meyerowitz, E.M. Genome-Wide Analysis of Gene Expression during Early *Arabidopsis* Flower Development. *PLoS Genet.* **2006**, *2*, e117. [CrossRef]

56. Ferrándiz, C.; Gu, Q.; Martienssen, R.; Yanofsky, M.F. Redundant regulation of meristem identity and plant architecture by *FRUITFULL, APETALA1* and *CAULIFLOWER*. *Development* **2000**, *127*, 725–734.

57. Tsubomura, M.; Kurita, M.; Watanabe, A. Determination of male strobilus developmental stages by cytological and gene expression analyses in Japanese cedar (*Cryptomeria japonica*). *Tree Physiol.* **2016**, *36*, 653–666. [CrossRef]

58. Hartweck, L.M. Gibberellin signaling. *Planta* **2008**, *229*, 1–13. [CrossRef]

59. Weiss, D.; Ori, N. Mechanisms of Cross Talk between Gibberellin and Other Hormones. *Plant Physiol.* **2007**, *144*, 1240–1246. [CrossRef]

60. Sun, T.P. The molecular mechanism and evolution of the review GA–GID1–DELLA signaling module in plants. *Curr. Biol.* **2011**, *21*, R338–R345. [CrossRef]

61. Yamaguchi, N.; Wu, M.-F.; Winter, C.M.; Berns, M.C.; Nole-Wilson, S.; Yamaguchi, A.; Coupland, G.; Krizek, B.A.; Wagner, D. A Molecular Framework for Auxin-Mediated Initiation of Flower Primordia. *Dev. Cell* **2013**, *24*, 271–282. [CrossRef]

62. Yamaguchi, N.; Jeong, C.W.; Nole-Wilson, S.; Krizek, B.A.; Wagner, D. AINTEGUMENTA and AINTEGUMENTA-LIKE6/PLETHORA3 Induce *LEAFY* Expression in Response to Auxin to Promote the Onset of Flower Formation in *Arabidopsis*. *Plant Physiol.* **2016**, *170*, 283–293. [CrossRef]

63. Fukaki, H.; Taniguchi, N.; Tasaka, M. PICKLE is required for SOLITARY-ROOT/IAA14-mediated repression of ARF7 and ARF19 activity during *Arabidopsis* lateral root initiation. *Plant J.* **2006**, *48*, 380–389. [CrossRef] [PubMed]

64. Hashizume, H. Chemical regulation of flower-bud formation in conifers. *J. Jpn. For. Soc.* **1968**, *50*, 14–16. (In Japanese)

65. Kohorn, B.D.; Johansen, S.; Shishido, A.; Todorova, T.; Martinez, R.; Defeo, E.; Obregon, P. Pectin activation of MAP kinase and gene expression is WAK2 dependent. *Plant J.* **2009**, *60*, 974–982. [CrossRef] [PubMed]

66. Gan, S.; Amasino, R.M. Making Sense of Senescence. *Plant Physiol.* **1997**, *113*, 313–319. [CrossRef]

67. Uauy, C.; Distelfeld, A.; Fahima, T.; Blechl, A.; Dubcovsky, J. A NAC Gene Regulating Senescence Improves Grain Protein, Zinc, and Iron Content in Wheat. *Science* **2006**, *314*, 1298–1301. [CrossRef]

68. Miao, Y.; Laun, T.; Zimmermann, P.; Zentgraf, U. Targets of the WRKY53 transcription factor and its role during leaf senescence in *Arabidopsis*. *Plant Mol. Biol.* **2004**, *55*, 853–867. [CrossRef]

69. Hu, Y.; Jiang, Y.; Han, X.; Wang, H.; Pan, J.; Yu, D. Jasmonate regulates leaf senescence and tolerance to cold stress: Crosstalk with other phytohormones. *J. Exp. Bot.* **2017**, *68*, 1361–1369. [CrossRef]

70. Zhang, H.; Zhou, C. Signal transduction in leaf senescence. *Plant Mol. Boil.* **2012**, *82*, 539–545. [CrossRef]

71. Jibran, R.; Hunter, D.; Dijkwel, P. Hormonal regulation of leaf senescence through integration of developmental and stress signals. *Plant Mol. Boil.* **2013**, *82*, 547–561. [CrossRef]

72. Zhang, K.; Xia, X.; Zhang, Y.; Gan, S.-S. An ABA-regulated and Golgi-localized protein phosphatase controls water loss during leaf senescence in *Arabidopsis*. *Plant J.* **2011**, *69*, 667–678. [CrossRef] [PubMed]

Article

Transcriptomic Profiling of *Cryptomeria fortunei* Hooibrenk Vascular Cambium Identifies Candidate Genes Involved in Phenylpropanoid Metabolism

Junjie Yang [1,2,3], Zhenhao Guo [1,2,3], Yingting Zhang [1,2,3], Jiaxing Mo [1,2,3], Jiebing Cui [1,2,3], Hailiang Hu [1,2,3], Yunya He [1,2,3] and Jin Xu [1,2,3,*]

[1] Key Laboratory of Forest Genetics and Biotechnology of Ministry of Education, Nanjing Forestry University, Nanjing 210037, China; yangjj@njfu.edu.cn (J.Y.); 18915019168@163.com (Z.G.); ytzhang@njfu.edu.cn (Y.Z.); jx_mo@njfu.edu.cn (J.M.); cuijiebing@163.com (J.C.); hhl1198264682@163.com (H.H.); 13611576662@163.com (Y.H.)

[2] Co-Innovation Center for Sustainable Forestry in Southern China, Nanjing Forestry University, Nanjing 210037, China

[3] College of Forestry, Nanjing Forestry University, Nanjing 210037, China

* Correspondence: xjinhsh@njfu.edu.cn; Tel.: +86-25-8542-7319

Received: 1 July 2020; Accepted: 15 July 2020; Published: 17 July 2020

Abstract: *Cryptomeria fortunei* Hooibrenk (Chinese cedar) is a coniferous tree from southern China that has an important function in landscaping and timber production. Lignin is one of the key components of secondary cell walls, which have a crucial role in conducting water and providing mechanical support for the upward growth of plants. It is mainly biosynthesized via the phenylpropanoid metabolic pathway, of which the molecular mechanism remains so far unresolved in *C. fortunei*. In order to obtain further insight into this pathway, we performed transcriptome sequencing of the *C. fortunei* cambial zone at 5 successive growth stages. We generated 78,673 unigenes from transcriptome data, of which 45,214 (57.47%) were successfully annotated in the non-redundant protein database (NR). A total of 8975 unigenes were identified to be significantly differentially expressed between Sample_B and Sample_A after analyzing their expression profiles. Of the differentially expressed genes (DEGs), 6817 (75.96%) and 2158 (24.04%) were up- and down-regulated, respectively. 83 DEGs were involved in phenylpropanoid metabolism, 37 DEGs that encoded v-Myb avian myeloblastosis viral oncogene homolog (MYB) transcription factor (TF), and many candidates that encoded lignin synthesizing enzymes. These findings contribute to understanding the expression pattern of *C. fortunei* cambial zone transcriptome. Furthermore, our results provide additional insight towards understanding the molecular mechanisms of wood formation in *C. fortunei*.

Keywords: *C. fortunei*; transcriptome; differentially expressed genes; phenylpropanoid metabolism; candidate genes

---

1. Introduction

The *Cryptomeria* genus consists of the species *Cryptomeria fortunei* Hooibrenk and *Cryptomeria japonica* (L.f.) D.Don (Japanese cedar). *Cryptomeria fortunei* Hooibrenk is an important coniferous timber species native to China. This species is a monoecious coniferous species which is widely planted in southern China due to its strong adaptability. *C. fortunei* has excellent properties that allow for efficient timber production, including a straight bole, soft texture, rapid growth, and ease of processing. Thus, its wood is widely used to construct wooden houses, barrels, and a large number of industrial materials. Additionally, as a photosynthesizing plant, *C. fortunei* is an important plant species in carbon storage and ecological restoration and is also a suitable landscaping tree due to its attractive appearance [1].

A transcriptome is a collection of all transcripts of a certain tissue or organ at a specific period or stage, including coding RNA and non-coding RNA. Based on transcriptome analysis, the molecular mechanisms of secondary growth have been elucidated in model plant [2]. In recent years, with rapid advances in RNA sequencing, it has been applied to non-model plants, for example, to develop simple sequence repeat (SSR) markers [3].

Lateral growth of tree stems occurs through cell divisions in the vascular cambium. Towards the inside, the cambium forms the secondary xylem, also called wood, while towards the outside, secondary phloem cells appear in the growing stem through the proliferation and differentiation. Wood formation is a complex biological process, including cambium cell division, cell extension, secondary cell wall formation, lignification, and finally, programmed cell death [4]. The formation of secondary cell walls is an important event during wood formation. Secondary cell walls are mainly composed of three polymers, lignin is one of the most important compounds that determine the properties of wood [5]. The main pathway of lignin biosynthesis is phenylpropanoid metabolism, which has been well described in *Populus trichocarpa* Torr. & A.Gray ex. Hook. [6] and Norway spruce [7]. However, the molecular mechanisms underlying this biosynthetic pathway are still uncertain in *C. fortunei*.

In plants, the phenylpropanoid metabolic pathway synthesizes a number of key components, including flavonoids, lignin, and others, all of which have been crucial for plants. Lignins contain 3 different components, all of which are synthesized into 3 monomers, including S-lignin, G-lignin, and H-lignin, respectively [8]. Lignins are mainly composed of G-H-lignin in conifers [9]. As a complex synthesis process, lignin biosynthesis can be roughly divided into 3 steps. Firstly, the synthesis of aromatic amino phenylalanine from photosynthetic assimilation products. Secondly, called phenylpropanoid metabolism, phenylalanine is synthesized into separate components, including phenylalanine ammonia-lyase (PAL), p-coumarate 3-hydroxylase (C3H), caffeoyl CoA O-methyltransferase (CCoAOMT), 4-(hydroxy) cinnamoyl CoA ligase (4CL). Thirdly, a specific metabolic pathway which contributes to lignin monomers biosynthesis, including enzymes such as cinnamoyl CoA reductase (CCR) and cinnamyl alcohol dehydrogenase (CAD) [10,11]. Lignins are polymerized from 3 monomers through laccase and peroxidase [9].

Previous reports have used transcriptomics to identify lignin biosynthetic genes in other woody plant species [7,12]. In this study, we conducted transcriptome sequencing of the *C. fortunei* cambial zone at 5 successive growth stages. Differentially expressed genes (DEGs) were identified and analyzed through analyzing the transcriptome data. Subsequently, we aimed to screen DEGs involved in phenylpropanoid metabolism and see how their expression levels would correlate to lignin deposition activity in *C. fortunei*. Our work lays the foundation for functionally elucidating the gene-regulated phenylpropanoid biosynthesis and molecular regulation of lignin biosynthesis in *C. fortunei*.

2. Materials and Methods

2.1. Plant Materials

We acquired the samples from *C. fortunei* trees, aged around 60 years, with no obvious presence of insect pests or disease, in the arboretum of Nanjing Forestry University, Nanjing City, Jiangsu Province, China. The exact dates of sampling were 4 April, 18 May, 10 July, 15 September, and 12 November in 2018, corresponding to 5 different growth stages. The letters A, B, C, D, and E represent these 5 successive stages, respectively. For each growth stage, three samples were taken as biological replicates, labeled as A–1, A–2, and A–3. We obtained the cambium region through scratching the stem by a sharp knife, then collected samples and immediately stored them in liquid nitrogen at $-80\,^{\circ}\mathrm{C}$ until use.

2.2. RNA Extraction and Transcriptome Sequencing

Total RNA was isolated from the samples of *C. fortunei* vascular cambium by a RNeasy Plant Mini Kit (Qiagen, Hilden, Germany). The integrity and concentration of total RNA were assessed by an Agilent 2100 Bioanalyzer (Agilent Technologies, Santa Clara, CA, USA) and a Thermo Scientific NanoDrop 2000 (Thermo Fisher Scientific, Wilmington, DE, USA). Library preparation and sequencing experiments were performed in accordance with the standard procedure provided by Illumina. Sequencing was then performed using an Illumina HiSeq™ 2500 system by OE biotech Co., Ltd. (Shanghai, China), generating 150 bp paired-end reads. The accession number of this project is PRJNA 644276.

2.3. Assembly and Functional Annotation

We obtained millions of clean reads after removing adaptor and low-quality sequences. Clean reads were then assembled into expressed sequence tag clusters (contigs), which were then de novo assembled into transcripts using Trinity and the paired-end method [13]. Subsequently, we described the longest transcript as a unigene using CD-HIT [14]. Unigene was chosen for subsequent analysis. Unigenes were aligned by diamond [15] and HMMER [16] to public databases NR, KOG, GO, Swiss-Prot, eggNOG, KEGG, and Pfam with the highest sequence similarity for protein functional annotation and classification.

2.4. Differential Expression Analysis

Unigene expression was quantified according to the fragments per kb per million reads (FPKM) method [17], using bowtie2 [18] and eXpress [19]. Through pairwise comparisons, DEGs of different stages were identified by DESeq [20]. A threshold of $p < 0.05$ and a greater than two-fold change were set [21]. To explore expression patterns, we performed a sample to sample distances cluster analysis [22]. GO and KEGG enrichment analysis of DEGs were performed using R based on the hypergeometric distribution [21].

2.5. Verification of Gene Expression Using qRT-PCR

8 unigenes were chosen for validation through qRT-PCR. RNA was extracted from the cambium region then reverse-transcribed using a HiScript III RT SuperMix (Vazyme Biotech Co., Ltd., Nanjing, China). We designed the primers using Primer Premier 5.0 software (Premier Biosoft International, Palo Alto, CA, USA) (Supplementary Table S1). Three biological replicates were run at a final volume of 20 μL, which consisted of 6 μL of ddH$_2$O, 1 μL of primers, 2 μL of cDNA, and 10 μL of 2× ChamQ SYBR qPCR Master Mix (Vazyme Biotech Co., Ltd., Nanjing, China). The *C. fortunei* β-actin gene was used as reference [11]. The primers used were F: GCCATCTTTGATTGGAATGG and R: GGTGCCACAACCTTGACTT. The qRT-PCR reaction was performed on an ABI 7500 Step One Plus Real-time PCR System (Applied Biosystems, Foster City, CA, USA). Reactions were performed at 95 °C for 30 s, followed by 40 cycles of 95 °C for 10 s, and 60 °C for 30 s. The delta-delta-Ct method was used to assess the amplification results [23].

2.6. Determination of Lignin Content

The lignin content of stems of the same year was determined according to Reference [24]. We acquired the data by a GeneQuant pro ultraviolet spectrophotometer (Biochrom Ltd., England, UK).

3. Results

3.1. Statistics of Transcriptome Sequencing Results and De Novo Assembly

In order to obtain candidate genes involved in phenylpropanoid metabolism, in this study, we performed transcriptome sequencing of the *C. fortunei* cambial zone at 5 successive growth stages. As a result, we obtained a total of 724,452,816 raw reads from *C. fortunei* cambium total RNA across all of our samples. From these, we assembled five complete transcriptomes, one for each growth stage (Table 1). The percentage of raw reads with a Q-value > 30 ranged from 92.33% to 94.89% for all samples, and the average GC content (the percentage of the total number of G's and C's in clean bases) was 44.01%. After removing low-quality reads and adaptor sequences, a total of 706,935,392 clean reads were obtained, which were used for de novo assembly. We obtained 78,673 unigenes using Trinity software, and found the average unigene length to be 957 bp, with an N50 length of 1576 bp. The sequence length distribution is shown in Supplementary Figure S1.

Table 1. Statistics of sequencing quality.

Sample	Raw_Reads	Clean_Reads	Clean_Bases	Valid_Bases (%)	Q30 [1] (%)	GC [2] (%)
A_1	46,571,826	45,283,706	6,697,931,982	95.88	93.64	43.53
A_2	53,925,378	52,724,482	7,839,827,405	96.92	93.96	43.81
A_3	53,242,134	51,808,658	7,669,074,710	96.03	93.55	43.83
B_1	46,927,200	45,482,626	6,725,348,970	95.54	93.11	44.00
B_2	52,320,204	50,588,050	7,464,083,436	95.11	92.91	43.84
B_3	47,901,258	46,465,580	6,850,549,711	95.34	92.33	44.10
C_1	48,353,956	47,494,412	7,067,068,800	97.44	94.45	44.15
C_2	49,948,000	49,090,122	7,312,542,131	97.60	94.33	44.56
C_3	48,035,404	47,186,328	7,004,509,885	97.21	94.30	44.44
D_1	46,718,046	45,434,522	6,720,860,268	95.91	93.01	44.09
D_2	48,454,586	47,792,530	7,068,774,362	97.26	94.85	44.03
D_3	43,553,002	42,426,072	6,280,307,490	96.13	92.99	43.71
E_1	47,230,764	46,004,016	6,795,292,538	95.92	93.20	44.04
E_2	47,250,800	45,758,216	6,745,283,335	95.17	92.51	44.04
E_3	44,020,258	43,396,072	6,443,702,336	97.59	94.89	44.03

[1] Q30 represents the percentage of bases whose phred number is greater than 30 in raw bases. [2] GC represents the percentage of the total number of G's and C's in clean bases.

3.2. Functional Annotation and Classification of All Unigenes

We next annotated all 78,673 unigenes using diamond software and HMMER software against seven public databases: NR, KOG, GO, Swiss-Prot, eggNOG, KEGG, and Pfam (Figure 1, Supplementary Table S2). Using these databases, 45,214 (57.47%), 26,866 (34.15%), 28,589 (36.34%), and 46,674 (59.33%) unigenes could be annotated in NR, Swiss-Prot, KOG, and eggNOG, respectively. Only 114 (0.14%) unigenes could be aligned to the Pfam database. We successfully annotated 24,312 (30.90%) unigenes into separate GO categories, including three functional categories: cellular component (CC), molecular function (MF), and biological process (BP), as well as 52 GO terms (Figure 2). In the 'cellular component' category, the most highly represented GO terms were 'cell' (20,396) and 'cell part' (20,364), while 'binding' (15,131) and 'catalytic activity' (13,131) were the two top GO terms in the 'molecular function' category. Additionally, regarding the 'biological process' category, these unigenes were clustered into 22 GO terms, with the three top terms being 'cellular process' (17,152), 'metabolic process' (14,158), and 'biological regulation' (6913). In total, 15,354 (19.52%) unigenes were assigned to 24 metabolic pathways (Figure 3). 'Translation' (1591), 'Signal transduction' (1515), and 'Carbohydrate metabolism' (1411) were the top three metabolic pathways.

Following this functional categorization, we continued analyzing transcription factor (TF) categorization across the different known plant TF families. As a result, we identified a total of 1401 TFs, which could be further classified into 66 TF families, such as WRKY, NAC, bZIP, and others (Supplementary Figure S2). We found the C2H2 family to be most abundant, with 233 unigenes, followed by AP2/ERF-ERF (115) and bHLH (100). We also found 71 and 34 unigenes encoding MYB and NAC TFs in this study.

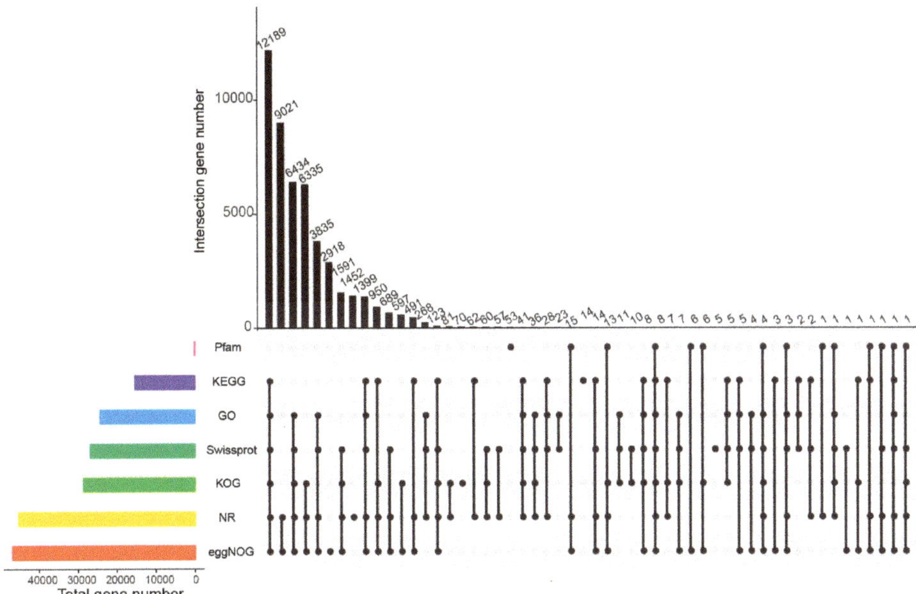

Figure 1. Unigenes functional annotation. The numbers on the top bar represent the results of the intersection of databases with black dots in the matrix below, and the columns on the left represent the total number of genes annotated to each database.

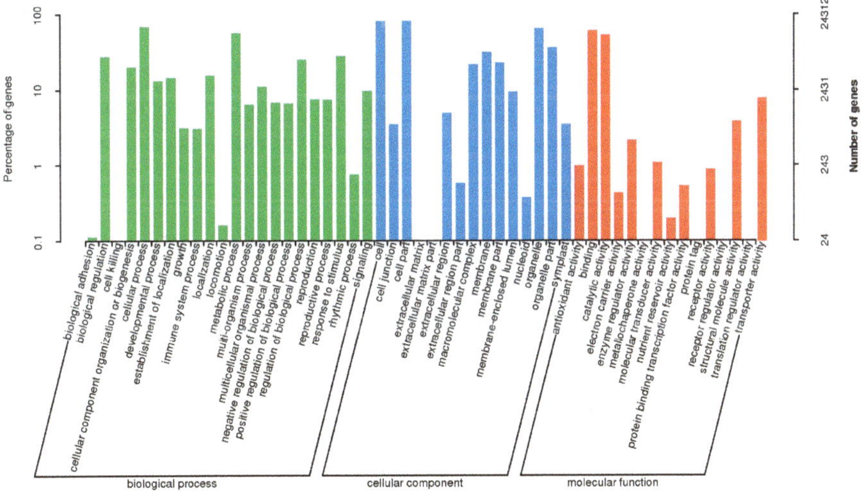

Figure 2. Functional distribution of Gene Ontology (GO) annotation. The *x*-axis represents the different GO functional classifications, the *y*-axis on the left and right respectively represent the percentage and absolute number of unigenes being annotated with each classification.

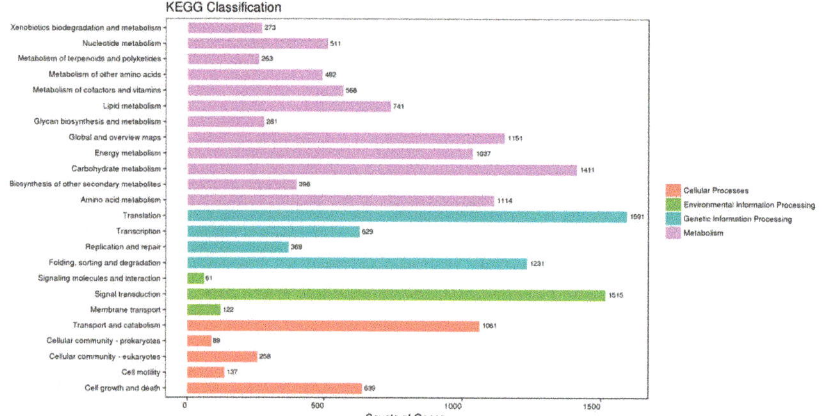

Figure 3. Functional distribution of Kyoto Encyclopedia of Genes and Genomes (KEGG) annotation. The *x*-axis represents the number of genes, and the *y*-axis represents the name of the KEGG metabolic pathway.

3.3. Identification of DEGs and the GO and KEGG Enrichment Analysis

In our study, we calculated unigene expression levels using the FPKM method and conducted annotation and enrichment analysis of DEGs. Firstly, we performed sample clustering analysis to obtain gene expression patterns of *C. fortunei* vascular cambium. The gene expression patterns of samples collected in September and November clustered together, while those from April and July did as well, making the samples collected in May their own cluster (Figure 4). These results indicate a higher sample similarity between these samples. In addition, the three biological replicates were clustered together, indicating the reliability of our transcriptome data.

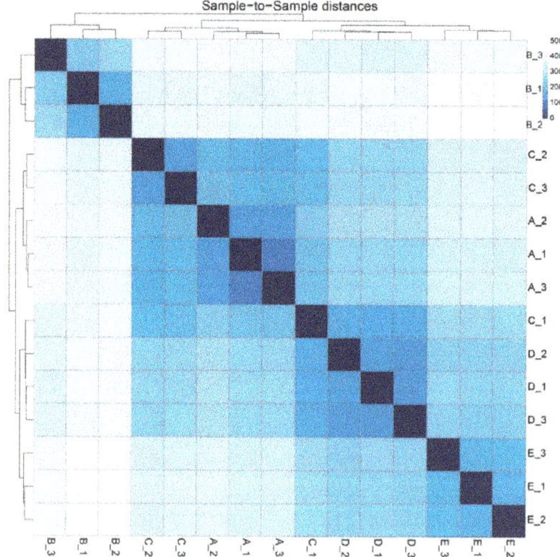

Figure 4. Gene expression analysis. Sample clustering of 5 growth stages. Sample correlation is based on their gene expression profile. The color scale represents the correlation coefficient. The sample clustering was used to investigate the expression patterns of genes at 5 successive growth stages in *C. fortunei* vascular cambium.

We then identified all *C. fortunei* vascular cambium DEGs through pairwise comparisons. The amount of DEGs for each pairwise comparison is shown in Supplementary Figure S3. With A as a reference, we compared B vs. A, C vs. A, D vs. A, and E vs. A: 8975, 4432, 11,683, and 17,774 unigenes were differentially expressed in all four pairwise comparisons, respectively.

In a previous study, we observed the development of *C. fortunei* vascular cambium by studying their morphology using paraffin sections [25]. We found that cellular growth and development were most vigorous in May. Here, we chose B vs. A (May vs. April) as an example to explain the DEGs' functionality. A total of 4165 DEGs were found through GO enrichment analysis in B vs. A, of which 3230 and 935 were up- and down-regulated, respectively (Figure 5A). To describe our GO annotation results, we constructed a directed acyclic graph (DAG) using topGO [26] (Figure 5B). The most significant enrichment in the 'biological process' category is 'secondary metabolic process' (GO: 0019748) and 'phenylpropanoid metabolic process' (GO: 0009698).

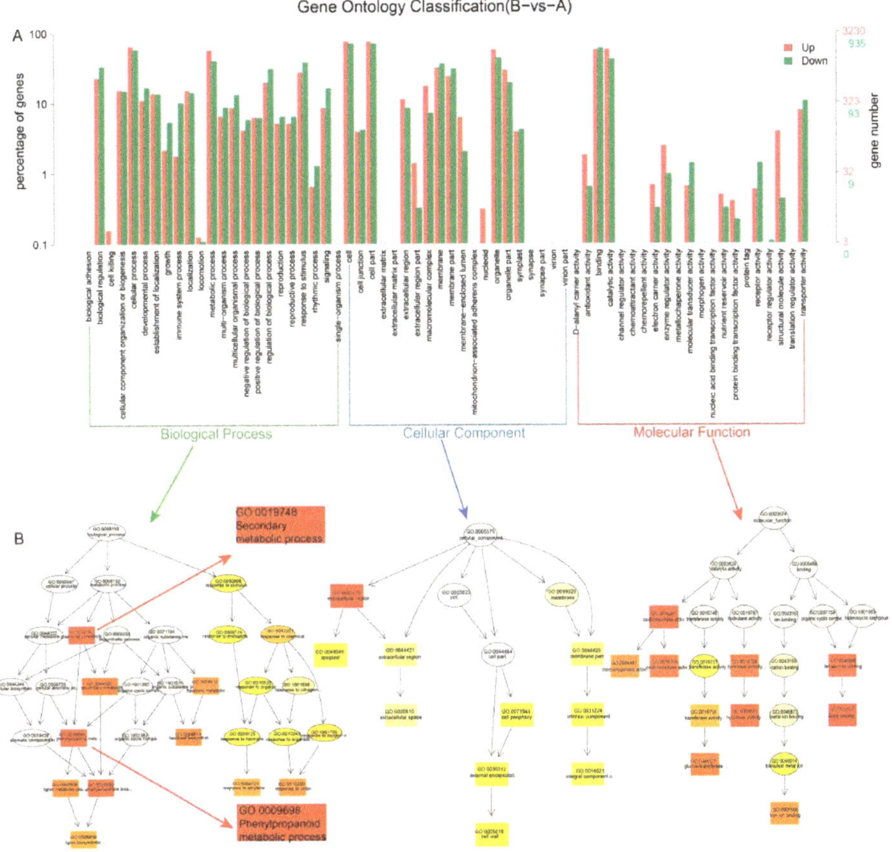

Figure 5. Gene Ontology (GO) functional classification of Differentially Expressed Genes (DEGs). (**A**) The *x*-axis represents the enriched GO terms. The *y*-axis represents the number and percentage of up- and down-regulated DEGs. (**B**) Directed acyclic graphs (DAGs) of Biological Process (BP), Cellular Component (CC), and Molecular Function (MF). The nodes are colored based on the *q*-value, and red indicates a high confidence level. The GO terms are presented at the horizontal node position. The red arrows represent 'secondary metabolic process' (GO: 0019748) and 'phenylpropanoid metabolic process' (GO: 0009698), respectively.

To further explore DEGs biological function, we performed a KEGG enrichment analysis: 2628 DEGs were successfully annotated into 24 pathways in B vs. A (Figure 6). We found 146 DEGs, of which 134 and 12 were up- and down-regulated, annotated to the secondary metabolism pathway, which indicates that these DEGs are involved in secondary metabolism.

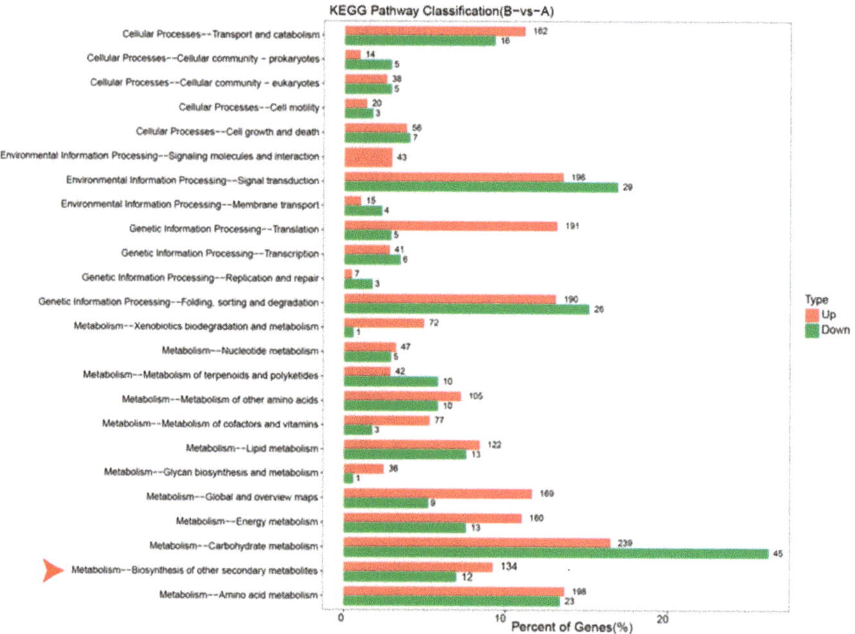

Figure 6. KEGG enrichment analysis of DEGs. The *x*-axis represents the percentage of genes. The *y*-axis represents the name of the KEGG metabolic pathway. Red arrow indicates 'Biosynthesis of other secondary metabolites'.

3.4. Identification of Candidate Genes Involved in Phenylpropanoid Metabolism

In order to understand how gene activity changes between growth stages, we continued by comparing KEGG pathway enrichment of DEGs (Figure 7). One of the main KEGG pathways undergoing enrichment dynamics was phenylpropanoid metabolism (ko00940). We identified 83 DEGs involved in this pathway in B vs. A, of which 76 DEGs were upregulated.

Lignin is mainly synthesized through the phenylpropanoid metabolic pathway (Figure 8), phenylalanines are converted to monolignols by the enzymes phenylalanine ammonia-lyase (PAL) (4.3.1.24), shikimate O-hydroxycinnamoyl transferase (HCT) (2.3.1.133), p-coumarate 3-hydroxylase (C3H) (1.14.13.36), 4-(hydroxy) cinnamoyl CoA ligase (4CL) (6.2.1.12), caffeoyl CoA O-methyltransferase (CCoAOMT) (2.1.1.104), cinnamoyl CoA reductase (CCR) (1.2.1.44), and cinnamyl alcohol dehydrogenase (CAD) (1.1.1.195). In B vs. A, we identified 8, 5, 2, 9, 4, and 4 unigenes encoding PAL, HCT, C3H, 4CL, CCoAOMT, and CCR respectively, most of which were upregulated.

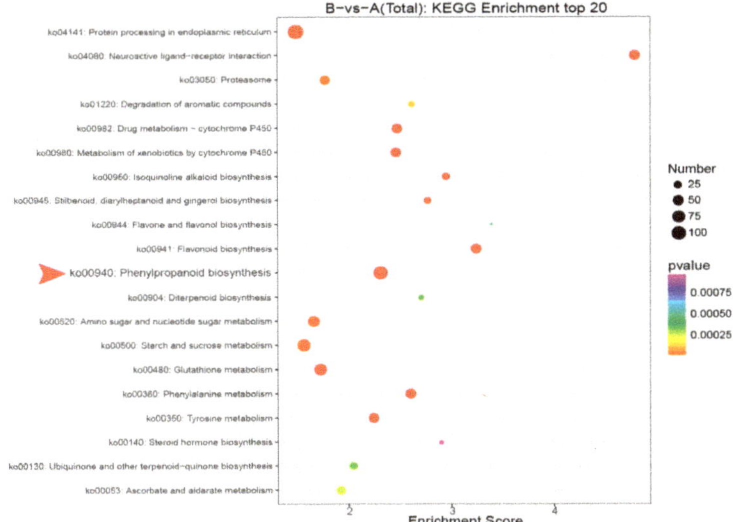

Figure 7. Bubble chart of KEGG enrichment. The *x*-axis represents the enrichment score. The *y*-axis represents the name of the KEGG metabolic pathway. The dot size indicates the number of DEGs, and the dot color indicates the *p*-value; the smaller the *p*-value, the greater the significance. Red arrow indicates the 'ko00940: Phenylpropanoid biosynthesis' pathway.

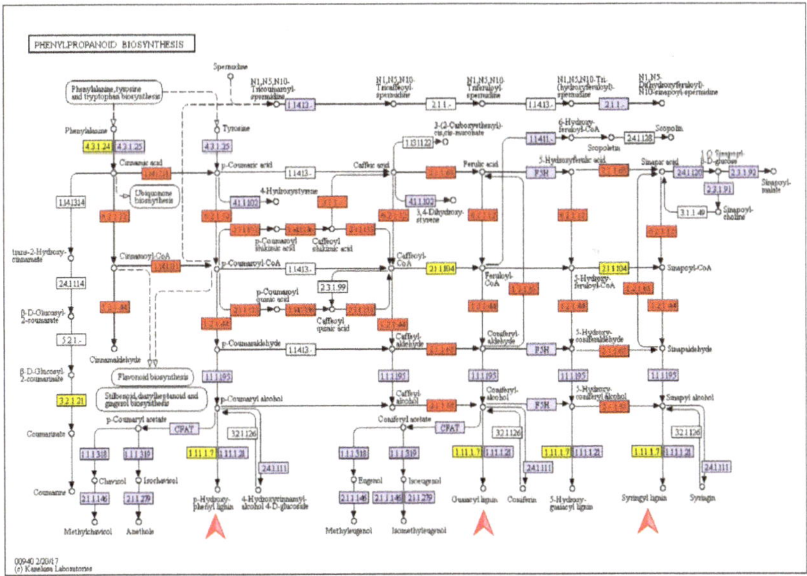

Figure 8. Pathway assignments based on the Kyoto Encyclopedia of Genes and Genomes (KEGG). A schematic representation of the phenylpropanoid biosynthesis pathway. The number in the rectangle indicates the corresponding enzyme. Red indicates upregulated unigenes, green indicates downregulated, yellow indicates unigenes that were both up- and down-regulated, and gray indicates no DEGs. Red arrows indicate H-type lignin, G-type lignin, and S-type lignin, respectively.

We then analyzed the expression of key enzymes involved in phenylpropanoid biosynthesis (Figure 9). All of them had different expression patterns at different stages. Five enzymes, including C3H, CCR, 4CL, PAL, and CCoAOMT, were all present at higher expression levels at stage_May. Two enzymes, including PAL and HCT, showed higher expression levels at stage_November, which indicates that these enzymes could play roles in response to cold stress. Most enzymes displayed lower expression levels at stage_April and November than at other stages, which could be caused by the seasonally cyclical pattern of dormancy and activity. Furthermore, the expression of these enzymes increased from July to September and decreased again from September to November. This finding is consistent with the general trend of lignin content. In the present study, the lignin content increased gradually from April (10.88%) to September (34.56%) and stabilized from September to November (34.75%) (Supplementary Figure S4). These phenomena revealed that key enzymes involved in phenylpropanoid biosynthesis might be responsible for the seasonal change in wood formation activity.

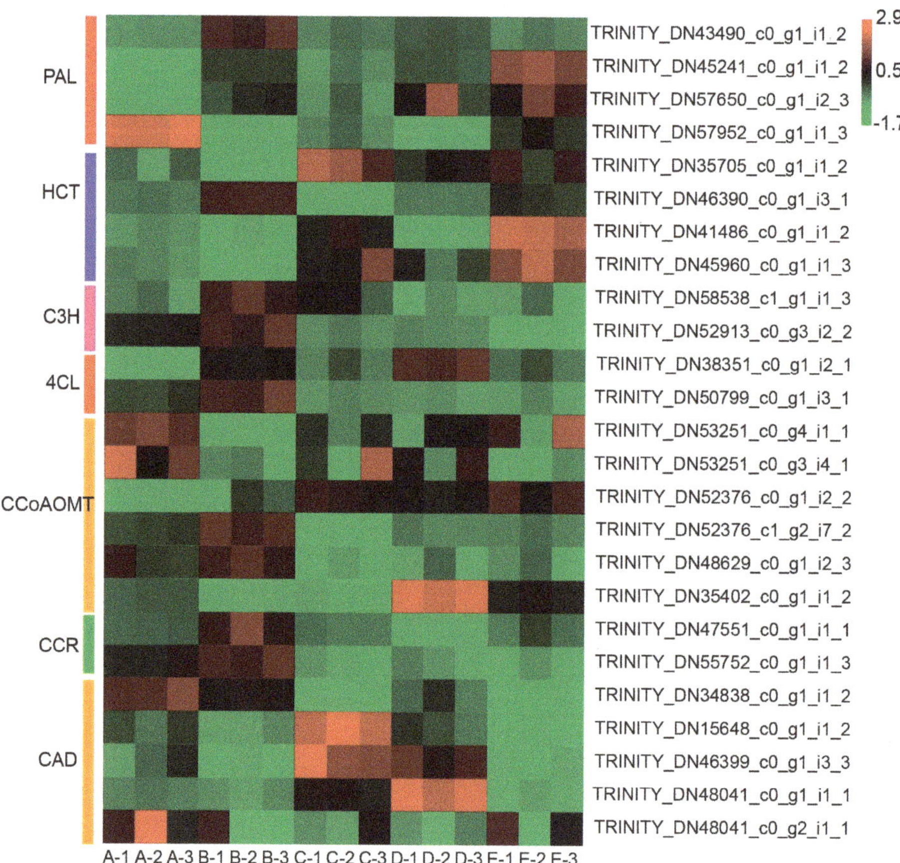

Figure 9. Expression levels of candidate genes associated with phenylpropanoid biosynthesis. PAL: Phenylalanine ammonia-lyase, HCT: Shikimate O-hydroxycinnamoyl transferase, C3H: Coumaroylquinate-3′-monooxygenase, 4CL: 4-coumarate-CoA ligase, CCoAOMT: Caffeoyl-CoA O-methyltransferase, CCR: Cinnamoyl-CoA reductase, CAD: Cinnamyl-alcohol dehydrogenase.

3.5. Quantitative Real-Time PCR Validation of Candidate Genes Involved in Phenylpropanoid Biosynthesis

In this study, we randomly selected eight unigenes involved in phenylpropanoid biosynthesis and examined their expression levels using qRT-PCR. We firstly performed the melt curve analysis. We found that all samples (including 3 replicates) have a single peak and the temperature is between 80 and 90 °C (Supplementary Figure S5), which indicates that the data is reliable. The expression profiles of these candidates are shown in Supplementary Figure S6. Although the exact fold changes between stages of each unigene varied somewhat between RNA-seq and qRT-PCR data, the trends between the different stages were overall similar. We could find just one candidate gene (TRINITY_DN57952_c0_g1_i1_3) of which the expression values were inconsistent with our RNA-seq data. Therefore, these results confirm the accurate assembly of the transcript sequences and reliability of our RNA-seq data.

4. Discussion

Wood is an important raw material with a rapidly increasing worldwide demand. As a result, more research is being devoted to analyzing the genetic regulation of wood formation. An important tool for such research is transcriptome sequencing, which can be used to discover genes that control economic traits. In a previous study, we identified the different expression pattern of reproductive genes in two conifer species through transcriptome sequencing [27]. In this study, approximately 72.44 million paired-end reads were obtained. After assembly, we obtained 78,673 unigenes, with an average length of 957 bp, significantly longer than has been reported previously for *Cunninghamia lanceolata* (Lamb.) Hook (449 bp) [28], *Camellia sinensis* (L.) O. Ktze. (355 bp) [29], and *Porphyra yezoensis* (Rhodophyta) (419 bp) [30], and slightly shorter than *C.japonica* (1069 bp) [31], and thereby providing more abundant genetic information to understand the mechanism of lignin biosynthesis.

Lignin is mainly synthesized through the phenylpropanoid metabolic pathway. We aimed to find DEGs involved in lignin biosynthesis. As a result, we obtained 83, 29, 49, and 74 DEGs in four pairwise comparisons, of which 76, 22, 43, and 46 DEGs were upregulated, respectively. We found most DEGs when comparing the growth stages May vs. April and November vs. April, which is most likely because of the seasonally cyclical pattern of dormancy and activity in wood formation. In this study, we found 8, 3, 4, and 8, and 9, 2, 1, and 1 DEGs encoding PAL and 4CL respectively, in four pairwise comparisons, most of which were upregulated. It is consistent with the expression patterns in Figure 9. Similarly, Mishima et al. [31] found the homologues, *PAL4* and *4CL3*, and showed an increasing expression pattern during cessation of growth. This finding indicates that PAL and 4CL might be regulated to the cold stress. We also found most DEGs encoding CAD, CCoAOMT, and CCR to be upregulated. Most enzymes were induced in April, and expression level gradually decreased from August to October in *C. japonica* [31]. The expression pattern corresponded to our previous study about anatomical observation of cambium cells [25]. Previous reports have found that CAD [32], CCoAOMT [11], and CCR [33] promote lignin synthesis, which is consistent with the expression patterns of these enzymes (Figure 9) and the associated corresponding changes in lignin content.

Previous studies have found that temperature plays an important role in dormancy development [34]. During our study, the daily maximum and minimum temperatures increased from 4 April to 18 May and peaked at 10 July (Supplementary Table S3). Subsequently, temperatures steadily declined from July to November. Similarly, from our previous work, we found cambium cells undergoing peri-planar divisions, with the largest number of layers, full cells, cytoplasm, and vigorous divisions in May, while the number of cell layers decreased in July and September compared to May, and with the number of cells being the least and division stopping in November [25]. Furthermore, we found that the expression of most lignin-synthesizing enzymes in *C. fortunei* vascular cambium changed in tandem with rising and falling temperatures, consistent with the growth season and previous studies [31]. Analogously, the growing season of Japanese cedar, another gymnosperm, runs from March until October, with lignin synthesis activity sharply increasing from March to June, then declining until dormancy in October [31]. Interestingly, Sato et al. [35] analyzed the diurnal periodicity of expression of lignin synthesizing genes in *C. japonica* and found that most enzymes showed different expression abundance at different times on the same day. This paper provides a good research direction for our future study.

Previous reports have found TFs involved in the regulation of lignin biosynthesis [36–38], response to the abiotic stress [39–43], and regulation of growth and developmental processes [44–48]. MYB TFs could regulate lignin synthesis by binding to the corresponding regulatory elements of lignin-synthesizing enzyme genes, whereas NAC TFs could regulate the corresponding MYB TFs to regulate lignin synthesis. During this study, we obtained 37, 16, 33, and 37 DEGs encoding MYB TFs in all four pairwise comparisons, of which 26, 10, 26, and 20 were upregulated. Similarly, Mishima et al. [31] found that 34 MYB were upregulated during the peak activity of xylem formation. These results are consistent with our findings in this paper. In addition, we found that these TFs regulated some metabolic pathways, including the phenylpropanoid metabolic pathway. We also analyzed the expression patterns of NAC TFs and obtained 15, 4, 12, and 12 NAC TFs showing differential expression in all four pairwise comparisons. Similarly, most TFs were upregulated. Mishima et al. [31] found a VND6 homolog and its expression was moderately decreased during peak xylem formation. According to the expression patterns of lignin-synthesizing enzymes in this study, these results imply a potential role of TFs in the regulation of lignin biosynthesis.

5. Conclusions

C. fortunei is a plant tree species that has a large number of excellent qualities, such as rapid growth, a straight bole, and ease of processing for wood production. In this study, we performed transcriptome sequencing on *C. fortunei* vascular cambium for 5 successive growth stages. We identified candidate genes involved in phenylpropanoid metabolism and analyzed expression patterns of lignin-synthesizing enzymes. Finally, we found the correlation of enzyme expression with different growth stages. Thus, our findings contribute to a better understanding of the molecular mechanisms underlying the phenylpropanoid biosynthesis pathway. Importantly, these results may be useful for molecular breeding of *C. fortunei* to improve wood characteristics.

Supplementary Materials: The following are available online at http://www.mdpi.com/1999-4907/11/7/766/s1: Figure S1: De novo transcriptome assembly, Figure S2: Distribution of TFs according to their TF family, Figure S3: Identification of DEGs. The *x*-axis represents the comparison groups, and the *y*-axis represents the number of DEGs, Figure S4: The lignin content of 5 growth stages, error bars represent standard deviation of three biological replicates, Figure S5: The melt curve of qRT-PCR, Figure S6: qPCR validation of RNA-seq data. The *x*-axis represents the growth stages, and the *y*-axis represents the fold change (log2). Table S1: Primers used for qRT-PCR, Table S2: Functional annotation of unigenes, Table S3: The daily maximum and minimum temperatures at the sampling date.

Author Contributions: Conceptualization, J.X.; formal analysis, J.Y., Y.Z., Z.G., J.M., J.C., H.H., and Y.H.; data curation, J.Y.; writing—original draft, J.Y.; writing—review and editing, J.Y. and J.M.; supervision, J.X.; funding acquisition, J.X.; project administration, J.X. All authors have read and agreed to the published version of the manuscript.

Funding: This research was funded by National Forestry and Grassland Administration of China, forestry public welfare industry research (201304104), and the Priority Academic Program Development of Jiangsu Higher Education Institutions (PAPD).

Acknowledgments: We would like to give thanks to Remco A. Mentink of Plant Research International, Wageningen University and Research Centre, for his careful scientific revision on the manuscript, and Tianci Yan of China Agricultural University for his careful revision of the figures. We also thank reviewers for insightful comments on this article.

Conflicts of Interest: The authors declare no conflict of interest.

References

1. Zhao, H.; Mo, J.X.; Hua, H.; Guo, Z.H.; Xu, J. Meiosis process and abnormal behavior of pollen mother cells in *Cryptomeria fortunei*. *J. Nanjing For. Univ. (Nat. Sci. Ed.)* **2019**, *43*, 45–50. (In Chinese)
2. Růžička, K.; Ursache, R.; Hejátko, J.; Helariutta, Y. Xylem development—From the cradle to the grave. *New Phytol.* **2015**, *207*, 519–535. [CrossRef] [PubMed]

3. Yi, M.; Zhang, L.; Lei, L.; Cheng, Z.S.; Sun, S.W.; Lai, M. Analysis of SSR information in transcriptome and development of EST-SSR molecular markers in *Pinus elliottii* Engelm. *J. Nanjing For. Univ. (Nat. Sci. Ed.)* **2020**, *44*, 75–83. (In Chinese)

4. Fagard, M.; Höfte, H.; Vernhettes, S. Cell wall mutants. *Plant Physiol. Biochem.* **2000**, *38*, 15–25. [CrossRef]

5. Soler, M.; Plasencia, A.; Larbat, R.; Pouzet, C.; Jauneau, A.; Rivas, S.; Pesquet, E.; Lapierre, C.; Truchet, I.; Grima-Pettenati, J. The Eucalyptus linker histone variant EgH1.3 cooperates with the transcription factor EgMYB1 to control lignin biosynthesis during wood formation. *New Phytol.* **2016**, *213*, 287–299. [CrossRef]

6. Shi, R.; Sun, Y.-H.; Li, Q.Z.; Heber, S.; Sederoff, R.; Chiang, V.L. Towards a systems approach for lignin biosynthesis in *Populus trichocarpa*: Transcript abundance and specificity of the monolignol biosynthetic genes. *Plant Cell Physiol.* **2009**, *51*, 144–163. [CrossRef]

7. Jokipii-Lukkari, S.; Delhomme, N.; Schiffthaler, B.; Mannapperuma, C.; Prestele, J.; Nilsson, O.; Street, R.N.; Tuominen, H. Transcriptional Roadmap to Seasonal Variation in Wood Formation of Norway Spruce. *Plant Physiol.* **2018**, *176*, 2851–2870. [CrossRef]

8. Boerjan, W.; Ralph, J.; Baucher, M. Ligninbiosynthesis. *Annu. Rev. Plant Boil.* **2003**, *54*, 519–546. [CrossRef] [PubMed]

9. Li, L.; Cheng, X.F.; Leshkevich, J.; Umezawa, T.; Harding, S.A.; Chiang, V.L. the last step of syringyl monolignol biosynthesis in angiosperms is regulated by a novel gene encoding sinapyl alcohol dehydrogenase. *Plant Cell* **2001**, *13*, 1567–1586. [CrossRef] [PubMed]

10. Zhang, X.-H.; Chiang, V.L. Molecular cloning of 4-coumarate:coenzyme A ligase in loblolly pine and the roles of this enzyme in the biosynthesis of lignin in compression wood. *Plant Physiol.* **1997**, *113*, 65–74. [CrossRef] [PubMed]

11. Guo, Z.; Hua, H.; Xu, J.; Mo, J.; Zhao, H.; Yang, J. Cloning and Functional Analysis of Lignin Biosynthesis Genes Cf4CL and CfCCoAOMT in *Cryptomeria fortunei*. *Genes* **2019**, *10*, 619. [CrossRef]

12. Qiu, Z.; Wan, L.; Chen, T.; Wan, Y.; He, X.; Lu, S.; Wang, Y.; Lin, J. The regulation of cambial activity in Chinese fir (*Cunninghamia lanceolata*) involves extensive transcriptome remodeling. *New Phytol.* **2013**, *199*, 708–719. [CrossRef] [PubMed]

13. Grabherr, M.G.; Haas, B.J.; Yassour, M.; Levin, J.Z.; Thompson, D.A.; Amit, I.; Adiconis, X.; Fan, L.; Raychowdhury, R.; Zeng, Q.; et al. Full-length transcriptome assembly from RNA-Seq data without a reference genome. *Nat. Biotechnol.* **2011**, *29*, 644–652. [CrossRef] [PubMed]

14. Li, W.; Jaroszewski, L.; Godzik, A. Clustering of highly homologous sequences to reduce the size of large protein databases. *Bioinformatics* **2001**, *17*, 282–283. [CrossRef] [PubMed]

15. Buchfink, B.; Xie, C.; Huson, D.H. Fast and sensitive protein alignment using DIAMOND. *Nat. Methods* **2014**, *12*, 59–60. [CrossRef]

16. Mistry, J.; Finn, R.D.; Eddy, S.R.; Bateman, A.; Punta, M. Challenges in homology search: HMMER3 and convergent evolution of coiled-coil regions. *Nucleic Acids Res.* **2013**, *41*, e121. [CrossRef]

17. Roberts, A.; Trapnell, C.; Donaghey, J.; Rinn, J.L.; Pachter, L. Improving RNA-Seq expression estimates by correcting for fragment bias. *Genome Boil.* **2011**, *12*, R22. [CrossRef]

18. Langmead, B.; Salzberg, S.L. Fast gapped-read alignment with Bowtie *Nat. Methods* **2012**, *9*, 357–359. [CrossRef]

19. Roberts, A. Ambiguous Fragment Assignment for High-Throughput Sequencing Experiments. Ph.D. Thesis, University of California, Berkeley, CA, USA, 2013.

20. Anders, S.; Huber, W. *Differential Expression of RNA-Seq Data at the Gene Level–the DESeq Package*; European Molecular Biology Laboratory (EMBL): Heidelberg, Germany, 2012.

21. Wu, Y.; Guo, J.; Zhou, Q.; Xin, Y.; Wang, G.; Xu, L.-A. De novo transcriptome analysis revealed genes involved in flavonoid biosynthesis, transport and regulation in *Ginkgo biloba*. *Ind. Crop. Prod.* **2018**, *124*, 226–235. [CrossRef]

22. Wang, L.; Feng, Z.; Wang, X.; Wang, X.; Zhang, X. DEGseq: An R package for identifying differentially expressed genes from RNA-seq data. *Bioinformatics* **2009**, *26*, 136–138. [CrossRef]

23. Livak, K.J.; Schmittgen, T.D. Analysis of relative gene expression data using real-time quantitative PCR and the $2^{-\Delta\Delta CT}$ method. *Methods* **2001**, *25*, 402–408. [CrossRef]

24. Rodrigues, J.; Faix, O.; Pereira, H. Improvement of the acetylbromide method for lignin determination within large scale screening programmes. *Holz als Roh-und Werkst.* **1999**, *57*, 341–345. [CrossRef]

25. Cao, Y.T. Transcriptome Sequencing and Profiling of Expression in Vascular Cambium of *Cryptomeria fortunei*. Master's Thesis, Nanjing Forestry University, Nanjing, China, 2016. (In Chinese)

26. Alexa, A.; Rahnenfuhrer, J. *Enrichment Analysis for Gene Ontology*; R Package Version 2.8; Bioconductor: Buffalo, NY, USA, 2010.

27. Mo, J.; Xu, J.; Cao, Y.; Yang, L.; Yin, T.; Hua, H.; Zhao, H.; Guo, Z.; Yang, J.; Shi, J. *Pinus massoniana* Introgression Hybrids Display Differential Expression of Reproductive Genes. *Forests* **2019**, *10*, 230. [CrossRef]

28. Huang, H.; Xu, L.-L.; Tong, Z.; Lin, E.; Liu, Q.; Cheng, L.J.; Zhu, M.-Y. De novo characterization of the Chinese fir (*Cunninghamia lanceolata*) transcriptome and analysis of candidate genes involved in cellulose and lignin biosynthesis. *BMC Genom.* **2012**, *13*, 648. [CrossRef] [PubMed]

29. Shi, C.-Y.; Yang, H.; Wei, C.; Yu, O.; Zhang, Z.; Jiang, C.; Sun, J.; Li, Y.; Chen, Q.; Xia, T.; et al. Deep sequencing of the *Camellia sinensis* transcriptome revealed candidate genes for major metabolic pathways of tea-specific compounds. *BMC Genom.* **2011**, *12*, 131. [CrossRef] [PubMed]

30. Yang, H.; Mao, Y.; Kong, F.; Yang, G.; Ma, F.; Wang, L. Profiling of the transcriptome of *Porphyra yezoensis* with Solexa sequencing technology. *Chin. Sci. Bull.* **2011**, *56*, 2119–2130. [CrossRef]

31. Mishima, K.; Fujiwara, T.; Iki, T.; Kuroda, K.; Yamashita, K.; Tamura, M.; Fujisawa, Y.; Watanabe, A. Transcriptome sequencing and profiling of expressed genes in cambial zone and differentiating xylem of Japanese cedar (*Cryptomeria japonica*). *BMC Genom.* **2014**, *15*, 219. [CrossRef] [PubMed]

32. Zhang, C.H.; Xiong, Z.H.; Wu, W.L.; Li, W.L. CAD enzyme activity and gene expression in connection with lignin synthesis during fruit development and ripening process of blackberry. *J. Nanjing For. Univ. (Nat. Sci. Ed.)* **2018**, *42*, 141–148. (In Chinese)

33. Douglas, C.J. Phenylpropanoid metabolism and lignin biosynthesis: From weeds to trees. *Trends Plant Sci.* **1996**, *1*, 171–178. [CrossRef]

34. Heide, O.M. Growth and dormancy in norway spruce ecotypes (*Picea abies*) i. interaction of photoperiod and temperature. *Physiol. Plant.* **1974**, *30*, 1–12. [CrossRef]

35. Sato, S.; Yoshida, M.; Ashizaki, Y.; Yamamoto, H. Diurnal periodicity of the expression of genes involved in monolignol biosynthesis in differentiating xylem of cryptomeria japonica. *Am. J. Plant Sci.* **2016**, *7*, 2457–2469. [CrossRef]

36. Zhong, R.; Ye, Z. Transcriptional regulation of lignin biosynthesis. *Plant Signal. Behav.* **2009**, *4*, 1028–1034. [CrossRef] [PubMed]

37. Lv, Y.Z.; Zheng, J.; Chen, J.H.; Shi, J.S. Cloning transcription factor ClMYB4 involving in secondary cell wall biosynthesis from *Cunninghamia lanceolata* (Lamb) Hook and expressing in *E. coli*. *Mol. Plant Breed.* **2012**, *10*, 512–519. (In Chinese)

38. Cassan-Wang, H.; Goué, N.; Saidi, M.N.; Legay, S.; Sivadon, P.; Goffner, D.; Grima-Pettenati, J. Identification of novel transcription factors regulating secondary cell wall formation in Arabidopsis. *Front. Plant Sci.* **2013**, *4*, 189. [CrossRef]

39. Movahedi, A.; Zhang, J.; Yin, T.; Zhuge, Q. Functional Analysis of Two Orthologous NAC Genes, CarNAC3, and CarNAC6 from *Cicer arietinum*, Involved in Abiotic Stresses in Poplar. *Plant Mol. Boil. Rep.* **2015**, *33*, 1539–1551. [CrossRef]

40. Wang, J.; Xu, M.; Gu, Y.; Xu, L. Differentially expressed gene analysis of *Tamarix chinensis* provides insights into NaCl-stress response. *Trees* **2016**, *31*, 645–658. [CrossRef]

41. Liu, X.; Yu, W.; Zhang, X.; Wang, G.; Cao, F.-L.; Cheng, H. Identification and expression analysis under abiotic stress of the R2R3-MYB genes in *Ginkgo biloba* L. *Physiol. Mol. Boil. Plants* **2017**, *23*, 503–516. [CrossRef]

42. Wang, Y.; Qiu, W.M.; Li, H.; He, X.L.; Liu, M.Y.; Han, X.J.; Qu, T.B.; Zhuo, R.Y. Research on the response of SaWRKY7 gene to cadmium stress in *Sedum alfredii* Hance. *J. Nanjing For. Univ. (Nat. Sci. Ed.)* **2019**, *43*, 59–66. (In Chinese)

43. Tian, X.Y.; Zhou, J.; Wang, B.S.; He, K.Y.; He, X.D. Cloning and expression pattern analysis of NAC genes in Salix. *J. Nanjing For. Univ. (Nat. Sci. Ed.)* **2020**, *44*, 119–124. (In Chinese)

44. Yang, C.; Xu, M.; Xuan, L.; Jiang, X.; Huang, M. Identification and expression analysis of twenty ARF genes in Populus. *Gene* **2014**, *544*, 134–144. [CrossRef]

45. Bi, C.; Xu, Y.; Ye, Q.; Yin, T.; Ye, N. Genome-wide identification and characterization of WRKY gene family in *Salix suchowensis*. *PeerJ* **2016**, *4*, e2437. [CrossRef] [PubMed]

46. Wang, P.; Cheng, T.; Lu, M.; Liu, G.; Li, M.; Shi, J.; Lu, Y.; Laux, T.; Chen, J. Expansion and Functional Divergence of AP2 Group Genes in Spermatophytes Determined by Molecular Evolution and Arabidopsis Mutant Analysis. *Front. Plant Sci.* **2016**, *7*, 1585. [CrossRef] [PubMed]

47. Mo, Z.H.; Li, F.D.; Su, W.C.; Cao, F.; Peng, F.R.; Li, Y.R. Cloning and expression analysis of CiMYB46 in the graft healing process of *Carya illinoinensis*. *J. Nanjing For. Univ. (Nat. Sci. Ed.)* **2019**, *43*, 156–162. (In Chinese)

48. Chen, G.; Wang, L.; Li, L.; Li, Y.; Li, H.; Ding, W.; Shi, T.; Yang, X.; Wang, L. Genome-Wide Identification of the Auxin Response Factor (ARF) Gene Family and Their Expression Analysis during Flower Development of *Osmanthus fragrans*. *Forests* **2020**, *11*, 245. [CrossRef]

 *forests*

Article

Profiling of Widely Targeted Metabolomics for the Identification of Secondary Metabolites in Heartwood and Sapwood of the Red-Heart Chinese Fir (*Cunninghamia Lanceolata*)

Sen Cao [1], Zijie Zhang [1], Yuhan Sun [1], Yun Li [1,*] and Huiquan Zheng [2,*]

[1] College of Biological Sciences and Technology, National Engineering Laboratory for Tree Breeding, Beijing Advanced Innovation Center for Tree Breeding by Molecular Design, Beijing Forestry University, Beijing 100083, China; caosen0419@bjfu.edu.cn (S.C.); zijiezhang@bjfu.edu.cn (Z.Z.); syh831008@163.com (Y.S.)

[2] Guangdong Provincial Key Laboratory of Silviculture, Protection and Utilization, Guangdong Academy of Forestry, Guangzhou 510520, China

* Correspondence: yunli@bjfu.edu.cn (Y.L.); zhenghq@sinogaf.cn (H.Z.); Tel./Fax: +86-10-6233-6094 (Y.L.)

Received: 20 July 2020; Accepted: 15 August 2020; Published: 18 August 2020

Abstract: The chemical composition of secondary metabolites is important for the quality control of wood products. In this study, the widely targeted metabolomics approach was used to analyze the metabolic profiles of heartwood and sapwood in the red-heart Chinese fir (*Cunninghamia lanceolata*), with an ultra-performance liquid chromatography-electrospray ionization tandem mass spectrometry system. A total of 224 secondary metabolites were detected in the heartwood and sapwood, and of these, flavonoids and phenolic acids accounted for 36% and 26% of the components, respectively. The main pathways appeared to be differentially activated, including those for the biosynthesis of phenylpropanoids and flavonoids. Moreover, we observed highly significant accumulation of naringenin chalcone, dihydrokaempferol, pinocembrin, hesperetin, and other important secondary metabolites in the flavonoid biosynthesis pathway. Our results provide insight into the flavonoid pathway associated with wood color formation in Chinese fir that will be useful for further breeding programs.

Keywords: Chinese fir; heartwood; secondary metabolites; widely targeted metabolomics; flavonoids

---

1. Introduction

The xylem of most woody plants can be divided into three parts: lighter sapwood (SW) on the periphery, darker heartwood (HW) on the inside, and the transition zone (TZ) at the color junction. HW contributes most to the value of the wood, and in most cases, it has a deep color that is likely associated with secondary metabolites. SW is often referred to as living tissue (5–25% of the components) and HW is mostly considered dead cells. The associated regulatory genes for metabolite production have been difficult to clarify in these tissues, particularly in HW. Some tree species have a TZ that produces secondary metabolites on a temporary basis, and includes live parenchyma cells; it is generally believed that this metabolic activity peaks rapidly in the TZ and produces a large number of secondary metabolites that accumulate in the HW [1].

Given the value of HW, some studies have specifically focused on its color. Miyamoto et al. [2] found that the moisture and potassium contents differed significantly between two groups of reddish and blackish HW in Sugi (*Cryptomeria japonica*). The color of HW in Scots pine (*Pinus sylvestris*) and Norway spruce (*Picea abies*) significantly changes after heat treatment, which may be due to changes in the chemical properties of secondary metabolites [3]. Gierlinger et al. [4] concluded that the red hue (+a*) of Japanese larch (*Larix kaempferi*) HW was strongly correlated with the number of phenols (r = 0.84) and decay resistance (r = 0.63), suggesting that color measurements of larch HW could be valuable in tree breeding programs to create an optimized larch timber tree. In addition, Deklerck et al. [5] predicted the resistance of wood against fungal decay, and as such for wood quality by using DART-TOFMS, secondary metabolites, and Partial Least Squares. It is also worth noting that heritability of chemical composition is high, which implies that it may be possible to improve chemical composition through the genetic breeding of forest trees [6,7].

In plant species, color-related metabolites are referred to as polyphenols, and they include multiple phenolic groups that have strong antioxidant effects. Some studies have found that the color determination of wood flour is a good indicator of phenols. Higher phenol content and higher corrosion resistance can be achieved through breeding [6,7]. Flavonoids, which are polyphenols, mainly include anthocyanins, flavans, flavones, flavanones, flavonols, and chalcone and are also important components of plant secondary metabolites. They are synthesized from phenylalanine via the phenylpropanoid and flavonoid pathways in the cytoplasm, representing the most-studied pathways [8]. Flavonoids are widely found in colored fruits and flowers [9], and play important roles in a variety of plant functions, such as pigmentation, prevention of dormancy, fertility, prevention of damage by ultraviolet rays and plant pathogens, and prevention of biotic and abiotic stress [10]. They function as antioxidants that can prevent chronic diseases, including cardiovascular disease, certain types of cancer [11], and inflammation [12].

Widely targeted metabolomics based on multiple reaction monitoring (MRM) is a highly sensitive and accurate method to measure targeted metabolites and has the advantages of high throughput, high sensitivity, and wide coverage. This method has been commonly used to analyze metabolites in crop species, such as rice [13], black sesame [14], and Chrysanthemum morifolium [15], and has successfully identified many valuable metabolites. Widely targeted metabolomics has also been used to successfully identify some flavonoid compounds in crop species [16–18].

Chinese fir (*Cunninghamia lanceolata* (Lamb.) Hook. is a fast-growing conifer tree species native to China that is well known as a quality wood used in the timber industry. This conifer has been used as a major breeding subject of the tree improvement programs of China for over 50 years [19], and numerous varieties were highlighted for their breeding value. Among them, the red-heart Chinese fir is particularly valuable because of its reddish HW [20]. Previous studies on red-heart Chinese fir have mainly focused on the collection and preservation of superior trees, the variation in the growth material based on provenance and family, the construction of seed orchards, and the breeding of seedlings [21–23]; few studies have investigated the mechanism underlying the color change in the wood. Fan et al. [24] found that contents of cold water, hot water, benzyl alcohol, ash, and 1% NaOH extracts in HW were higher than SW in the red-heart Chinese fir. Analyses of the chemical constituents of ethanol extracts of red-heart Chinese fir have demonstrated that two compounds, namely, cedrol and sclareol, are more concentrated in HW than in SW of the same age, and that the content increases with age [25]. In this context, we used a widely targeted metabolomics method to identify the main differences in metabolites between the HW and SW of the red-heart Chinese fir and described a metabolite base for further breeding programs.

2. Materials and Methods

2.1. Plant Materials

We used wood samples of 9-year-old red-heart Chinese fir belonging to clone cx746, collected from Xiaokeng State Forest Farm (Guangdong, China, 24°70′ N, 113°81′ E, 328–339 m above sea level). Three cores were collected from each individual tree from the main stems at breast height (1.3 m) using a tree growth cone and three random individuals were selected in clone cx746. The wood samples were split into two groups, based on color, and included SW in the outer wood, which was light yellow, and HW in the inner wood with a deep red color. All samples were flash-frozen in liquid nitrogen and maintained at −80 °C for further use.

2.2. Extraction and Separation of Flavonoid Secondary Metabolites

The three red-heart Chinese fir wood core samples were mixed and freeze-dried under a vacuum, ground into a fine powder in liquid nitrogen, mixed thoroughly, and then crushed for 1.5 min at 30 Hz using a mixer mill (MM 400, Retsch) with zirconia beads. The samples (100 mg, per sample) were extracted with 1 mL 70% methanol on a rotating wheel at 4 °C in the dark for 12 h. After centrifugation at 10,000 g for 10 min at 4 °C, the extracts were filtered (SCAA-104, 0.22 μm pore size; ANPEL) and then analyzed by LC–MS/MS (ThermoFisher Scientific, SanJose, CA, USA). Quality control samples were prepared by mixing all the samples equally. During the analyses, the quality control samples were run every 10 injections to monitor the stability of the analysis conditions.

The samples (5 μL, per sample) were analyzed using an ultra-performance liquid chromatography electrospray ionization tandem mass spectrometry (UPLC-ESI-MS/MS; hereafter, UEMS; SHIMADZU, Kyoto, Japan) system (Shim-pack UFLC SHIMADZU CBM30A, http://www.shimadzu.com.cn/; MS, Applied Biosystems 4500 QTRAP, http://www.appliedbiosystems.com.cn/) equipped with a C18 column (Waters ACQUITY UPLC HSS T3, 1.8 μm, 2.1 mm × 100 mm). The mobile phases consisted of ultra-pure water containing 0.04 acetic acid as mobile phase A, and acetonitrile containing 0.04% acetic acid as mobile phase B. The A:B (v/v) gradient was as follows: 95:5 at 0 min, 5:95 at 11.0 min, 5:95 at 12.0 min, 95:5 at 12.1 min, and 95:5 at 15.0 min. The flow rate was maintained at 0.40 mL/min, and the column temperature was kept at 40 °C. All eluents were pure grades from Merck. The samples obtained above were also connected to an ESI-triple quadrupole-linear ion trap mass spectrometer (Applied Biosystems, Framingham, MA, USA) equipped with an atmospheric pressure chemical ionization (APCI) turbo-ion spray interface, operating in negative ion mode and controlled by Analyst 1.6.3 software (Sciex, Framingham, MA). The operating parameters of the APCI ion source were as follows: ESI temperature, 550 °C; Mass spec voltage, 5500 V; and curtain gas, 25 psi.

2.3. Metabolite Identification and Quantification

The identification and structural analyses of the primary and secondary spectral data of the metabolites detected by mass spectrometry were based on the MWDB database of Wuhan (China) Mettewell Biotechnology Co., Ltd (Wuhan, China). and public databases, including MassBank (http://www.massbank.jp/), KNAPSAcK (http://kanaya.naist.jp/KNApSAcK/) [26], HMDB (http://www.hmdb.ca/), MoToDB (http://www.ab.wur.nl/moto/), and METLIN (http://metlin.scripps.edu/index.php), and followed the standard metabolic operating procedures. Metabolomics data were processed using System Software Analyst (Version 1.6.3). Metabolite quantification was performed using MRM. Orthogonal partial least squares discriminant analysis (OPLS-DA) was performed on the identified metabolites. Metabolites with significant content differences were set as thresholds of variable importance in projection (VIP) ≥1 and fold change ≥2 or ≤0.5.

2.4. Data Analyses

Principal component analysis (PCA) was employed to compress the original data into several principal components to describe the characteristics of the original data set [27]. The PC represents the combination of variables that explains most of the variance, in descending order. The main parameter obtained from PCA, R2X, represents the interpretation rate of the original data after dimensionality reduction and can be used to judge the quality of the model.

Partial least squares-discriminant analysis (PLS-DA) was used to help identify the differential metabolites. We also combined orthogonal signal correction (OSC) with PLS-DA being an OPLS-DA that further decomposes the X matrix into two types of information: related or unrelated to Y. OPLS-DA could filter differential variables by removing unrelated differences. The prediction parameters for evaluating the OPLS-DA model are R2X, R2Y, and Q2, where R2X and R2Y represent the interpretation rate of the model built to the X and Y matrices, and Q2 represents the prediction ability of the model. The closer these three are to 1, the more stable the model. When Q2 > 0.5, the model can be considered effective, and when Q2 > 0.9, it can be considered excellent [28]. The OPLS-DA results enable a preliminary screening of differential metabolites between different tissues. In this study, a combination of the fold-change value and the VIP value was used to screen for differential metabolites. For the experiments, we used a completely randomized block design, and repeated the experiments three times. Microsoft Excel 2016 (Microsoft, Seattle, WA, USA) and IBM SPSS Statistics 24.0 (IBM Corporation, Armonk, NY, USA) were also used to evaluate the data assessed differential secondary metabolites. Differences in metabolites in HW and SW were evaluated using analysis of variance (ANOVA) with Duncan's multiple range tests for multiple comparisons. A p-value $\leq$ 0.05 for the ANOVA F-test was considered statistically significant.

3. Results

3.1. Metabolic Profiling of Heartwood and Sapwood Based on LC-MS/MS

To investigate differences in the composition of secondary metabolites between the HW and SW of the red-heart Chinese fir, widely targeted metabolomics assay was performed, which analyzed the metabolic spectrum using UEMS. Metabolites were quantitatively analyzed following the collection of secondary data using the MRM model. Collectively, a total of 224 secondary metabolites were covered in our assay (HW and SW; Figure 1A), including 80 flavonoids, 58 phenolic acids, 14 alkaloids, 9 lignans and coumarins, 7 tannins, 4 terpenoids, and 52 others (Figure 1B). Among these secondary metabolites, the most abundant were flavonoids.

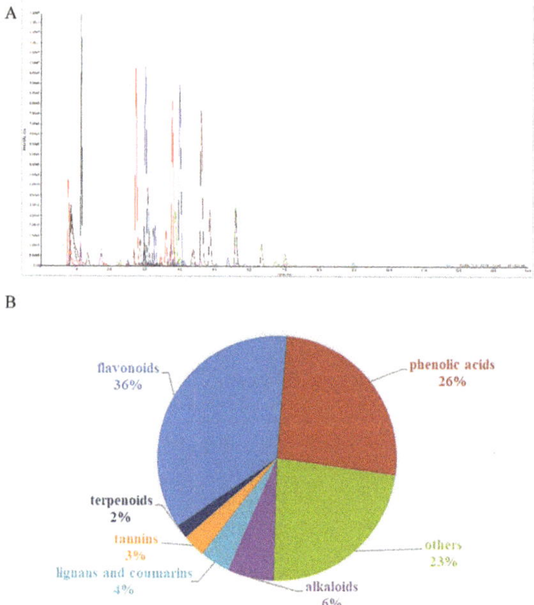

Figure 1. (**A**) The extracting ion current (XIC) chromatogram for the heart- and sapwood mix samples in a preliminary assay. (**B**) Types and proportions of differential secondary metabolites between heartwood and sapwood in the red-heart Chinese fir.

3.2. PCA

To compare the secondary metabolites in the HW and SW, the dataset obtained from the LC-MS/MS in ESI⁻ mode was subjected to PCA. In the PC1 × PC2 value plot (Figure 2B), the samples were separated in the first principal component (PC1), which reached 60.01%. The model of PC1 and PC2 explained 79.04% of the variance in total and showed different metabolites between the heartwood and sapwood. The results showed a distinct grouping of HW and SW samples into two distinct areas in the plot, indicating that each cultivar had a relatively distinct metabolic profile. In addition, multivariate statistics were used to assess further the differences in metabolic profiles between HW and SW, and hierarchical cluster analysis (HCA) of the secondary metabolites in HW and SW was also performed. This showed two main clusters based on the relative differences in the accumulation patterns in the different tissues (Figure 2C). The secondary metabolites belonging to cluster 1 accumulated at higher levels in SW and in contrast, the metabolites in cluster 2 accumulated at higher levels in HW. Taken together, the plot shows that there were significant differences in the secondary metabolites detected in HW and SW.

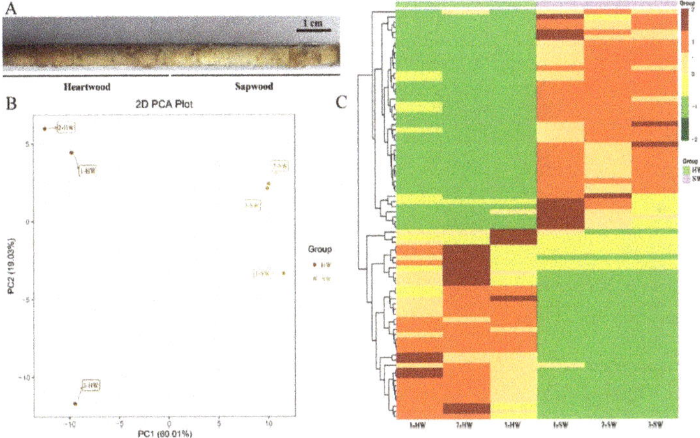

Figure 2. (**A**) A core of red-heart Chinese fir wood, showing red-colored tissue (heartwood, HW) and light-yellow tissue (sapwood, SW). (**B**) PCA score plot of the metabolites in the heartwood and sapwood. (**C**) Heatmap of the metabolites in the heartwood and sapwood.

3.3. OPLS-DA

The OPLS-DA model was used to screen the differential compounds in both groups of samples and analyze metabolic differences between HW and SW (Figure 3B). We observed high predictability (Q2) and strong goodness of fit (R2X, R2Y) between HW and SW (Q2Y = 0.998, R2X = 0.898, R2Y = 1.000). The HW was separated from SW, indicating major differences in the metabolic profiles between the two different colored tissues. Moreover, fold-change scores of ≥2 or ≤0.5 among the metabolites with a VIP value ≥1 was used to identify different metabolites. The screening results were analyzed using volcano plots (Figure 3A). A total of 80 significant differences between metabolites were identified, of which 37 were significantly up-regulated (the relative content of heartwood is larger than that of sapwood) and 43 were significantly down-regulated in HW and SW, respectively. The top 10 metabolites that were significantly up-regulated in HW are listed in Table 1, and among them, 6 flavonoids and 2 phenolic acids were significantly different between the HW and SW. These metabolites were mainly flavonoids and they can be considered the representative differential metabolites of HW and SW, influencing the properties and particularly the color of the wood.

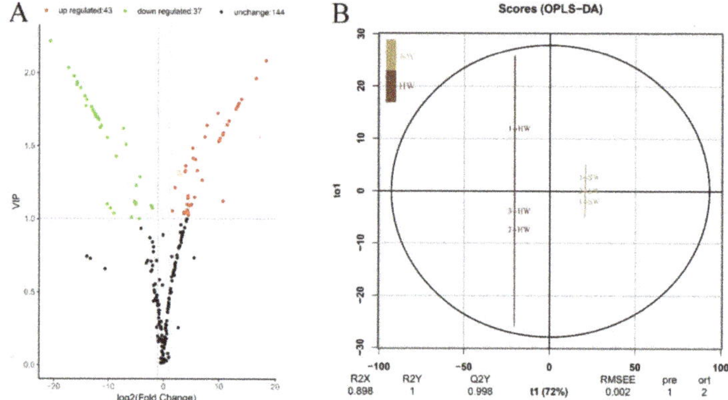

Figure 3. (**A**) The volcano plot of the differential metabolites in the heartwood and sapwood. Green dots represent down-regulated metabolites, red spots represent up-regulated metabolites, and gray dots represent insignificant differences in metabolites. (**B**) OPLS-DA model plot of the differential metabolites in the heartwood and sapwood.

Table 1. The top 10 metabolites that were significantly up-regulated in heartwood compared to sapwood.

Compounds	Formula	Class	VIP	Fold-Change	Log^{2}FC
Naringenin chalcone	$C_{15}H_{12}O_5$	Flavonoids	2.22	1.65×10^6	20.65
Dihydrokaempferol	$C_{15}H_{12}O_6$	Flavonoids	2.03	1.73×10^5	17.40
1-Feruloyl-sn-glycerol	$C_{13}H_{16}O_6$	Phenolic acids	1.98	8.43×10^4	16.36
Ailantinol E	$C_{21}H_{24}O_7$	Others	1.92	6.04×10^4	15.88
Pinocembrin (Dihydrochrysin)	$C_{15}H_{12}O_4$	Flavonoids	1.94	5.66×10^4	15.79
Z-6,7-Epoxyligustilide	$C_{12}H_{14}O_3$	Others	1.90	3.69×10^4	15.17
3-*O*-Acetylpinobanksin	$C_{17}H_{14}O_6$	Flavonoids	1.84	2.20×10^4	14.43
benzoyltartaric acid	$C_{11}H_{10}O_7$	Phenolic acids	1.84	2.12×10^4	14.37
Sexangularetin	$C_{16}H_{12}O_7$	Flavonoids	1.77	1.85×10^4	14.18
Quercetin galloglucoside	$C_{28}H_{24}O_{16}$	Flavonoids	1.82	1.66×10^4	14.02

3.4. Putative Metabolic Pathways for Signal Metabolites

To elucidate the pathways of the differential metabolites, we mapped the metabolites from the HW and SW to the Kyoto Encyclopedia of Genes and Genomes (KEGG) database (http://www.genome.jp/kegg/). The results are shown in Figure 4A. These differential metabolites are mainly involved in metabolic pathways and flavonoid biosynthesis, and others, such as phenylpropanoids, flavone, and flavonol also showed differences. Flavonoids synthesized via the general phenylpropanoid pathway constitute a group of secondary metabolites in plants that contribute to the key characteristics of HW in woody plants, such as decay resistance and insect decay [29,30]. Further enrichment analyses of the main secondary metabolites showed that they were highly enriched in the flavonoid synthesis pathway, as shown in Figure 4B, and that flavonoids constituted the main secondary metabolite difference between HW and SW. Cellulose, hemicellulose, and lignin, which are the major components of wood, generally do not exhibit color; wood color is due to the presence of colored extractives contained in the wood. These colored extractives turn from pale to dark due to oxidation and polymerization over time.

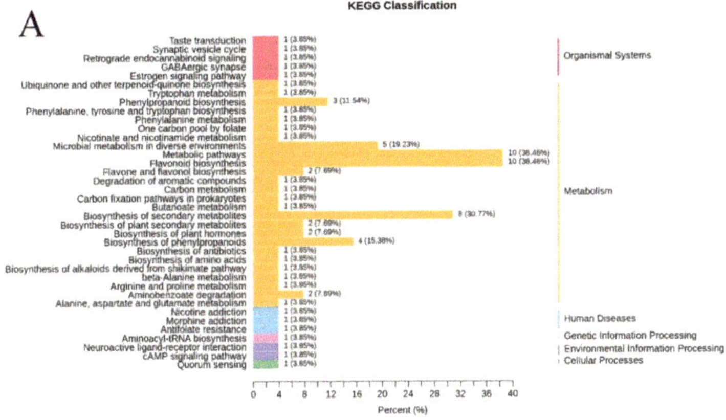

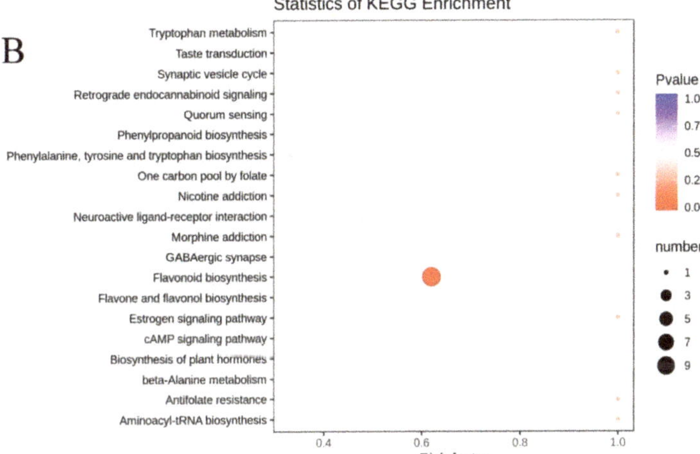

Figure 4. (**A**) Enrichment of the differential metabolites to highlight the KEGG pathway. Differential metabolites in the heartwood compared to those in the sapwood were mapped to distinct metabolic pathways. (**B**) Overview of pathway analyses of the heartwood and sapwood of the red-heart Chinese fir. The flavonoid biosynthesis metabolites are the most significantly different.

3.5. Secondary Metabolites Identified in the Flavonoid Pathway in HW and SW

Eleven significantly different secondary metabolites in the HW and SW of red-heart Chinese fir were selected for further analyses of the flavonoid pathway, and the differences are shown in Figure 5. The contents of naringenin chalcone, dihydrokaempferol, pinocembrin, hesperetin, pinobanksin, eriodictyol, luteolin, catechin, and apigenin were significantly higher in the HW than in SW, while the contents of epicatchin and myricetin were significantly higher in the SW than in HW. The annotation of the differential secondary metabolites to the flavonoid pathway, as shown in Figure 6, revealed that the Log$_2$FC values of naringenin chalcone, dihydrokaempferol, pinocembrin, and hesperetin were higher than 10 and were mainly concentrated in the upstream pathway. Myricetin, a downstream compound of dihydromyricetin, was more prevalent in SW than in HW, with a Log$_2$FC value of 10.6.

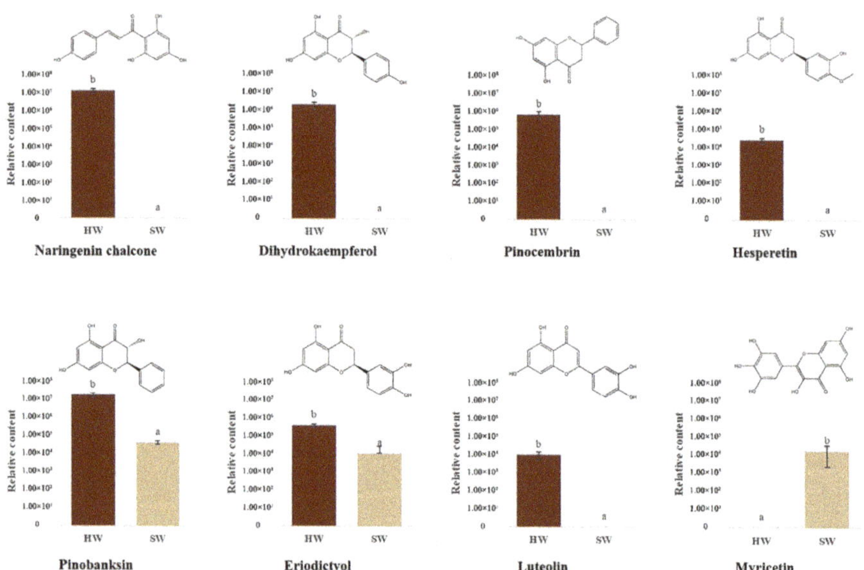

Figure 5. Chemical structure and contents of 8 significant differential flavonoids metabolites in the heartwood and sapwood of the red-heart Chinese fir. Means with different lowercase letters are significantly different at $p \leq 0.05$.

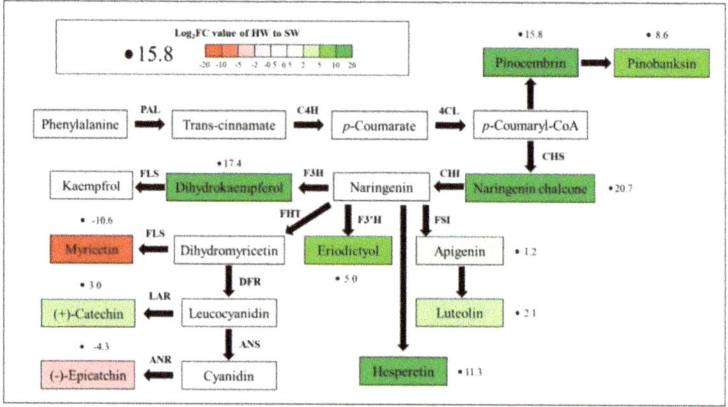

Figure 6. Metabolic profiling of the heartwood and sapwood in the flavonoid biosynthetic pathways in the red-heart Chinese fir. Grids with green and red color-scales from light to dark represent the Log_2FC values of HW to SW 0.5–2, 2–5. 5–10, 10–20, and over 20, respectively. PAL, phenylalanine ammonia-lyase; C4H, cinnamic acid 4-hydroxylase; 4CL, 4-coumarate CoA ligase; CHS, chalcone synthase; CHI, chalcone isomerase; F3H, flavanone 3-hydroxylase; F3′H, flavanoid 3′-hydroxylase; FHT, flacanone hydroxylase; FSI, flavonol synthase; DFR, dihydroflavonol 4-reductase; ANS, anthocyanidin synthase; FLS, flavonol synthesis; LAR, leucocyanidin reductase; ANR, anthocyanin reductase.

4. Discussion

We compared the secondary metabolites of the HW and SW in the red-heart Chinese fir by widely targeted metabolomics analyses, based on the UEMS system. There was a significant difference in the presence of 80 secondary metabolites. Metabolic pathway analyses revealed that the flavonoid biosynthesis pathway was significantly different between HW and SW. Furthermore, HW contained significantly more flavonoids than SW.

There are considerably more metabolites in the plant kingdom than in the animal kingdom, with the number estimated to exceed 200,000 [31]. New varieties of woody plants are deliberately selected for the creation of high-quality wood. However, differences have only recently been discovered in the distribution of secondary metabolites in the stems of the red-heart Chinese fir. A metabolomics study can provide new information on the different secondary metabolic compound profiles. Detailed metabolite profiling of thousands of plant samples has great potential to elucidate metabolic processes. However, it is difficult to simultaneously undertake both comprehensive and high-throughput analyses in plants due to the wide diversity of metabolites present. Widely targeted metabolomics approach can monitor both the specific precursor ions and the product ions of each metabolite using MRM and tandem quadrupole mass spectrometry (TQMS), as they enable high sensitivity, reproducibility, and a broad dynamic range [32]. In general, such analyses can provide a large data set that will aid understanding of plant metabolism, and which can be used in combination with other omics approaches to achieve a deeper understanding of the biological processes involved. Herein, the UEMS system-based widely targeted metabolomics analyses approach appeared be effective as it helps us to rapidly obtain information on 80 significantly different metabolites between HW and SW and to classify them accurately in Chinese fir.

Color differences are widely studied in horticulture and agriculture, and particularly in plants that are usually propagated vegetatively, such as most fruit trees. Cho et al. [33] analyzed the metabolites of three colored potato cultivars and found differences in the anthocyanin content. Xue et al. [34] analyzed the molecular and metabolic bases of pigmentation in *Lonicera japonica* flowers at different developmental stages and constructed regulatory networks of anthocyanin biosynthesis, chlorophyll metabolism, and carotenoid biosynthesis by weighted gene co-expression network analysis. In addition, there have been color-related studies on fig (*Ficus carica* L.), loquat (*Eriobotrya japonica* Lindl), pomegranate (*Punica granatum* L.), and other fruit trees [35–39]. Anthocyanins and flavonoids affect the colors and tastes of fruits; their antioxidant capacities confer health properties and reduce the risk for cardiovascular morbidity and mortality [40,41]. As with many secondary metabolite pathways, the flavonoid pathway is regulated by multiple genes. First, phenylalanine ammonia-lyase (PAL) catalyzes the conversion of phenylalanine into cinnamic acid. Then, dihydrokaempferol is converted from cinnamic acid by a series of enzymes, such as cinnamate 4 hydroxylase (C4H), 4-coumaroyl: CoA-ligase (4CL), chalcone synthase (CHS), chalcone isomerase (CHI), flavanone 3-hydroxylase (F3H), flavanone 3'-hydroxylase (F3'H), and flavanone 3'-5'-hydroxylase (F3'5'H). Subsequently, dihydroflavonol 4-reductase (DFR) catalyzes the conversion of dihydrokaempferol into unmodified and colorless anthocyanins [42,43]. Then, the conversion of colorless anthocyanins into colored anthocyanins is catalyzed by anthocyanidin synthase (ANS). Finally, the unstable colored anthocyanins are converted into blue-purple, brick red, or magenta glycosides by UDP glucose: flavonoid 3-O glucosyltransferase (UFGT) [44–46]. In this study, we identified the flavonoids in the HW of the red-heart Chinese fir, determined which substances differed from those in the SW, and more importantly, revealed the highly significant accumulation of naringenin chalcone, dihydrokaempferol, pinocembrin, hesperetin, and other important secondary metabolites in the flavonoid biosynthesis pathway. We found that flavonoids were the secondary metabolites that differed most between the SW and HW in the red-heart Chinese fir. In addition, HW with a natural red-orange color differed from SW due to the presence of pigments that proved difficult to extract but which were formed by the enzymatic reduction of dihydroquercetin, the main extractive of Douglas fir HW [47,48]. Pâques et al. [49] found that taxifolin and dihydrokaempferol showed significant differences among different varieties of larch by studying the distribution of HW extractives in hybrid larches and in their related European and Japanese larch parents. Takahashi and Mori [50] concluded that Sugi HW turns black because sequirin-C, a type of norlignan that is readily soluble in alkaline solution and can form a large intramolecular conjugation system when alkalized, was converted into products that had a deep purple color as the HW was basified. In another study, the metabolic profiles of HW and SW in *Taxus chinensis* were analyzed using widely targeted metabolomics assay, and a total of 607 metabolites were detected, including 71 flavonoids and isoflavones that were significantly different between HW and SW [51]. The differences in the secondary metabolites of HW are mainly a result of species differences. This is due to the great diversity of metabolic pathways that each plant species has evolved to survive under varying environmental conditions.

In our study, flavonoids were the main secondary metabolites that differed between the HW and SW in the red-heart Chinese fir, but it is not known if they underlie the differences in color, as this requires additional evidence. Anthocyanins, as downstream products in the flavonoid biosynthesis pathway, are widely distributed in flowers and fruits, where they play a role in coloring, but there are few reports of anthocyanins in wood. In the transition zone, the ray parenchyma cells synthesize flavonoids before programmed death and eventually become a part of the heartwood [52,53]. This process increases the wood quality and corrosion resistance, while giving the heartwood a more attractive color. In future studies, we will compare the secondary metabolites of HW between the red-heart Chinese fir and the common Chinese fir at a population scale, in the hope that it will address this issue more intuitively. In addition, metabolomics, in combination with genomics, transcriptomics, or proteomics, may help us to analyze molecules with similar chemical properties that are related to metabolites (namely, DNA, RNA, and proteins).

5. Conclusions

This study was the first to investigate the differential secondary metabolites of the red-heart Chinese fir using a widely targeted metabolomics method. We found significant differences between HW and SW. This study successfully identified components in the HW of the red-heart Chinese fir that are consistent with specific characteristics of HW in other conifers. The results provide critical metabolite inventory information for the study of specific metabolites in the HW of the red-heart Chinese fir, and provide a base study for the improvement of wood quality regarding to HW color on the metabolite aspect.

Author Contributions: Y.L. and H.Z. conceived and designed the experiments, S.C. wrote the paper, S.C. and Z.Z. performed the experiments and analyzed the data, Y.S. participated in and help to complete the experiments. All authors have read and agreed to the published version of the manuscript.

Funding: This research was supported by the Key-Area Research and Development Program of Guangdong Province (No. 2020B020215001), the Science and Technology Research Project of Beijing Forestry University (No. 2018WS01) and the National Natural Science Foundation of China (No. 31972956).

Conflicts of Interest: The authors declare no conflict of interest.

References

1. Spicer, R. Senescence in secondary xylem: Heartwood formation as an active developmental program. In *Vascular Transport in Plants*; Elsevier: Amsterdam, The Netherland, 2005; pp. 457–475.
2. Miyamoto, N.; Iizuka, K.; Nasu, J.; Yamada, H. Genetic effects on heartwood color variation in *Cryptomeria japonica*. *Silvae Genet.* **2016**, *65*, 80–87. [CrossRef]
3. Sundqvist, B. Color response of Scots pine (*Pinus sylvestris*), Norway spruce (*Picea abies*) and birch (*Betula pubescens*) subjected to heat treatment in capillary phase. *Eur. J. Wood Wood Prod.* **2002**, *60*, 106–114. [CrossRef]
4. Gierlinger, N.; Jacques, D.; Grabner, M.; Wimmer, R.; Schwanninger, M.; Rozenberg, P.; Pâques, L.E. Colour of larch heartwood and relationships to extractives and brown-rot decay resistance. *Trees-Struct. Funct.* **2004**, *18*, 102–108. [CrossRef]
5. Deklerck, V.; De Ligne, L.; Espinoza, E.; Beeckman, H.; Van den Bulcke, J.; Van Acker, J. Assessing the natural durability of xylarium specimens: Mini-block testing and chemical fingerprinting for small-sized samples. *Wood Sci. Technol.* **2020**, *54*, 981–1000. [CrossRef]
6. Fries, A.; Ericsson, T.; Gref, R. High heritability of wood extractives in *Pinus sylvestris* progeny tests. *Can. J. For. Res.* **2000**, *30*, 1707–1713. [CrossRef]

7. Ericsson, T.; Fries, A.; Gref, R. Genetic correlations of heartwood extractives in *Pinus sylvestris* progeny tests. *For. Genet.* **2001**, *8*, 73–80.

8. Jeong, S.T.; Goto-Yamamoto, N.; Kobayashi, S.; Esaka, M. Effects of plant hormones and shading on the accumulation of anthocyanins and the expression of anthocyanin biosynthetic genes in grape berry skins. *Plant Sci.* **2004**, *167*, 247–252. [CrossRef]

9. Gould, K.; Davies, K.M.; Winefield, C. *Anthocyanins: Biosynthesis, Functions, and Applications*; Springer Science & Business Media: Berlin/Heidelberg, Germany, 2008.

10. Jia, L.; Wu, Q.; Ye, N.; Liu, R.; Shi, L.; Xu, W.; Zhi, H.; Rahman, A.R.B.; Xia, Y.; Zhang, J. Proanthocyanidins inhibit seed germination by maintaining a high level of abscisic acid in *Arabidopsis thaliana* F. *J. Integr. Plant Biol.* **2012**, *54*, 663–673. [CrossRef]

11. Sehitoglu, M.H.; Farooqi, A.A.; Qureshi, M.Z.; Butt, G.; Aras, A. Anthocyanins: Targeting of signaling networks in cancer cells. *Asian Pac. J. Cancer Prev.* **2014**, *15*, 2379–2381. [CrossRef]

12. Lee, S.L.; Chin, T.Y.; Tu, S.C.; Wang, Y.J.; Hsu, Y.T.; Kao, M.C.; Wu, Y.C. Purple sweet potato leaf extract induces apoptosis and reduces inflammatory adipokine expression in 3T3-L1 differentiated adipocytes. *Evid. Based Complement. Altern. Med.* **2015**, 126302. [CrossRef]

13. Wei, C.; Liang, G.; Guo, Z.; Wang, W.; Zhang, H.; Liu, X.; Yu, S. A novel integrated method for large-scale detection, identification, and quantification of widely targeted metabolites: Application in the study of rice metabolomics. *Mol. Plant* **2013**, *6*, 1769–1780. [CrossRef]

14. Wang, D.; Zhang, L.; Huang, X.; Xiao, W.; Yang, R.; Jin, M.; Wang, X.; Wang, X.; Qi, Z.; Li, P. Identification of nutritional components in black sesame determined by widely targeted metabolomics and traditional Chinese Medicines. *Molecules* **2018**, *23*, 1180. [CrossRef]

15. Wang, T.; Zou, Q.; Guo, Q.; Yang, F.; Zhang, W. Widely targeted metabolomics analysis reveals the effect of flooding stress on the synthesis of flavonoids in *Chrysanthemum morifolium*. *Molecules* **2019**, *24*, 3695. [CrossRef]

16. Wu, X.; Prior, R.L. Systematic identification and characterization of anthocyanins by HPLC-ESI-MS/MS in common foods in the United States: Fruits and Berries. *J. Agric. Food Chem.* **2005**, *53*, 2589–2599. [CrossRef]

17. Lin, L.Z.; Sun, J.; Chen, P.; Harnly, J.A. LC-PDA-ESI/MS(n) Identification of New Anthocyanins in purple bordeaux radish (*Raphanus sativus* L. variety). *J. Agric. Food Chem.* **2011**, *59*, 6616–6627. [CrossRef]

18. Sun, J.; Lin, L.Z.; Pei, C. Study of the mass spectrometric behaviors of anthocyanins in negative ionization mode and its applications for characterization of anthocyanins and non-anthocyanin polyphenols. *Rapid Commun. Mass Spectrom. RCM* **2012**, *26*, 1123–1133. [CrossRef]

19. Zheng, H.; Hu, D.; Wei, R.; Yan, S.; Wang, R. Chinese fir breeding in the high-throughput sequencing era: Insights from SNPs. *Forests* **2019**, *10*, 681. [CrossRef]

20. Duan, H.; Hu, R.; Wu, B.; Chen, D.; Huang, K.; Dai, J.; Chen, Q.; Wei, Z.; Cao, S.; Sun, Y. Genetic characterization of red-colored heartwood genotypes of Chinese fir using simple sequence repeat (SSR) markers. *Genet. Mol. Res.* **2015**, *14*, 18552–18561. [CrossRef]

21. Qiu, F.; Sun, S.; Xiao, F.; Zeng, Z.; Liang, C.; Liang, X. Growth analysis of half-sib families from first generation seed orchard of *Cunninghamia lanceolata*. *South China For. Sci.* **2016**, *44*, 1–4. (In Chinese) [CrossRef]

22. Kui-Peng, L.I.; Wei, Z.C.; Huang, K.Y.; Dong, L.J.; Huang, H.F.; Chen, Q.; Dai, J.; Tan, W.J. Research on variation pattern of wood properties of red-heart Chinese fir plus trees, a featured provenance from Rongshui of Guangxi. *For. Res.* **2017**, *30*, 424–429. (In Chinese) [CrossRef]

23. Wen, H.; Deng, X.; Zhang, Y.; Wei, X.; Zhu, N. *Cunninghamia lanceolata* variant with red-heart wood: A mini-review. *Dendrobiology* **2018**, *79*, 156–167. [CrossRef]

24. Fan, G.R.; Zhang, W.Y.; Xiao, F.M.; Zhu, X.Z.; Jiang, X.M.; Yan-Shan, L.I. Studies on longitudinal variation of main chemical components in Chenshan red-heart Chinese-fir. *Acta Agric. Univ. Jiangxiensis* **2015**, *37*, 212–217. (In Chinese) [CrossRef]

25. Yang, H.; Wen, S.; Xiao, F.; Fan, G. Study on the chemical compositions of ethanol extraction from Chenshan Red-heart Chinese fir. *South China For. Sci.* **2016**, *44*, 35–37. (In Chinese) [CrossRef]

26. Afendi, F.M.; Okada, T.; Yamazaki, M.; Hirai-Morita, A.; Nakamura, Y.; Nakamura, K.; Ikeda, S.; Takahashi, H.; Altaf-Ul-Amin, M.; Darusman, L.K.; et al. KNApSAcK family databases: Integrated metabolite-plant species databases for multifaceted plant research. *Plant Cell Physiol.* **2012**, *53*, e1. [CrossRef]

27. Eriksson, L.; Kettanehwold, N.; Trygg, J.; Wikström, C.; Wold, S. Part I: Basic Principles and Applications. In *Multi- and Megavariate Data Analysis*, 2nd ed.; Umetrics Academy: Umea, Sweden, 2006.

28. Thevenot, E.A.; Roux, A.; Xu, Y.; Ezan, E.; Junot, C. Analysis of the human adult urinary metabolome variations with age, body mass index, and gender by implementing a comprehensive workflow for univariate and opls statistical analyses. *J. Proteome Res.* **2015**, *14*, 3322–3335. [CrossRef]

29. Mounguengui, S.; Tchinda, J.B.S.; Ndikontar, M.K.; Dumarçay, S.; Attéké, C.; Perrin, D.; Gelhaye, E.; Gérardin, P. Total phenolic and lignin contents, phytochemical screening, antioxidant and fungal inhibition properties of the heartwood extractives of ten Congo Basin tree species. *Ann. For. Sci.* **2016**, *73*, 287–296. [CrossRef]

30. Gierlinger, N.; Jacques, D.; Schwanninger, M.; Wimmer, R.; Pâques, L.E. Heartwood extractives and lignin content of different larch species (*Larix* sp.) and relationships to brown-rot decay-resistance. *Trees* **2004**, *18*, 230–236. [CrossRef]

31. Fiehn, O.; Kloska, S.; Altmann, T. Integrated studies on plant biology using multiparallel techniques. *Curr. Opin. Biotechnol.* **2001**, *12*, 82–86. [CrossRef]

32. Sawada, Y. Widely targeted metabolomics based on large-scale MS/MS Data for elucidating metabolite accumulation patterns in plants. *Plant Cell Physiol.* **2009**, *50*, 37–47. [CrossRef]

33. Cho, K.; Cho, K.; Sohn, H.; Ha, I.J.; Hong, S.; Lee, H.; Kim, Y.; Nam, M.H. Network analysis of the metabolome and transcriptome reveals novel regulation of potato pigmentation. *J. Exp. Bot.* **2016**, *67*, 1519–1533. [CrossRef]

34. Xue, Q.; Fan, H.; Yao, F.; Cao, X.; Liu, M.; Sun, J.; Liu, Y. Transcriptomics and targeted metabolomics profilings for elucidation of pigmentation in *Lonicera japonica* flowers at different developmental stages. *Ind. Crop. Prod.* **2020**, *145*, 111981. [CrossRef]

35. Hadjipieri, M.; Georgiadou, E.C.; Marin, A.; Diazmula, H.M.; Goulas, V.; Fotopoulos, V.; Tomasbarberan, F.A.; Manganaris, G.A. Metabolic and transcriptional elucidation of the carotenoid biosynthesis pathway in peel and flesh tissue of loquat fruit during on-tree development. *BMC Plant Biol.* **2017**, *17*, 102. [CrossRef] [PubMed]

36. Wang, Z.; Cui, Y.; Vainstein, A.; Chen, S.; Ma, H. Regulation of fig (*Ficus carica* L.) fruit color: Metabolomic and transcriptomic analyses of the flavonoid biosynthetic pathway. *Front. Plant Sci.* **2017**, *8*, 1990. [CrossRef] [PubMed]

37. Li, Y.; Fang, J.; Qi, X.; Lin, M.; Zhong, Y.; Sun, L.; Cui, W. Combined analysis of the fruit metabolome and transcriptome reveals candidate genes involved in flavonoid biosynthesis in *Actinidia arguta*. *Int. J. Mol. Sci.* **2018**, *19*, 1471. [CrossRef] [PubMed]

38. Paolo, B.; Saverio, O.; Mirko, M.; Matteo, B.; Lara, G.; Azeddine, S.A. Gene expression and metabolite accumulation during strawberry (*Fragaria × ananassa*) fruit development and ripening. *Planta* **2018**, *248*, 1143–1157. [CrossRef]

39. Zhang, Q.; Wang, L.; Liu, Z.; Zhao, Z.; Zhao, J.; Wang, Z.; Zhou, G.; Liu, P.; Liu, M. Transcriptome and metabolome profiling unveil the mechanisms of *Ziziphus jujuba* Mill. peel coloration. *Food Chem.* **2020**, *312*, 125903. [CrossRef]

40. Holt, R.R.; Lazarus, S.A.; Sullards, M.C.; Zhu, Q.Y.; Schramm, D.D.; Hammerstone, J.F.; Fraga, C.G.; Schmitz, H.H.; Keen, C.L. Procyanidin dimer B2 [epicatechin-(4β-8)-epicatechin] in human plasma after the consumption of a flavanol-rich cocoa. *Am. J. Clin. Nutr.* **2002**, *76*, 798–804. [CrossRef]

41. Wu, S.; Dastmalchi, K.; Long, C.; Kennelly, E.J. Metabolite profiling of Jaboticaba (*Myrciaria cauliflora*) and other dark-colored fruit juices. *J. Agric. Food Chem.* **2012**, *60*, 7513–7525. [CrossRef]

42. Boss, P.K.; Davies, C.H.J.; Robinson, S.P. Analysis of the expression of anthocyanin pathway genes in developing *Vitis vinifera* L. cv Shiraz Grape Berries and the implications for pathway regulation. *Plant Physiol.* **1996**, *111*, 1059–1066. [CrossRef]

43. Ferreyra, M.L.F.; Rius, S.P.; Casati, P. Flavonoids: Biosynthesis, biological functions and biotechnological applications. *Front. Plant Sci.* **2012**, *3*, 222. [CrossRef]

44. Pelletier, M.K.; Murrell, J.R.; Shirley, B.W. Characterization of flavonol synthase and leucoanthocyanidin dioxygenase genes in arabidopsis (further evidence for differential regulation of "early" and "late" genes). *Plant Physiol.* **1997**, *113*, 1437–1445. [CrossRef] [PubMed]

45. Dick, C.A.; Buenrostro, J.D.; Butler, T.; Carlson, M.L.; Kliebenstein, D.J.; Whittall, J.B. Arctic mustard flower color polymorphism controlled by petal-specific downregulation at the threshold of the anthocyanin biosynthetic pathway. *PLoS ONE* **2011**, *6*, e18230. [CrossRef] [PubMed]

46. Saito, K.; Yonekurasakakibara, K.; Nakabayashi, R.; Higashi, Y.; Yamazaki, M.; Tohge, T.; Fernie, A.R. The flavonoid biosynthetic pathway in Arabidopsis: Structural and genetic diversity. *Plant Physiol. Biochem.* **2013**, *72*, 21–34. [CrossRef] [PubMed]

47. Dellus, V.; Mila, I.; Scalbert, A.; Menard, C.; Michon, V.; Du Penhoat, C.L.M.H. Douglas-fir polyphenols and heartwood formation. *Phytochemistry* **1997**, *45*, 1573–1578. [CrossRef]
48. Dellus, V.; Scalbert, A.; Janin, G. Polyphenols and colour of douglas fir heartwood. *Holzforschung* **1997**, *51*, 291–295. [CrossRef]
49. Pâques, L.E.; Garciacasas, M.D.C.; Charpentier, J. Distribution of heartwood extractives in hybrid larches and in their related European and Japanese larch parents: Relationship with wood colour parameters. *Eur. J. For. Res.* **2013**, *132*, 61–69. [CrossRef]
50. Takahashi, K.; Mori, K. Relationships between blacking phenomenon and norlignans of sugi (*Cryptomeria japonica*) heartwood III: Coloration of norlignans with alkaline treatment. *J. Wood Sci.* **2006**, *52*, 134–139. [CrossRef]
51. Shao, F.; Zhang, L.; Guo, J.; Liu, X.; Ma, W.; Wilson, I.W.; Qiu, D. A comparative metabolomics analysis of the components of heartwood and sapwood in *Taxus chinensis* (Pilger) Rehd. *Sci. Rep.* **2019**, *9*, 1–8. [CrossRef]
52. Chen, S.; Yen, P.; Chang, T.; Chang, S.; Huang, S.; Yeh, T. Distribution of living ray parenchyma cells and major bioactive compounds during the heartwood formation of Taiwania cryptomerioides Hayata. *J. Wood Chem. Technol.* **2018**, *38*, 84–95. [CrossRef]
53. Nakaba, S.; Arakawa, I.; Morimoto, H.; Nakada, R.; Bito, N.; Imai, T.; Funada, R. Agatharesinol biosynthesis-related changes of ray parenchyma in sapwood sticks of Cryptomeria japonica during cell death. *Planta* **2016**, *243*, 1225–1236. [CrossRef]

Article

Physiological Characterization and Transcriptome Analysis of *Camellia oleifera* Abel. during Leaf Senescence

Shiwen Yang [1], Kehao Liang [1], Aibin Wang [1], Ming Zhang [1], Jiangming Qiu [2] and Lingyun Zhang [1,*]

[1] Key Laboratory of Forest Silviculture and Conservation of the Ministry of Education, The College of Forestry, Beijing Forestry University, 35 Qinghua East Road, Beijing 100083, China; ShiwenYang1995@hotmail.com (S.Y.); lkhbjfu@163.com (K.L.); wangaibin126@126.com (A.W.); 19801202580@163.com (M.Z.)

[2] Jiangxi Environmental Engineering Vocational College, Ganzhou 341000, China; 18270730388@163.com

[*] Correspondence: lyzhang@bjfu.edu.cn; Tel.: +86-10-62336044; Fax: +86-10-62338197

Received: 10 July 2020; Accepted: 25 July 2020; Published: 28 July 2020

Abstract: *Camellia (C.) oleifera* Abel. is an evergreen small arbor with high economic value for producing edible oil that is well known for its high level of unsaturated fatty acids. The yield formation of tea oil extracted from fruit originates from the leaves, so leaf senescence, the final stage of leaf development, is an important agronomic trait affecting the production and quality of tea oil. However, the physiological characteristics and molecular mechanism underlying leaf senescence of *C. oleifera* are poorly understood. In this study, we performed physiological observation and de novo transcriptome assembly for annual leaves and biennial leaves of *C. oleifera*. The physiological assays showed that the content of chlorophyll (Chl), soluble protein, and antioxidant enzymes including superoxide dismutase, peroxide dismutase, and catalase in senescing leaves decreased significantly, while the proline and malondialdehyde concentration increased. By analyzing RNA-Seq data, we identified 4645 significantly differentially expressed unigenes (DEGs) in biennial leaves with most associated with flavonoid and phenylpropanoid biosynthesis and phenylalanine metabolism pathways. Among these DEGs, 77 senescence-associated genes (SAGs) including *NOL*, *ATAF1*, *MDAR*, and *SAG12* were classified to be related to Chl degradation, plant hormone, and oxidation pathways. The further analysis of the 77 SAGs based on the Spearman correlation algorithm showed that there was a significant expression correlation between these SAGs, suggesting the potential connections between SAGs in jointly regulating leaf senescence. A total of 162 differentially expressed transcription factors (TFs) identified during leaf senescence were mostly distributed in MYB (myeloblastosis), ERF (Ethylene-responsive factor), WRKY, and NAC (NAM, ATAF1/2 and CUCU2) families. In addition, qRT-PCR analysis of 19 putative SAGs were in accordance with the RNA-Seq data, further confirming the reliability and accuracy of the RNA-Seq. Collectively, we provide the first report of the transcriptome analysis of *C. oleifera* leaves of two kinds of age and a basis for understanding the molecular mechanism of leaf senescence.

Keywords: *Camellia oleifera*; leaf senescence; transcriptome analysis; senescence-associated genes; physiological characterization

---

1. Introduction

Camellia (C.) oleifera Abel. is an evergreen small arbor within the genus *Camellia* of the family Theaceae. It has been cultivated and utilized in China for more than 2000 years and is one of the four major woody oil plants in the world [1]. The tea oil extracted from *C. oleifera* is mainly composed of unsaturated fatty acids including oleic acid and linoleic acid, which can improve the antioxidant capacity and protect the liver, among other benefits [2–4]. Moreover, the United Nations Food and Agriculture Organization recommended the tea oil as a high-quality, healthy vegetable oil due to its nutritional value and excellent storage quality [5]. However, with the growing demand for high-quality edible oil, the low production of tea oil can no longer meet the market demand, so increasing the output of *C. oleifera* has always been one of the most urgent problems to be solved in the research of this species [6].

Many researches have shown that leaf senescence is crucial to limit the yield and quality of tree species [7]. This process is tightly regulated by both endogenous factors such as leaf age, the state of plant hormones, and the stages of growth and development as well as external factors mainly including diverse environmental conditions like drought and salinity [8]. During the process of leaf senescence, some synergetic changes occurred at the leaf physiological, biochemical, and molecular processes including chlorophyll (Chl) degradation, hydrolysis of macromolecules, accumulation of malondialdehyde (MDA), phytohormone changes, some redox processes, and the responses of senescence-associated genes (SAGs) [9]. To date, many advancements have been made in the research on leaf senescence at the molecular level, and a large number of SAGs have been identified including *SAG* family genes, *PAO*, *S40*, *ATAF1*, *GBF1*, etc. [10–17]. These genes participate in the process of leaf senescence through a complex regulatory network.

Chl degradation is one of the most significant characteristics and the genes related to it can be induced during leaf senescence [18]. For example, in rice (*Oryza sativa* L.), *NOL* and *SGR (STAY GREEN)*, the key genes responsible for catalyzing Chl degradation, are significantly induced in older leaves [19,20]. Phytohormones are also involved in response to leaf senescence, with abscisic acid (ABA), ethylene, jasmonic acid (JA), and salicylic acid (SA) promoting leaf senescence, auxin, and cytokinin delaying senescence [21]. In arabidopsis, ABA receptor *PYL9* (*promoter-pyrabactin resistance 1-like*) transcripts are obviously more abundant in older leaves, and *PYL9* accelerates leaf senescence of the ABA pathway [22]. *AHK2*, known as a cytokinin receptor, mediates cytokinin-dependent Chl retention; in the leaves of *ahk2* mutant, Chl content reduced and the precocious senescence phenotype was observed [23]. In addition, recent studies have demonstrated that redox signaling in leaf changes accordingly during leaf senescence including a reduction of antioxidant enzyme activity, accumulation of reactive oxygen species (ROS), and expression changes in relevant genes [24]. In kiwifruit (*Actinidia chinensis*), the accumulation of monodehydroascorbate reductase (*MDAR*) reduces H_2O_2 content, thus delaying leaf senescence [25]. These studies indicate that leaf senescence is a very complicated life process, and involved in multiple layers and signaling pathways. In addition to these functional genes or key enzymes, more recently, there has been considerable progress in understanding the importance of transcription factors (TFs), which could be responsible for the onset of leaf senescence [26]. In arabidopsis, *WRKY75* acts as a positive regulator of leaf senescence, and overexpression of *WRKY75* increases the expression of *SID2* (*SA INDUCTION-DEFICIENT2*), thus promoting SA production; moreover, *WRKY75* suppresses H_2O_2 scavenging, and can form three interlinking positive feedback loops with SA and ROS [27]. *ORE1*, a member of NAC TF family, is involved in the direct regulation of Chl catabolic genes, and can accelerate ethylene production by inducing the expression of *ACS2*, a vital ethylene biosynthesis gene, thus positively regulating leaf senescence in arabidopsis [28].

The mechanism of senescence has been intensively studied in model plants and crops such as arabidopsis and rice, however, the genes involved in the molecular mechanism driving leaf senescence in *C. oleifera* remain elusive. Therefore, the main objective of this study was to determine the physiological changes and gene expression profiling between *C. oleifera* leaves of two types of ages. In the present study, physiological analysis of two leaf types was performed measuring Chl content, soluble protein concentration, antioxidant enzyme activity, and accumulation of MDA and proline. The gene expression profiles were determined by RNA-Seq and qRT-PCR (Quantitative Real-time PCR). The putative SAGs and pathways associated with leaf senescence were identified and analyzed by means of multiple bioinformatics tools. Our study provides an initial analysis for understanding the underlying molecular mechanisms of leaf senescence and is a valuable resource for further identification of genes related to leaf senescence in *C. oleifera*.

2. Materials and Methods

2.1. Plant Materials and Growth Conditions

The plant material used in this study was 6-year-old *C. oleifera* 'Cenxiruanzhi No. 3'. Trees were originally grown in the Fengsheng seedling field in Cenxi City, Guangxi Zhuang Autonomous Region, and transplanted into the greenhouse at Sanqingyuan, Beijing Forestry University, Beijing, with the maximum air temperature of 28 °C in the daytime and a minimum of 21 °C at night, and the humidity of 72%–85%. All plants were watered and fertilized under the same conditions. Seedlings displaying uniform growth and with no signs of disease or insects were selected for further research. The intact annual leaves (AL) and biennial leaves (BL) in the same position on the sunny side were collected in August 2019. All samples were immediately frozen in liquid nitrogen and stored at −80 °C for further analysis. More than three seedlings were pooled as one biological replicate, and three biological replicates were included for our collection. The samples were marked with serial numbers: AL_1, AL_2, AL_3, BL_1, BL_2, and BL_3.

2.2. Physiological Analysis of C. oleifera Leaves

The physiological changes differing between the two leaf types *C. oleifera* were determined before RNA-Seq. The total Chl content was determined by the ethanol–acetone method, with minor modifications [29]. The soluble protein content was measured by using the Coomassie brilliant blue G250 dye method [30]. Malondialdehyde (MDA) concentration was measured by the thiobarbituric acid method, as proposed by a previous study [31]. The proline content was determined according to the sulfosalicylic acid–acid ninhydrin method, with slight modifications [32]. The activities of superoxide dismutase (SOD) was measured by the Total Superoxide Dismutase Assay Kit (hydroxylamine method) and the activity of peroxidase (POD) was determined by the Peroxidase Assay Kit. The activity of catalase (CAT) was measured by the CATalase Assay Kit (Visible light). These assay kits were provided by the Jiancheng Bioengineering Institute (Nanjing, China).

2.3. Total RNA Extraction

C. oleifera leaves were ground into powder in liquid nitrogen with a sterilized mortar and pestle for RNA extraction. The total RNA was extracted from the leaves using FastPure Plant Total RNA Isolation Kits (polysaccharides and polyphenolics-rich) (Vazyme, Nanjing, China) according to the manufacturer's instructions. Agilent 2100 and NanoDrop were used to quantify and evaluate the purified RNA. Each assay included three biological replicates.

2.4. Construction of the cDNA Library and RNA Sequencing

The extracted RNA was sent to the Beijing Genomics Institute (BGI) (Shenzhen, China), where the library was constructed and sequenced. The mRNA was purified using magnetic beads attached to Oligo (dT) and then purified mRNA was fragmented into small pieces. The first-strand cDNA was synthesized by random hexamer-primed reverse transcription, and subsequently, second strand cDNA was generated. Afterward, the A-tail adaptor was added through incubation, the cDNA fragments obtained after end repair were amplified by PCR, and then were purified by Ampure XP Beads. The purified product was tested on an Agilent Technologies 2100 bioanalyzer. Subsequently, the double-stranded PCR products were denatured and circularized to obtain the final library. The constructed library was finally sequenced on the BGIseq500 platform (BGI, Shenzhen, China).

2.5. Transcriptome Assembly and Annotation

SeqPrep (https://github.com/jstjohn/SeqPrep) and Sickle (https://github.com/najoshi/sickle) software were used to filter the raw sequencing reads. After filtering, the remaining reads were considered as clean reads and used for further bioinformatics analysis. Afterward, the clean reads were used for de novo assembly using Trinity (https://github.com/trinityrnaseq/trinityrnaseq) [33]. TransRate (http://hibberdlab.com/transrate/) [34] and CD-HIT (http://weizhongli-lab.org/cd-hit/) [35] were utilized to optimize and filter the initial assembly sequence, and BUSCO (Benchmarking Universal Single-Copy Orthologs, http://busco.ezlab.org) [36] was used to evaluate the assembly results. After assembly, all unigenes were aligned to the public protein databases including NR (NCBI non-redundant) (ftp://ftp.ncbi.nlm.nih.gov/blast/db/) [37], Swiss-Prot (http://web.expasy.org/docs/swiss-prot_guideline.html) [38], Pfam (http://pfam.xfam.org/) [39], COG (Clusters of Orthologous Groups of proteins, http://www.ncbi.nlm.nih.gov/COG/) [40], GO (Gene Ontology, http://www.geneontology.org) [41] and KEGG (Kyoto Encyclopedia of Genes and Genomes, http://www.genome.jp/kegg/) [42] databases.

2.6. Analysis and Enrichment of Differentially Expressed Genes (DEGs)

The data were analyzed on the free online platform of Majorbio Cloud Platform (www.majorbio.com). RSEM (RNA-Seq by Expectation Maximization) was utilized to calculate gene expression, and TPM (Transcripts Per Million reads) was used to quantify the expression level of genes [43]. DESwq2 (http://bioconductor.org/packages/stats/bioc/DESeq2/) [44] was used for statistical analysis of raw counts, and the DEGs were defined with p-adjust < 0.05 and $|\log2FC| \geq 1$, p-adjust is the p-value after multiple test correction using BH (fdr correction with Benjamini/Hochberg). GO enrichment analysis of DEGs were performed by using Goatools (https//github.com/tanghaibao/GOatools) [45]. When the corrected p-value < 0.05, the DEGs were defined to be significantly enriched. The KEGG pathway enrichment analysis of DEGs was performed by KOBAS (http://kobas.cbi.pku.edu.cn/home.do) [46]. The DEGs were considered to be significantly enriched when the corrected p-value < 0.05.

2.7. Screening of Key Senescence-Associated Genes (SAGs)

We initially screened 77 SAGs on the Majorbio Cloud Platform. By using the BLAST (Basic Local Alignment Search Tool) program of the NCBI (National Center for Biotechnology Information) and LSD 3.0 (Leaf Senescence Database, https://bigd.big.ac.cn/lsd/index.php) [47], we selected some SAGs with high homology to important SAGs in other species for further analysis.

2.8. Analysis of Expression Correlation

The Spearman algorithm was used to obtain the correlation coefficient between genes, and the expression correlation was defined with a correlation coefficient ≥ 0.5 and p-adjust < 0.05 [48]. According to the expression correlation between genes, the visual network diagram was constructed.

2.9. Quantitative Real-Time PCR (qRT-PCR) Analysis of Gene Expression

To evaluate the quality of RNA-Seq data, 19 unigenes were selected for qRT-PCR analysis. The extracted RNA was used to synthesize first-strand cDNAs using a FastQuant cDNA First-Strand Synthesis Kit (Tiangen Biotechnology, Beijing, China). We used the StepOne Plus Real-Time PCR System (ABI, Vernon, CA, USA) for qRT-PCR, according to the manufacturer's instructions of 2 × Fast SYBR Green qPCR Master Mix (High ROX) (Servicebio, Wuhan, China). *GAPDH* was used as a reference gene. The primer sequences of selected genes are shown in Table S4. All qRT-PCR assays were performed for three biological replicates and three technical repeats.

3. Results

3.1. Morphological and Physiological Observation of Leaves at Different Leaf Age

We observed the phenotype and measured physiological changes of two different kinds of *C. oleifera* leaves in the rapid fruit growth period. Our results showed that biennial leaves were significantly yellower than annual leaves (Figure 1A). The visible morphological changes we observed are consistent with previous reports [49] where yellowing due to Chl degradation is one of the most obvious characteristics of leaf senescence. Accordingly, in contrast to annual leaves, the Chl content in biennial leaves was significantly reduced (by 34.6%), suggesting that the degradation rate of Chl in biennial *C. oleifera* leaves was higher than the synthesis rate (Figure 1B). Additionally, the soluble protein content of biennial leaves was only 48.5% of that in annual leaves (Figure 1C), suggesting that the hydrolysis of soluble protein dominates during the senescence process of *C. oleifera*. In addition, we also examined the accumulation of MDA and proline and found that the contents of MDA and proline in biennial leaves were 1.48-fold higher and 2.57-fold higher than that in annual leaves, respectively (Figure 1D,E). Moreover, compared with annual leaves, the activity of SOD, POD, and CAT in senescent leaves decreased by 9.8%, 44.5%, and 49.8% (Figure 1F–H), respectively, indicating that the activities of antioxidant enzymes were reduced in *C. oleifera* biennial leaves, thus repressing the scavenging ability of ROS (reactive oxygen species).

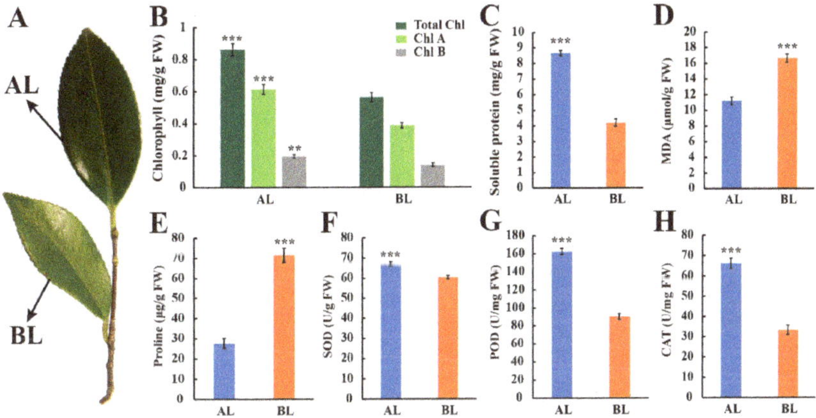

Figure 1. Morphological and physiological observation of different leaves of *Camellia Oleifera* Abel. (**A**) The phenotype of annual leaves and biennial leaves (AL: annual leaves; BL: biennial leaves). (**B**) Chl (chlorophyll) content. (**C**) Soluble protein content. (**D**) MDA (malondialdehyde) content. (**E**) Proline content. (**F**) SOD (superoxide dismutase) activity. (**G**) POD (peroxidase) activity. (**H**) CAT (catalase) activity. *: $0.01 \leq p < 0.05$, **: $0.001 \leq p < 0.01$, ***: $p < 0.001$.

3.2. RNA Sequencing, De Novo Assembly, and Annotation

In order to explore the molecular mechanisms underlying the leaf senescence of *C. oleifera* at the transcriptomic level, six RNA libraries from *C. oleifera* leaves (AL1, AL2, AL3; BL1, BL2, and BL3) were sequenced and 37.16 Gb high-quality clean data were obtained. The clean data of each sample reached more than 6.08 Gb, and the clean reads of each sample ranged from 40,602,122 to 41,859,058 with the Q30 percentage over 90.32% (Table S1). After de novo assembly of the *C. oleifera* transcriptome, a total of 78,860 unigenes were generated with an N50 length of 1374 bp and an average length of 865.96 bp. The GC content was approximately 39.19% (Table 1).

Table 1. Summary of sequence assembly.

Item	Unigene
Total number	78,860
Total base	68,289,653
Average length (bp)	865.96
N50 length (bp)	1374
GC percent (%)	39.19

The assembled unigenes were searched against the NR (NCBI non-redundant protein sequences), Swiss-Prot (Annotated protein sequence database), Pfam (Protein family), COG (Clusters of Genes), GO (Gene Ontology), and KEGG (Kyoto Encyclopedia of Genes and Genomes) databases, with 51.3% (40,425 of 78,860) of unigenes annotated. Among them, 26,486, 13,831, 29,317, 39,848, 24,214, and 23,566 annotated unigenes were obtained from the GO, KEGG, COG, NR, Swiss-Prot, and Pfam databases, respectively (Figure 2). The results indicated that the NR database had the highest proportion of annotation and 39,848 unigenes were annotated. Due to the lack of a reference genome for *C. oleifera*, the NR database can provide us with information about the most similar matches to gene sequences from other species, and according to the NR database, the species with the greatest number of *C. oleifera* unigenes are *Camellia sinensis* (87.70%), followed by *Actinidia chinensis* (2.02%) and *Vitis vinifera* (1.22%) (Figure S1). These results provide a basis for the future research of the potential function of these annotated unigenes.

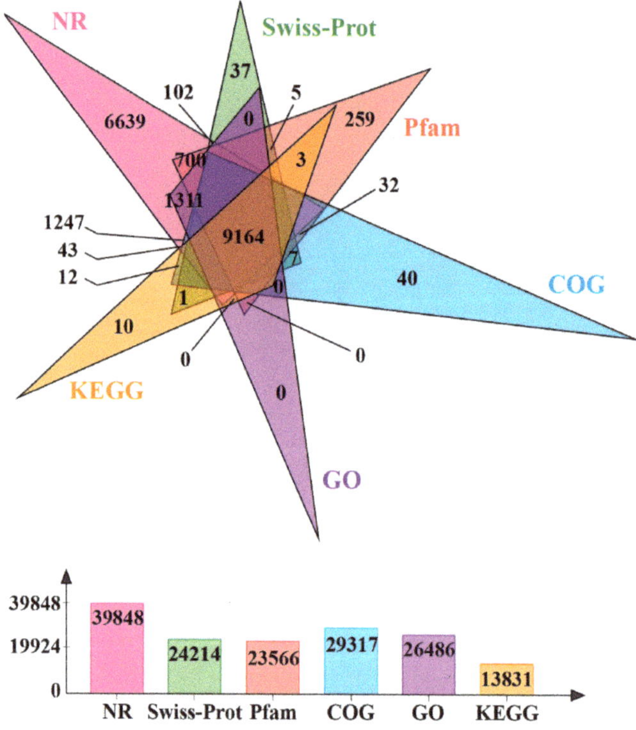

Figure 2. Venn diagram of unigenes annotated in NR (NCBI non-redundant), Swiss-Prot, Pram (Protein family), COG (Clusters of Genes), GO (Gene Ontology), and KEGG (Kyoto Encyclopedia of Genes and Genomes) protein databases.

3.3. Analyses of Gene Expression and Identification of Differentially Expressed Genes (DEGs)

The TPM values were used to calculate the expression level of unigenes in each tissue, and unigenes with TPM ≥1 were defined as expressed [50]. The number of unigense expressed in annual leaves was 59,212, and that in biennial leaves was 47,700. In addition, 40,421 unigenes were co-expressed in both leaves (Figure S2). Moreover, correlation analyses were performed to verify the consistency among the samples, and our results demonstrated that the correlation among three biological replicates was very high (Figure S3). The DEGs were defined with the adjusted *p*-value < 0.05 and fold-change ≥ 2, and our results showed that there were 4645 DEGs between *C. oleifera* leaves of two types of ages. Taking the annual leaves as a control group, the number of upregulated genes and downregulated genes in biennial leaves were 2729 and 1916, respectively (Figure 3A). To facilitate the visualization, a heatmap was produced on the basis of the clustering analysis of expression patterns for all DEGs (Figure 3B). These findings demonstrate that a large number of genes are responsive to the senescence process.

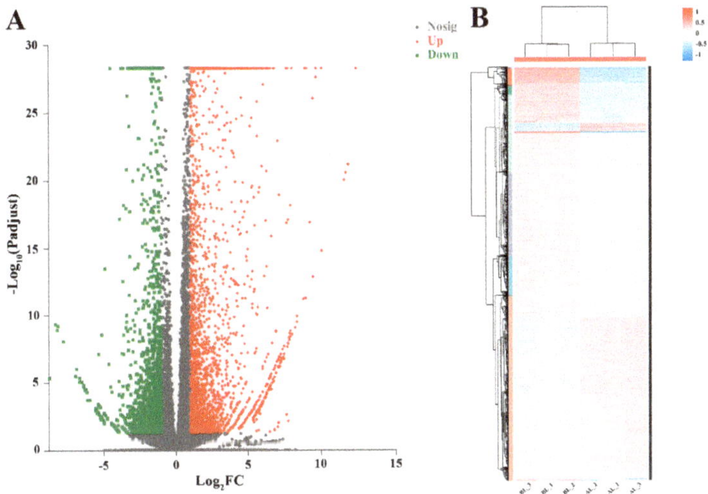

Figure 3. Analysis of gene expression in two kinds of leaves. (**A**) Volcanic map of differentially expressed gene. Each point represents an expressed unigene, the abscissa represents the fold change value of gene expression difference in the two samples, the vertical axis represents the statistical test value of the difference in gene expression, *p*-value. Compared with annual leaves, red shows that the gene expression was upregulated in biennial leaves, green represents the gene expression that was downregulated in biennial leaves, and grey represents non-differentially expressed genes. (**B**) Heat map of differentially expressed genes. AL_1, AL_2 and AL_3 represent three biological replicates of annual leaves. BL_1, BL_2, and BL_3 represent three biological replicates of biennial leaves.

3.4. Gene Ontology (GO) Enrichment Analysis of DEGs

To further elucidate the function of DEGs, a GO enrichment analysis was performed. In terms of all DEGs, there were 298 enriched GO terms (*p*-value ≤ 0.05). The most significantly enriched GO category was the extracellular region (GO ID: 0005576), followed by the intrinsic component of the membrane (GO ID: 0031224), nucleic acid binding transcription (GO ID: 0001071), transcription factor activity, sequence-specific DNA binding (GO ID: 0003700), and oxidoreductase activity (GO ID: 0016491) (Figure 4A). In terms of upregulated DEGs, there were 292 significantly enriched GO terms including the extracellular region (GO ID: 0005576), dioxygenase activity (GO ID: 0051213), membrane part (GO ID: 0044425), etc (Figure 4B). As for downregulated DEGs, there were 219 terms such as beta-amyrin synthase activity (GO ID: 0042300), oxidosqualene cyclase activity (GO ID: 0031559), and lanosterol synthase activity (GO ID: 0000250) (Figure 4C).

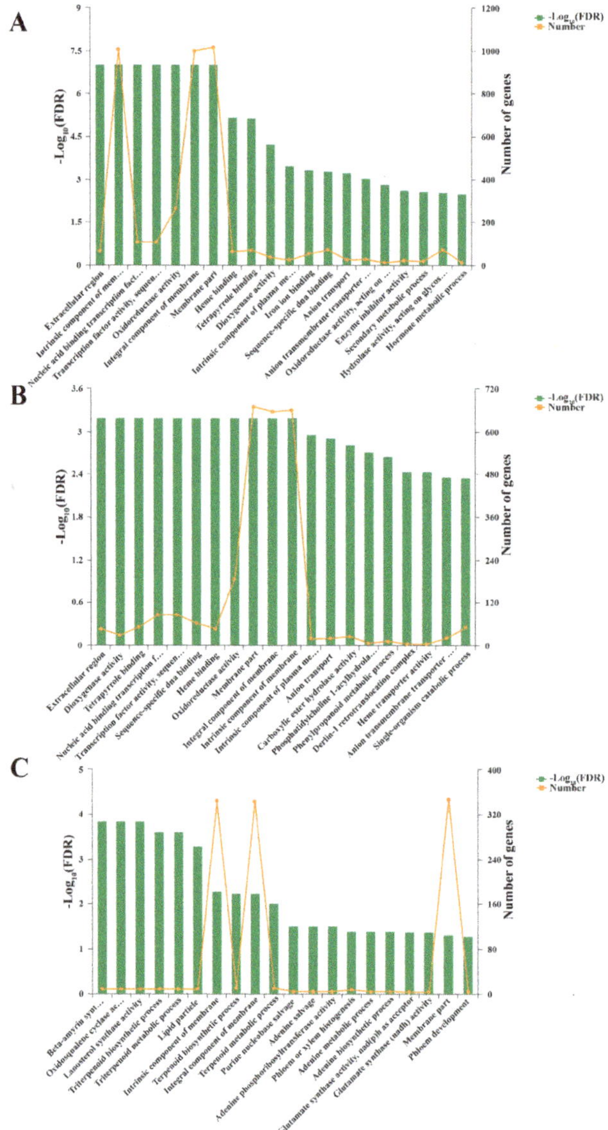

Figure 4. GO (Gene Ontology) enrichment analysis of DEGs (differentially expressed unigenes). (**A**) GO enrichment of all DEGs. (**B**) GO enrichment of upregulated DEGs. (**C**) GO enrichment of downregulated DEGs. The abscissa represents different GO terms, the vertical axis on the left represents the significance level of enrichment, corresponding to the height of the column, and the vertical axis on the right represents the number of unigenes of each GO term, corresponding to different points on the polyline.

3.5. Kyoto Encyclopedia of Genes and Genomes (KEGG) Pathway Analysis of DEGs

To further discover the leaf senescence pathway of *C. oleifera* leaves, KEGG analysis of DEGs was performed. A total of 126 KEGG pathways were significantly enriched, among which the phenylpropanoid biosynthesis (Pathway id: map00940), plant hormone signal transduction (Pathway id: map04075), and phenylalanine metabolism (Pathway id: map00360) pathways were the most obviously enriched pathways (Figure 5A). The upregulated DEGs were assigned to the KEGG pathways involved in 125 pathways, among them, phenylpropanoid biosynthesis (Pathway id: map00940), phenylalanine metabolism (Pathway id: map00360), and arginine and proline metabolism (Pathway id: map00330) were the most highly represented (Figure 5B). The downregulated DEGs were assigned to 90 KEGG pathways, among which the sesquiterpenoid and triterpenoid biosynthesis (Pathway id: map00909), plant hormone signal transduction (Pathway id: map04075), and ascorbate and aldarate metabolism (Pathway id: map00053) pathways were the most significantly enriched pathways (Figure 5C).

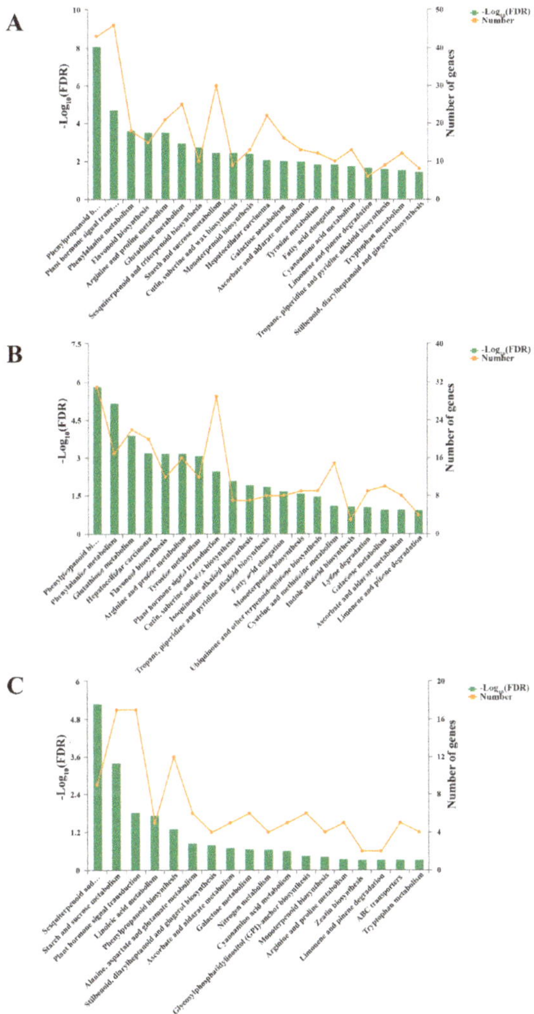

Figure 5. Enriched KEGG (Kyoto Encyclopedia of Genes and Genomes) pathway analysis of DEGs (differentially expressed unigenes). (**A**) KEGG pathway analysis of all DEGs. (**B**) KEGG pathway analysis of upregulated DEGs. (**C**) KEGG pathway analysis of downregulated DEGs. The abscissa represents different KEGG pathways, the vertical axis on the left represents significance level of enrichment, corresponding to the height of the column, and the vertical axis on the right represents the number of unigenes of each KEGG pathway, corresponding to different points on the polyline.

3.6. SAGs Differentially Expressed in Two Kinds of Leaves

To identify the important gene expressions of *C. oleifera leaves* during leaf senescence, we analyzed DEGs associated with Chl degradation, plant hormones, oxidation pathways, and found 77 unigenes homologous to the key SAGs that have been reported in other species (Table S2). Many genes involved in the Chl degradation pathway were upregulated in biennial leaves including *PAO*, *NOL*, *SGR*, etc. (Figure 6A, Figure S4A). In addition, genes related to ABA, ethylene, JA, and SA were mostly upregulated in biennial leaves such as *MYBL*, *ATAF1*, and *PYL9* (Figure S4B). Genes associated with auxin and cytokinin were also differentially expressed during leaf senescence. For example, the gene homologous to cytokinin receptor gene *AHK2* in arabidopsis was downregulated, and the gene homologous to auxin-related gene *SAUR36* was upregulated. During foliar senescence, evident changes for oxidation-related genes have been observed, for instance, *WRKY75*, encoding a transcription factor inhibiting the elimination of ROS, was upregulated, whereas *MDAR*, encoding a monodehydroascorbate reductase, was downregulated (Figure 6C, Figure S4C). Furthermore, we found some other SAGs were also responsive to the senescence of *C. oleifera* leaves such as *SAG12*, which was evidently upregulated in biennial leaves (Figure 6D). These findings suggest that there are multiple signaling pathways involved in the regulation of *C. oleifera* leaf senescence.

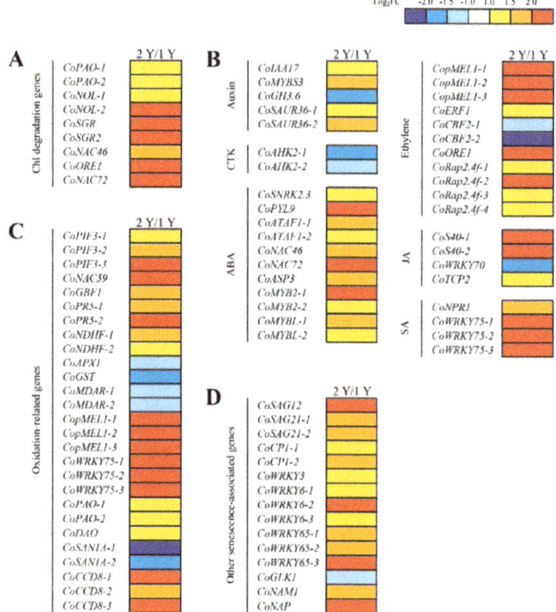

Figure 6. The ratios of expression levels (biennial leaves/annual leaves) for putative SAGs. (**A**) The genes involved in the Chl (chlorophyll) degradation pathway. (**B**) Genes associated with plant hormone pathway. (**C**) Genes associated with oxidation pathway. (**D**) Other senescence-associated genes. The relative expression value was normalized to the annual leaf expression level.

3.7. Correlation Analysis of SAGs

Previous studies have shown that leaf senescence is a complicated physiological process regulated by multiple signal pathways. To better reveal the potential co-expression relationship between genes, we constructed an expression correlation network of the 77 SAGs on the basis of the Spearman correlation algorithm. The results indicated that there was a possible correlation among genes involved in different pathways such as Chl degradation, phytohormones, and oxidation pathways. Among these

SAGs, *CoWRKY75-2*, *CoPAO-2*, *CoSAG12*, *CoWRKY65-2*, *CoMYBL-1*, *CoSAUR36-1*, *CoNOL-1*, *CoNAC59*, *CoGBF1*, and *CoS40-2* were significantly correlated with the expression of other SAGs (Figure 7). For instance, there is a significant expression correlation between *CoSAG12* and many other SAGs such as *CoNOL-2*, *CoORE1*, *CoMYBL-1*, *CoS40-2*, and *CoWRKY75-2*, and these genes probably play important roles in Chl degradation, hormones, and oxidation pathways.

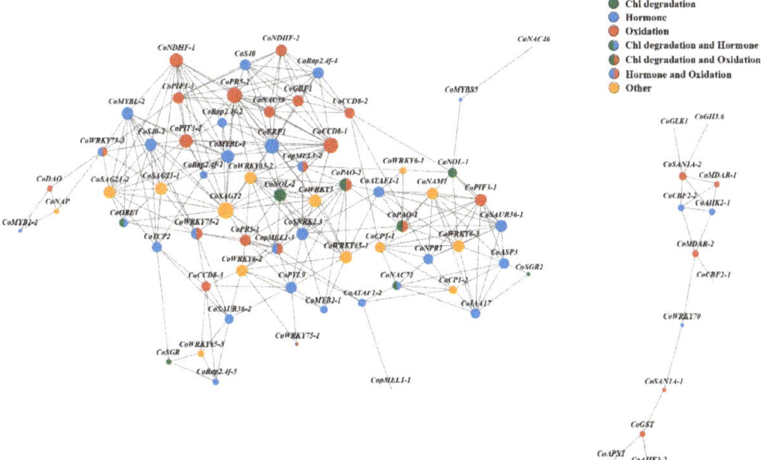

Figure 7. Expression correlation analysis of putative SAGs. Each node represents a unigene, and the connection between nodes represents the expression correlation of genes. Large node suggests that this gene has expression correlation with a large number of other genes.

3.8. Transcription Factors (TFs) Responding to Leaf Senescence

Transcription factors (TFs) exert vital function in the process of leaf senescence. Therefore, we analyzed the expression levels of TFs in *C. oleifera* leaves of two types of ages. Among all the DEGs, we identified 162 TFs (123 upregulated TFs and 39 downregulated TFs) (Table S3). Among these different TF families, the genes encoding MYB proteins (16.67%) accounted for the largest proportion of these genes, followed by the genes encoding AP2/ERF proteins (16.05%), WRKY (12.35%), and NAC proteins (9.3%) (Figure 8). Notably, many TFs we have identified are homologous to SAGs in other species genome such as MYBL, ERF1, WRKY75, ATAF1, etc. Among these potential SAGs in *C. oleifera*, the genes that exhibited the most significant changes were selected for further qRT-PCR analysis.

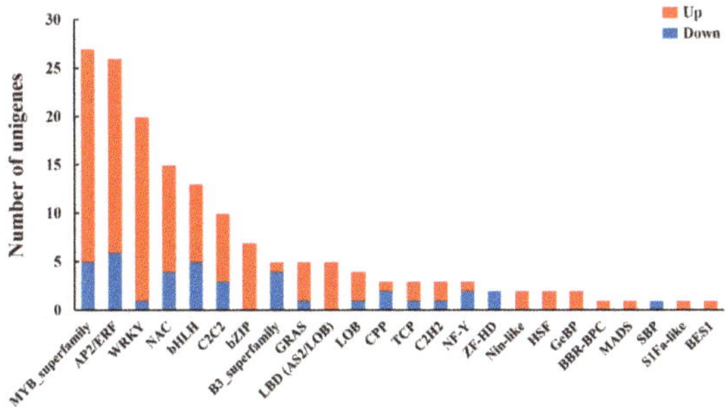

Figure 8. Distribution of differentially expressed transcription factors (TFs). Compared with annual leaves, red represents up-regulated TFs in biennial leaves, blue represents down-regulated TFs in biennial leaves.

3.9. Validation of RNA-Seq Data by qRT-PCR

To verify the accuracy and reliability of the transcriptome data, 19 putative SAGs, which were speculated to be highly correlated with other SAGs by expression correlation analysis, were selected for further analysis using qRT-PCR. Our findings indicated that the results of qRT-PCR analysis were in accordance with the RNA-Seq data. Compared with annual leaves, the expression level of all predicted upregulated SAGs was significantly increased in biennial leaves, and two downregulated SAGs were correspondingly inhibited in senescent leaves (Figure 9). Therefore, these results suggest that our RNA-Seq data were reliable and consistent with the qRT-PCR analysis.

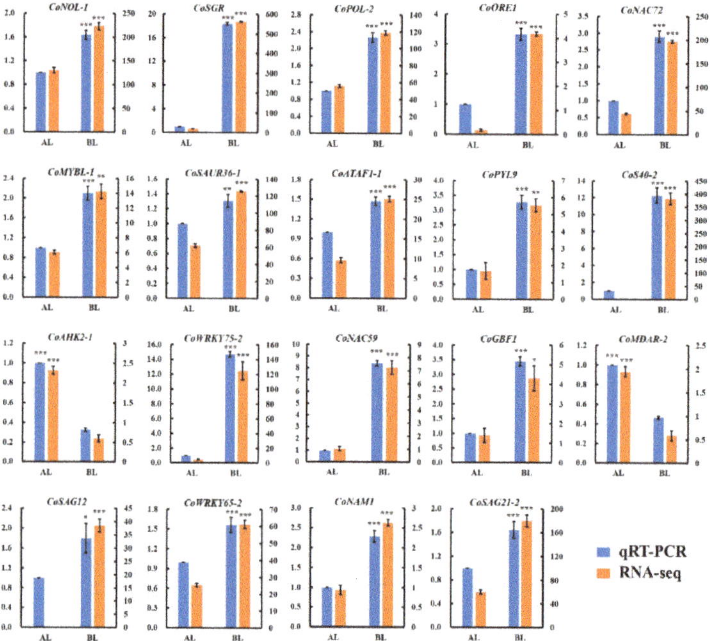

Figure 9. The validation of transcriptome by qRT-PCR (Quantitative Real-time PCR). The vertical axis on the left represents the relative expression level of genes by qRT-PCR analysis, and the vertical axis on the right represents the TPM (Transcripts Per Million reads) value analyzed by RNA-Seq. *: $0.01 \leq p < 0.05$, **: $0.001 \leq p < 0.01$, ***: $p < 0.001$.

4. Discussion

In this study, we demonstrated that some important physiological differences occurred between younger and older leaves of *C. oleifera*. For example, compared with the younger annual leaves, the older biennial leaves showed less Chl concentration and antioxidant enzyme activities, and more accumulation of MDA. In fact, similar phenomena were also observed in other species. For instance, *Gossypium hirsutum* displays Chl breakdown and increased production of MDA during leaf senescence [51]. In *Triticum aestivum*, the activities of antioxidant enzymes such as SOD and CAT were inhibited in the process of leaf senescence, suggesting that the leaf senescence of the plant is generally accompanied by physiological changes such as the degradation of Chl and some macromolecules, and the reduction of antioxidant enzyme activity [52]. Our findings indicated that compared with the annual leaves, the biennial leaves showed obvious senescence phenotypes and physiological change, suggesting that these two kinds of leaves are suitable for the study of *C. oleifera* leaf senescence.

To gain more insight into the mechanism of leaf senescence in *C. oleifera*, a transcriptomic analysis was performed in both annual leaves and biennial leaves. In our study, a total of 4645 DEGs were identified, and we further interpreted the potential biological functions of DEGs from the gene function and signaling pathway through GO enrichment analysis and KEGG enrichment analysis, respectively. The results of GO enrichment showed that with the duration of leaf senescence of *C. oleifera*, 'extracellular region', 'intrinsic component of membrane', and 'oxidoreductase activity' were significantly enriched, which were consistent with the physiological changes we measured. During leaf senescence of *C. oleifera*, the activity of antioxidant enzyme and MDA concentration changed, and MDA is one of the most important products associated with membrane lipid peroxidation, and can further cause damage to cell membranes [53]. Therefore, on one hand, our results suggest that ROS plays a vital role in the process of *C. oleifera* leaf senescence, and on the other hand, these findings demonstrate that the physiological changes we determined and the analysis of transcriptome are reliable. Furthermore, based on KEGG enrichment, we found that 'phenylpropanoid biosynthesis' and 'plant hormone signal transduction' were significantly enriched. In fact, previous study also showed that during the leaf senescence of *Sorghum bicolor*, 'phenylpropanoid biosynthesis' and 'plant hormone signal transduction' exhibited the most DEGs [54].

Leaf senescence is a highly complex process, which is controlled by multiple pathways [49]. Based on previous studies and our findings on the physiological changes and transcriptomic analysis of *C. oleifera*, we speculated that leaf senescence may cause changes in pathways such as Chl degradation, plant hormones, and oxidation, etc. In order to reveal these changes, we performed differential expression statistics on the paramount pathways and found 77 essential SAGs in the process of *C. oleifera* leaf senescence, many of which are associated with Chl degradation, phytohormone signaling, and oxidation. For instance, compared with annual leaves, *CoNOL* and other Chl degradation related genes, ABA receptor gene *CoPYL9* and senescence-related *SAG12* were upregulated in biennial leaves, while cytokinin receptor *CoAHK2* and antioxidant enzyme gene *CoMDAR-2* were downregulated. Similar results were also observed in other species. For example, in *Brassica napus*, the *SAG12* transcript evidently accumulates in older leaves [17]. In addition, in arabidopsis, during dark-induced senescence, the expression of *SGR*, *NOL*, and *PAO* also significantly increased to degrade Chl [55]. These results indicate that a similar senescence mechanism exists in the process of leaf senescence induced by environmental factor or age.

Notably, in this study, based on the analysis of these SAGs, we found that *CoORE1* is simultaneously involved in Chl degradation and plant hormone signaling pathways, suggesting that SAGs may participate in more than one pathway to regulate leaf senescence. In fact, previous study in *Brassica rapa* demonstrated that *BrNAC055* accelerates leaf senescence by activating the transcription of *RBOH* (ROS-producing enzymes respiratory burse oxidase homologue) and two Chl catabolic genes (*BrNYC1* and *BrNYE1*), indicating that *BrNAC055* is involved in Chl degradation and oxidation pathways [56]. To identify the potential correlation between these important pathways, we further constructed an expression correlation network for the 77 SAGs identified and found that there is an expression correlation between genes in different pathways. For example, *CoORE1*, which is involved in both Chl degradation and plant hormone pathways, correlated with *CoWRKY75-2*, a WRKY transcription factor that participates in both oxidation and hormone pathways. Furthermore, *CoORE1* also correlated with other key SAGs such as *CoSAG21-2* and *CoSAG12*. Expression correlation implies, to a certain extent, potential connections between genes. For instance, there is an expression correlation between senescence-related gene *ONAC054* and ABA signaling associated gene *OsABI5*, and further experimental results indicated that *ONAC054* directly upregulates the expression of *OsABI5*, thus regulating leaf senescence of rice [57]. Therefore, by constructing the expression correlation network, we predicted the possible connection between SAGs from a bioinformatics perspective. Nevertheless, the accurate regulatory relationship needs further experimental verification in future research.

TFs have important functions in transmitting and amplifying leaf senescence signals, thus activating a cascade of downstream genes [9]. Overexpression of *Cucumis melo* TF CmNAC60 in arabidopsis can promote leaf senescence of transgenic plants [58]. In this study, we performed statistical analysis on TFs and found that the differentially expressed TFs were mostly distributed in the MYB, WRKY, NAC families, etc. Expression correlation analysis showed that the functions of these senescence-associated TFs are complex. For instance, *CoWRKY75-2* is highly correlated with the expression of several significant senescence-associated genes such as *CoSAG12* and *CoSAG21-1*. *CoNAC72* not only participates in Chl degradation, but also plays an important role in the hormone pathway. Notably, there is an expression correlation between *CoNAC72* and *CoATAF1-2*. It is reported that homodimerization and heterodimerization among NAC TFs are crucial mechanisms in controlling NAC transcription factor mediated plant developmental processes [59,60], suggesting that these NAC TFs may form homodimers or heterodimers to function in regulating leaf senescence of *C. oleifera*. This phenomenon can also be found in rice, in which *ONAC020* and *ONAC026* share similar expression patterns during seed development, and *ONAC020* can interact with *ONAC026* [61]. In fact, in addition to NAC TFs, there have been similar reports on the study of WRKY TFs. In arabidopsis, *WRKY54* and *WRKY70* have been identified as important in leaf senescence, and they share a similar expression pattern during leaf senescence and co-operate as negative regulators [62]. Furthermore, in the present study, by using qRT-PCR, we found that the expression levels of key transcription factors such as *CoMYBL-1* and *CoNAC59* changed significantly, which was in good accordance with the transcriptome data, suggesting that TFs play important roles in *C. oleifera* leaf senescence based on the transcriptional level. In future research, *CoMYBL-1*, *CoNAC72*, *CoWRKY75-2*, and other TFs can be chosen as important candidates for the potential association between TFs and other SAGs to elucidate the leaf senescence mechanism. In addition to participating in senescence, TFs such as NAC72, ATAF1, and WRKY75 have also been widely reported to play important roles in the process of abiotic stress response in plants such as drought and salinity [63–65]. Our physiological results demonstrated that the accumulation of ROS in the annual leaves and biennial leaves was significantly different. Previous reports have shown that it is a common function of TFs to participate in the regulation of plants in response to abiotic stress through ROS pathways. Our qRT-PCR analysis indicated that the stress-responsive genes were differentially expressed between the two leaf types. Therefore, we speculated that beyond senescence, the two types of leaves also have tolerance differences in response to drought and other abiotic stresses, and TFs such as CoNAC72 and CoATAF1 may also be involved in the process of *C. oleifera* in response to abiotic stress. However, future research should acquire more data to verify this information.

5. Conclusions

To summarize, our results present a transcriptome analysis that combined physiological data and phenotypic observation, which contributes to the understanding of gene expression profiling underlying leaf senescence of *C. oleifera*. Compared with the annual leaves, a large number of DEGs were identified in the biennial leaves including 162 differentially expressed TFs. The present study explored 77 putative SAGs, which may be involved in the regulation of leaf senescence through Chl degradation, plant hormones, and oxidation pathways. There was a significant expression correlation between these SAGs, suggesting that these SAGs may jointly regulate leaf senescence. qRT-PCR analysis of 19 putative SAGs were in accordance with the RNA-Seq data, and in future research, these SAGs may serve as candidate genes for delaying leaf senescence in molecular breeding programs, thus increasing the yield of *C. oleifera*.

Supplementary Materials: The following are available online at http://www.mdpi.com/1999-4907/11/8/812/s1, Figure S1: NR annotated homologous species distribution map, Figure S2: Number of expressed unigenes in two kinds of *C. oleifera* leaves, Figure S3: Correlation analysis between different samples, Figure S4: Heat map of SAGs associated with important pathways, Table S1: Summary of sequencing data, Table S2: The details of putative SAGs, Table S3: The detail of differentially expressed TFs, Table S4: Primer sequences used in the qRT-PCR experiment.

Author Contributions: Conceptualization, S.Y. and L.Z.; Methodology, S.Y. and K.L.; Software, S.Y. and K.L.; Validation, S.Y. and L.Z.; Formal analysis, S.Y. and K.L.; Investigation, S.Y., M.Z., A.W., and J.Q.; Resources, L.Z.; Data curation, S.Y. and K.L.; Writing—original draft preparation, S.Y. and K.L.; Writing—review and editing, L.Z.;

Visualization, S.Y.; Supervision, L.Z.; Project administration, L.Z.; Funding acquisition, L.Z. All authors have read and agreed to the published version of the manuscript.

Funding: This research was funded by a grant from the National Key R&D Program Project, grant number 2018YFD1000603 to Lingyun Zhang.

Conflicts of Interest: The authors declare no conflicts of interest.

References

1. Zhuang, R. *Camellia Oleifera in China*; China Forestry Press: Beijing, China, 2008; p. 366.
2. Yang, C.; Liu, X.; Chen, Z.; Lin, Y.; Wang, S. Comparison of Oil Content and Fatty Acid Profile of Ten New Camellia oleifera Cultivars. *J. Lipids* **2016**, *2016*, 3982486. [CrossRef]
3. Lee, C.P.; Yen, G.C. Antioxidant activity and bioactive compounds of tea seed (Camellia oleifera Abel.) oil. *J. Agric. Food Chem.* **2006**, *54*, 779–784. [CrossRef]
4. Lee, C.P.; Shih, P.H.; Hsu, C.L.; Yen, G.C. Hepatoprotection of tea seed oil (Camellia oleifera Abel.) against CCl4-induced oxidative damage in rats. *Food Chem. Toxicol.* **2007**, *45*, 888–895. [CrossRef]
5. Chen, Y.; Deng, S.; Chen, L.; Ma, L.; He, H.; Wang, X.; Peng, S.; Liu, C.; Wang, R.; Xu, Y.; et al. A new view on the development of oil tea camellia industry. *J. Nanjing For. Univ. Nat. Sci. Edit.* **2020**, *44*, 1–10. [CrossRef]
6. Lin, P.; Wang, K.; Zhou, C.; Xie, Y.; Yao, X.; Yin, H. Seed Transcriptomics Analysis in Camellia oleifera Uncovers Genes Associated with Oil Content and Fatty Acid Composition. *Int. J. Mol. Sci.* **2018**, *19*, 118. [CrossRef]
7. Guo, Y.; Gan, S.S. Translational researches on leaf senescence for enhancing plant productivity and quality. *J. Exp. Bot.* **2014**, *65*, 3901–3913. [CrossRef] [PubMed]
8. Lim, P.O.; Kim, H.J.; Nam, H.G. Leaf senescence. *Annu. Rev. Plant Biol.* **2007**, *58*, 115–136. [CrossRef] [PubMed]
9. Podzimska-Sroka, D.; O'Shea, C.; Gregersen, P.L.; Skriver, K. NAC Transcription Factors in Senescence: From Molecular Structure to Function in Crops. *Plants (Basel)* **2015**, *4*, 412–448. [CrossRef] [PubMed]
10. Woo, H.R.; Kim, H.J.; Lim, P.O.; Nam, H.G. Leaf Senescence: Systems and Dynamics Aspects. *Annu. Rev. Plant Biol.* **2019**, *70*, 347–376. [CrossRef]
11. Miller, J.D.; Arteca, R.N.; Pell, E.J. Senescence-associated gene expression during ozone-induced leaf senescence in Arabidopsis. *Plant Physiol.* **1999**, *120*, 1015–1024. [CrossRef]
12. Pruzinska, A.; Tanner, G.; Anders, I.; Roca, M.; Hortensteiner, S. Chlorophyll breakdown: Pheophorbide a oxygenase is a Rieske-type iron-sulfur protein, encoded by the accelerated cell death 1 gene. *Proc. Natl. Acad. Sci. USA* **2003**, *100*, 15259–15264. [CrossRef] [PubMed]
13. Jehanzeb, M.; Zheng, X.; Miao, Y. The Role of the S40 Gene Family in Leaf Senescence. *Int. J. Mol. Sci.* **2017**, *18*, 2152. [CrossRef] [PubMed]
14. Garapati, P.; Xue, G.P.; Munne-Bosch, S.; Balazadeh, S. Transcription Factor ATAF1 in Arabidopsis Promotes Senescence by Direct Regulation of Key Chloroplast Maintenance and Senescence Transcriptional Cascades. *Plant Physiol.* **2015**, *168*, 1122–1139. [CrossRef] [PubMed]
15. Smykowski, A.; Zimmermann, P.; Zentgraf, U. G-Box binding factor1 reduces CATALASE2 expression and regulates the onset of leaf senescence in Arabidopsis. *Plant Physiol.* **2010**, *153*, 1321–1331. [CrossRef] [PubMed]
16. Zheng, X.; Jehanzeb, M.; Habiba; Zhang, Y.; Li, L.; Miao, Y. Characterization of S40-like proteins and their roles in response to environmental cues and leaf senescence in rice. *BMC Plant Biol.* **2019**, *19*, 174. [CrossRef]
17. Gombert, J.; Etienne, P.; Ourry, A.; Le Dily, F. The expression patterns of SAG12/Cab genes reveal the spatial and temporal progression of leaf senescence in Brassica napus L. with sensitivity to the environment. *J. Exp. Bot.* **2006**, *57*, 1949–1956. [CrossRef]
18. Kuai, B.; Chen, J.; Hortensteiner, S. The biochemistry and molecular biology of chlorophyll breakdown. *J. Exp. Bot.* **2018**, *69*, 751–767. [CrossRef]
19. Ramkumar, M.K.; Senthil Kumar, S.; Gaikwad, K.; Pandey, R.; Chinnusamy, V.; Singh, N.K.; Singh, A.K.; Mohapatra, T.; Sevanthi, A.M. A Novel Stay-Green Mutant of Rice with Delayed Leaf Senescence and Better Harvest Index Confers Drought Tolerance. *Plants (Basel)* **2019**, *8*, 375. [CrossRef]

20. Sato, Y.; Morita, R.; Katsuma, S.; Nishimura, M.; Tanaka, A.; Kusaba, M. Two short-chain dehydrogenase/reductases, NON-YELLOW COLORING 1 and NYC1-LIKE, are required for chlorophyll b and light-harvesting complex II degradation during senescence in rice. *Plant J.* **2009**, *57*, 120–131. [CrossRef]

21. Jan, S.; Abbas, N.; Ashraf, M.; Ahmad, P. Roles of potential plant hormones and transcription factors in controlling leaf senescence and drought tolerance. *Protoplasma* **2019**, *256*, 313–329. [CrossRef]

22. Zhao, Y.; Chan, Z.; Gao, J.; Xing, L.; Cao, M.; Yu, C.; Hu, Y.; You, J.; Shi, H.; Zhu, Y.; et al. ABA receptor PYL9 promotes drought resistance and leaf senescence. *Proc. Natl. Acad. Sci. USA* **2016**, *113*, 1949–1954. [CrossRef] [PubMed]

23. Riefler, M.; Novak, O.; Strnad, M.; Schmulling, T. Arabidopsis cytokinin receptor mutants reveal functions in shoot growth, leaf senescence, seed size, germination, root development, and cytokinin metabolism. *Plant Cell* **2006**, *18*, 40–54. [CrossRef] [PubMed]

24. Rogers, H.; Munne-Bosch, S. Production and Scavenging of Reactive Oxygen Species and Redox Signaling during Leaf and Flower Senescence: Similar but Different. *Plant Physiol.* **2016**, *171*, 1560–1568. [CrossRef] [PubMed]

25. Liang, D.; Shen, Y.; Ni, Z.; Wang, Q.; Lei, Z.; Xu, N.; Deng, Q.; Lin, L.; Wang, J.; Lv, X.; et al. Exogenous Melatonin Application Delays Senescence of Kiwifruit Leaves by Regulating the Antioxidant Capacity and Biosynthesis of Flavonoids. *Front Plant Sci.* **2018**, *9*, 426. [CrossRef]

26. Bengoa Luoni, S.; Astigueta, F.H.; Nicosia, S.; Moschen, S.; Fernandez, P.; Heinz, R. Transcription Factors Associated with Leaf Senescence in Crops. *Plants (Basel)* **2019**, *8*, 411. [CrossRef]

27. Guo, P.; Li, Z.; Huang, P.; Li, B.; Fang, S.; Chu, J.; Guo, H. A Tripartite Amplification Loop Involving the Transcription Factor WRKY75, Salicylic Acid, and Reactive Oxygen Species Accelerates Leaf Senescence. *Plant Cell* **2017**, *29*, 2854–2870. [CrossRef]

28. Qiu, K.; Li, Z.; Yang, Z.; Chen, J.; Wu, S.; Zhu, X.; Gao, S.; Gao, J.; Ren, G.; Kuai, B.; et al. EIN3 and ORE1 Accelerate Degreening during Ethylene-Mediated Leaf Senescence by Directly Activating Chlorophyll Catabolic Genes in Arabidopsis. *PLoS Genet.* **2015**, *11*, e1005399. [CrossRef]

29. Zhang, X. Determination of chlorophyll content in plants – Acetone ethanol mixture method. *Liaoning Agric. Sci.* **1986**, *3*, 28–30.

30. Sedmak, J.J.; Grossberg, S.E. A rapid, sensitive, and versatile assay for protein using Coomassie brilliant blue G250. *Anal. Biochem.* **1977**, *79*, 544–552. [CrossRef]

31. Dhindsa, R.S.; Plumb-Dhindsa, P.; Thorpe, T.A. Leaf senescence: Correlation with increased levels of membrane permeability and lipid peroxidation and increased levels of superoxide dismutase and catalase. *J. Exp. Bot.* **1981**, *32*, 93–101. [CrossRef]

32. Bates, L.S.; Waldren, R.P.; Teare, I.D. Rapid Determination of Free Proline for Water-Stress Studies. *Plant & Soil* **1973**, *39*, 205–207.

33. Grabherr, M.G.; Haas, B.J.; Yassour, M.; Levin, J.Z.; Thompson, D.A.; Amit, I.; Xian, A.; Lin, F.; Raychowdhury, R.; Zeng, Q.; et al. Full-length transcriptome assembly from RNA-Seq data without a reference genome. *Nat. Biotech.* **2011**, *29*, 644–652. [CrossRef] [PubMed]

34. Smith-Unna, R.; Boursnell, C.; Patro, R.; Hibberd, J.M.; Kelly, S. TransRate: Reference-free quality assessment of de novo transcriptome assemblies. *Genome Res.* **2016**, *26*, 1134–1144. [CrossRef] [PubMed]

35. Fu, L.; Niu, B.; Zhu, Z.; Wu, S.; Li, W. CD-HIT: Accelerated for clustering the next-generation sequencing data. *Bioinformatics* **2012**, *28*, 3150–3152. [CrossRef]

36. Seppey, M.; Manni, M.; Zdobnov, E.M. BUSCO: Assessing Genome Assembly and Annotation Completeness. *Methods Mol. Biol.* **2019**, *1962*, 227.

37. Deng, Y.Y.; Li, J.Q.; Wu, S.F.; Zhu, Y.P.; He, F.C. Integrated nr Database in Protein Annotation System and Its Localization. *Comput. Eng.* **2006**, *32*, 71–72.

38. Rolf, A.; Amos, B.; Wu, C.H.; Barker, W.C.; Brigitte, B.; Serenella, F.; Elisabeth, G.; Huang, H.; Rodrigo, L.; Michele, M. UniProt: The Universal Protein knowledgebase. *Nucleic Acids Res.* **2004**, *32*, D115–D119.

39. Finn, R.D. *Pfam: The Protein Families Database*; American Cancer Society: Atlanta, GA, USA, 2005.

40. Tatusov, R.L.; Galperin, M.Y.; Natale, D.A.; Koonin, E.V. The COG database: A tool for genome-scale analysis of protein functions and evolution. *Nucleic Acids Res.* **2000**, *28*, 33. [CrossRef]

41. Ashburner, M.; Ball, C.A.; Blake, J.A.; Botstein, D.; Butler, H.; Cherry, J.M.; Davis, A.P.; Dolinski, K.; Dwight, S.S.; Eppig, J.T.; et al. Gene Ontology: Tool for the unification of biology. *Nat. Genet.* **2000**, *25*, 25–29. [CrossRef]

42. Minoru, K.; Yoko, S.; Masayuki, K.; Miho, F.; Mao, T. KEGG as a reference resource for gene and protein annotation. *Nuclc Acids Res.* **2016**, *44*, D457–D462.

43. Wagner, G.P.; Kin, K.; Lynch, V.J. Measurement of mRNA abundance using RNA-seq data: RPKM measure is inconsistent among samples. *Theory Biosci.* **2012**, *131*, 281–285. [CrossRef] [PubMed]

44. Love, M.I.; Huber, W.; Anders, S. Moderated estimation of fold change and dispersion for RNA-seq data with DESeq2. *Genome Biol.* **2014**, *15*, 550. [CrossRef] [PubMed]

45. Klopfenstein, D.V.; Zhang, L.; Pedersen, B.S.; Fidel, R.; Alex, W.V.; Aurélien, N.; Mungall, C.J.; Yunes, J.M.; Olga, B.; Mark, W. GOATOOLS: A Python library for Gene Ontology analyses. *Sci. Rep.* **2018**, *8*, 10872. [CrossRef] [PubMed]

46. Xie, C.; Mao, X.; Huang, J.; Ding, Y.; Wu, J.; Dong, S.; Kong, L.; Gao, G.; Li, C.Y.; Wei, L. KOBAS 2.0: A web server for annotation and identification of enriched pathways and diseases. *Nuclc Acids Res.* **2011**, *39*, 316–322. [CrossRef]

47. Li, Z.; Zhang, Y.; Zou, D.; Zhao, Y.; Wang, H.; Zhang, Y.; Xia, X.; Luo, J.; Guo, H.; Zhang, Z. LSD 3.0: A comprehensive resource for the leaf senescence research community. *Nucleic Acids Res.* **2019**, *48*, D1069–D1075. [CrossRef]

48. Liang, K.; Wang, A.; Sun, Y.; Yu, M.; Zhang, L. Identification and Expression of NAC Transcription Factors of Vaccinium corymbosum L. in Response to Drought Stress. *Forests* **2019**, *10*, 1088. [CrossRef]

49. Jeongsik, K.; Hee, K.J.; Il, L.J.; Ryun, W.H.; Ok, L.P. New insights into the regulation of leaf senescence in Arabidopsis. *J. Exp. Bot.* **2017**, *4*, 787–799.

50. Landis, J.B.; Soltis, D.E.; Soltis, P.S. Comparative transcriptomic analysis of the evolution and development of flower size in Saltugilia (Polemoniaceae). *BMC Genom.* **2017**, *18*, 475. [CrossRef]

51. Lin, M.; Pang, C.; Fan, S.; Song, M.; Wei, H.; Yu, S. Global analysis of the Gossypium hirsutum L. Transcriptome during leaf senescence by RNA-Seq. *BMC Plant Biol.* **2015**, *15*, 43. [CrossRef]

52. Srivalli, S.; Khanna-Chopra, R. Delayed wheat flag leaf senescence due to removal of spikelets is associated with increased activities of leaf antioxidant enzymes, reduced glutathione/oxidized glutathione ratio and oxidative damage to mitochondrial proteins. *Plant Physiol. Biochem.* **2009**, *47*, 663–670. [CrossRef]

53. Moore, K.; Roberts, L.J. Measurement of Lipid Peroxidation. *Free Radic. Res. Commun.* **1998**, *28*, 659–671. [CrossRef]

54. Wu, X.; Hu, W.; Luo, H.; Xia, Y.; Zhao, Y.; Wang, L.; Zhang, L.; Luo, J.; Jing, H. Transcriptome profiling of developmental leaf senescence in sorghum (Sorghum bicolor). *Plant Mol. Biol.* **2016**, *92*, 555–580. [CrossRef] [PubMed]

55. Zhang, Y.; Liu, Z.; Chen, Y.; He, J.; Bi, Y. PHYTOCHROME-INTERACTING FACTOR 5 (PIF5) positively regulates dark-induced senescence and chlorophyll degradation in Arabidopsis. *Plant Sci.* **2015**, *237*, 57–68. [CrossRef] [PubMed]

56. Fan, Z.Q.; Tan, X.L.; Chen, J.W.; Liu, Z.L.; Kuang, J.F.; Lu, W.J.; Shan, W.; Chen, J.Y. BrNAC055, a Novel Transcriptional Activator, Regulates Leaf Senescence in Chinese Flowering Cabbage by Modulating Reactive Oxygen Species Production and Chlorophyll Degradation. *J. Agric. Food Chem.* **2018**, *66*, 9399–9408. [CrossRef]

57. Sakuraba, Y.; Kim, D.; Han, S.-H.; Kim, S.-H.; Paek, N.-C. Multilayered Regulation of Membrane-Bound ONAC054 Is Essential for Abscisic Acid-Induced Leaf Senescence in Rice. *Plant Cell* **2020**, *32*, 630–649. [CrossRef] [PubMed]

58. Cao, S.; Zhang, Z.; Wang, C.; Li, X.; Guo, C.; Yang, L.; Guo, Y. Identification of a Novel Melon Transcription Factor CmNAC60 as a Potential Regulator of Leaf Senescence. *Genes* **2019**, *10*, 584. [CrossRef]

59. Jin, S.J.; Park, Y.T.; Jung, H.; Park, S.H.; Kim, J.K. Rice NAC proteins act as homodimers and heterodimers. *Plant Biotechnol. Rep.* **2009**, *3*, 127–134.

60. Wang, X.; Culver, J.N. DNA binding specificity of ATAF2, a NAC domain transcription factor targeted for degradation by Tobacco mosaic virus. *BMC Plant Biol.* **2012**, *12*, 1–14. [CrossRef]

61. Mathew, I.E.; Das, S.; Mahto, A.; Agarwal, P. Three Rice NAC Transcription Factors Heteromerize and Are Associated with Seed Size. *Front Plant Sci.* **2016**, *7*, 1638. [CrossRef]

62. Besseau, S.; Li, J.; Palva, E.T. WRKY54 and WRKY70 co-operate as negative regulators of leaf senescence in Arabidopsis thaliana. *J. Exp. Bot.* **2012**, *63*, 2667–2679. [CrossRef]

63. Liu, C.; Sun, Q.; Zhao, L.; Li, Z.; Peng, Z.; Zhang, J. Heterologous Expression of the Transcription Factor EsNAC1 in Arabidopsis Enhances Abiotic Stress Resistance and Retards Growth by Regulating the Expression of Different Target Genes. *Front Plant Sci.* **2018**, *9*, 1495. [CrossRef] [PubMed]

64. Lopez-Galiano, M.J.; Gonzalez-Hernandez, A.I.; Crespo-Salvador, O.; Rausell, C.; Real, M.D.; Escamilla, M.; Camanes, G.; Garcia-Agustin, P.; Gonzalez-Bosch, C.; Garcia-Robles, I. Epigenetic regulation of the expression of WRKY75 transcription factor in response to biotic and abiotic stresses in Solanaceae plants. *Plant Cell Rep.* **2018**, *37*, 167–176. [CrossRef] [PubMed]

65. Wu, Y.; Deng, Z.; Lai, J.; Zhang, Y.; Yang, C.; Yin, B.; Zhao, Q.; Zhang, L.; Li, Y.; Yang, C.; et al. Dual function of Arabidopsis ATAF1 in abiotic and biotic stress responses. *Cell Res.* **2009**, *19*, 1279–1290. [CrossRef]

Article

Marker-Assisted Selection of Trees with *MALE STERILITY 1* in *Cryptomeria japonica* D. Don

Yoshinari Moriguchi [1,*], Saneyoshi Ueno [2], Yoichi Hasegawa [2], Takumi Tadama [1], Masahiro Watanabe [1], Ryunosuke Saito [1], Satoko Hirayama [3], Junji Iwai [4] and Yukinori Konno [5]

[1] Graduate School of Science and Technology, Niigata University, 8050, Igarashi 2-Nocho, Nishi-ku, Niigata 950-2181, Japan; tama52178@gmail.com (T.T.); f20e024j@mail.cc.niigata-u.ac.jp (M.W.); f19e013a@mail.cc.niigata-u.ac.jp (R.S.)

[2] Department of Forest Molecular Genetics and Biotechnology, Forestry and Forest Products Research Institute, Forest Research and Management Organization, 1 Matsunosato, Tsukuba, Ibaraki 305-8687, Japan; saueno@ffpri.affrc.go.jp (S.U.); yohase@ffpri.affrc.go.jp (Y.H.)

[3] Niigata Prefecture Niigata Regional Promotion Bureau, 2009 Akiha-ku Hodojima, Niigata 956-8635, Japan; hirayama.satoko@pref.niigata.lg.jp

[4] Niigata Prefectural Forest Research Institute, 2249-5 Unotoro, Murakami, Niigata 958-0264, Japan; iwai.junji@pref.niigata.lg.jp

[5] Miyagi Prefectural Forest Research Institute, 3-8-1 Aoba-ku Honmachi, Sendai 980-8570, Japan; konno-yu943@pref.miyagi.lg.jp

* Correspondence: chimori@agr.niigata-u.ac.jp; Tel.: +81-25-262-6861

Received: 25 May 2020; Accepted: 3 July 2020; Published: 6 July 2020

Abstract: The practical use of marker-assisted selection (MAS) is limited in conifers because of the difficulty with developing markers due to a rapid decrease in linkage disequilibrium, the limited genomic information available, and the diverse genetic backgrounds among the breeding material collections. First, in this study, two families were produced by artificial crossing between two male-sterile trees, 'Shindai11' and 'Shindai12', and a plus tree, 'Suzu-2' (*Ms1/ms1*) (S11-S and S12-S families, respectively). The segregation ratio between the male-sterile and male-fertile trees did not deviate significantly from the expected 1:1 ratio in either family. These results clearly suggested that the male-sterile gene of 'Shindai11' and 'Shindai12' is *MALE STERILITY 1* (*MS1*). Since it is difficult to understand the relative positions of each marker, due to the lack of a linkage map which all the closely linked markers previously reported are mapped on, we constructed a partial linkage map of the region encompassing *MS1* using the S11-S and S12-S families. For the S11-S and S12-S families, 19 and 18 markers were mapped onto the partial linkage maps of the *MS1* region, respectively. There was collinearity (conserved gene order) between the two partial linkage maps. Two markers (CJt020762_*ms1-1* and reCj19250_2335) were mapped to the same position as the *MS1* locus on both maps. Of these markers, we used CJt020762 for the MAS in this study. According to the MAS results for 650 trees from six prefectures of Japan (603 trees from breeding materials and 47 trees from the Ishinomaki natural population), five trees in Niigata Prefecture and one tree in Yamagata Prefecture had heterozygous *ms1-1*, and three trees in Miyagi Prefecture had heterozygous *ms1-2*. The results obtained in this study suggested that *ms1-1* and *ms1-2* have different geographical distributions. Since MAS can be used effectively to reduce the labor and time required for selection of trees with a male-sterile gene, the research should help ensure that the quantity of breeding materials will increase to assist future tree-breeding efforts.

Keywords: conifer; linkage map; male sterility; marker-assisted selection

1. Introduction

Molecular marker-assisted selection (MAS), which can reduce the time required for a breeding cycle, is an attractive method for conifers, which have longer generation times than those of most crop species [1]. However, in conifers, the practical use of MAS is limited because it is difficult to develop markers for MAS due to a rapid decrease in linkage disequilibrium, the limited genomic information available, and the diverse genetic backgrounds among the breeding material collections. Nevertheless, the progress with genome analysis technologies has recently accelerated, producing an enormous volume of sequences and the subsequent development of markers linked to a particular target gene.

Sugi (*Cryptomeria japonica* D. Don) is an important forestry species that occupies nearly 4.5 million hectares of planted forest in Japan, which corresponds to approximately 44% of all the planted forest area in the country [2]. The increase in area covered by planted *C. japonica* has triggered pollinosis. *C. japonica* pollinosis is one of the most serious allergies in Japan, affecting 26.5% of the Japanese population [3]. As a countermeasure against *C. japonica* pollinosis, the use of male-sterile individuals is effective. The first male-sterile tree ('Toyama1') in *C. japonica*, which produced no pollen grains, was found in Toyama Prefecture in 1992 [4]. Genetic male sterility was found to be conferred by a major recessive gene [5], *MALE STERILITY 1* (*MS1*) [6]. Hasegawa et al. [7] reported that *MS1* was different from a gene reported in *Arabidopsis* (GenBank Accession AJ344210, [8]); however, its protein structure showed functional and structural similarities to wheat male-sterile genes (*ms1* and *ms5*). Since the discovery of this individual, six male-sterile trees homozygous for *MS1* (*ms1/ms1*) have been selected ('Shindai3', 'Fukushima-funen1', 'Fukushima-funen2', 'Tahara-1', 'Sosyun', and 'Mie-funen1') [6,9–14]. The frequency of these male-sterile trees in the forest is considered to be very low, because Igarashi et al. [9] identified only two male-sterile trees in a screening of 8700 trees distributed across a 19-ha planted forest. Male-sterile trees are generally identified by observing pollen release and/or by the direct inspection of the male strobili using a magnifying glass or microscope. In the selected male-sterile trees, the confirmation of the male-sterile gene *MS1* was made based on the results of test crossings. These test crossings led to the discovery of three other male-sterile genes: *MS2*, *MS3*, and *MS4* [5,6,15,16]. In some male-sterile trees such as 'Shindai11' and 'Shindai12', male-sterile genes have not yet been investigated.

Mutations in the *MS1* gene lead to the collapse of microspores after the separation of pollen tetrads [17], whereas that of the *MS2* gene lead to the formation of microspore clumps after normal microsporogenesis [15]. On the other hand, mutations in the *MS3* and *MS4* genes lead to the formation of microspores of various sizes after normal microsporogenesis [15,16]. The four male-sterile genes *MS1*, *MS2*, *MS3*, and *MS4* have been mapped to different linkage groups: the ninth (referred to as LG9 hereafter), fifth, first, and fourth linkage groups, respectively [17–19]. Only one tree with *ms2*, *ms3*, and *ms4* was selected, respectively. Therefore, trees with *ms1* have generally been used for tree improvement and seedling production. Both male-sterile trees and also trees heterozygous for the male-sterile gene are important for tree improvement and seed production as the maternal and paternal parents, respectively. Currently, seven trees heterozygous for *MS1* (*Ms1/ms1*), 'Suzu-2', 'Naka-4', 'Ooi-7', 'Ohara-13', 'Zasshunbo', 'Kamiukena-16', and 'Kurihara-4', have been selected ([4,6,14,20,21], Konno, personal communication). For the precise selection of trees heterozygous for *MS1*, it is generally necessary to produce F_1 trees by artificial crossing and to confirm whether these F_1 trees are male-sterile or -fertile trees. Confirmation is performed by the direct inspection of male strobili using a magnifying glass (or a microscope) or by observing the pollen release.

Due to the large amount of labor required for selection, the number of trees with the male-sterile gene is not sufficient. To reduce the labor of screening, the MAS of trees with the male-sterile gene is necessary. Recently, some markers closely linked to the *MS1* gene or derived from a putative *MS1* gene have been developed [22–25]. Moriguchi et al. [22] and Ueno et al. [25] reported that estSNP04188 and dDcontig_3995-165 were 1.8 cM and 0.6 cM from *MS1* in the T5 family (173 trees), respectively. Hasegawa et al. [23] reported that 15 markers were 0 cM from *MS1* in the F1O7 family (84 trees). Among these, AX-174127446 showed a high rate of predicting trees with *ms1*. Mishima et al. [24] reported two markers from contig "reCj19250" that can be used to select trees with *ms1*. On the other hand, Hasegawa et al. [7] reported a candidate male-sterile gene CJt020762 at the *MS1* locus, and all the breeding materials with the allele *ms1* had either a 4-bp or 30-bp deletion in the gene (they defined these alleles as *ms1-1* and *ms1-2*, respectively). Both of these were expected to result in faulty gene transcription and function; therefore, they developed two markers [26] from contig "CJt020762". The lack of a linkage map for these markers constructed from the same family makes it difficult to understand the relative position of each marker.

Therefore, in this study, we (1) checked whether the male-sterile gene of Shindai11 and Shindai12 was *MS1*, based on the results of test crossings, (2) constructed a partial linkage map of the region encompassing *MS1* using 46 markers, and (3) selected the trees with *ms1* by MAS. As there are few studies pertaining to the practicable applications of MAS in conifers, this study should provide a valuable model.

2. Materials and Methods

2.1. Phenotyping of Male Sterility and Single Nucleotide Polymorphism (SNP) Genotyping for Linkage Analysis

We used two families, S11-S and S12-S, in this study. These families were produced by artificial crossing between two male-sterile trees, 'Shindai11' and 'Shindai12', and a plus tree, 'Suzu-2' (*Ms1/ms1*), during March of 2016. These trees are considered to be diploid ($2n = 22$), because two or fewer alleles have been obtained from microsatellite analyses and they are able to produce an amount of seeds. Strobili production was promoted by spraying the trees with gibberellin-3 (100 ppm) in July 2018. Approximately five male strobili were sampled from each individual from early November to early December 2018. Each sampled male strobilus was bisected vertically with a razor, and male sterility was determined using a microscope (SZ-ST, Olympus, Tokyo, Japan). Individuals without male strobili and individuals in whom it was difficult to discriminate male sterility were excluded from further analysis. Finally, 130 individuals from S11-S and 138 individuals from S12-S were used to construct a linkage map. Needle tissue was collected from three parent trees ('Shindai11', 'Shindai12', and 'Suzu-2') and all the F$_1$ trees (268 trees) of two mapping populations. Genomic DNA was extracted from these needles using a modified hexadecyltrimethylammonium bromide (CTAB) method [26,27].

Single nucleotide polymorphism (SNP) markers from contigs "reCj19250" and "CJt020762" [7,24] and SNP markers mapped to LG9 [23,25,28] were used to construct a partial linkage map of the region encompassing the *MS1* locus for each of the two families (because the gene was located in LG9) [18]. For estSNP00204 [22], AX-174127446 [23], and CJt020762 [7], the SNaPshot assay, which extends primers by a single base, was used for genotyping. The primer sequences used to target the three markers in the SNaPshot assay (estSNP00204 [22], AX-174127446 [23], and CJt020762 [7]) are shown in Table S1. Although CJt020762 contained a 4-bp and 30-bp deletion, we used the 4-bp deletion for primer design because there is no polymorphism associated with the 30-bp deletion between the parents of the mapping populations. Multiplex polymerase chain reaction (PCR) was performed using three primer pairs and the Multiplex PCR Kit (QIAGEN, Hilden, Germany). Each reaction contained 2× QIAGEN multiplex PCR master mix, 1 μL primer mix (2.5 μM for each primer), and 40 ng genomic DNA in a total volume of 6 μL. Amplification was performed in the Takara PCR Thermal Cycler (Takara, Tokyo, Japan) using an initial denaturation step at 95 °C for 15 min, followed by 30 cycles of denaturation at 94 °C for 30 s, annealing at 57 °C for 1.5 min, and extension at 72 °C for 1 min, with a final extension at 60 °C for 30 min. To remove any primers and dNTPs, 5.0 μL of the PCR products were treated with 2.0 μL ExoSAP-IT reagent (Thermo Fisher Scientific, Waltham, MA, USA), followed by incubation at 37 °C for 30 min and then 80 °C for 15 min to inactivate the enzyme. Single-base extension reactions were carried out in a 5.0 μL final volume containing 0.5 μL SNaPshot Multiplex Ready Mix (Thermo Fisher Scientific), 1 μL primer mix (1.0 μM for each primer), and 2.0 μL of the treated PCR products. The reactions were performed in the Takara PCR Thermal Cycler (Takara) with 25 cycles of denaturation at 96 °C for 10 s and annealing and elongation at 60 °C for 30 s. The final extension products were treated with 1 U shrimp alkaline phosphatase (Thermo Fisher Scientific) and incubated at 37 °C for 1 h, followed by enzyme inactivation at 80 °C for 15 min. The PCR products (1.0 µL) were mixed with 0.2 µL GeneScan 120 LIZ size standard and 8 µL Hi-Di formamide prior to electrophoresis. Capillary electrophoresis was performed on the 3130xl genetic analyzer using POP-7 (Thermo Fisher Scientific), and the alleles were analyzed using GeneMaker v2.4.0 software (SoftGenetics, State College, PA, USA). For the other 43 SNP markers mapped to LG9, genotyping was performed using the 48.48 Dynamic Array (Fluidigm, South San Francisco, CA, USA). For the 48.48 Dynamic Array, 6.25 ng genomic DNA per sample (at a concentration of 5 ng/µL) were used for specific target amplification. The assays were performed following the protocol provided by the manufacturer. The data obtained were analyzed using Fluidigm SNP Genotyping Analysis software (ver. 4.5.1). The primer information is provided in Table S2.

Chi-square tests were performed for each locus to assess the deviation from the expected Mendelian segregation ratio. Loci showing an extreme segregation distortion ($p < 0.01$) and with many missing data points (more than five individuals) were excluded from further linkage analysis. The linkage analyses were performed using the maximum likelihood mapping algorithm in JoinMap ver. 4.1 software (Kyazma, Wageningen, The Netherlands) with a cross pollination-type population (hk × hk, lm × ll, and nn × np) and two rounds of map calculation [29]. The markers were assigned to the LG9 linkage group using a logarithm of odds ratio threshold of 8.0, which was the same value as in previous reports on *C. japonica* [18,19,22,28]. The maximum likelihood mapping algorithm was used to determine the marker order in the linkage group. The map distance was calculated using the Kosambi mapping function [30]. The default settings were used for the recombination frequency threshold and ripple value.

2.2. MAS of Trees with ms1

The needles for MAS selection were collected from breeding materials in Niigata (Tohoku breeding region), Yamagata (Tohoku breeding region), Miyagi (Tohoku breeding region), Shizuoka (Kanto breeding region), Tottori (Kansai breeding region), and Kumamoto (Kyushu breeding region) Prefectures with sample numbers of 238, 163, 30, 34, 72, and 66, respectively. In the samples from Miyagi Prefecture, Kurihara-4, a tree heterozygous for *MS1*, was included. Genomic DNA was extracted from these needles using a modified CTAB method [26,27]. In addition, we also performed MAS selection using previously extracted DNA from 47 *C. japonica* trees in the Ishinomaki natural population of Miyagi Prefecture, where clonal analysis was performed in 2017 [31].

Based on the sequence information of CJt020762, Hasegawa et al. [26] developed two primer pairs that sandwiched the two deletions, respectively. These two markers were used for MAS selection in this study. PCR amplifications were performed in 10 µL reaction volumes containing 5 ng of genomic DNA, 1× PCR Kapa2G buffer with 1.5 mM MgCl$_2$, 0.2 µL of 25 mM MgCl$_2$, 0.2 µL of 10 mM each dNTP mix, 0.4 µL of 5 µM forward primers labeled with dye (CJt020762_*ms1-1*_F and CJt020762_*ms1-2*_F), 0.2 µL 5 µM reverse primers (CJt020762_*ms1-1*_R and CJt020762_*ms1-2*_R), 5 ng template DNA, and 0.5 U KAPA2G Fast PCR enzyme (KAPA2G Fast PCR kit; KAPA Biosystems, Wilmington, USA). Amplification was performed on the Takara PCR Thermal Cycler (Takara) under the following conditions: initial denaturation for 3 min at 95 °C, followed by 35 cycles of denaturation for 15 s at 95 °C, annealing for 15 s at 60 °C, extension for 1 s at 72 °C, and a final extension for 1 min at 72 °C. The PCR products and the DNA size marker (LIZ600; Thermo Fisher Scientific) were separated by capillary electrophoresis on the ABI 3130 Genetic Analyzer (Applied Biosystems, Tokyo, Japan). The DNA fragments were detected using the GeneMarker software (ver. 2.4.0; SoftGenetics, State College, PA, USA).

3. Results and Discussion

3.1. Linkage Maps of the MS1 Region

Of the 130 progenies in S11-S family produced by artificial crossing between 'Shindai11' and 'Suzu-2', 75 were male-fertile and 55 male-sterile. On the other hand, of the 138 progenies in S12-S family produced by artificial crossing between 'Shindai12' and 'Suzu-2', 65 were male-fertile and 73 male-sterile. The segregation ratio between the male-sterile and male-fertile trees in the S11-S and S12-S progenies did not deviate significantly from the expected ratios of 1:1 ($X^2 = 0.31$ ($p = 0.08$) and 0.46 ($p = 0.50$), respectively). These results clearly suggested that the male-sterile gene of 'Shindai11' and 'Shindai12' was *MS1*. Based on observations using a microscope, Miura et al. [32] reported that the male-sterile phenotype of 'Shindai11' and 'Shindai12' was similar to those of 'Fukushima-funen1', 'Fukushima-funen2', and 'Shindai3', which are regulated by the *MS1* gene [6,9]. These previous observational results obtained by microscopy were consistent with the results in this study.

The 19 and 18 markers were mapped onto the partial linkage maps of the region encompassing *MS1* for the S11-S and S12-S families, respectively (Figure 1). There was collinearity (conserved gene order) among the two partial linkage maps. Two markers (CJt020762_ms1-1 and reCj19250_2335) were mapped to the same position as the *MS1* locus in both maps. Of these markers, reCj19250_2335 could not be selected 'Ooi-7' heterozygous for *MS1* (*ms1-2/Ms1*), suggesting that reCj19250 was not the causative gene of *MS1*; the marker did not select trees with *ms1* with 100% accuracy [7,23]. As genome sequencing has now been completed in *C. japonica*, the question of whether these markers are located close to each other within the genome will probably be investigated in the near future. The gene CJt020762 was found to code for a lipid transfer protein, and the loss of protein function was predicted in breeding materials with *ms1* due to the 4-bp and the 30-bp deletions in the coding region [7]. CJt020762 also showed functional and structural similarity with wheat male-sterile genes; the mRNA sequence from CJt020762 showed a two-fold higher expression in male-fertile strobili than in male-sterile strobili [7]. Therefore, we used CJt020762 for the MAS in this study.

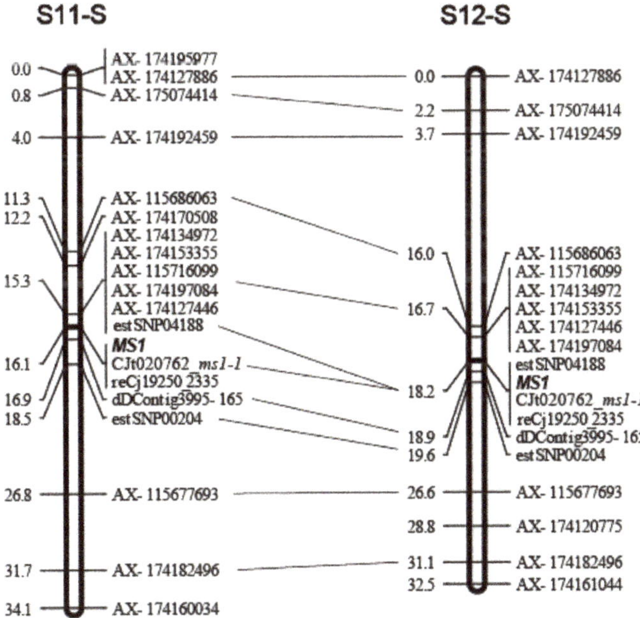

Figure 1. Partial linkage maps of the region encompassing *MS1* in the S11-S and S12-S *C. japonica* families.

3.2. MAS of Trees with ms1

In the MAS results of this study, we found that five trees in Niigata Prefecture ('Kashiwazakishi-1', 'Setsugai Niigata-6', 'Setsugai Murakami-2', 'Setsugai Aikawa-8', and 'Kamikiri Niigata-55') and one tree in Yamagata Prefecture ('Taisetsu Yamagata-8') had heterozygous *ms1-1*, and three trees in Miyagi Prefecture ('Kurihara-4' and two trees in the natural population) had heterozygous *ms1-2*. Two male-sterile trees in Niigata Prefecture ('Shindai11' and 'Shinadai-12') used as the mother trees of the mapping families had homozygous *ms1-1*. The two trees with *ms1-2* in the Ishinomaki natural forest (Ishinomaki_J284 and Ishinomaki_J278) were considered to have a parent–child relationship according to their genotypes. Because Hasegawa et al. [7] also found trees with *ms1-2* in this forest, trees with *ms1-2* may be distributed at a high frequency in this forest. Through further selections from this natural forest, it may be possible to obtain more breeding materials for male sterility.

Because half of the offspring in the mapping family 'Fukushima-funen1' (*ms1-1/ms1-1*) × 'Ooi-7' (*Ms1/ms1-2*) [23] showed male sterility, both of the trees with *ms1-1* and *ms1-2* can be used in a breeding program. Therefore, MAS should target both the *ms1-1* and *ms1-2* alleles. In Niigata Prefecture, where three male-sterile trees ('Shindai3' (*ms1-1/ms1-1*), 'Shindai11' (*ms1-1/ms1-1*), 'Shindai12' (*ms1-1/ms1-1*)) have been found thus far [6,7,32], five trees heterozygous for *MS1* (*Ms1/ms1-1*) were newly found among the 238 trees. This higher number of trees with *ms1* may be due to the large number of samples analyzed in this study. However, considering the achievements attained in the past selection, the rate of trees with *ms1* in Niigata Prefecture appears to be higher than the rates of other prefectures. In Miyagi Prefecture, where one tree heterozygous for *MS1* ('Kurihara-4', *Ms1/ms1-2*) was found thus far [Konno, personal communication], two trees with *ms1-2* (they have a parent–child relationship) were newly found among the 77 trees. Thus, the deletion mutations detected in Niigata Prefecture (*ms1-1*) and Miyagi Prefecture (*ms1-2*) were different. These results suggest that *ms1-1* and *ms1-2* may have different geographical distributions. Among the four breeding regions in *C. japonica*, the Tohoku breeding region has a relatively large amount of breeding materials for male sterility (Figure 2). However, the breeding materials for male sterility in the Kanto and Kansai breeding regions are still fewer than those in the Tohoku breeding region, and there are no breeding materials for male sterility in the Kyushu breeding region.

In this study, 650 trees were examined for male-sterile alleles. The precise selection of trees heterozygous for *MS1* using a magnifying glass or a microscope requires considerable labor, time (approximately 5 years: 1 year to promote flowering, 1 year for seed production, and 3 years to confirm male sterility), and space (approximately 56 seedlings per 1 m^2). Labor includes gibberellin treatment, artificial crossing (i.e., pollen collection, male strobilus removal, pollination bag setting and pollen application), seed collection, field plowing, seed sowing, watering, fertilizer application, weeding, pesticide application, gibberellin treatment, and the observation of male strobili. For the examination of 650 trees using this method, firstly, we have to prepare 19,500 F_1 seedlings (age, 3 years) by 650 artificial crossings (30 F_1 seedlings per clone); in contrast, the MAS performed in this study (consisting of fragment analysis using a sequencer) requires approximately two weeks to complete, including DNA extraction (approximately 100 samples per day) and genotyping (approximately 200 samples per day). Since MAS is effective for reducing the amount of labor and time required to select trees with the male-sterile allele, the research should help ensure that the quantity of breeding materials will increase to assist future tree-breeding efforts. Hasegawa et al. [26] developed allele specific PCR (ASP) and amplified length polymorphism (ALP) markers to analyze *ms1-1* and *ms1-2*, respectively, without a sequencer. In a laboratory without a sequencer, these markers are effective for performing MAS. If both *ms1-1* and *ms1-2* must be analyzed, the total estimated cost of fragment analysis by sequencer (approximately 236.7 US\$ for 96 samples) is very similar to that of the ASP and ALP marker-based analysis (approximately 241.6 US\$ for 96 samples) (Table S3). These estimates may be further reduced through the development of DNA extraction methods with greater efficiency and lower cost.

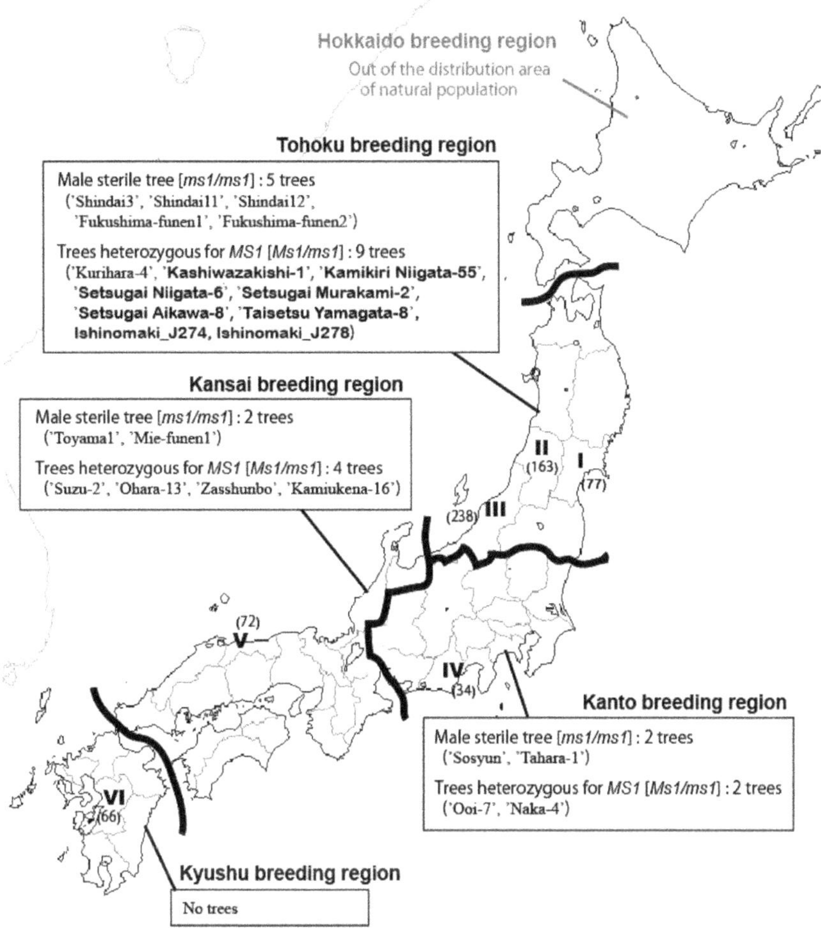

Figure 2. Breeding materials with *MALE STERILITY 1* in four *C. japonica* breeding regions. The bold font shows the selected trees in this study. Roman numerals indicate the prefectures where marker-assisted selection (MAS) was performed (I: Miyagi, II: Yamagata, III: Niigata, IV: Shizuoka, V: Tottori, VI: Kumamoto). Numbers in parentheses indicate the sample size.

4. Conclusions

In this study, we performed MAS for 650 trees from six prefectures of Japan using CJt020762_*ms1-1* markers and found that five trees in Niigata Prefecture ('Kashiwazakishi-1', 'Setsugai Niigata-6', 'Setsugai Murakami-2', 'Setsugai Aikawa-8', and 'Kamikiri Niigata-55') and one tree in Yamagata Prefecture ('Taisetsu Yamagata-8') had heterozygous *ms1-1*, and three trees in Miyagi Prefecture ('Kurihara-4' and two trees in the natural population) had heterozygous *ms1-2*. The results obtained in this study suggested that a difference in geographical distribution between *ms1-1* and *ms1-2*. In this study, we selected trees with *ms1* among 650 trees, without producing 19,500 F_1 seedlings by 650 artificial crossing (30 F_1 seedlings per clone). This MAS is able to complete within approximately two weeks. Because MAS can effectively reduce the labor and time for the selection of trees with the male-sterile gene, research should help ensure that the quantity of breeding materials will increase to assist future tree-breeding efforts.

Supplementary Materials: The following are available online at http://www.mdpi.com/1999-4907/11/7/734/s1, Table S1: Primer sequence of the SNaPshot assay, Table S2: Primer sequence for a 48.48 Dynamic Array, Table S3: Cost comparison of ASP and ALP markers by means of agarose gel and the fragment analysis using a sequencer in marker-assisted selection (MAS) of 96 samples.

Author Contributions: Conceptualization, Y.H., S.U., S.H. and Y.M.; material preparation and phenotype data curation, T.T., S.H., J.I., Y.K. and Y.M.; marker development and genotype data collection, S.U., Y.H., T.T., M.W., R.S. and Y.M.; funding acquisition, Y.M.; writing—original draft, Y.M.; writing—review and editing, Y.H., S.U., M.W. and T.T. All authors have read and agreed to the published version of the manuscript.

Funding: This research was supported by the grants from Ministry of Agriculture, Forestry and Fisheries of Japan (MAFF) and NARO Bio-oriented Technology Research Advancement Institution (BRAIN) (the Science and technology research promotion program for agriculture, forestry, fisheries and food industry (No. 28013B)) and the grants from NARO Bio-oriented Technology Research Advancement Institution (BRAIN) (Research program on development of innovative technology (No. 28013BC)).

Acknowledgments: The authors would like to thank Y. Abe, Y. Komatsu for assistance with laboratory works. We also thank Y. Sato for artificial crossing. We also thank Y. Ito, S. Ikemoto, M. Sonoda, K. Yokoo, T. Hakamata and T. Miyashita for providing samples.

Conflicts of Interest: The authors declare no conflict of interest.

References

1. Neale, D.B.; Kremer, A. Forest tree genomics: Growing resources and applications. *Nat. Rev.* **2011**, *12*, 111–122. [CrossRef]

2. Forestry Agency. *Statistical Handbook of Forest and Forestry*; Forestry Agency, Ministry of Agriculture, Forestry and Fisheries: Tokyo, Japan, 2014; pp. 8–9. (In Japanese)

3. Baba, K.; Nakae, K. The national epidemiological survey of nasal allergy 2008 (compared with 1998) in otolaryngologists and their family members. *Prog. Med.* **2008**, *28*, 2001–2012. (In Japanese)

4. Taira, H.; Teranishi, H.; Kenda, Y. A case study of male sterility in sugi (*Cryptomeria japonica*). *J. Jpn. For. Soc.* **1993**, *75*, 377–379, (In Japanese with English summary).

5. Taira, H.; Saito, M.; Furuta, Y. Inheritance of the trait of male sterility in *Cryptomeria japonica*. *J. For. Res.* **1999**, *4*, 271–273. [CrossRef]

6. Saito, M. Breeding strategy for the pollinosis preventive cultivars of *Cryptomeria japonica* D Don. *J. Jpn. For. Soc.* **2010**, *92*, 316–323, (In Japanese with English summary). [CrossRef]

7. Hasegawa, Y.; Ueno, S.; Wei, F.J.; Matsumoto, A.; Uchiyama, K.; Ujino-Ihara, T.; Hakamata, T.; Fujino, T.; Kasahara, M.; Bino, T.; et al. Identification and genetic diversity analysis of a male-sterile gene (*MS1*) in Japanese cedar (*Cryptomeria japonica* D. Don). *Sci. Rep.* **2020**. under review. [CrossRef]

8. Wilson, Z.A.; Morroll, S.M.; Dawson, J.; Swarup, R.; Tighe, P. The *Arabidopsis MALE STERILITY1* (*MS1*) gene is a transcriptional regulator of male gametogenesis, with homology to the PHD-finger family of transcription factors. *Plant J.* **2001**, *28*, 27–39. [CrossRef]

9. Igarashi, M.; Watanabe, J.; Ozawa, H.; Saito, Y.; Taira, H. The male sterile sugi (*Cryptomeria japonica* D. Don) was found in Fukushima Prefecture (I): Selection of the search ground and identification of the male sterility. *Tohoku J. For. Sci.* **2004**, *9*, 86–89, (In Japanese with English abstract).

10. Takahashi, M.; Iwaizumi, M.G.; Hoshi, H.; Kubota, M.; Fukuda, Y.; Fukatsu, E.; Kondo, T. Survey of male sterility on sugi (*Cryptomeria japonica* D. Don) clones collected from Kanto Breeding Region and characteristics of two male-sterile clones. *Bull. Natl. For. Tree Breed. Cent.* **2007**, *23*, 11–36, (In Japanese with English abstract).

11. Fujisawa, T.; Saito, H.; Fujimiya, T.; Taira, H.; Saito, M. Selection and practical application of male-fertile cedar from elite trees in Kanagawa prefecture. In Proceedings of the 120th Japanese Forest Society Meeting, Kyoto, Japan, 26 March 2009. Abstract Number Pb01-06 (In Japanese).

12. Ueuma, H.; Yoshii, E.; Hosoo, Y.; Taira, H. Cytological study of a male-sterile *Cryptomeria japonica* that does not release microspores from tetrads. *J. For. Res.* **2009**, *14*, 123–126. [CrossRef]

13. Yamada, H.; Yamaguchi, K. Survey of male sterility of sugi (*Cryptomeria japonica* D. Don) clones collected from Kansai Breeding Region and characteristics of the male-sterile clone. *Appl. For. Sci.* **2009**, *8*, 33–36. (In Japanese)

14. Kawai, K.; Kubota, M.; Endo, K.; Isoda, K. Trial of efficient method for screening *Cryptomeria japonica* trees heterozygous for male-sterile gene by segregation of male-sterile seedlings derived from self-pollinated progeny. *Bull. FFPRI* **2017**, *444*, 265–266. (In Japanese)

15. Yoshii, E.; Taira, H. Cytological and genetical studies on male sterile sugi (*Cryptomeria japonica* D. Don), Shindai 1 and Shindai 5. *J. Jpn. For. Soc.* **2007**, *89*, 26–30, (In Japanese with English summary). [CrossRef]

16. Miyajima, D.; Yoshii, E.; Hosoo, Y.; Taira, H. Cytological and genetic studies on male sterility in *Cryptomeria japonica* D. Don (Shindai 8). *J. Jpn. For. Soc.* **2010**, *92*, 106–109, (In Japanese with English summary). [CrossRef]

17. Saito, M.; Taira, H.; Furuta, Y. Cytological and genetical studies on male sterility in *Cryptomeria japonica* D. Don. *J. For. Res.* **1998**, *3*, 167–173. [CrossRef]

18. Moriguchi, Y.; Ujino-Ihara, T.; Uchiyama, K.; Futamura, N.; Saito, M.; Ueno, S.; Matsumoto, A.; Tani, N.; Taira, H.; Shinohara, K.; et al. The construction of a high-density linkage map for identifying SNP markers that are tightly linked to a nuclear-recessive major gene for male sterility in *Cryptomeria japonica* D. Don. *BMC Genom.* **2012**, *13*, 95. [CrossRef] [PubMed]

19. Moriguchi, Y.; Ueno, S.; Higuchi, Y.; Miyajima, D.; Itoo, S.; Futamura, N.; Shinohara, K.; Tsumura, Y. Establishment of a microsatellite panel covering the sugi (*Cryptomeria japonica*) genome, and its application for localization of a male sterile gene (*ms-2*). *Mol. Breed.* **2014**, *33*, 315–325. [CrossRef]

20. Isoda, K.; Kawai, K.; Yamaguchi, K.; Kubota, M.; Yamada, H. Screening of trees heterozygous for male-sterile gene in Kansai breeding region. *Ann. Rep. Tree Breed. Cent.* **2013**, *2013*, 55–59.

21. Saito, M.; Aiura, H.; Kato, A.; Matsuura, T. Characterization of the cutting cultivar of *Cryptomeria japonica* D. Don, Zasshunbo, with a heterozygous male-sterility gene in a heterozygous state, selected from Toyama Prefecture. *Jpn. For. Genet. Tree Breed.* **2015**, *4*, 45–51, (In Japanese with English abstract).

22. Moriguchi, Y.; Uchiyama, K.; Ueno, S.; Ujino-Ihara, T.; Matsumoto, A.; Iwai, J.; Miyajima, D.; Saito, M.; Sato, M.; Tsumura, Y. A high-density linkage map with 2,560 markers and its application for the localization of the male-sterile genes *ms3* and *ms4* in *Cryptomeria japonica* D. Don. *Tree Genet. Genomes* **2016**, *12*, 57. [CrossRef]

23. Hasegawa, Y.; Ueno, S.; Matsumoto, A.; Ujino-Ihara, T.; Uchiyama, K.; Totsuka, S.; Iwai, J.; Hakamata, T.; Moriguchi, Y. Fine mapping of the male-sterile genes (*MS1*, *MS2*, *MS3*, and *MS4*) and development of SNP markers for marker-assisted selection in Japanese cedar (*Cryptomeria japonica* D. Don). *PLoS ONE* **2018**, *13*, e0206695. [CrossRef] [PubMed]

24. Mishima, K.; Hirao, T.; Tsubomura, M.; Tamura, M.; Kurita, M.; Nose, M.; Hanaoka, S.; Takahashi, M.; Watanabe, A. Identification of novel putative causative genes and genetic marker for male sterility in Japanese cedar (*Cryptomeria japonica* D. Don). *BMC Genom.* **2018**, *19*, 277. [CrossRef] [PubMed]

25. Ueno, S.; Uchiyama, K.; Moriguchi, Y.; Ujino-Ihara, T.; Matsumoto, A.; Fu-Jin, W.; Saito, M.; Higuchi, Y.; Futamura, N.; Kanamori, H.; et al. Scanning RNA-Seq and RAD-Seq approach to develop SNP markers closely linked to *MALE STERILITY 1* (*MS1*) in *Cryptomeria japonica* D. Don. *Breed. Sci.* **2019**, *69*, 19–29. [CrossRef] [PubMed]

26. Hasegawa, Y.; Ueno, S.; Fu-Jin, W.; Matsumoto, A.; Ujino-Ihara, T.; Uchiyama, K.; Moriguchi, Y.; Kasahara, M.; Fujino, T.; Shigenobu, S.; et al. Development of diagnostic PCR and LAMP markers for *MALE STERILITY 1* (*MS1*) in *Cryptomeria japonica* D Don. *BMC Res. Note* **2020**. under review. [CrossRef]

27. Murray, M.; Thompson, W.F. Rapid isolation of high molecular weight plant DNA. *Nucleic Acids Res.* **1980**, *8*, 4321–4325. [CrossRef]

28. Moriguchi, Y.; Ueno, S.; Saito, M.; Higuchi, Y.; Miyajima, D.; Itoo, S.; Tsumura, Y. A simple allele-specific PCR marker for identifying male-sterile trees: Towards DNA marker-assisted selection Cryptomeria japonica breeding program. *Tree Genet. Genomes* **2014**, *10*, 1069–1077. [CrossRef]

29. Van Ooijen, J.W.; Voorrips, R.E. *JoinMap, Version 3.0: Software for the Calculation of Genetic Linkage Maps*; University and Research Center: Wageningen, The Netherlands, 2001.

30. Kosambi, D.D. The estimation of map distances from recombination values. *Ann. Eugen.* **1944**, *12*, 172–175. [CrossRef]

31. Doğan, G.; Tadama, T.; Kohama, H.; Matsumoto, A.; Moriguchi, Y. Evidence of clonal propagation in *Cryptomeria japonica* D. Don distributed on Pacific Ocean side in Japan. *Silvae Genet.* **2017**, *66*, 43–46. [CrossRef]
32. Miura, S.; Nameta, M.; Yamamoto, T.; Igarashi, M.; Taira, H. Mechanisms of male sterility in four *Cryptomeria japonica* individuals with obvious visible abnormality at the tetrad stage. *J. Jpn. For. Soc.* **2011**, *93*, 1–7, (In Japanese with English summary). [CrossRef]

 *forests*

MDPI

Article

SNP Genotyping with Target Amplicon Sequencing Using a Multiplexed Primer Panel and Its Application to Genomic Prediction in Japanese Cedar, *Cryptomeria japonica* (L.f.) D.Don

Soichiro Nagano [1,*], Tomonori Hirao [1,2], Yuya Takashima [1], Michinari Matsushita [1], Kentaro Mishima [1], Makoto Takahashi [1], Taiichi Iki [3], Futoshi Ishiguri [4] and Yuichiro Hiraoka [1,5]

[1] Forest Tree Breeding Center (FTBC), Forestry and Forest Products Research Institute (FFPRI), 3809-1 Ishi, Juo, Hitachi, Ibaraki 319-1301, Japan; hiratomo@affrc.go.jp (T.H.); ytakashima@ffpri.affrc.go.jp (Y.T.); matsushita@affrc.go.jp (M.M.); mishimak@affrc.go.jp (K.M.); makotot@affrc.go.jp (M.T.);

[2] Forest Bio-Research Center, FFPRI, 3809-1 Ishi, Juo, Hitachi, Ibaraki 319-1301, Japan

[3] Tohoku Regional Breeding Office, FTBC, FFPRI, 95 Osaki, Takizawa, Iwate 020-0621, Japan; iki@affrc.go.jp

[4] School of Agriculture, Utsunomiya University, 350 Mine-machi, Utsunomiya, Tochigi 321-8505, Japan; ishiguri@cc.utsunomiya-u.ac.jp

[5] Faculty of Production and Environmental Management, Shizuoka Professional University of Agriculture, 678-1 Tomigaoka, Iwata, Shizuoka 438-8577, Japan; hiraoka.yuichiro@spua.ac.jp

* Correspondence: snagano@ffpri.affrc.go.jp; Tel.: +81-294-39-7045

Received: 22 July 2020; Accepted: 13 August 2020; Published: 19 August 2020

Abstract: Along with progress in sequencing technology and accumulating knowledge of genome and gene sequences, molecular breeding techniques have been developed for predicting the genetic potential of individual genotypes and for selecting superior individuals. For Japanese cedar (*Cryptomeria japonica* (L.f.) D.Don), which is the most common coniferous species in Japanese forestry, we constructed a custom primer panel for target amplicon sequencing in order to simultaneously determine 3034 informative single nucleotide polymorphisms (SNPs). We performed primary evaluation of the custom primer panel with actual sequencing and *in silico* PCR. Genotyped SNPs had a distribution over almost the entire region of the *C. japonica* linkage map and verified the high reproducibility of genotype calls compared to SNPs obtained by genotyping arrays. Genotyping was performed for 576 individuals of the F_1 population, and genomic prediction models were constructed for growth and wood property-related traits using the genotypes. Amplicon sequencing with the custom primer panel enables efficient obtaining genotype data in order to perform genomic prediction, manage clones, and advance forest tree breeding.

Keywords: amplicon sequencing; AmpliSeq; genomic selection; Japanese cedar (*Cryptomeria japonica*); multiplexed SNP genotyping; spatial autocorrelation error

---

1. Introduction

Molecular breeding techniques for plants that predict phenotypes from individual genotypes have been developed to shorten the breeding period compared to conventional methods [1,2]. With advances using a large number of genome-wide DNA markers to predict genomic estimated breeding values [3], genomic selection (GS) has been performed in many plant species (reviewed by Lin et al. [4] and Desta and Ortiz [5]) since it has become possible to construct genomic and transcriptomic information with next-generation DNA sequencing (NGS) and to prepare genome-wide DNA markers with various genotyping techniques [6].

Forest trees are a plant group for which breeding can be accelerated by using molecular breeding techniques [7–9]. The timespan required for genetic improvement is generally longer in forest trees than in agricultural cultivars because forest trees require a longer period for evaluating their economically important traits and reaching reproductive maturity. For forest trees with large genomic sequences that are rich in repetitive segments, it is not easy to obtain the genomic information that is necessary for molecular breeding. In particular, conifers have large genome sizes [10,11] and low linkage disequilibrium (LD) due to being undomesticated [11–14], which hinders genome-wide studies. Both the identification of genome-wide DNA markers and using these markers in trials of genomic predictions are essential for implementing GS in conifers.

Despite these disadvantages, various genotyping platforms and genetic mapping methods have been developed and applied to conifers, as summarized by Ritland et al. [15]. For example, the most widespread electrophoresis-based methods, simple sequence repeat (SSR) markers, which are also called microsatellites, have been developed for many species such as pine (*Pinus strobus* L., [16]; *P. sylvestris* L., [17]; *P. teada* L., [18]; *P. pinaster* Aiton, [19]) and spruce *Picea abies* (L.) H. Karst., [20,21]). Kompetitive allele specific PCR (KASP) assays utilizing fluorescence detection, which is suitable for detecting a few to several hundred single nucleotide polymorphisms (SNPs) in large numbers of samples, have been developed for *Abies alba* Mill. [22]. These markers can be applied to genetic mapping, population genetics, or lineage management. In addition, for large numbers of SNPs identified through genomic and transcriptomic sequencing, microarray-based genotyping platforms have been developed, including GoldenGate (Illumina; e.g., for Japanese black pine, [23]), Infinium (Illumina; e.g., for white spruce, [24]), and Axiom (Applied Biosystems; e.g., for Douglas fir, [25]). The advantages of these microarray platforms are the larger number (3 to 1000 K) of loci and the cost per marker per assay, but the initial cost to design a custom array and the cost per sample are relatively high. Genotyping by sequencing (GBS), such as with restriction site-associated DNA sequencing (RAD-seq, reviewed by Parchman et al. [26]) or multiplexed inter-simple sequence repeat (ISSR) genotyping by sequencing (MIG-seq, [27]), is available to accomplish high-throughput SNP genotyping with NGS, although its limited marker density and linkage disequilibrium often compromise its utility [26], and genotype numbers among genotyping assays are not sufficiently reproducible to apply GS, which requires high marker density to cover the entire genome. It is necessary to select suitable genotyping methods that offer an appropriate yield of genotypes for the intended research purpose from various available analysis platforms.

Japanese cedar, *Cryptomeria japonica* (L.f.) D.Don, is the most common coniferous tree species in Japanese forestry, and various molecular information and markers have been prepared to evaluate its genetic diversity, to perform reliable lineage management, and to examine genetic demography in natural populations. Marker development in allozymes [28] and cleaved amplified polymorphic sequence (CAPS, [29]) drove early studies of population genetics in *C. japonica*. SSR markers have been developed [30–33] and used in various studies of genetic diversity [32], gene flow [34], and core collection [35]. The GoldenGate SNP genotyping platform (Illumina, San Diego, CA, USA) was used to detect over 1000 SNPs and conduct a genome-wide association study (GWAS) for wood properties and the quantity of male strobili [9]. Using even larger numbers of SNPs, Mishima et al. [36] constructed a comprehensive expressed sequence tag (EST) collection from multiple tissues and developed custom Axiom arrays that enabled the simultaneous genotyping of more than 70K SNPs. With Axiom arrays, they constructed a linkage map for the F_1 population that was capable of detecting significant male sterility-related SNPs [36]. By applying the genotypes acquired by the Axiom arrays, Hiraoka et al. [37] performed GWAS and made genomic predictions for economically and socially important traits in unrelated first-generation *C. japonica* plus trees; many SNPs detected in the arrays were significantly correlated with economically and socially important traits. They also clarified that the accuracies of genomic predictions were dependent on the traits and populations reflecting the genetic architecture and on the background of the traits [37]. Although massive numbers of SNPs provide greater analytical capabilities, the authors noted the high cost of genotyping and suggested reducing the SNP number as an effective way of cutting genotyping cost [37]. In particular, analysis cost per individual is an important consideration when performing genotyping on large numbers of individuals. Prediction accuracies using SNPs selected based on the results of GWAS were similar to those using all SNPs for several combinations of traits and populations [37]. Pre-selection SNPs could be crucial for improving the quality of genomic predictions [38]. An efficient SNP genotyping system is required to verify the practicality of these SNPs and to apply them to actual breeding populations.

However, there are not many choices for medium-scale (up to several thousand loci) genotyping methods that can be redesigned flexibly and applied to GS in conifers. It is necessary to construct a platform for medium-scale high reproducibility genotyping to perform genomic prediction in *C. japonica* with high reliability. AmpliSeq (Thermo Fisher Scientific Inc., Waltham, MA, USA), which is an NGS-based genotyping method using multiplexed primer solutions for targeted amplicon sequencing, can enable amplification of up to 6000 amplicons simultaneously with ultra-high multiplex PCR and the construction of a targeted amplicon sequencing library in 10 h. This method has been used for studying inherited cancer in humans (e.g., [39]) and flowering time in soybean [40].

In this study, we constructed a medium-scale SNP genotyping system for *C. japonica*. We adopted an AmpliSeq custom primer panel (Thermo Fisher Scientific Inc.) as the platform and performed primary evaluation of the custom primer panel with actual sequencing and variant calling and with *in silico* PCR. We also examined the applicability of the custom primer panel for genomic prediction in a F_1 population of *C. japonica* plus trees.

2. Materials and Methods

2.1. Primer Panel Design

Here, we selected 3034 target SNPs based on allelic effects of the SNPs on growth- (height and diameter at breast height, or DBH), wood property- (wood stiffness and wood density), and reproductive- (male fecundity) related traits as suggested by GWAS results [37] and their comprehensive distribution on the linkage map [36]. Primers for targeted amplicon sequencing on Ion Torrent platforms were designed with the online application program, AmpliSeq designer (https://www.ampliseq.com/login/login.action) with EST sequences containing target SNPs in *C. japonica* as reported by Mishima et al. [36] and bed files describing the position of target SNPs on contigs. A total of 3031 EST sequences and 3034 target SNPs (four SNPs were located in one sequence) were analyzed. A total of 3004 primer pairs for 99.0% of the target SNPs were designed in a multiplexed 2 × Ion AmpliSeq Primer Pool (Table S1). Primer pairs to amplify target sequences for the remaining 30 SNPs could not be designed.

2.2. Panel Evaluation Via Actual Genotyping

We used a total of 16 clones of *C. japonica* for primary evaluation of the custom primer panel; 11 of the 16 clones were genotyped with the Axiom custom genotyping array (Thermo Fisher Scientific Inc.) designed by Mishima et al. [36], and five clones had an unknown SNP genotype. Current year needles were collected and stored at −20 °C until DNA extraction. For DNA extraction, about 50 mg of frozen needles were transferred into 2.0 mL microtubes and ground with liquid nitrogen and beads in a Shake Master Auto Ver. 2.0 (Bio Medical Science, Tokyo, Japan). DNA was extracted from the pellet with a plant DNA extraction kit (Qiagen Inc., Hilden, Germany), and DNA was quantified with the Qubit 3.0 Fluorometer and the Qubit dsDNA BR Assay Kit (Thermo Fisher Scientific Inc.), according to the manufacturer's instructions.

Libraries for amplicon sequencing were constructed with the AmpliSeq Library Kit v2.0 (Thermo Fisher Scientific Inc.) using the following protocol. For multiplex-PCR amplification, 5 ng DNA of each sample was amplified with the custom primer pool (3004 primer pairs) per reaction. Each reaction mix contained 2 µL of 5 × Ion AmpliSeq HiFi Master Mix, 5 µL of 2 × Ion AmpliSeq Primer Pool, and 5 ng of DNA, and it was brought to a volume of 10 µL with nuclease-free water. The reaction mix was heated for 2 min at 99 °C for enzyme activation, followed by 13 two-step cycles at 99 °C for

15 s and at 60 °C for 8 min, and ending with a holding period at 10 °C. Following the PCR, 1 µL of FuPa enzyme regents per sample was added to the reaction mix, the reaction mix was incubated at 50 °C for 10 min and at 55 °C for 10 min to digest the primers of the amplicons, and the mixture was incubated at 60 °C for 20 min to inactivate the enzyme. To enable library multiplexing on a single semiconductor chip, 2 µL of Switch Solution, 1 µL of diluted unique barcode adapter mix containing Ion Xpress Barcode and Ion P1 Adapters at standard volumes, and 1 µL of DNA ligase were added to the digested reaction mix, and the reaction mix was incubated at 22 °C for 30 min, followed by ligase inactivation for 5 min at 68 °C and 5 min at 72 °C. The adapter-ligated AmpliSeq library was purified using 22.5 µL of Agencourt AMPure XP Reagent (Beckman Coulter Inc., Brea, CA, USA), followed by washing with 75 µL of 70% ethanol twice. After the magnetic beads were dry, the AmpliSeq library was dissolved in 20 µL of Tris-EDTA (TE) buffer.

AmpliSeq libraries were evaluated for size distribution with a Bioanalyzer 2100 and High Sensitivity DNA Kit (Agilent Technologies Inc., Santa Clara, CA, USA), and the quantity was evaluated with a Qubit 3.0 Fluorometer and Qubit dsDNA HS Assay Kit (Thermo Fisher Scientific Inc.). The libraries identified by Ion Xpress Barcodes (Thermo Fisher Scientific Inc.) were multiplexed into a group of 16 samples for sequencing with an Ion Gene Studio S5 semiconductor sequencer and an Ion 520 semiconductor chip (Thermo Fisher Scientific Inc.). Emulsion PCR, emulsion breaking, and enrichment for the template preparation of ion sphere particles were performed with the Ion Chef and the Ion 520 and 530 Kit Chef (Thermo Fisher Scientific Inc.), according to the manufacturer's instructions. Following preparation of the semiconductor tip, sequencing was performed with an Ion Gene Studio S5 semiconductor sequencer using the Ion 520 Chip (Thermo Fisher Scientific Inc.), according to the manufacturer's instructions. The sequenced data were mapped to the reference gene sequence of *C. japonica* on a Torrent Server. Plugins, variantCaller v5.10.0.18 (Thermo Fisher Scientific Inc.) and reformatGBSCov v3.1 (Thermo Fisher Scientific Inc.) were used to construct a genotype table for the sequencing.

2.3. Data Processing and Visualization

Read depth per amplicon, read quality score, number of variants per site, GC ratio, and marker position within the genotyping with the AmpliSeq custom primer panel were summarized for each amplicon and drawn with the linkage position of the published *C. japonica* linkage map.

Genotyping efficiency for the primer panel was calculated for the 16 clones of *C. japonica*. Linkage positions of the SNPs on the published SNP linkage map [36] were drawn with R 3.6.0 [41]. The SNPs were distinguished by color on the linkage map at an 80% genotype call rate threshold. In addition, the SNP genotypes of the 11 of 16 clones were genotyped with the Axiom custom array (Axiom_Cj_70K_ver. 1.0; 73,274 SNPs; Gene Expression Omnibus Dataset (GEO): GSE95616; [36]) and compared with the genotyped results. The reproducibility of alleles was summarized and graphed on the R platform [41].

2.4. In Silico Panel Evaluation

The designed custom primer panel was tested with an *in silico* PCR program, Simulate_PCR version 1.2 ([42], https://sourceforge.net/projects/simulatepcr/) under a Perl 5.16.3 environment. The fasta files containing the 3031 EST sequences used to design the AmpliSeq custom primer panel, and the 34,731 EST sequences for the reference gene [36] of *C. japonica* were used for the input template reference file. The default settings and the options regarding included amplicon sizes were as follows: -minlen 50 and -maxlen 1000. The numbers of expected PCR products were summarized as follows: (a) amplified products in *in silico* PCR, (b) amplicon size ≤300 bp in *in silico* PCR, (c) correct pairing (intended pair of forward and reverse primers), (d) correct amplicon (annealing to correct contig with expected amplicon size), (e) off-target amplification (unintended primers annealing to the wrong contig), (f) unintended amplicon size (shorter or longer than expected amplicon), (g) mismatched pairing (unintended pair of forward and reverse primers), and (h) others (possible missing detection; amplicon size ≤300 bp and off target, unintended amplicon size, or mismatched pairing primers).

2.5. Genotyping a F_1 Population

We used a total of 576 individuals of a F_1 population, consisting of 547 individuals from eight half-diallelic F_1 populations constructed with outcrossing between 32 plus trees originating from the Tokai breeding area (Table S2) and 29 individuals from open pollinated four maternal plus trees originating from the North-Kanto breeding area. The F_1 population were planted in the Forest Tree Breeding Center (36°69′ N, 140°69′ E, 49.5 m above sea level), Hitachi, Ibaraki, Japan in 1995 with a plantation density of 3000 individuals per ha in a random block design with 6 replicates. At the timing of thinning in 2015, the materials for genotyping and phenotyping were collected. For genotyping, needles were placed in a plastic bag and stored at −20 °C until DNA extraction as described above. DNA concentration was measured with the QuantStudio 5 Real-Time PCR System and the Qubit dsDNA BR Assay Kit, according to the manufacturer instructions.

AmpliSeq libraries were constructed with an AmpliSeq Library Kit v2.0 (Thermo Fisher Scientific Inc.) as described above. Then, 22.5 µL of Agencourt AMPure XP Reagent (Beckman Coulter Inc.) was added with epMotion 96 (Eppendorf, Hamburg, Germany), and the magnetic beads were washed with 50 µL ethanol three times using HydroFlex (Tecan Group Ltd., Männedorf, Zürich, Switzerland) to purify the adapter-ligated AmpliSeq library. After the magnetic beads were dry, the AmpliSeq library was dissolved in 20 µL of TE buffer.

The purified AmpliSeq libraries were evaluated for size distribution with the Bioanalyzer 2100 and the High Sensitivity DNA Kit (Agilent Technologies Inc.) with quantification with the Qubit 3.0 fluorometer (Thermo Fisher Scientific Inc.). The libraries were distinguished by Ion Xpress Barcodes (Thermo Fisher Scientific Inc.), and 96 samples were multiplexed for sequencing with the Ion Gene Studio S5 semiconductor sequencer and a total of six Ion 540 semiconductor chips (Thermo Fisher Scientific Inc.). The sequenced data were mapped to the reference gene sequence of *C. japonica* on the Torrent Server. Plugins, variantCaller v5.10.0.18 and reformatGBSCov v3.1 were used to construct genotype tables for each run. All six sets of genotype data were collected in one genotype table.

2.6. Phenotypic Data

Phenotypic data of growth- and wood property-related traits were obtained by the following methods. Tree height, diameter at breast height (DBH), stress wave velocity (SWV), and pilodyn penetration depth (PP) were measured on trees that were 18 years old before harvesting. Tree height was measured with a Vertex III ultrasound instrument system (Haglöf, Västernorrland, Sweden), and DBH was measured at 1.3 m above ground with diameter calipers. The stress-wave propagation time of the stem was measured using a stress-wave timer (TreeSonic; Fakopp, Agfalva, Hungary). Briefly, the start sensor and the stop sensor were set on the stem from 0.7 to 1.7 m above the ground, and the stress-wave propagation time was measured five times from two directions at right angles to the direction of the slope, and SWV was determined by dividing the distance between the sensors by the mean value of the stress-wave propagation time. PP was measured using a Pilodyn 6J Forest (PROCEQ, Zurich, Switzerland) with a 2.5 mm diameter pin without removing the bark from the same directions used to obtain the stress-wave propagation time. After cutting the trees, logs from 1.0 to 2.5 m above the ground were collected for measuring the dynamic Young's modulus (DMOE). DMOE was measured using a portable FFT analyzer AD-3527 (A&D, Tokyo, Japan), following the tapping methods described by Sobue et al. [43]. After measuring DMOE, discs (about 40 mm in thickness) and short logs (about 400 mm in length) were collected from the butt end of the logs used to measure DMOE.

Square specimens were prepared for determining basic density (BD) at every fifth annual ring from the pith. The BD calculated from oven-dried weight was divided by the green volume, which was measured by the water displacement method [44]. BD_1, BD_2, and BD_3 were defined as the BD of the segments containing the 1st to 5th annual rings, 6th to 10th annual rings, and 11th to 15th annual rings, respectively. BD was not measured outside the 16th annual ring. The average BD of the whole disc (BD_means) was estimated by the weighted average method using the area of segments BD_1, BD_2, and BD_3.

The heart wood color was measured using a color meter CR-300 (Minolta, Tokyo, Japan). Bark to bark radial boards (30 mm thickness) were prepared from the short logs. The surface was previously smoothed by a belt sander under air-dried conditions. Measurements were expressed in the $L^*a^*b^*$ color space. L^* indicates lightness, a^* indicates the red–green axis, and b^* indicates the yellow–blue axis. The average values for five scattered points within each heartwood sample were used for analysis.

To determine the modulus of elasticity (MOE) and modulus of rupture (MOR), static bending tests were conducted according to the Japanese Industrial Standard (JIS) Z 2101-2009 [45] using bark-to-bark radial boards (30 mm thickness) prepared from the short logs. The boards were air-dried under laboratory conditions. A small clear specimen (20 (R) × 20 (T) × 320 (L) mm) was prepared from each board. All specimens were prepared at the same radial position: cross-section centered on the 4th annual ring from the pith. Static bending tests were conducted using a universal testing machine MSC-5/500-2 (Tokyo Testing Machine, Tokyo, Japan). Load was applied to the center of the radial surface of the specimen at 5 mm/min over a span of 280 mm. Data regarding load and deflection were recorded using a personal computer. After static bending tests, a small block was collected from each specimen for measuring the air-dried density and moisture content of the small clear specimen.

The microfibril angle of the S_2 layer in latewood tracheid (MFA) was measured as the angle of the slit-like pit aperture of boarded pits in latewood tracheids [46–48]. Using a sliding microtome (ROM-710, Yamatokohki, Saitama, Japan) and small clear specimens after the static bending test, tangential sections of 20 μm in thickness containing latewood of the 4th annual ring from the pith were obtained. These sections were stained with 1% safranine and then dehydrated using graded ethanol. The dehydrated sections were dipped into xylene and mounted on slides with bioleit (Okenshoji, Tokyo, Japan). Photomicrographs of tangential sections were taken using a light microscope CX-41 (Olympus Corporation, Tokyo, Japan) equipped with a digital camera E-300 (Olympus Corporation, Tokyo, Japan). The angle of the slit-pit aperture in the bordered pits of latewood tracheids to longitudinal direction was measured as MFA using ImageJ (National Institute of Health, Bethesda, MD, USA). Thirty tracheids were measured in each tree.

For the phenotypic data of each individual, a spatial autocorrelation error was adjusted with the breedR package [49] of the R platform [41]. Coordinates of individuals in the plantation site and values for each trait were used to calculate the spatial autocorrelation error, and each error was subtracted from the raw value to calculate the adjusted value (Figure S1).

2.7. Genomic Prediction Within the F_1 Population

We performed the genomic prediction for the F_1 population and each trait using two methods: genomic best linear unbiased prediction (GBLUP) and Random Forest (RF). GBLUP and RF were performed using the "kin.BLUP" function of the rrBLUP package v 4.6.1 [50] and "randomForest" function of the randomForest package v 4.6-14 [51] of the R platform [41], respectively. For the methods of GBLUP and RF, raw trait and adjusted trait values with special autocorrelation errors were used independently to construct the genomic prediction model. Prediction accuracy was estimated using Pearson's correlation coefficient between the phenotypic value and the genomic prediction value obtained from the validation dataset in the 10-fold cross-validation. The correlation coefficients from the 10-time replications in the 10-fold cross-validations were summarized.

3. Results

3.1. Primary Evaluation of the Multiplex Primer Panel

In the primary Ion S5 sequence run with an Ion 520 tip, a total of 1.32 Gb corresponding to 6.30 M reads was generated. The mean, median, and mode of the total sequence reads were 210, 228, and

235 bp, respectively, and 95.3% of obtained reads (6.00 out of 6.30 M reads) were successfully aligned to the reference sequence (Figure S2).

SNPs with a call rate of more than 80% were distributed over 11 linkage groups covering the entire previously constructed linkage map of *C. japonica* [36] (Figure 1). SNPs with a call rate of less than 80% were scattered over the linkage map (Figure 1), and there were some open areas where SNP markers were not originally designed because of the low degree of polymorphism among clones.

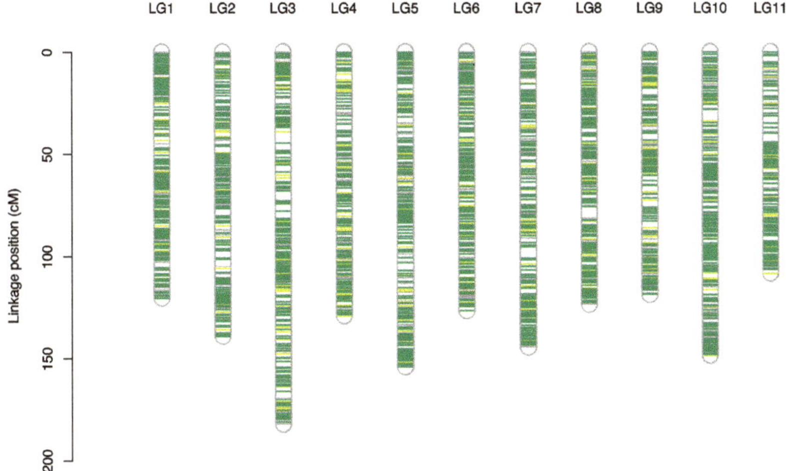

Figure 1. Linkage map of single nucleotide polymorphisms (SNP) distribution on the 11 linkage groups for *Cryptomeria japonica* (L.f.) D.Don published by Mishima et al. [36]. SNPs that are 80% or more genotyped, less than 80% genotyped, and ungenotyped with the custom primer panel are shown in deep green, light green, and gray, respectively.

Read depth on the mapped loci and relative read quality for amplicon sequencing varied among the loci (Figure 2a). Alignment of amplicons to the reference sequence with variant calls shows that novel variants aside from the targeted SNPs were detected with more than one variant per amplicon (Figure 2c) and with 18 variants per locus in the upper part of linkage group (LG) 1, 20 at the lower part of LG5, and 19 at the intermediate part of LG9 (Figure 2c). GC ratios of the sequenced amplicons were not largely biased (Figure 2d). The GC ratio and read depth were not correlated in this analysis.

The average number of called SNPs was 1990, corresponding to 68.5% of the SNP call rate through the genotyping for 16 clones of *C. japonica*. The obtained SNPs covered the linkage map (total of 1492.8 cm covering 11 linkage groups) of a previous study [36] with a mean distance between adjacent SNPs of 0.75 cm per SNP. For 11 clones, which were also previously genotyped with the Axiom arrays in a previous study [36] out of 16 clones, there were variations in the SNP call rate (Figure 3a) that are probably due to the low read depth (Table S3). The average SNP call rate of the 11 clones was 99.4% when genotyped with the Axiom arrays (Figure 3b). The comparison between the two genotyping platforms, i.e., Axiom and AmpliSeq, shows a genotype call rate of target SNPs with a custom primer panel for AmpliSeq that was less than that with Axiom (Figure 3c), although the average for the ratio consensus per genotyped SNP reached 94.9% (Figure 3d). Therefore, most of the called SNPs with the custom primer panel of AmpliSeq were consistent with results obtained with the custom genotyping array of Axiom, indicating the high reproducibility of these two genotyping systems.

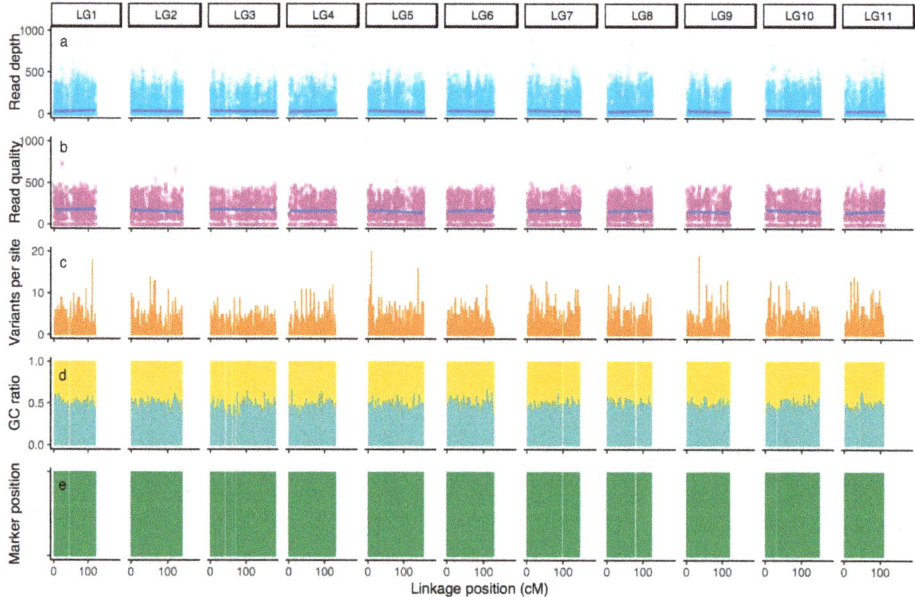

Figure 2. Parameters determined through genotyping for primary evaluation with the custom primer panel: (**a**) read depth; (**b**) read quality score; (**c**) number of variants per site; (**d**) GC ratio; and (**e**) marker position. The linkage positions of the amplicons on the linkage map in *C. japonica* published by Mishima et al. [36] are shown.

Of the set of 3031 EST sequences that were used to design the custom primer panel and were used as the template reference for *in silico* panel evaluations, 3004 amplicons were synthesized by *in silico* PCR (Table 1), and all of the intended SNP genotyping was obtained. The total number of amplified products by *in silico* PCR was 3157, as some unintended additional amplicons were produced. Out of 3157 amplicons, 3052 amplicons were intended PCR products generated with the correct primer pairs, and 105 amplicons were unintended PCR products generated with wrong primer pairings (Table 1). Among the amplicons generated with correct primer pairings, six were off-target amplicons due to unintended primers annealing to the wrong contigs, and 42 were of unexpected size. The remaining 63 PCR products had an amplicon size ≤300 bp due to the wrong annealing position, presumably resulting in missing alleles in the process of genotyping (Table 1). In contrast, when all 34,731 EST sequences [36] were used as a template reference for *in silico* PCR, 2747 out of 3004 targeted amplicons were amplified with *in silico* PCR (Table 1). The total number of amplified products by *in silico* PCR was 3395, consisting of 3265 PCR products amplified by the intended primer pairs and 130 amplified by the wrong primer pairs (Table 1). Among the PCR products with correct primer pairing, 340 products were off target amplicons and 178 were of unintended amplicon size (Table 1). The remaining 504 PCR products had an amplicon size ≤300 bp (Table 1). When sequence data used for designing the custom primer panel were used for the template reference of the *in silico* PCR, all the targeted SNPs were genotyped, although some additional unintended amplification occurred. When different sequence data (in this case, 34,731 contigs) were applied to the template reference for *in silico* PCR, the proportion of the targeted amplification decreased to 2747 amplicons (80.9%), suggesting the redundancy of gene sequences in the *C. japonica* genome.

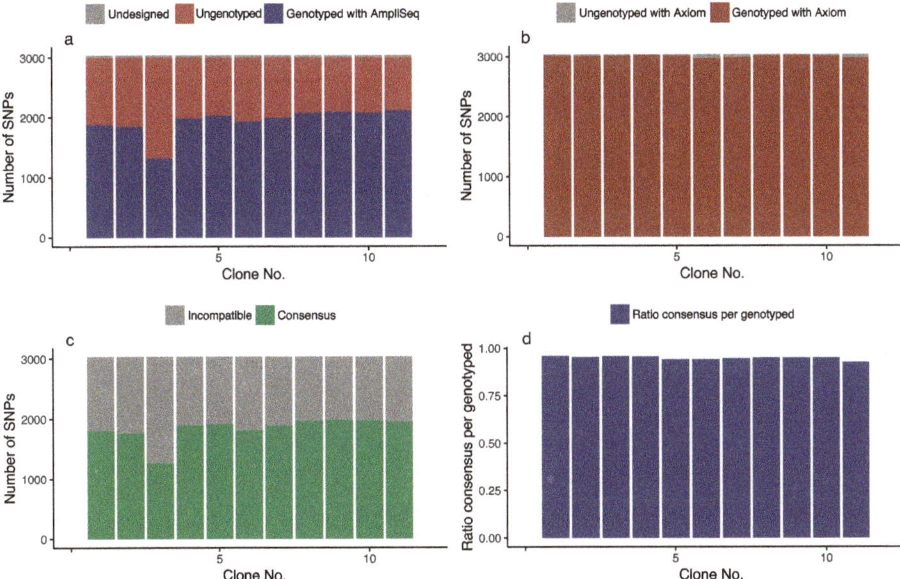

Figure 3. Genotyping summary and comparison between the AmpliSeq custom primer panel and Axiom custom genotyping array: (**a**) number of the SNPs detected with AmpliSeq; (**b**) number of SNPs detected with Axiom; (**c**) number of SNPs in consensus in a comparison between AmpliSeq and Axiom; (**d**) ratio consensus per genotyped SNP by AmpliSeq. Of a total of 16 clones of *C. japonica*, 16 were genotyped with AmpliSeq and 11 were genotyped with Axiom.

Table 1. Summary of *in silico* PCR evaluations of the custom primer panel for *Cryptomeria japonica* (L.f.) D.Don.

Reference Templates Used for *In Silico* PCR Amplification	3031 Contigs [1]		34,731 Contigs [2]	
Number of designed primer pairs		3004		
Amplified products in *in silico* PCR [3] (a; a = d + e + f + g)	3157		3395	
Amplicon size ≤ 300 bp in *in silico* PCR (b; b = d + h)	3067	(97.1)	3251	(95.8)
Correct pairing (intended pair of forward and reverse primers) (c; c = d + e + f)	3052	(96.7)	3265	(96.2)
Correct amplicon (annealing to correct contig with expected amplicon size) (d)	3004	(95.2)	2747	(80.9)
Off target amplification (unintended primers annealing to the wrong contig) (e)	6	(0.2)	340	(10.0)
Unintended amplicon size (shorter or longer than expected amplicon) (f)	42	(1.3)	178	(5.2)
Mismatched paring (unintended pair of forward and reverse primers) (g)	105	(3.3)	130	(3.8)
Others (possible missing detection) [4] (h)	63	(2.0)	504	(14.8)

[1] Contigs used for the custom amplicon panel design, which were selected from the total of the 34,731 contigs., [2] Cj_454_34731EST.fasta, reported by Mishima et al. [36], [3] Number of amplicons and percentage in brackets are shown., [4] Amplicon size ≤ 300 bp and off target, unintended amplicon size, or mismatched paring primers.

3.2. Genotyping of the F_1 Population

Ion S5 sequencing runs with the six Ion 540 tips generated a total of 62.2 Gb corresponding to 347.73 M reads. The mean, median, and mode of the total sequence reads were 178, 202, and 233 bp, respectively. Of these, 321.91 M reads (92.6%) were aligned to the reference sequence. Alignment accuracy of the reads to the reference sequence was high as well as the results of the primary evaluation of the custom primer panel.

Through genotyping 576 individuals of the F_1 population, the average number of read counts per sample was 563,800 ± 242,311 (mean ± SD) and the average number of called SNPs was 1963 ± 153, giving an average SNP call rate of 64.7% (Figure 4a). The remaining SNPs (34.3%) were not genotyped. The relationship between the acquired read count per sample and the proportion of genotype call rate per sample was examined, and the proportion of the SNP call rate was saturated when the read count per sample reached more than 250,000 reads (Figure 4b), suggesting that a sufficient number of reads was obtained for most genotyping samples.

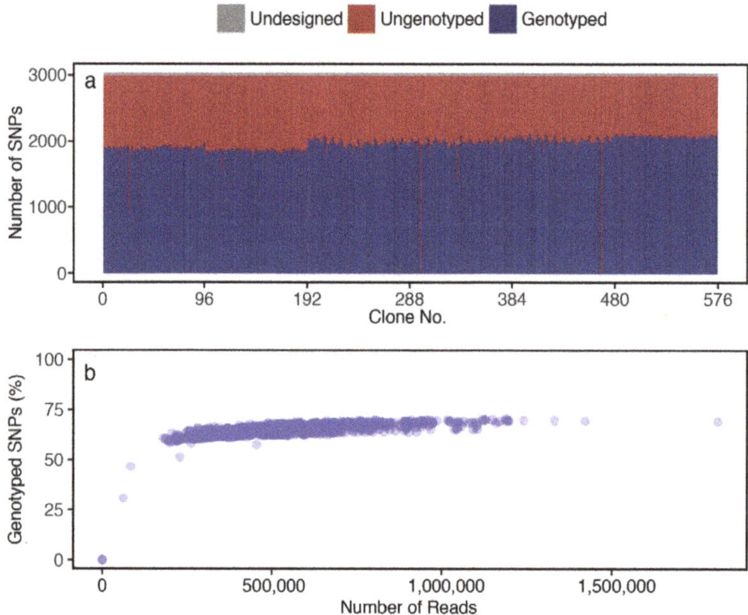

Figure 4. For genotyping the F_1 population, (**a**) number of SNPs detected by AmpliSeq; and (**b**) relationships between read count per sample and genotype call rate.

3.3. Construction of the Genomic Prediction Models with the F_1 Population

The prediction accuracy ranged from 0.166 to 0.555 depending on the type of data applied to the prediction, examined traits, and the models used for the genomic prediction (Table 2). Prediction accuracy was improved for all traits when the data were adjusted with spatial autocorrelation (Table 2, Figure 5 and Figure S3). In the models constructed with GBLUP, the prediction accuracy ranged from 0.166 to 0.454 when using raw trait values, but it was 0.185 to 0.544 for the adjusted trait values (Figure S3). In models constructed with RF, the prediction accuracy ranged from 0.176 to 0.450 for the raw trait values and from 0.195 to 0.555 for the adjusted trait values (Table 2, Figure 5). The prediction accuracy for traits related to wood properties (e.g., PP) was higher than for growth-related traits (e.g., height) (Table 2, Figure 5). The prediction accuracy showed a greater range of variation for the growth-related traits (0.197–0.418 for height and 0.236–0.408 for DBH) than for those of wood properties (0.445–0.487 for SWV, 0.456–0.555 for PP, and 0.409–0.544 for BD_means). For many wood properties (SWV, PP, DMOE, BD_2, BD_3, and BD_means), the prediction accuracies were higher than 0.40 regardless of the applied models (Table 2), although prediction accuracies for heart wood color (L^*, a^*, and b^*) were lower than 0.30 (Table 2). For DBH, DMOE, MOE, MOR, MFA, BD_1, BD_2, BD_3, and BD_means, the prediction accuracies were higher for the model with GBLUP than in the model with RF, but for tree height, L^*, a^*, b^*, SWV, and PP, prediction accuracy was higher in the model with RF than in the model with GBLUP (Table 2).

Table 2. Correlation coefficients for the genomic prediction models in the F_1 population. BD: basic density, DBH: diameter at breast height, DMOE: dynamic Young's modulus, GBLUP: genomic best linear unbiased prediction, MFA: microfibril angle, MOE: modulus of elasticity, MOR: modulus of rupture, PP: pilodyn penetration depth, SWV: stress wave velocity.

Model	GBLUP		Random Forest	
Trait	Raw Data	Adjusted Data	Raw Data	Adjusted Data
Height	0.197 ± 0.006	0.392 ± 0.003	0.302 ± 0.005	**0.418 ± 0.004**
DBH	0.236 ± 0.005	**0.408 ± 0.004**	0.255 ± 0.005	0.383 ± 0.004
L^*	0.225 ± 0.004	0.241 ± 0.006	0.251 ± 0.006	**0.251 ± 0.007**
a^*	0.212 ± 0.005	0.236 ± 0.003	0.232 ± 0.007	**0.252 ± 0.006**
b^*	0.166 ± 0.005	0.185 ± 0.005	0.189 ± 0.005	**0.195 ± 0.004**
SWV	0.445 ± 0.004	0.481 ± 0.002	0.449 ± 0.003	**0.487 ± 0.004**
PP	0.436 ± 0.004	0.542 ± 0.002	0.450 ± 0.003	**0.555 ± 0.001**
DMOE	0.410 ± 0.004	**0.445 ± 0.004**	0.408 ± 0.003	0.436 ± 0.003
MOE	0.283 ± 0.007	**0.372 ± 0.006**	0.252 ± 0.010	0.324 ± 0.008
MOR	0.248 ± 0.008	**0.358 ± 0.006**	0.228 ± 0.007	0.299 ± 0.007
MFA	0.192 ± 0.015	**0.246 ± 0.007**	0.176 ± 0.008	0.226 ± 0.008
BD_1	0.395 ± 0.004	**0.521 ± 0.003**	0.388 ± 0.004	0.515 ± 0.002
BD_2	0.437 ± 0.003	**0.516 ± 0.003**	0.437 ± 0.004	0.507 ± 0.004
BD_3	0.423 ± 0.003	**0.505 ± 0.004**	0.407 ± 0.002	0.487 ± 0.002
BD_means	0.454 ± 0.009	**0.544 ± 0.004**	0.409 ± 0.004	0.506 ± 0.004

$n = 10$, Mean $\pm$ SE are shown. Highest averaged value in each of the traits is in bold.

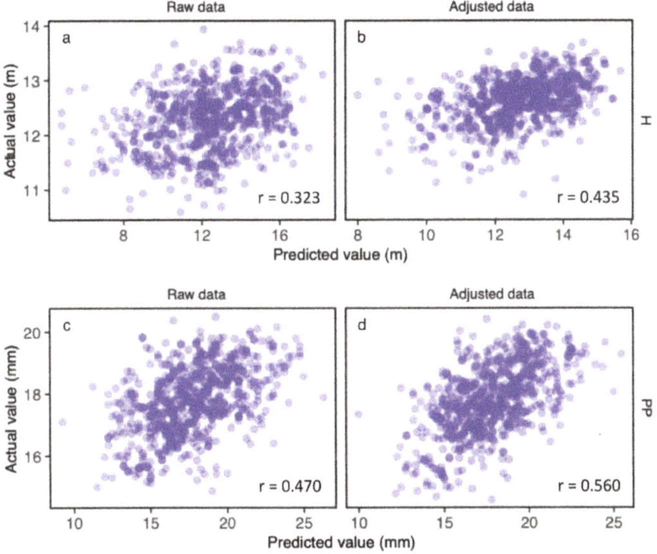

Figure 5. Relationships between actual and predicted trait values for tree height (H) for (**a**) raw data and (**b**) adjusted data, and for pyridine penetration depth (PP) for (**c**) raw data and (**d**) adjusted data. Genomic predictions within the F_1 population are shown. Predicted values were estimated by the genomic prediction model with data from random forest (RF). The relationships with the highest correlation efficient (r) were estimated by a round of 10-fold cross-validation, as shown for each trait without or with adjustment.

4. Discussion

In this study, we constructed and evaluated a multiplexed custom primer panel for amplicon sequencing in order to perform genomic prediction in *C. japonica*. We verified the high reproducibility of genotype calls by comparing results for two different methods: the custom genotyping array (Axiom) and the massive amplicon sequencing based genotyping (AmpliSeq). Genotyped SNPs by these two methods were in consensus for almost 2000 of the 3034 targeted SNPs (94.9%). Genotyped SNPs were distributed over the entire linkage map (1492.8 cm covering 11 LGs) of *C. japonica* without regional bias, as presented in a previous study [36] with a mean distance between adjacent SNPs of 0.75 cm per SNP. An unbiased distribution of the dense marker is essential for performing an accurate estimation of breeding value [3]. We also conducted genomic prediction with the F_1 population using the genotypes acquired with the custom primer panel, and we therefore created the first platform for middle-scale genotyping with amplicon sequencing to archive genomic prediction in *C. japonica*.

A comparison with other genotyping platforms indicates the usefulness and availability of the custom primer panel for targeted amplicon sequencing. Amplicon sequencing with a custom primer panel is characterized by the high reproducibility of genotype calls and short processing time. In routine genotyping for breeding, NGS-based techniques need to meet several criteria, e.g., short processing time until interpretation of genotyping results for selection, limited requirements for DNA, sufficient read depth to accurately detect variants [40], and completeness of the genotype call. In lodgepole pine (*Pinus contorta* Douglas) and white spruce (*Picea glauca* (Moench) Voss), 17,765 and 17,845 SNPs were obtained for the pine and spruce, respectively, through GBS [52], and those were greater than the currently targeted SNP numbers by AmpliSeq. However, other crops have shown that genotyping with GBS shows a low completeness of SNPs called in those experiments, especially regarding low read depth [53]. Low completion rates of the total detected variants among samples leads to an increase in the number of genotypes that are treated as dominant markers rather than co-dominant markers. Stochastic molecular reactions at the condensation stage of genomic DNA at the step of constructing the sequencing library, such as cleavage by restriction enzymes in RAD-seq or binding less specific primers on short SSR sequences in PCR in MIG-seq, may produce different null alleles between experiments. These genotyping methods are suitable for mutation extraction, e.g., SNP discovery and/or experiments that do not assume repetition. In fact, Ueno et al. [54] detected hundreds of thousands of candidate SNPs in SNP discovery with RNA-Seq and RAD-Seq in *C. japonica*, although when analyzing the mapping population using them, they used the Fluidigm (South San Francisco, CA, USA) SNPType assay with 129 candidate SNPs and showed that 75 SNPs, representing 58.1%, were available as markers [54]. On the other hand, genotyping with AmpliSeq is suitable for breeding applications such as genomic selection with effective SNPs, owing to its high reproducibility among experiments.

In amplicon sequencing with the custom primer panels used in this study, around 34.3% of 3034 target SNPs were not detected (Figure 3a), even though primers for the target amplicon were designed and included in the custom primer panel. The relationship between acquired read count per sample and genotype call rate per sample shows a saturation curve (Figure 4), and it is suggested that a sufficient number of reads is obtained from most of the samples for genotyping. The most likely cause of why more than 30% of the target SNPs were not genotyped is that the amplicons were not synthesized as designed in the first PCR step of library construction. One of the following may prevent the successful synthesis of amplicons: (1) interception of the primers by other homologous gene loci, (2) synthesis of chimerical products, and (3) insertion of large introns in the genomic sequence. In the present study, we employed partial sequences of expressed genes [36] as a reference to design the custom primer panel. For the first scenario, duplicated gene loci, which were not included among the applied 3031 sequences, may interrupt binding to the specific loci of the designed amplicons. An *in silico*-based evaluation of the custom primer panel suggests the possibility that designed primers would anneal to off-targeted gene sequences (Table 1). For the second scenario, chimeric PCR products synthesized between different pairs of forward and reverse primers also consumed primers and seemed to have a low alignment ratio to the reference, although the primary evaluation of the custom primer panel showed a high alignment ratio against the reference (Figure S2). The third scenario is also likely to be an obstacle to sequencing and sequence alignments when designing a custom primer panel that does not target genome sequencing, because the expressed gene sequence after RNA splicing by spliceosomes does not include introns in the genomic sequence. However, the size distribution of the product lengths of the synthesized library was matched to the requirements of sequencing, and excess product length was not observed in the synthesized library. Therefore, the first scenario is the most likely cause to prevent the successful synthesis of amplicons. These considerations suggest that re-designing the custom primer panel would be necessary to improve the genotyping efficiency. Using redundant gene sequences or, if possible, genome sequences as a reference, it would be possible to construct a more accurate panel for genotyping in *C. japonica*.

In the F_1 population of *C. japonica*, although results depend on traits and applied models, we confirmed that moderate accuracies (>0.5) were obtained for some wood properties in the genomic prediction modeling with the SNP genotypes (Table 2). Among the traits, genomic prediction accuracies for wood properties (e.g., SWV, PP, and basic densities) were higher than for growth traits (height and DBH). The ranges of prediction accuracies were more variable in growth traits (0.197–0.418 for height, and 0.236–0.408 for DBH) than in the wood property-related traits (0.445–0.487 for SWV, 0.456–0.555 for PP, and 0.409–0.544 for BD_means). Previous studies suggest that trait heritability is an important factor for the accuracy of genomic prediction [6]. The ranges of broad-sense heritability, which were previously reported for the associated traits in this study, were as follows: 0.37–0.72 for height [55,56], 0.21–0.52 for DBH [55–57], 0.65 for wood stiffness [57], and 0.78–0.88 for wood density [55,56]. In addition, higher prediction accuracies were observed for each trait when the trait values were adjusted by the spatial autocorrelation residuals that were employed for genomic predictions. This suggests that the F_1 individuals at the plantation site are affected by local micro-environmental factors for both growth traits and wood properties; the growth traits were more sensitive to environmental heterogeneity than the wood property-related traits. Furthermore, this suggests that accurate individual phenotyping is important for accurate genomic prediction modeling.

In previous genomic prediction studies performed by Hiraoka et al. [37], prediction accuracy differed among populations and unrelated plus trees of *C. japonica*; the prediction accuracies in the Kyushu population were generally the highest, followed in order by those in North Kanto and South Kanto populations [37]. For example, prediction accuracies in DBH were 0.523, 0.299, and 0.033 in Kyushu, North Kanto, and South Kanto populations, respectively, when the prediction model was constructed based on all 32,036 SNPs [37]. In this study, the F_1 population was mostly constructed by artificial crossings between the plus trees originating from the South Kanto population. Although the prediction model was constructed based on around 2000 SNPs, prediction accuracies in the F_1 population were higher than the results of trait prediction modeling in unrelated plus trees in the South Kanto population. Genomic relationships arising from population structures could influence the prediction accuracy as well as linkage disequilibrium [58], and the genetic structure of the F_1 population may show improved prediction accuracies compared to unrelated plus trees. Extended linkage disequilibrium in *C. japonica* [59] may have a positive effect on increased prediction accuracies. In addition, large numbers of individuals ($n = 576$) may positively affect prediction accuracies because the number of F_1 individuals used in modeling was greater than that of the unrelated plus trees ($n = 159$ in the South Kanto population [37]). In a genomic prediction study in *Pinus pinaster* Aiton using 661 individuals and 2500 markers [14], the prediction accuracy for tree height and DBH was 0.47 and 0.43, respectively, which was comparable to the prediction accuracy in the present study. In *Eucalyptus*, the prediction accuracy for basic wood density was 0.67 in genomic prediction with 768 individuals and 24,806 SNPs [60]. Therefore, the population structure and the number of individuals may affect prediction accuracy. Assuming that the medium-scale genotyping is a prerequisite, it is necessary to increase the number of individuals in order to improve the prediction accuracies.

In this study, we performed genomic prediction for the F_1 population, which was mostly constructed by artificial crossing between plus trees originating from the South Kanto population with

AmpliSeq, a medium-scale genotyping platform. Genomic estimated breeding values generated by prediction models with moderate accuracies may be usable as a threshold for selecting individuals that have not been virtually phenotyped as candidate or superior trees. We consider that the GS procedure with the medium-scale genotyping applied in this study is more practical than a large-scale genotyping such as Axiom when adapted to a large number of individuals. In a simulation study of GS in *C. japonica*, Iwata et al. [8] showed that GS breeding with model updating based on a realistic number of markers (e.g., one in every 1 cm) outperformed phenotypic selection breeding over 60-year periods, even for a low-heritability polygenic trait. In the present study, we developed and used a comparable number of markers (0.75 cm intervals) as that assumed by Iwata et al. [8] for the genomic prediction. This indicates that the breeding scheme proposed by Iwata et al. [8] is one option for future genome-based breeding in this species. GS breeding with model updating particularly for traits with high heritability such as wood properties may obtain higher genetic gain than for a low-heritability polygenic trait. In-house genotyping using the custom primer panel allows for flexible model improvement.

Further trials of genomic prediction in progenies of other populations, e.g., North Kanto and Kyusyu populations, would make it possible to verify the effectiveness of these prediction models. In addition, an application of models constructed for first-generation plus trees would be required and useful for predictions in subsequent generations and further applications of GS in *C. japonica* breeding.

5. Conclusions

In this study, we constructed a custom primer panel for amplicon sequencing for *C. japonica*, and we evaluated the custom primer panel with actual sequencing and with *in silico* PCR. Although genotyping efficiency could be improved through redesign of the custom panel, based on the trials for the genotyping and the genomic prediction modeling in the F_1 population, we demonstrated that the custom panel is useful for genomic prediction in *C. japonica*. In addition, we showed that prediction accuracy was improved when considering special autocorrelation errors arising from environmental heterogeneities. Since models are considered to be constructed for the first-generation plus trees and would also be useful for predictions in subsequent generations, further considerations are necessary for applying genomic prediction to *C. japonica*. Amplicon sequencing with the custom panel enables us to obtain genotype data efficiently in order to perform genomic prediction, to manage clones, and to advance forest tree breeding.

Supplementary Materials: The following are available online at http://www.mdpi.com/1999-4907/11/9/898/s1, Figure S1: Example of the effect of trait-value adjustment with special autocorrelation error for tree height, Figure S2: Alignment of amplicon sequences obtained by a primary evaluation of the custom primer panel for *C. japonica*, Figure S3: Relationships between actual and predicted trait value for tree height (H, a and b) and for pyridine penetration rate (PP, c and d) with best linear unbiased prediction (GBLUP), Table S1: List for the probe ID in the Axiom custom genotyping array, corresponding amplicon name, contig, SNP position, and primer sequence in the AmpliSeq custom panel, Table S2: Contents of half diallelic F_1 populations, Table S3: Summary statistics in the primary evaluation for the custom primer panel.

Author Contributions: Conceptualization, Y.H. and T.H.; amplicon sequencing and SNP genotyping, S.N. and T.H.; *in silico* panel evaluation, S.N.; wood property analysis, T.I., Y.T., and F.I.; genomic prediction, Y.H., S.N., and M.M.; writing—original draft preparation, S.N. and Y.T.; writing—review and editing, M.T., K.M., T.H., and Y.H. All authors have read and agreed to the published version of the manuscript.

Funding: This work was supported in part by 'Development of adaptation techniques to the climate change in the sectors of agriculture, forestry, and fisheries' (Ministry of Agriculture, Forestry and Fisheries of Japan).

Acknowledgments: We are grateful to E. Fukatsu for introducing analysis software; K. Kato, T. Kaminaga, and M. Shibata for their assistance with laboratory experiment; N. Kuramoto for coordination of the research project; and other members of the Forest Tree Breeding Center for help with field investigations.

Conflicts of Interest: The authors declare no conflict of interest.

References

1. Crossa, J.; Pérez-Rodríguez, P.; Cuevas, J.; Montesinos-López, O.; Jarquin, D.; Campos, G.; Burgueño, J.; Camacho-González, J.; Perez-Elizalde, S.; Beyene, Y.; et al. Genomic selection in plant breeding: Methods, models, and perspectives. *Trends Plant Sci.* **2017**, *22*, 961–997. [CrossRef]

2. Lenaerts, B.; Collard, B.C.Y.; Demont, M. Review: Improving global food security through accelerated plant breeding. *Plant Sci.* **2019**, *287*, 110207. [CrossRef]

3. Meuwissen, T.H.E.; Hayes, B.J.; Goddard, M.E. Prediction of total genetic value using genome-wide dense marker maps. *Genetics* **2001**, *157*, 1819–1829.

4. Lin, Z.; Hayes, B.J.; Daetwyler, H.D. Genomic selection in crops, trees and forages: A review. *Crop Pasture Sci.* **2014**, *65*, 1177–1191. [CrossRef]

5. Desta, Z.A.; Ortiz, R. Genomic Selection: Genome-wide prediction in plant improvement. *Trends Plant Sci.* **2014**, *19*, 592–601. [CrossRef]

6. Davey, J.W.; Hohenlohe, P.A.; Etter, P.D.; Boone, J.Q.; Catchen, J.M.; Blaxter, M.L. Genome-wide genetic marker discovery and genotyping using next-generation sequencing. *Nat. Rev. Genet.* **2011**, *12*, 499–510. [CrossRef]

7. Grattapaglia, D.; Resende, M.D.V. Genomic selection in forest tree breeding. *Tree Genet. Genomes* **2011**, *7*, 241–255. [CrossRef]

8. Iwata, H.; Hayashi, T.; Tsumura, Y. Prospects for genomic selection in conifer breeding: A simulation study of *Cryptomeria japonica*. *Tree Genet. Genomes* **2011**, *7*, 747–758. [CrossRef]

9. Uchiyama, K.; Iwata, H.; Moriguchi, Y.; Ujino-Ihara, T.; Ueno, S.; Taguchi, Y.; Tsubomura, M.; Mishima, K.; Iki, T.; Watanabe, A.; et al. Demonstration of genome-wide association studies for identifying markers for wood property and male strobili traits in *Cryptomeria japonica*. *PLoS ONE* **2013**, *8*, e79866. [CrossRef]

10. Chagné, D.; Brown, G.; Lalanne, C.; Madur, D.; Pot, D.; Neale, D.; Plomion, C. Comparative genome and qtl mapping between maritime and loblolly pines. *Mol. Breed.* **2003**, *12*, 185–195. [CrossRef]

11. Neale, D.B.; Wegrzyn, J.L.; Stevens, K.A.; Zimin, A.V.; Puiu, D.; Crepeau, M.W.; Cardeno, C.; Koriabine, M.; Holtz-Morris, A.E.; Liechty, J.D.; et al. Decoding the massive genome of loblolly pine using haploid DNA and novel assembly strategies. *Genome Biol.* **2014**, *15*, R59. [CrossRef]

12. Neale, D.B.; Savolainen, O. Association genetics of complex traits in conifers. *Trends Plant Sci.* **2004**, *9*, 325–330. [CrossRef]

13. Plomion, C.; Chancerel, E.; Endelman, J.; Lamy, J.-B.; Mandrou, E.; Lesur, I.; Ehrenmann, F.; Isik, F.; Bink, M.C.A.M.; van Heerwaarden, J.; et al. Genome-wide distribution of genetic diversity and linkage disequilibrium in a mass-selected population of maritime pine. *BMC Genomics* **2014**, *15*, 171. [CrossRef]

14. Isik, F.; Bartholomé, J.; Farjat, A.; Chancerel, E.; Raffin, A.; Sanchez, L.; Plomion, C.; Bouffier, L. Genomic selection in maritime pine. *Plant Sci.* **2016**, *242*, 108–119. [CrossRef]

15. Ritland, K.; Krutovsky, K.V.; Tsumura, Y.; Pelgas, B.; Isabel, N.; Bousquet, J. Genetic Mapping in Conifers. In *Genetics, Genomics and Breeding of Conifers*; Plomion, C., Bousquet, J., Eds.; CRC Press: Boca Raton, FL, USA, 2011; pp. 196–238.

16. Echt, C.S.; May-Marquardt, P.; Hseih, M.; Zahorchak, R. Characterization of microsatellite markers in eastern white pine. *Genome* **1996**, *39*, 1102–1108. [CrossRef]

17. Soranzo, N.; Provan, J.; Powell, W. Characterization of microsatellite loci in *Pinus sylvestris* L. *Mol. Ecol.* **1998**, *7*, 1260–1261.

18. Elsik, C.G.; Minihan, V.T.; Hall, S.E.; Scarpa, A.M.; Williams, C.G. Low-copy microsatellite markers for *Pinus Taeda* L. *Genome* **2000**, *43*, 550–555. [CrossRef]

19. Mariette, S.; Chagne, D.; Decroocq, S.; Giovanni Giuseppe, V.; Lalanne, C.; Madur, D.; Plomion, C. Microsatellite markers for *Pinus Pinaster* Ait. *Ann. For. Sci.* **2001**, *58*. [CrossRef]

20. Paglia, G.P.; Olivieri, A.M.; Morgante, M. Towards second-generation STS (sequence-tagged sites) linkage maps in conifers: A genetic map of Norway spruce (*Picea Abies* K.). *Mol. Gen. Genet.* **1998**, *258*, 466–478. [CrossRef]

21. Scotti, I.; Magni, F.; Paglia, G.; Morgante, M. Trinucleotide microsatellites in Norway spruce (*Picea abies*): Their features and the development of molecular markers. *Theor. Appl. Genet.* **2002**, *106*, 40–50. [CrossRef]

22. Roschanski, A.M.; Csilléry, K.; Liepelt, S.; Oddou-Muratorio, S.; Ziegenhagen, B.; Huard, F.; Ullrich, K.; Postolache, D.; Giovanni Giuseppe, V.; Fady, B. Evidence of divergent selection for drought and cold tolerance

at landscape and local scales in *Abies alba* Mill. in the French Mediterranean Alps. *Mol. Ecol.* **2016**, *25*, 776–794. [CrossRef]

23. Hirao, T.; Matsunaga, K.; Hirakawa, H.; Shirasawa, K.; Isoda, K.; Mishima, K.; Tamura, M.; Watanabe, A. construction of genetic linkage map and identification of a novel major locus for resistance to pine wood nematode in Japanese black pine (*Pinus thunbergii*). *BMC Plant Biol.* **2019**, *19*. [CrossRef]

24. Pavy, N.; Gagnon, F.; Rigault, P.; Blais, S.; Deschênes, A.; Boyle, B.; Pelgas, B.; Deslauriers, M.; Clément, S.; Lavigne, P.; et al. Development of high-density SNP genotyping arrays for white spruce (*Picea glauca*) and transferability to subtropical and Nordic congeners. *Mol. Ecol. Res.* **2013**, *13*, 324–336. [CrossRef]

25. Howe, G.T.; Jayawickrama, K.; Kolpak, S.E.; Kling, J.; Trappe, M.; Hipkins, V.; Ye, T.; Guida, S.; Cronn, R.; Cushman, S.A.; et al. An Axiom SNP genotyping array for douglas-fir. *BMC Genom.* **2020**, *21*, 9. [CrossRef]

26. Parchman, T.L.; Jahner, J.P.; Uckele, K.A.; Galland, L.M.; Eckert, A.J. RADseq Approaches and applications for forest tree genetics. *Tree Genet. Genomes* **2018**, *14*, 39. [CrossRef]

27. Suyama, Y.; Matsuki, Y. MIG-seq: An effective PCR-Based method for genome-wide single-nucleotide polymorphism genotyping using the next-generation sequencing platform. *Sci. Rep.* **2015**, *5*, 16963. [CrossRef]

28. Tsumura, Y.; Ohba, K. Allozyme variation of five natural populations of *Cryptomeria japonica* in western Japan. *Jpn. J. Genet.* **1992**, *67*, 299–308. [CrossRef]

29. Tsumura, Y.; Kado, T.; Takahashi, T.; Tani, N.; Ujino-Ihara, T.; Iwata, H. Genome scan to detect genetic structure and adaptive genes of natural populations of *Cryptomeria japonica*. *Genetics* **2007**, *176*, 2393. [CrossRef]

30. Moriguchi, Y.; Iwata, H.; Ujino-Ihara, T.; Yoshimura, K.; Taira, H.; Tsumura, Y. Development and characterization of microsatellite markers for *Cryptomeria japonica* D. Don. *Theor. Appl. Genet.* **2003**, *106*, 751–758. [CrossRef]

31. Tani, N.; Takahashi, T.; Ujino-Ihara, T.; Iwata, H.; Yoshimura, K.; Tsumura, Y. Development and characteristics of microsatellite markers for sugi (*Cryptomeria japonica* D. Don) derived from microsatellite-enriched libraries. *Ann. For. Sci.* **2004**, *61*, 569–575. [CrossRef]

32. Takahashi, T.; Tani, N.; Taira, H.; Tsumura, Y. Microsatellite markers reveal high variation in natural populations of *Cryptomeria japonica* near refugial areas of the last glacial period. *J. Plant Res.* **2005**, *118*, 83–90. [CrossRef]

33. Moriguchi, Y.; Ueno, S.; Higuchi, Y.; Miyajima, D.; Itoo, S.; Futamura, N.; Shinohara, K.; Tsumura, Y. Establishment of a microsatellite panel covering the sugi (*Cryptomeria japonica*) genome, and its application for localization of a male-sterile gene (*ms-2*). *Mol. Breed.* **2014**, *33*, 315–325. [CrossRef]

34. Moriguchi, Y.; Tani, N.; Itoo, S.; Kanehira, F.; Tanaka, K.; Yomogida, H.; Taira, H.; Tsumura, Y. Gene flow and mating system in five *Cryptomeria japonica* D. Don seed orchards as revealed by analysis of microsatellite markers. *Tree Genet. Genomes* **2005**, *1*, 174–183. [CrossRef]

35. Miyamoto, N.; Ono, M.; Watanabe, A. Construction of a core collection and evaluation of genetic resources for *Cryptomeria japonica* (Japanese cedar). *J. For. Res.* **2015**, *20*, 186–196. [CrossRef]

36. Mishima, K.; Hirao, T.; Tsubomura, M.; Tamura, M.; Kurita, M.; Nose, M.; Hanaoka, S.; Takahashi, M.; Watanabe, A. Identification of novel putative causative genes and genetic marker for male sterility in Japanese cedar (*Cryptomeria japonica* D. Don). *BMC Genom.* **2018**, *19*, 277. [CrossRef]

37. Hiraoka, Y.; Fukatsu, E.; Mishima, K.; Hirao, T.; Teshima, K.M.; Tamura, M.; Tsubomura, M.; Iki, T.; Kurita, M.; Takahashi, M.; et al. Potential of genome-wide studies in unrelated plus trees of a coniferous species, *Cryptomeria japonica* (Japanese cedar). *Front. Plant Sci.* **2018**, *9*. [CrossRef]

38. Croiseau, P.; Legarra, A.; Guillaume, F.; Fritz, S.; Baur, A.; Colombani, C.; Robert-Granié, C.; Boichard, D.; Ducrocq, V. Fine tuning genomic evaluations in dairy cattle through SNP pre-selection with the elastic-net algorithm. *Genet. Res.* **2011**, *93*, 409–417. [CrossRef]

39. Liu, S.; Wang, H.; Zhang, L.; Tang, C.; Jones, L.; Ye, H.; Ban, L.; Wang, A.; Liu, Z.; Lou, F.; et al. Rapid detection of genetic mutations in individual breast cancer patients by next-generation DNA sequencing. *Hum. Genom.* **2015**, *9*, 2. [CrossRef]

40. Ogiso-Tanaka, E.; Shimizu, T.; Hajika, M.; Kaga, A.; Ishimoto, M. Highly multiplexed AmpliSeq technology identifies novel variation of flowering time-related genes in soybean (*Glycine Max*). *DNA Res.* **2019**, *26*, 243–260. [CrossRef]

41. R Core Team. *A Language and Environment for Statistical Computing*; R Foundation for Statistical Computing: Vienna, Austria; Available online: https://www.R-project.org/2019 (accessed on 24 May 2019).

42. Gardner, S.N.; Slezak, T. Simulate_PCR for Amplicon prediction and annotation from multiplex, degenerate primers and probes. *BMC Bioinform.* **2014**, *15*, 2–7. [CrossRef]

43. Sobue, N. Measurement of Young's modulus by transient longitudinal vibration of wooden beams using a fast fourier transformation spectrum analyzer. *Mokuzai Gakkaishi* **1986**, *32*, 744–747, (In Japanese with English Summary).

44. Panshin, A.J.; de Zeeuw, C. *Text Book of Wood Technology*; McGraw-Hill Book, Co.: New York, NY, USA, 1980.

45. Japanese Industrial Standards. *Methods of Testing for Woods JIS Z 2101-2009*; Japanese Standards Association: Tokyo, Japan, 2009.

46. Cookrell, R.A. A Comparison of latewood pits, fibril orientation, and shrinkage of normal and compression wood of giant sequoia. *Wood Sci. Technol.* **1974**, *8*, 197–206. [CrossRef]

47. Donaldson, L.A. The Use of pit apertures as windows to measure microfibril angle in chemical pulp fibers. *Wood Fiber Sci.* **1991**, *23*, 290–295.

48. Hirakawa, Y.; Fujisawa, Y. The Relationships between microfibril angles of the S_2 layer and latewood tracheid lengths in elite sugi tree (*Cryptomeria japonica*) Clones. *Mokuzai Gakkaishi* **1995**, *41*, 123–131, (In Japanese with English Summary).

49. Muñoz, F.; Sanchez, L. breedR: Statistical Methods for Forest Genetic Resources Analysis. Available online: https://github.com/famuvie/breedR/2018 (accessed on 1 November 2018).

50. Endelman, J. Ridge regression and other kernels for genomic selection with R package rrBLUP. *Plant Genome* **2011**, *4*, 250–255. [CrossRef]

51. Liaw, A.; Wiener, M. Classification and regression by randomForest. *R News* **2002**, *2*, 18–22.

52. Chen, C.; Mitchell, S.E.; Elshire, R.J.; Buckler, E.S.; El-Kassaby, Y.A. Mining conifer's mega-genome using rapid and efficient multiplexed high-throughput genotyping-by-sequencing (GBS) SNP discovery platform. *Tree Genet. Genomes* **2013**, *9*, 1537–1544. [CrossRef]

53. Huang, Y.; Poland, J.; Wight, C.; Jackson, E.; Tinker, N. Using genotyping-by-sequencing (GBS) for genomic discovery in cultivated oat. *PLoS ONE* **2014**, *9*. [CrossRef]

54. Ueno, S.; Uchiyama, K.; Moriguchi, Y.; Ujino-Ihare, T.; Matsumoto, A.; Wei, F.; Saito, M.; Higuchi, Y.; Futamura, N.; Kanamori, H.; et al. Scanning RNA-Seq and RAD-Seq approach to develop SNP markers closely linked to *MALE STERILITY 1* (*MS1*) in *Cryptomeria japonica* D. Don. *Breed. Sci.* **2019**, *69*, 19–29. [CrossRef]

55. Tamura, A.; Kurinobu, S.; Fukatsu, E.; Iizuka, K. An investigation on the allocation of selection weight on growth and wood basic density to maximize carbon storage in the stem of sugi (*Cryptomeria japonica* D. Don) plus-tree clones. *J. Jpn. For. Soc.* **2006**, *88*, 15–20. [CrossRef]

56. Fukatsu, E.; Tamura, A.; Takahashi, M.; Fukuda, Y.; Nakada, R.; Kubota, M.; Kurinobu, S. Efficiency of the indirect selection and the evaluation of the genotype by environment interaction using pilodyn for the genetic improvement of wood density in *Cryptomeria japonica*. *J. For. Res.* **2011**, *16*, 128–135. [CrossRef]

57. Fujisawa, Y.; Ohta, S.; Nishimura, K.; Toda, T.; Tajima, M. Wood characteristics and genetic variations in sugi (*Cryptomeria japonica*). 3. Estimation of variance-components of the variation in dynamic modulus of elasticity with plus-tree clones. *Mokuzai Gakkaishi* **1994**, *40*, 457–464.

58. Daetwyler, H.; Kemper, K.; van der Werf, J.; Hayes, B. Components of the accuracy of genomic prediction in a multi-breed sheep population. *J. Anim. Sci.* **2012**, *90*, 3375–3384. [CrossRef]

59. Moritsuka, E.; Hisataka, Y.; Tamura, M.; Uchiyama, K.; Watanabe, A.; Tsumura, Y.; Tachida, H. Extended linkage disequilibrium in noncoding regions in a conifer, *Cryptomeria japonica*. *Genetics* **2012**, *190*, 1145–1148. [CrossRef]

60. Resende, R.; Resende, M.; Silva, F.; Azevedo, C.; Takahashi, E.; Silva, O.; Grattapaglia, D. Assessing the expected response to genomic selection of individuals and families in *Eucalyptus* breeding with an additive-dominant model. *Heredity* **2017**, *119*, 245–255. [CrossRef]

MDPI

St. Alban-Anlage 66

4052 Basel

Switzerland

Tel. +41 61 683 77 34

Fax +41 61 302 89 18

www.mdpi.com

Forests Editorial Office

E-mail: forests@mdpi.com

www.mdpi.com/journal/forests

CPSIA information can be obtained
at www.ICGtesting.com
Printed in the USA
LVHW070925300821
696448LV00002B/30